Multivariate Analysis

MULTIVARIATE ANALYSIS

TECHNIQUES FOR EDUCATIONAL AND PSYCHOLOGICAL RESEARCH

SECOND EDITION

Maurice M. Tatsuoka
University of Illinois at Urbana-Champaign

with contributions by
Paul R. Lohnes
State University of New York at Buffalo

Macmillan Publishing Company
New York
Collier Macmillan Publishers
London

Printed in the United States of America

Macmillan Publishing Company
866 Third Avenue, New York, New York 10022

Collier Macmillan Canada, Inc.

Library of Congress Cataloging in Publication Data

Tatsuoka, Maurice M.
Multivariate analysis.

Bibliography: p.
Includes index.
1. Multivariate analysis. I. Title.
QA278.T37 1988 519.5'35 86-33127
ISBN 0-02-419120-5

ISBN 0-02-419120-5

Printing: 2 3 4 5 6 7 8 Year: 0 1 2 3 4 5 6 7

PREFACE

Fifteen years have elapsed since the first edition of this book was published. A number of new multivariate techniques have appeared in the interim, and a computer revolution has overtaken virtually all phases of our culture, including statistical data analysis. Computers were, of course, very much in evidence in the latter field for many decades prior to 1970. But the recent proliferation of microcomputers and statistical software has made it a relatively simple matter for anyone with a minimal knowledge of statistical theory to carry out quite complicated data analyses including those that involve multivariate techniques. This has both desirable and undesirable consequences. The former are obvious; the latter are less so, and indeed may seem nonexistent to some people. It is my conviction, however, that unrestrained use of "high-powered" techniques without adequate understanding of their rationale can lead to various kinds of abuse and misuse and misinterpretation of the results. For this reason, although the second edition makes many references to computer packages and includes exercises that are intended to be done with the aid of computers (often with hints for computer programming), I have stopped short of urging wholesale use of packaged programs for "one-stroke" execution of complete analyses. Instead, I encourage students, insofar as possible, to develop their own programs through the use of matrix operations. It is believed that this will lead to a better understanding of what would otherwise be a black box.

Other changes from the first edition include the addition of chapters on multiple regression analysis, factor analysis, and the general linear model. The first two of these topics were excluded in the first edition in accordance with the somewhat narrow definition of "multivariate analysis" adopted there, as a body of statistical techniques that involve multiple dependent variables; multiple regression has but one and factor analysis, none. On the other hand, the chapter on covariance analysis with multiple covariates has been deleted. It was included in the first edition mainly to introduce the concept of sums-of-squares-and-cross-products (SSCP) matrices in a context that was assumed to be familiar to the reader. With the addition of the chapter on multiple regression, however, the covariance analysis chapter no longer seemed necessary for that purpose.

In addition to my early mentors in multivariate analysis, who are acknowledged in the preface to the first edition, I now wish to thank also the several hundred students to whom I, in turn, taught this subject. They have—some explicitly and the rest implicitly—conveyed to me wherein there were shortcomings in the pedagogy used in my first edition. These weak points have been amended in the present edition.

I am also indebted to a number of colleagues who have used the first edition as a text—in particular, Professors Ledyard R. Tucker and Robert L. Linn of the University of Illinois—and have suggested a number of improvements. Another colleague, who I know only by name, Dr. Juan C. Rosales of the University of Laguna, Spain, has been kind enough to compile a detailed list of numerical and typographical errors, many of which had eluded my detection throughout the decade-and-a-half I had been teaching from the book!

Professor Paul R. Lohnes and another reviewer read the manuscript of the present edition with extreme care and made many valuable suggestions for improvement, most of which I have incorporated. Some of Paul's suggestions, however—in particular, those that urged the addition of more real-life research examples—were not easy for me to comply with. I therefore invited him to contribute such examples, and he graciously agreed. As a result, each chapter from the third one on now contains a final section by Paul Lohnes that discusses one or more examples, drawn from the literature, to illustrate actual applications of the technique(s) discussed in that chapter. Of course, by its very nature, real-life research does not compartmentalize itself into neat categories such as "applications of technique A," "applications of technique B," and so on. Each example, while primarily invoking one technique, will almost always utilize other related (and sometimes not closely related) techniques as well. Thus for example, one of the studies illustrating the use of multivariate significance tests of Chapter 4 (Section 4.7) refers also to discriminant functions and classification procedures, which are discussed in Chapters 7 and 10, respectively. Also, I have not attempted to modify his personal style to conform to my more formal style. I believe that Paul's contributions have added greatly to the usefulness of the book as a teaching aid as well as a reference source.

I am indebted to the Literary Executor of the late Sir Ronald A. Fisher, F.R.S., to Dr. Frank Yates, F.R.S., and to Oliver and Boyd, Edinburgh, for permission to reprint Tables E.2 and E.4 from their book, *Statistical Tables for Biological, Agricultural and Medical Research.* Table E.3 was abridged from a table prepared by Catherine M. Thompson, and has been reproduced with the permission of the *Biometrika* Trustees. Data for Example 10.6 were partly taken from Table B.1 of the book *Multivariate Statistics for Personnel Classification*, by Rulon, Tiedeman, Tatsuoka, and Langmuir, with the permission of John Wiley & Sons, Inc.

To Ms. Karen Cattell, Director of the Institute for Personality and Ability Testing, I extend grateful thanks for permission to use as Chapter 9 my booklet

The General Linear Model: A ''New'' Trend in Analysis of Variance, published and copyrighted by IPAT.

I gratefully acknowledge the work of Dr. Menucha Birenbaum and Ms. Lih-Shing Wang, who prepared the solutions of all the exercises that were newly added in the present edition.

Finally, I take this opportunity to right a past wrong by acknowledging the painstaking work of my wife, Kikumi Tatsuoka, who wrote the long (10 pages) and complicated computer program for the Monte Carlo study to validate the multivariate omega-squared statistic discussed in Section 4.5. This program appeared in the Final Report of my project titled ''An Examination of the Statistical Properties of a Multivariate Measure of Strength of Relationship,'' funded by the U.S. Office of Education, Bureau of Research. But I failed to acknowledge my wife's contribution in that report, under the misguided notion that it was ''all in the family.''

M.M.T.

PREFACE TO THE FIRST EDITION

This book developed from a series of mimeographed class notes distributed in advanced-statistics and multivariate analysis courses, which I taught in the Department of Educational Psychology and the Department of Psychology, respectively, at the University of Illinois at Urbana-Champaign. I am indebted to the many students in these courses for having pointed out ways in which my presentation could be improved, and for detecting and correcting numerical errors in some of the examples.

However, I am most grateful to the late Phillip J. Rulon of the Graduate School of Education, Harvard University, under whose stimulating tutelage I was first introduced to multivariate analysis. His untimely death in June 1968 has been a source of great dismay and sadness for me. I had planned to dedicate this book to Professor Rulon as a token of my deep gratitude to him, and to show him that his wayward former student had returned to multivariate statistics after teaching science and mathematics at the undergraduate level for several years. Now, alas, I can only dedicate it to his memory.

I am also grateful to many other former teachers of mine, among whom Professors John B. Carroll, Frederick Mosteller, and David V. Tiedeman have given me constant encouragement and moral support by expressing interest in my work and by offering suggestions for topics to be included. My colleagues at the University of Illinois Psychology Department, Ledyard R. Tucker and Arthur Whimbey, each read parts of my manuscript and offered valuable suggestions.

I sincerely thank Edward H. Wolf, Rutgers University, and George P.H. Styan, McGill University, who reviewed the manuscript thoroughly and made many excellent suggestions. I followed most of these suggestions and made revisions. I alone am responsible for any remaining defects and inaccuracies.

M.M.T.

CONTENTS

1

INTRODUCTION

Multivariate statistical analysis, or *multivariate analysis* for short, is that branch of statistics which is devoted to the study of multivariate (or multidimensional) distributions and samples from those distributions. This, at least, is how the mathematical statistician would characterize this discipline. For applied statisticians and researchers who use statistics as a tool, however, this characterization—although technically correct—would leave much to be desired in communicative value; it may even sound circular.

1.1 SCOPE OF THIS BOOK AND PREREQUISITES FOR ITS STUDY

In applied contexts, particularly in educational and psychological research, multivariate analysis is concerned with a group (or several groups) of individuals, each of whom possesses values or scores on two or more variables, such as tests or other measures. We are interested in studying the interrelations among these variables, in looking for possible group differences in terms of these variables, and in drawing inferences relevant to these variables concerning the populations from which the sample groups were chosen.

The above, admittedly rather loose description of the subject matter of this book is very broad. The variables involved in a multivariate technique may be of various scale types: interval, ordinal, or nominal. Most often, the variables would also constitute two distinct sets playing opposite roles: a set of p independent variables and a set of q dependent variables. Double-classifying the variables according to the three scale types and two roles gives us six cells. When each cell is further subdivided in accordance with the number of independent and dependent variables ($p = 0, 1, \geq 2$; $q = 0, 1, \geq 2$), we get a table with a total of $6 \times 5 = 30$ cells. [The number is not 6×9 because three of the nine (p, q) combinations—(0, 0), (0, 1), and (1, 0)—cannot or do not occur, and the (1, 1) case does not represent a multivariate technique.] This type of table provides a convenient way to organize the various multivariate techniques and was used by Tatsuoka and Tiedeman (1963) for this purpose. We shall not display such a table here, but it may be helpful for the reader to keep the implied classification system in mind when studying this book.

As for the extent of statistical background required, a two-semester graduate (or advanced undergraduate) course in applied statistics, using such texts as Glass and Hopkins (1984) or Hays (1981) is presumed. Thus the reader is expected to be familiar with statistical inference in the one-dependent-variable ($q = 1$) case. Included in this rubric are such matters as interval estimation, univariate significance tests based on the normal, chi-square, t-, and F-distributions. It is also assumed that the reader has a fair mastery of bivariate correlation and regression analyses.

The question of how much mathematical background is necessary for tackling any textbook in applied statistics is always a difficult one to answer accurately. Many authors like to assure their prospective readers that "a working knowledge of high school algebra" is all that is required. This is usually true in the sense that nothing beyond high school algebra is involved in the actual development presented. Nevertheless, it often turns out, unhappily, that most students who have not had at least one or two college mathematics courses simply lack the mathematical "sophistication" or "way of thinking" to follow the text adequately. The same condition applies, perhaps with greater force than usual, to this book.

Although extensive use is made of matrix algebra, its elements are discussed in considerable detail in Chapter 2, and its slightly more advanced aspects are treated in Chapter 5. Once the basic rules of matrix algebra are learned, there is actually little beyond the skills of high school algebra that is needed to achieve, with practice, a working competence in matrix operations. However, to understand fully the geometric implications of those aspects of matrix algebra treated in Chapter 5, some familiarity with analytic geometry is necessary.

It is true that some differential calculus is used on a few occasions, but this is solely for the benefit of those who are familiar with the calculus. Those who are not may take the results on faith, and the chain of reasoning will not be broken, since differentiation is involved only in one step in each instance of its use. Thus it is probably fair to say that one or two college mathematics courses and an ability to follow mathematical thinking (or the willingness to acquire this ability) are the mathematical prerequisites for studying this book.

1.2 EXPOSITORY APPROACH AND ORGANIZATION OF THIS BOOK

Pointing out the analogy between a given multivariate technique and the corresponding univariate method is one of the principal didactic strategies used throughout this book, especially in Chapters 3, 4, and 8. Readers should soon come to appreciate the utility of matrix notation in highlighting this analogy and will thus be amply rewarded for the trouble they may have to take in learning the elements of matrix algebra in Chapter 2. Besides the drawing of analogies, other plausible arguments are of course presented. But rigorous mathematical proofs and derivations are held at a minimum except in the two chapters specifically devoted to supplying the necessary background in matrix algebra. The reader who is interested in pursuing more of the mathematics of multivariate analysis should consult such books as those by Anderson (1958) and Rao (1965). A recent book by Morrison (1976) is less mathematical than these, but somewhat more advanced than the present book.

It has already been mentioned that an introduction to the elements of matrix algebra (through matrix inversion) is presented in Chapter 2. Readers who are familiar with matrix algebra and the matrix algebraic solution of systems of linear equations may skim or omit this chapter.

Chapter 3 is concerned with multiple regression analysis, which is probably the most widely used multivariate technique. It starts with a review of simple (i.e., bivariate) regression analysis and draws heavily on an analogy with it to describe the rationale of multiple regression analysis. Here the reader is introduced to the close parallels between a univariate technique and its multivariate counterpart, which is highlighted by the use of matrix notation. Grasping the "correspondence principle" involving quadratic forms will facilitate an understanding of several of the subsequent chapters.

In Chapter 4 we discuss the multivariate counterparts of the familiar t-test for the significance of the difference between two group means (Hotelling's T^2-test) and of the F-test for the significance of overall differences among several group means (simplest case of Wilks' Λ-ratio test). As a preliminary to understanding the multivariate significance tests, the first half of the chapter is devoted to a study of the multivariate normal distribution and the problem of establishing confidence regions (corresponding to univariate confidence intervals) for the population centroid when the population variance–covariance matrix $\mathbf{\Sigma}$ is known. Once this is thoroughly grasped, it is a short step to understand the rationale of the T^2-test, even though the proof that this statistic follows an F-distribution is beyond our scope. The transition from the univariate test when σ is known to that when σ is unknown is very closely paralleled by the transition, in the multivariate situation, from the case when $\mathbf{\Sigma}$ is known to that when $\mathbf{\Sigma}$ is unknown. This parallelism is clearly shown by the use of matrix notation. The jump to the Λ-test for more than two groups is over a greater gap that cannot be bridged by elementary means. We therefore only show that a simple function of Λ is a natural multivariate extension of the ratio of between-groups to within-groups mean squares used in the univariate situation and point out that when the number of groups is two, the Λ-ratio reduces to a simple function of Hotelling's T^2. The final section is devoted to Box's test for the equality of population covariance matrices.

A further discourse on matrix algebra is the subject matter of Chapter 5. Here, the reader is introduced to the problem of determining the eigenvalues and eigenvectors of a matrix in a manner not usually found in books on matrix algebra per se. This is because the problem is developed in the statistical context of determining a linear combination of a given set of variables that has a larger variance than any other linear combination; it is also tied in with the geometric problem of axis rotation. Thus this chapter actually presents the essential mathematics of principal components analysis, so it prepares the reader for the brief introduction to factor analysis that is given in the following chapter; it may also be studied profitably in conjunction with a course in factor analysis. The first half of Section 5.6, in particular, is devoted to showing how principal-components analysis fits into the broader context of factor analysis, and the points made are illustrated by a numerical example in Appendix D. Some of the proofs in this chapter may be a little difficult to follow for readers with limited mathematical background. However, the results are stated in the form of numbered theorems and properties, so that students may skim or omit the proofs if desired, at least on a first reading, yet absorb the content of the theorems that are utilized later.

As intimated above, a brief introduction to factor analysis is presented in Chapter 6. It makes no pretense of being a complete discourse on this interesting but very complicated topic; rather, we attempt to give a simple conceptual overview of factor analysis in order to enable the reader to better understand such treatises as those of Harman (1967) and Mulaik (1972), which are devoted to this subject.

In Chapter 7 the mathematically related techniques of discriminant analysis and canonical correlation analysis are discussed. It should be emphasized that their relationship is mathematical rather than substantive, for their purposes are quite distinct. In discriminant analysis we seek linear combinations of a set of variables that best differentiate among several groups, while in canonical analysis we determine a linear combination of each of two sets of variables such that correlate most highly with each other. The relationship between the two techniques lies in the fact that the problem of discriminant analysis can be formally recast into a canonical correlation problem (although no computational benefit is gained by such recasting). It is with the hope that understanding either one of the two techniques will facilitate an understanding of the other that they are treated in the same chapter.

Chapter 8 is a short one on multivariate analysis of variance (MANOVA) for factorial experiments. The brevity of this chapter does not imply that this technique is any simpler than those discussed in the other chapters, only that the basic statistical principles involved have been sufficiently developed by this time that a brief exposition suffices. A detailed numerical example supplements the brief description, since the actual calculations are probably the most difficult aspect of MANOVA, once the rationale of Wilks' Λ-criterion is thoroughly understood—as we hope it will be after its repeated applications in Chapters 4 and 7.

Chapter 9 differs from the rest of this book in that it deals almost exclusively with a *univariate* technique, ANOVA, by the general linear model approach. The multivariate extension to MANOVA is merely outlined. This strategy was adopted because it was assumed that very few readers would have been exposed to the general linear model (GLM) even in the univariate case (although this may become less the case as more textbooks in intermediate applied statistics are coming to introduce GLM). It was therefore decided to go into considerable detail in describing the rationale of GLM, rather than making a half-hearted attempt at explaining GLM for MANOVA after a cursory description of its use for ANOVA. Readers who wish to learn more about GLM in its multivariate applications are referred to the excellent textbook by Lunneborg and Abbott (1983).

The final chapter deals with a rather special application of multivariate analysis, namely, to classification problems. It is addressed primarily to those readers who are concerned with educational or vocational guidance, although the versatile reasearcher will no doubt find its contents applicable, with some modification, to other fields of endeavor, such as clinical diagnosis and personality research. A brief discussion of the fairly recent nonparametric classification method called the ''*k*-nearest-neighbor'' rule is included.

Computer programs as such are not included in this book because good books on multivariate analysis in behavioral research that do list them are already in existence—such as those by Cooley and Lohnes (1971), Overall and Klett (1972), and Harris (1985). However, at least two exercises that require the use of a computer are included in each chapter. Data for 600 subjects, selected from a tape of the two-year follow-up study of the U.S. Department of Education–sponsored longitudinal project ''High School and Beyond'' (initiated in 1980) are given in Appendix F for use in these exercises. Individual instructors who teach multivariate statistics in the behavioral and social sciences will no doubt have their own preferences among the widely used packages that include multivariate procedures—such as BMD, BMDP, OSIRIS, SAS, SPSS, and SPSSx.

The present author holds the belief (perhaps outmoded by most standards) that much is to be gained by the student's going through the calculations by hand (i.e., by using only a hand-held electronic calculator) for at least a few small exercises. Next, for exercises of ''real-life'' magnitude, the student should—for relatively simple analyses such as multiple regression and significance tests for two-sample and K-sample problems—use matrix algebra procedures to carry out the analyses step by step instead of using a "canned" program that does everything automatically. (For more complicated analyses such as stepwise procedures in multiple regression analysis or discriminant analysis, it would, of course be unreasonable to expect students to develop their own programs, so canned programs will have to be used.) Students who have undergone this sort of learning experience will be more likely to develop a thorough understanding of the major steps involved in a sequence of computations than will those who, from the outset, leave all the "busy work" to the computer. They will consequently be in a better position later, when dealing with a real-life research problem, to make flexible and efficient use of available computer programs for the successive steps of its solution in the event that a tailor-made program for the complete solution in one fell swoop is unavailable. The point may be brought home by the true story of a graduate student whose thesis work was held up for several days while he had someone write a program for computing point-biserial correlation coefficients, a large number of which he had to calculate. All the while, a program for the usual product-moment correlation coefficient was available at the computer center to which he had access.

2

MATHEMATICAL PRELIMINARIES: SOME MATRIX ALGEBRA

Readers probably already have a general idea of what a *matrix* is. They have heard of a correlation matrix, a matrix of sociometric choices, and the like. They will have associated the term "matrix" with a rectangular (most often square) arrangement of numbers in rows and columns. Perhaps, however, they have thought of a matrix as being no more than a tabular display of certain types of data, and have not regarded it as a mathematical entity amenable to manipulations or operations (like addition and multiplication) analogous to those with which they are familiar in the realm of numbers themselves.

As a starting point, the notion of a matrix as a convenient rectangular display of numerical data is convenient and useful; it will remind us of the concrete meaning of the abstract entity with which we shall deal more and more as we proceed in this book. To be able to understand and apply the several multivariate statistical techniques described later, however, we must go beyond the mere tabular-display notion and develop some facility with the algebra of matrices. Although this chapter and Chapter 5 should provide sufficient understanding of matrix algebra for the purposes of this book, those who wish to pursue more advanced phases of multivariate analysis will find it profitable to consult such texts as Horst (1963), Searle (1966), and Green and Carroll (1976). A computer-oriented presentation of matrix algebra (among other things) is given by Ralston (1978).

2.1 DEFINITION AND BASIC TERMINOLOGY

A *matrix* is a rectangular arrangement of numbers in several *rows* (horizontal arrays) and several *columns* (vertical arrays). In applicational contexts, each row and each column has a specific referent in the field of interest, such as a particular test or experimental variable in educational or psychological measurement.

A familiar example is a *correlation matrix,* which is a square matrix (i.e., one in which the number of rows and the number of columns are equal) displaying the coefficients of correlation between pairs of tests. For instance, if three tests, say an algebra test (A), an English test (E), and a physics test (P), are given to a group of college applicants, the coefficients of correlation between pairs of these tests, for this group, may be displayed in a square matrix with three rows and three columns:

$$\begin{bmatrix} 1.00 & .32 & .47 \\ .32 & 1.00 & .35 \\ .47 & .35 & 1.00 \end{bmatrix}$$

This is called a 3×3 matrix (read ''three by three''); the first numeral refers to the number of rows and the second, to the number of columns. In this example the first row refers to test A, the second row to test E, and the third row to test P; the three columns also refer to these three tests in the same order.

The numerical entries are called the *elements* of the matrix. A particular element is specified by the number of the row and the number of the column in which it stands. Thus the ''.35'' found in the second row and third column of the correlation matrix above is called its (2, 3)-element; again, the first numeral refers to the row, and the second, to the column.

Often it is necessary to denote a matrix and its elements without specifying actual numerical values for the latter. It is then customary to denote the matrix by an uppercase letter in boldface type, like **A, B,** or **R** (often used for correlation matrices); its elements are then denoted by the corresponding lowercase letter with double subscripts indicating the row and column numbers. Thus a 4×3 matrix **A** may be denoted as

$$\mathbf{A} = \begin{bmatrix} a_{11} & a_{12} & a_{13} \\ a_{21} & a_{22} & a_{23} \\ a_{31} & a_{32} & a_{33} \\ a_{41} & a_{42} & a_{43} \end{bmatrix}$$

Sometimes this denotation is abbreviated to

$$\mathbf{A} = (a_{ij}) \qquad [i = 1, 2, 3, 4; \quad j = 1, 2, 3]$$

and the statement in brackets, indicating the range of i and the range of j, may be omitted if these are clear from the context.

In many contexts it becomes necessary simultaneously to consider a given matrix and an associated matrix obtained by writing the elements of each *row* of the former in the corresponding *column* of the latter. The matrix thus generated is called the *transpose* of the first matrix, and is denoted by suffixing a prime (′) to the symbol denoting the first matrix. For example, if

$$\mathbf{B} = \begin{bmatrix} 3 & -1 & 2 \\ -5 & 0 & 4 \end{bmatrix}$$

its transpose is

$$\mathbf{B}' = \begin{bmatrix} 3 & -5 \\ -1 & 0 \\ 2 & 4 \end{bmatrix}$$

(The elements 3, -1, 2 of the first row of **B** constitute, in this order, the elements of the first *column* of $\mathbf{B}'$; and similarly with -5, 0, 4.)

From the definition of a transpose, it is evident that the transpose of an $m \times n$ matrix is an $n \times m$ matrix. It is also easy to see that the transpose *of the transpose* of a given matrix is identical to the given matrix itself. Thus $(\mathbf{B}')'$ would be formed by writing 3 and -5 in its first column, -1 and 0 in its second, and 2 and 4 in its third column, so that we have

$$(\mathbf{B}')' = \begin{bmatrix} 3 & -1 & 2 \\ -5 & 0 & 4 \end{bmatrix}$$

which is none other than the original matrix **B** itself. We state this simple, but often useful fact as an equation for future reference: For any matrix **A**,

$$(\mathbf{A}')' = \mathbf{A} \tag{2.1}$$

Implicit in this use of the equality sign relating two indicated matrices is the definition of equality of two matrices: They are said to be *equal* if and only if they are identical—that is, if each element of the first matrix is equal to the corresponding element of the second matrix. Thus when we assert that

$$\mathbf{C} = \mathbf{B}$$

where **B** is the matrix (2.1), we are saying that

$$\begin{array}{lll} c_{11} = 3 & c_{12} = -1 & c_{13} = 2 \\ c_{21} = -5 & c_{22} = 0 & c_{23} = 4 \end{array}$$

A single matrix equation involving $m \times n$ matrices is therefore equivalent to mn ordinary algebraic equations. Symbolically, if **A** and **B** are $m \times n$ matrices,

$$\mathbf{A} = \mathbf{B} \Leftrightarrow a_{ij} = b_{ij} \qquad (\text{for } i = 1, 2, \ldots, m; \;\; j = 1, 2, \ldots, n) \tag{2.2}$$

(The double arrow "⇔" is read "if and only if.") The power of matrix algebra lies in allowing compact expressions of sets of equations and thereby facilitating their further manipulation.

A correlation matrix such as the one displayed at the beginning of this section exemplifies a square matrix, with the additional special feature that it is equal to its own transpose. Such a matrix, called a *symmetric matrix,* is characterized by the fact that for each i and j, the (i, j)-element and the (j, i)-element are equal. (In a correlation matrix, this is so because the two elements r_{ij} and r_{ji} both represent the coefficient of correlation between the ith and jth variables.) Most of the matrices that we encounter in multivariate statistical analysis will be square matrices, and many of these will be symmetric matrices—for reasons that will soon become evident.

Another type of matrix that will frequently be encountered later is the opposite extreme from a square matrix: namely, a matrix with only one row or one column, such as

$$[3, \ -1, \ 2, \ 4] \qquad \text{or} \qquad \begin{bmatrix} -2 \\ 1 \\ 5 \end{bmatrix}$$

It is quite proper to refer to such matrices as $1 \times n$ or $n \times 1$ matrices, but they are also called *vectors*. More specifically, a $1 \times n$ matrix is called an n-dimensional row vector and an $n \times 1$ matrix is called an n-dimensional column vector. (The reason for this geometrical nomenclature will be explained later.)

It is customary to denote a *column* vector by a lowercase letter in boldface type, and a *row* vector by a letter followed by a prime (since it is the transpose of a column vector). Thus an n-dimensional row vector with numerically unspecified elements will be indicated as, for example,

$$\mathbf{u}' = [u_1, u_2, u_3, \ldots, u_n]$$

while an n-dimensional column vector will be written as

$$\mathbf{v} = \begin{bmatrix} v_1 \\ v_2 \\ \vdots \\ v_n \end{bmatrix}$$

2.2 ADDITION OF MATRICES; MULTIPLICATION OF A MATRIX BY A SCALAR

Suppose that each of five sixth-grade pupils is given an arithmetic test which yields two subtest scores: one for computational skills and one for the ability to solve word problems. The 10 scores thus obtained may be arranged in a 5 × 2 matrix, called a *score matrix,* thus:

$$\mathbf{X} = \begin{bmatrix} 30 & 15 \\ 25 & 10 \\ 28 & 12 \\ 32 & 14 \\ 22 & 13 \end{bmatrix}$$

where each row shows the scores of a particular pupil, the first column lists the computational-skills scores, and the second column lists the word-problem scores.

Next, suppose that after a period of instruction, these same five pupils are given an alternative form of the arithmetic test. The second set of scores may be presented in another score matrix:

$$\mathbf{Y} = \begin{bmatrix} 34 & 17 \\ 27 & 12 \\ 30 & 16 \\ 35 & 17 \\ 26 & 15 \end{bmatrix}$$

where each row refers to the same pupil as did the corresponding row in the first score matrix **X**.

We may, for some reason, be interested in displaying the total score earned by each pupil for each part of the two forms of the test. We would then report the matrix

$$\mathbf{T} = \begin{bmatrix} 64 & 32 \\ 52 & 22 \\ 58 & 28 \\ 67 & 31 \\ 48 & 28 \end{bmatrix}$$

which is an element-by-element sum of the two matrices **X** and **Y** ($64 = 30 + 34$, $32 = 15 + 17$, $52 = 25 + 27$, etc.). This procedure for obtaining **T** from **X** and **Y** illustrates the definition of *matrix addition:* Each element of the sum matrix is the sum of the corresponding elements of two addend matrices. Symbolically, we may write

$$\mathbf{C} = \mathbf{A} + \mathbf{B} \Leftrightarrow c_{ij} = a_{ij} + b_{ij} \qquad \text{(for each } (i, j)\text{-pair} \qquad (2.3)$$

From this definition it should be clear that two matrices can be added if and only if they have the same number of rows *and* the same number of columns (briefly, if they are of the same *order*). Thus the expression **A** + **B** would be meaningless if, for example, **A** is a 3 × 4 matrix and **B** is a 3 × 3 matrix (or anything other than a 3 × 4 matrix).

Instead of the total score earned by each pupil on each part of the two forms of our arithmetic test, we may want to report the average score on each part for each pupil. Each average would, of course, be obtained by dividing the relevant element of **T** by 2—or, equivalently, by multiplying each element of **T** by $\frac{1}{2}$. The resulting "score" matrix (in which each "score" is really the average of two scores) is

$$\mathbf{A} = \begin{bmatrix} 32.0 & 16.0 \\ 26.0 & 11.0 \\ 29.0 & 24.0 \\ 33.5 & 15.5 \\ 24.0 & 14.0 \end{bmatrix}$$

In matrix notation, the relation between **A** and **T** is expressed as

$$\mathbf{A} = \tfrac{1}{2}\mathbf{T}$$

The rule for multiplying any matrix by a number (which, to emphasize its distinction from matrices and vectors, is called a *scalar*) is exemplified in the foregoing equation. The rule is simply to multiply each and every element of the matrix by that scalar. A general, symbolic statement of this is, therefore, as follows:

$$\mathbf{B} = k\mathbf{A} \Leftrightarrow b_{ij} = ka_{ij} \qquad \text{for each } (i, j)\text{-pair} \tag{2.4}$$

Finally, we may want to take the difference between corresponding scores on form Y (the form administered after the instructional period) and form X (the first-administered form) of our arithmetic test as a measure of the gain occurring between the pre- and postinstruction performances. An element-by-element subtraction of matrix **X** from matrix **Y** would give us the desired gain-score matrix, **G**:

$$\mathbf{G} = \mathbf{Y} - \mathbf{X} = \begin{bmatrix} 4 & 2 \\ 2 & 2 \\ 2 & 4 \\ 3 & 3 \\ 4 & 2 \end{bmatrix}$$

The rule for subtracting one matrix from another, just exemplified, is really a consequence for the addition rule (2.3) and a special case of the scalar-multiplication rule (2.4), in which k is given the value -1. That is because the difference, $a - b$, between any two numbers, a and b, is equivalent to the sum

$$a + (-1) \cdot b$$

When this fact is applied, element by element, in forming the difference, $\mathbf{A} - \mathbf{B}$, between two matrices, we are naturally led to the rule

$$\mathbf{C} = \mathbf{A} - \mathbf{B} \Leftrightarrow c_{ij} = a_{ij} - b_{ij} \qquad \text{for each } (i, j)\text{-pair} \qquad (2.3a)$$

EXERCISES

Given three matrices,

$$\mathbf{A} = \begin{bmatrix} 6 & 0 & -9 \\ 3 & -5 & 1 \end{bmatrix} \qquad \mathbf{B} = \begin{bmatrix} -1 & 0 & \frac{3}{2} \\ 2 & 1 & -\frac{1}{2} \end{bmatrix} \qquad \mathbf{C} = \begin{bmatrix} 0 & \frac{3}{2} \\ 0 & \frac{1}{6} \\ 0 & -\frac{1}{3} \end{bmatrix}$$

answer the following questions.

1. Which of the following matrix expressions are meaningful?

(a) $\mathbf{A} + \mathbf{B}$ (b) $\mathbf{B} + \mathbf{C}$ (c) $\mathbf{C}' + \mathbf{A}$ (d) $\mathbf{B}' - \mathbf{C}$ (e) $\mathbf{A} - \mathbf{C}$

Write each of the meaningful expressions as a single matrix, and state why the others are not meaningful.

2. Determine the matrices

$$\mathbf{D} = \tfrac{1}{3}\mathbf{A} \qquad \text{and} \qquad \mathbf{E} = 2\mathbf{B}$$

3. Verify that the matrix $\mathbf{D} + \mathbf{E}$ can be written as a scalar multiple of $\mathbf{C}'$.

4. By computing each side separately, verify the following equation:

$$(\mathbf{A} + \mathbf{B})' = \mathbf{A}' + \mathbf{B}'$$

2.3 MULTIPLICATION OF TWO MATRICES

So far, our operations with matrices have been straightforward "natural" extensions of the corresponding operations with numbers. The addition and subtraction of two matrices, and the multiplication of a matrix by a scalar, were each defined by an element-by-element operation using the corresponding operations on numbers, and nothing else.

In matrix multiplication we encounter, for the first time, an operation that goes beyond a "commonsense" extension of the corresponding operation with numbers. First, there is a rather peculiar restriction on the forms of pairs of matrices that can be multiplied to yield products. Moreover, this restriction is not that the two matrices must be of the same order (which is a rather "natural" restriction), as in the case of matrix addition. Instead, the condition to be satisfied by two matrices **A** and **B**, in order that their product **AB** can be formed, is that *the number of columns in* **A** be equal to *the number of rows in* **B**. For example, if **A** is a 2×3 matrix, then **B** must be a $3 \times n$ matrix for the product

AB to be defined (where n may be any natural number). An illustration will clarify both the procedure of matrix multiplication and the reason for the restriction.

Let

$$\mathbf{A} = \begin{bmatrix} 2 & -3 & 1 \\ -1 & 4 & 0 \end{bmatrix}$$

and

$$\mathbf{B} = \begin{bmatrix} 3 & 1 \\ 4 & 2 \\ 5 & -3 \end{bmatrix}$$

Then, since **A** has three columns and **B** has three *rows,* the product **AB** = **C** (say) can be formed. The procedure is as follows:

1. Each element of the first *row* of **A** is multiplied by the corresponding element of the first *column* of **B**, and the resulting products are added. This sum is c_{11}, the (1, 1)-element of the product matrix **C**. Thus for the present example,

$$c_{11} = (2)(3) + (-3)(4) + (1)(5) = 6 - 12 + 5 = -1$$

2. The elements of the first row of **A** are multiplied by the respectively corresponding elements of the second column of **B** and the products added, to obtain c_{12}:

$$c_{12} = (2)(1) + (-3)(2) + (1)(-3) = 2 - 6 - 3 = -7$$

3. Similar operations with the second-row elements of **A** and the first-column elements of **B** yield c_{21}:

$$c_{21} = (-1)(3) + (4)(4) + (0)(5) = -3 + 16 + 0 = 13$$

4. Similarly,

$$c_{22} = (-1)(1) + (4)(2) + (0)(-3) = -1 + 8 + 0 = 7$$

Arranging the four elements obtained above in a matrix, we have the product matrix

$$\mathbf{AB} = \mathbf{C} = \begin{bmatrix} -1 & -7 \\ 13 & 7 \end{bmatrix}$$

These operations for computing the elements of a product matrix may be summarized in the general statement

$$\mathbf{C} = \mathbf{AB} \Leftrightarrow c_{ij} = \sum_{k=1}^{s} a_{ik} b_{kj}$$

$$(\text{for } i = 1, 2, \ldots, m; \quad j = 1, 2, \ldots, n) \qquad (2.5)$$

where **A** is an $m \times s$ matrix and **B** is an $s \times n$ matrix; hence **C** is an $m \times n$ matrix.

Careful examination of the operations involved in matrix multiplication, illustrated above, should reveal several things:

1. The reason the number of *columns* in the first factor matrix (**A** in our example) must agree with the number of *rows* of the second factor (**B** in our example) is that the elements of each row of the former are "paired off" with elements of each *column*, in turn, of the latter.
2. The product matrix (**C**, above) has as many rows as does the first factor, and as many columns as does the second factor. Thus if a $p \times q$ matrix is multiplied by a $q \times r$ matrix, the product is a $p \times r$ matrix.
3. The (i, j)-element of the product matrix is the result of "pairing" the ith row of the first factor with the jth column of the second factor.
4. The order of multiplication is important. That is, unlike the multiplication of two numbers (where $ab = ba$), **AB** and **BA** are usually not equal to each other; in fact, even when one of these products exists, the other may be undefined. (Consider, for example, the case when **A** is a 3×4 matrix and **B** is 5×3. Which order of multiplication is possible in this case?)

The property of matrix multiplication noted above constitutes an important departure of matrix algebra from the algebra of numbers. Mathematicians describe this difference by saying that although the multiplication of numbers is *commutative* ($ab = ba$), the multiplication of matrices is *noncommutative* ($\mathbf{AB} \neq \mathbf{BA}$, in general). In contrast, it should be noted that the commutative law holds for matrix addition, as should be clear from definition (2.3). That is, for any two matrices of the same order, $\mathbf{A} + \mathbf{B} = \mathbf{B} + \mathbf{A}$, just as $a + b = b + a$ for any two numbers.

In view of the noncommutativity of matrix multiplication, it would be ambiguous to speak simply of "multiplying **A** and **B**." Reference must be made to the order in which the factors are taken. Thus, if the product **AB** is intended, we say that "**A** is *postmultiplied* by **B**" or that "**B** is *premultiplied* by **A**;" the letters "**A**" and "**B**" are reversed in both of these descriptions when the product **BA** is implied.

The Addition-Operator Vector

A vector of special utility in statistical operations is one whose elements are all equal to 1. Such a vector is denoted by $\mathbf{1}_a$ or $\mathbf{1}'_b$ depending on whether it has a column or a row form, respectively. The subscript in each case indicates the dimensionality and may be omitted when this is evident from the context. For example,

$$\mathbf{1}_3 = \begin{bmatrix} 1 \\ 1 \\ 1 \end{bmatrix} \qquad \text{and} \qquad \mathbf{1}'_4 = [1, 1, 1, 1]$$

From the definition of matrix multiplication given above, it is evident that postmultiplying an $n \times p$ score matrix by $\mathbf{1}_p$ will produce an n-dimensional column vector whose successive elements are the sums of the p test scores for the successive examinees. Similarly, when the same score matrix is premultiplied by $\mathbf{1}'_n$, the result is a p-dimensional row vector exhibiting the sums, over the n examinees, of the p tests. That is, if $\mathbf{X}$ is a score matrix, then

$$\mathbf{X1} \qquad \text{and} \qquad \mathbf{1}'\mathbf{X}$$

give the column of row totals of $\mathbf{X}$, and the row of column totals of $\mathbf{X}$, respectively. For this reason, $\mathbf{1}$ (or $\mathbf{1}'$) is called an addition-operator vector. Using the 5×2 score matrix $\mathbf{X}$ shown at the beginning of Section 2.2 as an example, we get

$$\mathbf{X}\,\mathbf{1}_2 = \begin{bmatrix} 30 & 15 \\ 25 & 10 \\ 28 & 12 \\ 32 & 14 \\ 22 & 13 \end{bmatrix} \begin{bmatrix} 1 \\ 1 \end{bmatrix} = \begin{bmatrix} 45 \\ 35 \\ 40 \\ 46 \\ 35 \end{bmatrix}$$

and

$$\mathbf{1}'_5\,\mathbf{X} = [1, 1, 1, 1, 1] \begin{bmatrix} 30 & 15 \\ 25 & 10 \\ 28 & 12 \\ 32 & 14 \\ 22 & 13 \end{bmatrix} = [137, 64]$$

An Example of Statistical Import

The matrix operations previously expounded may be applied to an example foreshadowing the important role of matrices in multivariate statistical analysis. Consider, once again, the 5×2 score matrix displayed at the beginning of Section 2.2:

$$\mathbf{X} = \begin{bmatrix} 30 & 15 \\ 25 & 10 \\ 28 & 12 \\ 32 & 14 \\ 22 & 13 \end{bmatrix}$$

whose elements are the scores earned by five pupils on two subtests (computation and word problems) of an arithmetic test. Now, the means of the five scores on the two subtests are 27.4 and 12.8, respectively. Let us define a 5×2 *mean-score matrix*

$$\bar{\mathbf{X}} = \begin{bmatrix} 27.4 & 12.8 \\ 27.4 & 12.8 \\ 27.4 & 12.8 \\ 27.4 & 12.8 \\ 27.4 & 12.8 \end{bmatrix}$$

The elements in the first column are all equal to the mean of the computations subtest, and those in the second column are equal to the mean of the word-problems subtest. By Eq. 2.3a, defining the difference between two matrices, the *deviation-score matrix* may be written as

$$\mathbf{x} = \mathbf{X} - \bar{\mathbf{X}} = \begin{bmatrix} 2.6 & 2.2 \\ -2.4 & -2.8 \\ 0.6 & -0.8 \\ 4.6 & 1.2 \\ -5.4 & 0.2 \end{bmatrix}$$

Next, let us form the product matrix $\mathbf{x}'\mathbf{x}$. Following the rules for matrix multiplication given earlier, we find

$$\mathbf{x}'\mathbf{x} = \begin{bmatrix} 63.2 & 16.4 \\ 16.4 & 14.8 \end{bmatrix}$$

The important point is that the elements of this product matrix are familiar statistical quantities, as may readily be verified by retracing how each element was computed. For example, the (1, 1)-element was obtained as

$$(2.6)^2 + (-2.4)^2 + (0.6)^2 + (4.6)^2 + (-5.4)^2 = 63.2$$

Thus this element is precisely the "sum of squares" (i.e., the sum of squared deviations from the mean) for the first subtest, which, following customary statistical notation, may be denoted by $\Sigma\, x_1^2$. (A more complete notation would require a second subscript, indicating the pupil number, after the "1" and we would write

$$\sum_{i=1}^{5} x_{1i}^2$$

but it is conventional to abbreviate this as $\Sigma\, x_1^2$ when the context specifies the number of scores involved.) Similarly, the (2, 2)-element gives the value of

Σx_2^2, the sum of squares for the second subtest. The two off-diagonal elements (which are equal in value) give the sum of cross-products Σx_1x_2 (and, equivalently, Σx_2x_1) between the two subtests.

Thus the elements of $\mathbf{x}'\mathbf{x}$ are such that when divided by the number of degrees of freedom (n.d.f. = 4, in this example), they yield the sample variances (along the diagonal) and the sample covariances (off-diagonal) for the tests involved. Symbolically, we may write for the general case of p tests,

$$\mathbf{x}'\mathbf{x} = \begin{bmatrix} \Sigma x_1^2 & \Sigma x_1x_2 & \cdots & \Sigma x_1x_p \\ \Sigma x_2x_1 & \Sigma x_2^2 & \cdots & \Sigma x_2x_p \\ \vdots & & \ddots & \vdots \\ \Sigma x_px_1 & \Sigma x_px_2 & \cdots & \Sigma x_p^2 \end{bmatrix} \tag{2.6}$$

Such a matrix will be referred to as a *sums of squares and cross-products,* or *SSCP, matrix.* It is evident that any SSCP matrix is symmetrical, just as is a correlation matrix (with which the former is closely related, as shown in Exercise 8 at the end of Section 2.4).

Another point worth mentioning in connection with an SSCP matrix is that it can be computed directly from the raw-score matrix and the mean-score matrix without first constructing the deviation-score matrix. This is accomplished by using the following identity, which may be easily verified by recalling a corresponding identity for the sum of squares of a single test.

$$\mathbf{x}'\mathbf{x} = \mathbf{X}'\mathbf{X} - \overline{\mathbf{X}}'\overline{\mathbf{X}} \tag{2.7}$$

The corresponding equation for the univariate case is

$$\Sigma x^2 = \Sigma X^2 - N(\overline{X})^2$$

where N is the sample size, that is, the number of scores at hand. The analogy becomes closer when this equation is rewritten as

$$\Sigma x^2 = \Sigma X^2 - \Sigma \overline{X}^2$$

No summation sign appears in the matrix counterpart Eq. 2.7 because the summation is already "built in" in the matrix products.

An alternative way to compute an SSCP matrix is to utilize the addition-operator matrix introduced above. Before we do this, however, it is necessary to discuss a special case of matrix multiplication, that of a row vector and a column vector of the same dimensionality. Multiplications in both orders are possible, but the results are quite different. For example, if

$$\mathbf{u}' = [1, \ -2, \ 3]$$

and

$$\mathbf{v} = \begin{bmatrix} 3 \\ 1 \\ -2 \end{bmatrix}$$

then

$$\mathbf{u}'\mathbf{v} = [1, \; -2, \; 3] \begin{bmatrix} 3 \\ 1 \\ -2 \end{bmatrix} = (1)(3) + (-2)(1) + (3)(-2) = -5$$

which is a scalar. For this reason, $\mathbf{u}'\mathbf{v}$ is called the *scalar product* of $\mathbf{u}$ and $\mathbf{v}$. It is also called the *inner product* or *dot product* of these two vectors.

On the other hand, multiplication in the reverse order yields what is called the *matrix product:*

$$\mathbf{v}\mathbf{u}' = \begin{bmatrix} 3 \\ 1 \\ -2 \end{bmatrix} [1, \; -2, \; 3]$$

$$= \begin{bmatrix} 3 \times 1 & 3 \times (-2) & 3 \times 3 \\ 1 \times 1 & 1 \times (-2) & 1 \times 3 \\ (-2) \times 1 & (-2) \times (-2) & (-2) \times (3) \end{bmatrix} = \begin{bmatrix} 3 & -6 & 9 \\ 1 & -2 & 3 \\ -2 & 4 & -6 \end{bmatrix}$$

a 3×3 matrix. (Note that a "degenerate case" of the rule for matrix multiplication is applied here: Each "row" of $\mathbf{v}$ contains just one element, as does each "column" of $\mathbf{u}'$.)

It is the matrix product of two vectors that we shall need to use in conjunction with the addition-operator vector $\mathbf{1}'$ to derive the alternative computational formula for an SSCP matrix. Recalling from Eq. 2.7 that

$$\mathbf{x}'\mathbf{x} = \mathbf{X}'\mathbf{X} - \overline{\mathbf{X}}'\overline{\mathbf{X}}$$

we now proceed to transform the second term $\overline{\mathbf{X}}'\overline{\mathbf{X}}$ on the right-hand side. It was shown that if $\mathbf{X}$ is an $n \times p$ score matrix, then

$$\mathbf{1}_n'\mathbf{X} = [T_1, T_2, \ldots, T_p]$$

where

$$T_j = \sum_{i=1}^{n} X_{ij} \qquad (j = 1, 2, \ldots, p)$$

is the total, for the group of n examinees, of the jth test, X_j. Abbreviating this row vector of totals by $\mathbf{t}_p'$ (and hence its transpose, a column vector, by $\mathbf{t}_p$), the matrix product of $\mathbf{t}_p$ and $\mathbf{t}_p'$ is readily seen to be

$$\mathbf{t}_p\mathbf{t}_p' = \begin{bmatrix} T_1^2 & T_1T_2 & \cdots & T_1T_p \\ T_2T_1 & T_2^2 & \cdots & T_2T_p \\ \vdots & & & \\ T_pT_1 & T_pT_2 & \cdots & T_p^2 \end{bmatrix}$$

$$= n(\overline{\mathbf{X}}'\overline{\mathbf{X}})$$

since the general element $(\overline{\mathbf{X}}'\overline{\mathbf{X}})_{jk}$ of $\overline{\mathbf{X}}'\overline{\mathbf{X}}$ is, as mentioned earlier, $n\overline{X}_j\overline{X}_k$, which is equivalent to

$$n(T_j/n)(T_k/n) = T_jT_k/n$$

Hence

$$\overline{\mathbf{X}}'\overline{\mathbf{X}} = \mathbf{t}_p\mathbf{t}_p'/n$$

Substituting this in Eq. 2.7 and dropping the subscript p for generality, we get

$$\begin{aligned} \mathbf{x}'\mathbf{x} &= \mathbf{X}'\mathbf{X} - \mathbf{t}\mathbf{t}'/n \\ &= \mathbf{X}'\mathbf{X} - (\mathbf{1}'\mathbf{X})'(\mathbf{1}'\mathbf{X})/n \end{aligned} \tag{2.8}$$

This formula is more convenient than Eq. 2.7 for computing an SSCP matrix, especially when the computation is to be done on a computer or microcomputer using a matrix-manipulation package. The step of obtaining $\mathbf{1}'\mathbf{X}$ need not be done literally in this way but may be accomplished by a single command for getting the row vector of column sums of the score matrix $\mathbf{X}$. The expression $\mathbf{1}'\mathbf{X}$ merely serves as the convenient mnemonic to issue this command.

Properties of Matrix Multiplication

Although we emphasized above a difference between the multiplication of matrices and the multiplication of numbers, there are also important points of similarity. One is the property of *associativity*. The reader has long been familiar with the fact that given any three numbers a, b, and c, their triple product abc may be formed equivalently as either $(ab)c$ or $a(bc)$. [For example, $(2 \times 3) \times 4 = 6 \times 4 = 24$; and $2 \times (3 \times 4) = 2 \times 12 = 24$]. This property of associativity also holds for matrix multiplication. That is, if $\mathbf{A}$, $\mathbf{B}$, and $\mathbf{C}$ are three matrices such that the product $\mathbf{AB}$ can be formed, and this product matrix, in turn, can be postmultiplied by $\mathbf{C}$ to yield a triple product $(\mathbf{AB})\mathbf{C}$, then $\mathbf{A}(\mathbf{BC})$ (i.e., postmultiplying $\mathbf{A}$ by the product $\mathbf{BC}$) is also a legitimate triple product, and the two triple products are identical. We state this property in the equation

$$(\mathbf{AB})\mathbf{C} = \mathbf{A}(\mathbf{BC}) \tag{2.9}$$

Because of this property, either triple product may, without ambiguity, be denoted as $\mathbf{ABC}$, omitting the parentheses. As an exercise, the reader should verify

Eq. 2.9 by computing both (**AB**)**C** and **A**(**BC**) and noting that they are equal, using the following three matrices:

$$\mathbf{A} = \begin{bmatrix} 3 & 1 \\ 2 & -1 \\ -1 & 2 \end{bmatrix} \qquad \mathbf{B} = \begin{bmatrix} 5 & 2 \\ 3 & 1 \end{bmatrix} \qquad \mathbf{C} = \begin{bmatrix} 1 & 2 & 3 & 4 \\ 1 & -1 & 0 & 2 \end{bmatrix}$$

A type of triple product that occurs very frequently in the formulas of multivariate analysis is that having the general form $\mathbf{u}'\mathbf{A}\mathbf{u}$, where $\mathbf{u}$ is an n-dimensional column vector, $\mathbf{u}'$ is its transpose, and $\mathbf{A}$ is an $n \times n$ square matrix. Since $\mathbf{u}'$ consists of only one row and $\mathbf{u}$ of only one column, the indicated product $\mathbf{u}'\mathbf{A}\mathbf{u}$ denotes a 1×1 "matrix," that is, a scalar. For example, if

$$\mathbf{u}' = [3 \quad -1 \quad 2]$$

and

$$\mathbf{A} = \begin{bmatrix} 5 & 2 & -1 \\ -3 & 4 & 2 \\ 1 & 2 & 3 \end{bmatrix}$$

then

$$\mathbf{u}'\mathbf{A}\mathbf{u} = [3 \quad -1 \quad 2] \begin{bmatrix} 5 & 2 & -1 \\ -3 & 4 & 2 \\ 1 & 2 & 3 \end{bmatrix} \begin{bmatrix} 3 \\ -1 \\ 2 \end{bmatrix}$$

$$= [20 \quad 6 \quad 1] \begin{bmatrix} 3 \\ -1 \\ 2 \end{bmatrix} = 56$$

It is interesting to note that when the elements of $\mathbf{u}$ and $\mathbf{A}$ are indicated by letters, the result of carrying out the indicated product $\mathbf{u}'\mathbf{A}\mathbf{u}$ can be written as a rather simple algebraic expression involving the u_i and a_{ij}:

$$\mathbf{u}'\mathbf{A}\mathbf{u} = a_{11}u_1^2 + a_{22}u_2^2 + a_{33}u_3^2 + (a_{12} + a_{21})u_1u_2 + (a_{13} + a_{31})u_1u_3 + (a_{23} + a_{32})u_2u_3 \tag{2.10}$$

An expression of this form is called a *quadratic form* (of the u_i). Observe that each term of this expression consists either of the square of an element of $\mathbf{u}$ or of the product of two different elements of $\mathbf{u}$, each multiplied by a coefficient that comes from $\mathbf{A}$ according to the following rules:

1. The coefficient of u_i^2 is a_{ii}, for each i. (An element such as a_{ii}—where the row and column indices are equal—of a square matrix is called a *diagonal element*. Such elements stand on the diagonal line extending from the upper left corner to the lower right corner of the matrix.)

2. The coefficient of $u_i u_j$ (where $i \neq j$) is $(a_{ij} + a_{ji})$, for each (i, j)-pair. [It may be mnemonically better to imagine the term $(a_{ij} + a_{ji})u_i u_j$ decomposed into $a_{ij}u_i u_j + a_{ji}u_j u_i$.]

These rules provide us with an alternative method for evaluating triple products of the stated form, without going through the two-stage matrix multiplication indicated. It is good computational policy to calculate quadratic forms in both ways, as an arithmetic check. As an exercise, verify the result $\mathbf{u'Au} = 56$ for the example given earlier, using the alternative computational scheme.

Another property common to the multiplication of matrices and the multiplication of numbers is that of *distributivity* of multiplication over addition. This principle, quite familiar to the reader in the realm of numbers, asserts that $a(b + c) = ab + ac$. We state this property for matrices in the equations

$$\mathbf{A(B + C) = AB + AC} \tag{2.11a}$$

$$\mathbf{(B + C)A = BA + CA} \tag{2.11b}$$

Two equations are needed to make it clear that distributivity holds both when the indicated sum $(\mathbf{B + C})$ occurs as the second factor and when it comes first.

One other principle of matrix algebra connected with products is the rule concerning the transpose of the product of two or more matrices. This rule may be stated, for the case of two matrices, as follows:

$$\mathbf{(AB)' = B'A'} \tag{2.12}$$

That is, the transpose of the product of two matrices is equal to the product of their respective transposes, multiplied in the *reverse order*. Verification of this rule is left to the reader. To remember that $(\mathbf{AB})'$ is *not* equal to $\mathbf{A'B'}$, as one might unwarily think it is, we need only recall the following. If, for example, $\mathbf{A}$ is 3×4 and $\mathbf{B}$ is 4×5, then $\mathbf{AB}$ is 3×5; so $(\mathbf{AB})'$ is 5×3. $\mathbf{B'A'}$ is $(5 \times 4) \times (4 \times 3) = 5 \times 3$, while $\mathbf{A'B'}$ is not even defined.

EXERCISES

Do the following problems with reference to the three matrices

$$\mathbf{A} = \begin{bmatrix} 3 & 1 \\ 2 & 0 \\ -1 & 2 \end{bmatrix} \qquad \mathbf{B} = \begin{bmatrix} 1 & -2 & 1 \\ -1 & 2 & 1 \end{bmatrix} \qquad \mathbf{C} = \begin{bmatrix} 2 & 3 & -1 \\ 1 & -2 & 1 \end{bmatrix}$$

and whatever other matrices that may be given in the respective problems.

1. Compute $\mathbf{A(B + C)}$ and $\mathbf{AB + AC}$, thereby verifying Eq. 2.11a.

2. Verify Eq. 2.11b by computing $\mathbf{(B + C)A}$ and $\mathbf{BA + CA}$.

3. Compute $\mathbf{(AB)C'}$ and $\mathbf{A(BC')}$. What law do the results verify?

4. Verify that $[(\mathbf{AB})\mathbf{C}']' = \mathbf{CB}'\mathbf{A}'$. Which two equations does this result verify?

5. Further, let

$$\mathbf{u}' = [1 \quad 2 \quad -1]$$

Compute the quadratic form $\mathbf{u}'(\mathbf{AB})\mathbf{u}$ in two ways.

6. Let

$$\mathbf{D} = \begin{bmatrix} 2 & 0 \\ 0 & 3 \end{bmatrix}$$

Compute **DC**. How is this product related to **C**? (That is, state a simple rule by which you could immediately write out **DC**, given **C**.)

7. Compute **AD**, where **D** is as given in Exercise 6. What do you notice about this product in relation to **A**?

8. Let

$$\mathbf{E} = \begin{bmatrix} 5 & 0 & 0 \\ 0 & 5 & 0 \\ 0 & 0 & 5 \end{bmatrix}$$

Compute **EA**. How does this result compare with **A**?

9. The following score matrix **Y** lists the scores on three subtests of an English test that were earned by a group of 10 high school juniors. (To save space, $\mathbf{Y}'$ instead of **Y** itself is displayed, so that each *column* shows the three subtest scores of one student.)

$$\mathbf{Y}' = \begin{bmatrix} 15 & 12 & 17 & 10 & 17 & 11 & 15 & 19 & 16 & 10 \\ 12 & 9 & 12 & 11 & 11 & 11 & 12 & 12 & 11 & 13 \\ 22 & 15 & 22 & 24 & 17 & 32 & 27 & 29 & 39 & 16 \end{bmatrix}$$

Use the addition-operator vector **1** or $\mathbf{1}'$, as the case may be, for indicating as well as actually calculating **(a)** the *column* vector of the total English test scores earned by the 10 students; and **(b)** the *row* vector showing the three subtest totals across the group. Be sure to indicate the dimensionality of the applicable addition-operator vector by subscripting the **1** or $\mathbf{1}'$ by 3 or 10, as appropriate.

10. Use Eqs. 2.7 and 2.8 (with the **X**'s replaced by **Y**'s, of course), in turn, to compute the SSCP matrix for the three English subtests for the sample of 10 students whose score matrix is given in Exercise 9.

11. Show that another way to write Eq. 2.8 is

$$\mathbf{x}'\mathbf{x} = \mathbf{X}'(\mathbf{I} - (\mathbf{1}\,\mathbf{1}')/n)\mathbf{X} \qquad (2.8a)$$

citing the property of matrix multiplication that you are using for each step in going from Eq. 2.8 to Eq. 2.8a. Although form (2.8a) is not particularly convenient for computational purposes, it is sometimes useful for theoretical purposes (i.e., in deriving new formulas), as we shall have occasion to see in Chapter 5.

2.4 THE IDENTITY MATRIX AND MATRIX INVERSION

Exercises 6–8 in Section 2.3 involved square matrices that were peculiar in that their elements were all 0's except along the diagonal from the upper left corner to the lower right corner (which is called the *main diagonal* of the matrix). Such a matrix is known as a *diagonal matrix,* and the reader has observed the effects of premultiplying and postmultiplying a given matrix by a diagonal matrix. Namely, premultiplying any $n \times m$ matrix $\mathbf{A}$ by an $n \times n$ diagonal matrix

$$\mathbf{D} = \begin{bmatrix} d_1 & 0 & 0 & \cdots & 0 \\ 0 & d_2 & 0 & \cdots & 0 \\ \vdots & & & \ddots & \vdots \\ 0 & 0 & 0 & \cdots & d_n \end{bmatrix}$$

has the following effect: Each element in the ith *row* of $\mathbf{A}$ gets multiplied by d_i the (i, i)-element of $\mathbf{D}$. Symbolically,

$$\mathbf{B} = \mathbf{DA} \Leftrightarrow b_{ij} = d_i a_{ij} \quad (i = 1, 2, \ldots, n; \quad j = 1, 2, \ldots, m) \qquad (2.13a)$$

Similarly, if $\mathbf{A}$ is an $m \times n$ matrix, then

$$\mathbf{B} = \mathbf{AD} \Leftrightarrow b_{ij} = d_j a_{ij} \qquad (2.13b)$$

That is, when $\mathbf{A}$ is postmultiplied by a diagonal matrix $\mathbf{D}$, the elements in each *column* of $\mathbf{A}$ get multiplied by the corresponding diagonal element of $\mathbf{D}$.

Exercise 8 involved a diagonal matrix $\mathbf{E}$ with the further special property that all its diagonal elements are equal. It follows from Eqs. 2.13a and 2.13b that either pre- or postmultiplying by a matrix like $\mathbf{E}$ gives rise to a matrix each of whose elements is equal to a constant multiple of the corresponding element of the original matrix, the multiple being the common value of the diagonal elements of $\mathbf{E}$ (which was 5 in Exercise 8). But referring to Eq. 2.4, we see that this is the same result as would be obtained when a matrix is multiplied by a scalar. For this reason, a diagonal matrix with all its diagonal elements equal is called a *scalar matrix*.

A special type of scalar matrix is one whose diagonal elements are all equal to unity. It should be clear from the foregoing discussion that the effect of pre- or postmultiplying any matrix by this special scalar matrix is to leave the given matrix unchanged. That is, the effect is the same as that of multiplying by 1 in the realm of numbers. ($x \cdot 1 = x$, for any number x.) A scalar matrix of the form

$$\begin{bmatrix} 1 & 0 & \cdots & 0 \\ 0 & 1 & \cdots & 0 \\ \vdots & \vdots & \ddots & \vdots \\ 0 & 0 & \cdots & 1 \end{bmatrix}$$

is, therefore, called an *identity matrix,* and is denoted by the letter **I**, with a subscript to indicate its order if necessary. Thus, if **A** is any $n \times n$ matrix,

$$\mathbf{I}_n\mathbf{A} = \mathbf{A}\mathbf{I}_n = \mathbf{A} \tag{2.14}$$

(If **A** is not a square matrix but an $n \times m$, then $\mathbf{I}_n\mathbf{A} = \mathbf{A}$ and $\mathbf{A}\mathbf{I}_m = \mathbf{A}$.)

Having established the matrix analog of the number 1, a natural question to ask next would be whether there exists an analog to the reciprocal of a number. That is, given a matrix **A**, can we find another matrix **X** such that $\mathbf{AX} = \mathbf{I}$ (just as, for any number $a \neq 0$, there exists a reciprocal $x = 1/a$ such that $ax = 1$)? It turns out that subject to certain restrictions (which will become clear later), the answer is affirmative. But the determination of such an **X** for a nonsquare matrix **A** requires more advanced techniques than we now have at our command. Its discussion will therefore be deferred to Chapter 5, and for the present we confine our attention to square matrices. For example, if

$$\mathbf{A} = \begin{bmatrix} 2 & 4 & 3 \\ -3 & 2 & 1 \\ -1 & 3 & 2 \end{bmatrix}$$

the reader may readily verify, by forming the product **AX**, that

$$\mathbf{X} = \begin{bmatrix} 1 & 1 & -2 \\ 5 & 7 & -11 \\ -7 & -10 & 16 \end{bmatrix}$$

is its "reciprocal" matrix, which is usually known as its *inverse matrix.* In analogy with the exponential notation for a reciprocal of a number ($1/a = a^{-1}$), the inverse of a given matrix **A**, when it exists, is denoted as $\mathbf{A}^{-1}$. It is true, only for square matrices, that either pre- or postmultiplying by the inverse yields the identity. That is, $\mathbf{A}^{-1}\mathbf{A} = \mathbf{A}\mathbf{A}^{-1} = \mathbf{I}$.

To compute the inverse of a given square matrix from first principles, one needs a working familiarity with the concept of a *determinant*. A brief summary of the rules for computing a determinant is given in Appendix A, which should suffice as a working guide. [However, the reader who has never before been introduced to determinants may want to consult a college algebra text, such as Beckenbach, Drooyan, and Wooton (1978, pp. 263–277), for a more detailed presentation. Here we shall assume a knowledge of determinants and illustrate the basic computational method for finding the inverse of a 3×3 matrix. The method will be shown to reduce almost to triviality for a 2×2 matrix. For practical computation of the inverse of a matrix of order greater than 3×3, however, a systematized routine, such as that described in Appendix B, becomes a near-necessity. Other computational routines may be found in Horst's *Matrix Algebra for Social Scientists*. With the widespread availability of computer software for matrix operations on both mainframes and microcomputers, it may be thought that such hand-computation routines are obsolete. Nevertheless, there are certain advantages to be gained from "suffering" through such computations for at least a few exercises—especially for those who wish to understand something about the theory of matrices rather than just having a conceptual familiarity.

Thus, Exercises 1, 2 and 5 on page 00 are intended for hand-calculation. On the other hand, Exercise 9 definitely requires a computer, while Exercise 10 may be done either way, following the algebraic work done in Exercise 8. For matrix calculations on a computer, readers who are familiar with some programming language such as FORTRAN, BASIC or PASCAL are encouraged to use these. Otherwise, using packaged matrix routines such as the SAS (1979) PROC MATRIX or the SAS/IML (1985) is recommended. Any one of these routines enables the user to carry out matrix operations by simply issuing commands that are very closely parallel to standard matrix formulas. For example, the command

```
G = A - B*(C** - 1)*B'
```

will cause the computer to calculate the matrix $\mathbf{A} - \mathbf{B}\mathbf{C}^{-1}\mathbf{B}'$ and assign to the result the symbol $\mathbf{G}$.

EXAMPLE 2.1 Find the inverse of

$$\mathbf{A} = \begin{bmatrix} 1 & -1 & 2 \\ 3 & 4 & -2 \\ -2 & 1 & 3 \end{bmatrix}$$

Step 1. Evaluate the determinant $|\mathbf{A}|$ of the matrix $\mathbf{A}$.

As stated in Appendix A, this may be done by expanding the determinant with respect to any one of its rows or columns. We illustrate by expanding along the first row. The required *cofactors* are as follows:

$$A_{11} = \begin{vmatrix} 4 & -2 \\ 1 & 3 \end{vmatrix} = (4)(3) - (-2)(1) = 14$$

$$A_{12} = -\begin{vmatrix} 3 & -2 \\ -2 & 3 \end{vmatrix} = -[(3)(3) - (-2)(-2)] = -5$$

$$A_{13} = \begin{vmatrix} 3 & 4 \\ -2 & 1 \end{vmatrix} = (3)(1) - (4)(-2) = 11$$

The value of the determinant $|\mathbf{A}|$ is then given by

$$\begin{aligned} |\mathbf{A}| &= a_{11}A_{11} + a_{12}A_{12} + a_{13}A_{13} \\ &= (1)(14) + (-1)(-5) + (2)(11) \\ &= 14 + 5 + 22 = 41 \end{aligned}$$

If the value of the determinant of the given matrix (41 in our example) happens to be 0, no further steps are taken; *an inverse matrix simply does not exist* in this case—for reasons that will become clear in Step 3.

Step 2. Compute the *adjoint* or *adjugate* **adj(A)** of the given matrix.

For this purpose we need the cofactor of each and every element of **A**. Since we have already (Step 1) computed the cofactors of the elements of some row or column (row 1 in this illustration), we need only to compute those of elements in the remaining rows or columns. For the present example, these are

$$A_{21} = -\begin{vmatrix} -1 & 2 \\ 1 & 3 \end{vmatrix} = 5 \qquad A_{22} = \begin{vmatrix} 1 & 2 \\ -2 & 3 \end{vmatrix} = 7$$

$$A_{23} = -\begin{vmatrix} 1 & -1 \\ -2 & 1 \end{vmatrix} = 1 \qquad A_{31} = \begin{vmatrix} -1 & 2 \\ 4 & -2 \end{vmatrix} = -6$$

$$A_{32} = -\begin{vmatrix} 1 & 2 \\ 3 & -2 \end{vmatrix} = 8 \qquad A_{33} = \begin{vmatrix} 1 & -1 \\ 3 & 4 \end{vmatrix} = 7$$

[As an arithmetic check, it should be verified that expansions along rows or columns other than that used in Step 1 yield the same value for $|\mathbf{A}|$. Thus, in our example,

$$(3)(5) + (4)(7) + (-2)(1) = 41$$

and

$$(-2)(-6) + (1)(8) + (3)(7) = 41]$$

We now collect the values of all the cofactors of **A** into a matrix, in this manner: The cofactors of the elements of the first *row* of **A** are written in the first *column* of this new matrix; those of the second *row* of **A** constitute the second *column* of the new matrix; and so on. The resulting matrix is called the *adjoint* or *adjugate* of **A**, and is denoted by **adj(A)**. That is,

$$\mathbf{adj(A)} = \begin{bmatrix} A_{11} & A_{21} & A_{31} \\ A_{12} & A_{22} & A_{32} \\ A_{13} & A_{23} & A_{33} \end{bmatrix}$$

[To repeat, A_{ij} is the jth *row*, ith *column* element of **adj(A)**, the two subscripts referring to row and column numbers in reverse order from the usual notation.] For our example,

$$\mathbf{adj(A)} = \begin{bmatrix} 14 & 5 & -6 \\ -5 & 7 & 8 \\ 11 & 1 & 7 \end{bmatrix}$$

is the adjoint of **A**.

Step 3. Divide each element of **adj(A)** by $|\mathbf{A}|$ (the determinant of **A**, obtained in Step 1); the resulting matrix is $\mathbf{A}^{-1}$, the desired inverse of **A**.

It should now be clear why $\mathbf{A}^{-1}$ fails to exist when $|\mathbf{A}| = 0$; division by 0 is forbidden. A matrix whose determinant is zero is called a *singular* matrix; only nonsingular matrices have inverses. For our example,

$$\mathbf{A}^{-1} = \mathbf{adj(A)}/|\mathbf{A}| = (1/|\mathbf{A}|)\mathbf{adj(A)}$$

$$= (1/41)\begin{bmatrix} 14 & 5 & -6 \\ -5 & 7 & 8 \\ 11 & 1 & 7 \end{bmatrix} \tag{2.15}$$

$$= \begin{bmatrix} \frac{14}{41} & \frac{5}{41} & -\frac{6}{41} \\ -\frac{5}{41} & \frac{7}{41} & \frac{8}{41} \\ \frac{11}{41} & \frac{1}{41} & \frac{7}{41} \end{bmatrix}$$

The reader should verify that when the original matrix **A** is either pre- or post-multiplied by the matrix just obtained, the identity matrix

$$\mathbf{I}_3 = \begin{bmatrix} 1 & 0 & 0 \\ 0 & 1 & 0 \\ 0 & 0 & 1 \end{bmatrix}$$

is obtained.

Let us now apply the computational steps, outlined above, to a 2×2 matrix in general notation; that is,

$$\mathbf{A} = \begin{bmatrix} a_{11} & a_{12} \\ a_{21} & a_{22} \end{bmatrix}$$

Its determinant is

$$|\mathbf{A}| = a_{11}a_{22} - a_{12}a_{21}$$

The cofactors of the four elements of **A** are simply

$$A_{11} = a_{22} \qquad A_{12} = -a_{21}$$
$$A_{21} = -a_{12} \qquad A_{22} = a_{11}$$

Hence the adjoint of **A** is

$$\mathbf{adj(A)} = \begin{bmatrix} a_{22} & -a_{12} \\ -a_{21} & a_{11} \end{bmatrix}$$

(Comparing this with **A** itself, we see that the rule for finding the adjoint of a 2×2 matrix may be stated as follows: Switch around the diagonal elements, and change the signs of the off-diagonal elements.) Then, provided that $a_{11}a_{22} - a_{12}a_{21} \neq 0$, the inverse of **A** is given by

$$\mathbf{A}^{-1} = 1/(a_{11}a_{22} - a_{12}a_{21})\begin{bmatrix} a_{22} & -a_{12} \\ -a_{21} & a_{11} \end{bmatrix} \tag{2.16}$$

If $a_{11}a_{22} - a_{12}a_{21} = 0$, **A** is singular and has no inverse.

EXERCISES

Find the inverse of each of the matrices given in Exercise 1 and 2 that are nonsingular.

1. **(a)** $\begin{bmatrix} 5 & 3 \\ 6 & 4 \end{bmatrix}$ **(b)** $\begin{bmatrix} 1 & -2 \\ -2 & 3 \end{bmatrix}$ **(c)** $\begin{bmatrix} -4 & 2 \\ -6 & 3 \end{bmatrix}$

(d) $\begin{bmatrix} 3 & 0 \\ 0 & -5 \end{bmatrix}$

2. **(a)** $\begin{bmatrix} 2 & 2 & -4 \\ 10 & 14 & -22 \\ -14 & -20 & 32 \end{bmatrix}$ **(b)** $\begin{bmatrix} 1 & 1 & 1 \\ 2 & -1 & -2 \\ 1 & -2 & -1 \end{bmatrix}$ **(c)** $\begin{bmatrix} 1 & -3 & 2 \\ 3 & 1 & 0 \\ 2 & -1 & 1 \end{bmatrix}$

(d) $\begin{bmatrix} 1 & 4 & 1 \\ 2 & -1 & -2 \\ 1 & 1 & -1 \end{bmatrix}$

3. Exercise 1(d) exemplifies the simple rule for finding the inverse of a diagonal matrix. State the rule and prove it for diagonal matrices of arbitrary order.

4. Comparing the solution of Exercise 2(a) with the matrix **A** given in the example on page 25, the reader should note two properties of matrix inversion:

$$(\mathbf{A}^{-1})^{-1} = \mathbf{A} \qquad (2.17)$$

$$(c\mathbf{A})^{-1} = (1/c)\mathbf{A}^{-1} \qquad (2.18)$$

where c is any scalar not equal to 0. Prove these properties algebraically.

5. Let

$$\mathbf{A} = \begin{bmatrix} 5 & 2 \\ 8 & 4 \end{bmatrix}$$

and

$$\mathbf{v}' = [-2 \quad 6]$$

Find the value of the quadratic form $\mathbf{v}'\mathbf{A}^{-1}\mathbf{v}$.

6. If **A** and **B** are both nonsingular square matrices of the same order, prove that

$$(\mathbf{AB})^{-1} = \mathbf{B}^{-1}\mathbf{A}^{-1} \qquad (2.19)$$

That is, prove that the inverse of the product of two matrices is equal to the product of their respective inverses, in *reverse order*.

7. **Square root of a diagonal matrix.** In certain applications, it is necessary to determine a square matrix **X** that has the property that **XX** = **A**, where **A** is a given square matrix. Such an **X** may be called the *square root* of **A**, and is denoted as $\mathbf{A}^{1/2}$. Determination of $\mathbf{A}^{1/2}$ in general is extremely difficult, but it becomes a simple matter in the special case when **A** is a diagonal matrix. Show that if

$$\mathbf{A} = \begin{bmatrix} a_1 & & & \\ & a_2 & & 0 \\ & 0 & \ddots & \\ & & & a_n \end{bmatrix}$$

then

$$\mathbf{A}^{1/2} = \begin{bmatrix} \sqrt{a} & & & \\ & \sqrt{a_2} & & 0 \\ & 0 & \ddots & \\ & & & \sqrt{a_p} \end{bmatrix}$$

8. **Relation between SSCP matrix and correlation matrix.** Let

$$\mathbf{S} = \begin{bmatrix} \Sigma x_1^2 & \Sigma x_1x_2 & \cdots & \Sigma x_1x_p \\ \Sigma x_2x_1 & \Sigma x_2^2 & \cdots & \Sigma x_2x_p \\ \vdots & & \ddots & \vdots \\ \Sigma x_px_1 & \Sigma x_px_2 & \cdots & \Sigma x_p^2 \end{bmatrix}$$

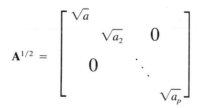

be the SSCP matrix, and

$$\mathbf{R} = \begin{bmatrix} 1 & r_{12} & \cdots & r_{1p} \\ r_{21} & 1 & \cdots & r_{2p} \\ \vdots & & \ddots & \\ r_{p1} & r_{p2} & \cdots & 1 \end{bmatrix}$$

be the correlation matrix among p variables. If we now define a diagonal matrix

$$\Delta = \begin{bmatrix} \Sigma x_1^2 & & & \\ & \Sigma x_2^2 & & 0 \\ & 0 & \ddots & \\ & & & \Sigma x_p^2 \end{bmatrix}$$

show that

$$\mathbf{S} = \Delta^{1/2}\mathbf{R}\,\Delta^{1/2} \tag{2.20}$$

If, furthermore, we let

$$(\Delta^{1/2})^{-1} = \Delta^{-1/2}$$

Eq. 2.20 may be solved for **R** to yield

$$\mathbf{R} = \Delta^{-1/2}\mathbf{S}\,\Delta^{-1/2} \tag{2.20a}$$

9. Using Eq. 2.8, compute **(a)** the SSCP matrix, and thence **(b)** the covariance matrix, of the four variables

X_1 = locus of control

X_2 = self-concept

X_3 = motivation

X_4 = writing score

in the sample of 600 students from the "High School and Beyond" project (hereafter abbreviated "HSB") given in Appendix F.

10. Use Eq. 2.20a to convert the SSCP matrix computed in Exercise 9 into a correlation matrix.

2.5 ANOTHER TYPE OF MATRIX MULTIPLICATION

(This section may be skipped on first reading without loss of continuity.)

Although the type of matrix multiplication that is about to be discussed will not be used in this book until Chapter 9, it is placed here because it is conceptually straightforward and hence would not fit in well with the second chapter on matrix algebra (Chapter 5), in which more advanced and mainly geometrically oriented material is discussed.

The multiplication in question yields what is called the *Kronecker product* or *direct product* of two matrices and is defined for any pair of matrices, regardless of their orders. The operation is denoted by $\otimes$ (or sometimes just by a boldface **x**), and it is defined as follows: Use *each* element of the *first* matrix as a scalar multiplier of the *second matrix in its entirety* and juxtapose the resulting product matrices. For example, if **A** is a 2 × 2 matrix and **B** is 2 × 3, their Kronecker product in the order in which the two matrices were mentioned is

$$\mathbf{A} \otimes \mathbf{B} = \left[\begin{array}{c|c} a_{11}\mathbf{B} & a_{12}\mathbf{B} \\ \hline a_{21}\mathbf{B} & a_{22}\mathbf{B} \end{array}\right]$$

$$= \left[\begin{array}{ccc|ccc} a_{11}b_{11} & a_{11}b_{12} & a_{11}b_{13} & a_{12}b_{11} & a_{12}b_{12} & a_{12}b_{13} \\ a_{11}b_{21} & a_{11}b_{22} & a_{11}b_{23} & a_{12}b_{21} & a_{12}b_{22} & a_{12}b_{23} \\ \hline a_{21}b_{11} & a_{21}b_{12} & a_{21}b_{13} & a_{22}b_{11} & a_{22}b_{12} & a_{22}b_{13} \\ a_{21}b_{21} & a_{21}b_{22} & a_{21}b_{23} & a_{22}b_{21} & a_{22}b_{22} & a_{22}b_{23} \end{array}\right]$$

As a numerical example (using matrices of orders different from those in the foregoing illustration), suppose that

$$\mathbf{A} = [-1, 2] \qquad \text{and} \qquad \mathbf{B} = \begin{bmatrix} 3 & 2 \\ 1 & 4 \\ -2 & 1 \end{bmatrix}$$

Then

$$\mathbf{A} \otimes \mathbf{B} = \left[(-1)\begin{bmatrix} 3 & 2 \\ 1 & 4 \\ -2 & 1 \end{bmatrix}, \ (2)\begin{bmatrix} 3 & 2 \\ 1 & 4 \\ -2 & 1 \end{bmatrix} \right]$$

$$= \begin{bmatrix} -3 & -2 & 6 & 4 \\ -1 & -4 & 2 & 8 \\ 2 & -1 & -4 & 2 \end{bmatrix}$$

In general, if **A** is of order $m \times n$ and **B** is $p \times q$ and we let $\mathbf{A} \otimes \mathbf{B} = \mathbf{C}$, then **C** is an $mp \times nq$ matrix, and we may symbolically write

$$\mathbf{A} \otimes \mathbf{B} = \mathbf{C} = [a_{ij}\mathbf{B}] \tag{2.21}$$

provided we understand that what is written inside the brackets is not the (i, j)-element of **C** in the ordinary sense. It may, however, be construed as the (i, j)-*submatrix* $(i = 1, 2, \ldots, m; j = 1, 2, \ldots, n)$ of a "supermatrix" **C**.

(We use the terminology of submatrices and supermatrices more meaningfully in Chapter 5.)

One use of Kronecker products, albeit a relatively trivial one, lies in writing a matrix of enormous size, but with a recurring pattern of elements, in a compact way. Recall that preparatory to obtaining Eq. 2.7 for the SSCP matrix, we defined and wrote out a 5 × 2 mean-score matrix which consisted of five identical rows, each listing the means on the two tests (27.4 and 12.8) for the five pupils. Suppose that the group size had been 50 instead of 5. Then the mean-score matrix $\bar{\mathbf{X}}$ would have been of order 50 × 2, and it would have been a formidable and tedious task to write this out. Actually, we would not have to write all the rows, but merely use the ellipsis sign (. . . , written vertically) with an indication that the number of rows is 50—*if* recording $\bar{\mathbf{X}}$ on paper was all we needed to do. However, if we needed to enter this 50 × 2 matrix $\bar{\mathbf{X}}$ into a computer for carrying out the operation of Eq. 2.7, we would be faced with a boring, time-consuming task, which using the Kronecker product allows us to circumvent. We need only write the computer-language equivalent of the statement "$\bar{\mathbf{X}} = \mathbf{1}_{50} \times [27.4, 12.8]$," which will produce the required matrix with 50 repeated rows containing 27.4 and 12.8 as their elements. (All matrix-manipulation packages will have the capability of generating $\mathbf{1}_n$ without our actually having to type in n 1's; Kronecker multiplication is also routinely available.)

Properties of Kronecker Products

Four useful properties of the Kronecker product are as stated in Eqs. 2.22–2.25.

Associativity: $$(\mathbf{A} \otimes \mathbf{B}) \otimes \mathbf{C} = \mathbf{A} \otimes (\mathbf{B} \otimes \mathbf{C}) \qquad (2.22)$$

Distributivity over addition: $$(\mathbf{A} + \mathbf{B}) \otimes \mathbf{C} = \mathbf{A} \otimes \mathbf{C} + \mathbf{B} \otimes \mathbf{C} \qquad (2.23)$$

Transposition: $$(\mathbf{A} \otimes \mathbf{B})' = \mathbf{A}' \otimes \mathbf{B}' \qquad (2.24)$$

The first of these are identical to the corresponding properties of ordinary matrix products. Note, however, that the third is different from the rule for getting the transpose of the ordinary product of two matrices given in Eq. 2.12 [i.e., $(\mathbf{AB})' = \mathbf{B}'\mathbf{A}'$]. There, the respective transposes of the two matrices are multiplied in the *reverse* order to get the transpose of the product; for the Kronecker product, the order of multiplication is preserved. To remember this, one need only recall that, by definition, $\mathbf{A}(m, n) \times \mathbf{B}(p, q) = \mathbf{C}(mp, nq)$, and hence $\mathbf{C}'$ is of order $(nq \times mp)$, as is $\mathbf{A}'(n, m) \times \mathbf{B}'(q, p)$. The verification of the three foregoing properties, using matrices of small order, is a simple matter and is left as an exercise.

The last property to be mentioned here is a rather peculiar one, with no counterpart in ordinary matrix products. It has to do with the (ordinary) *product of two Kronecker products* being equal to the Kronecker product of two ordinary products:

$$(\mathbf{A} \otimes \mathbf{B})(\mathbf{C} \otimes \mathbf{D}) = \mathbf{AC} \otimes \mathbf{BD} \qquad (2.25)$$

In this case there are constraints on the orders of the four matrices so that the ordinary products involved will be executable. The following four small matrices will qualify:

$$\mathbf{A} = \begin{bmatrix} 3 \\ 2 \end{bmatrix}, \qquad \mathbf{B} = [2, 1], \qquad \mathbf{C} = [1, 3], \qquad \mathbf{D} = \begin{bmatrix} 1 & 2 \\ 3 & 4 \end{bmatrix}$$

We then obtain

$$\mathbf{A} \otimes \mathbf{B} = \begin{bmatrix} 3[2, 1] \\ 2[2, 1] \end{bmatrix} = \begin{bmatrix} 6 & 3 \\ 4 & 2 \end{bmatrix}$$

and

$$\mathbf{C} \otimes \mathbf{D} = \left[1\begin{bmatrix} 1 & 2 \\ 3 & 4 \end{bmatrix}, \; 3\begin{bmatrix} 1 & 2 \\ 3 & 4 \end{bmatrix} \right] = \begin{bmatrix} 1 & 2 & 3 & 6 \\ 3 & 4 & 9 & 12 \end{bmatrix}$$

Hence

$$(\mathbf{A} \otimes \mathbf{B})(\mathbf{C} \otimes \mathbf{D}) = \begin{bmatrix} 6 & 3 \\ 4 & 2 \end{bmatrix} \begin{bmatrix} 1 & 2 & 3 & 6 \\ 3 & 4 & 9 & 12 \end{bmatrix} = \begin{bmatrix} 15 & 24 & 45 & 72 \\ 10 & 16 & 30 & 48 \end{bmatrix}$$

On the other hand, the two ordinary products on the right-hand side are found to be

$$\mathbf{AC} = \begin{bmatrix} 3 & 9 \\ 2 & 6 \end{bmatrix} \qquad \text{and} \qquad \mathbf{BD} = [5, 8]$$

and their Kronecker product is

$$\mathbf{AC} \otimes \mathbf{BD} = \begin{bmatrix} 3 & 9 \\ 2 & 6 \end{bmatrix} \times [5, 8] = \begin{bmatrix} 3[5, 8] & 9[5, 8] \\ 2[5, 8] & 6[5, 8] \end{bmatrix}$$

$$= \begin{bmatrix} 15 & 24 & 45 & 72 \\ 10 & 16 & 30 & 48 \end{bmatrix}$$

which is equal to the result just obtained for the left-hand side of Eq. 2.25.

EXERCISES

1. By definition of Kronecker products, no constraints are placed on the orders of the matrices **A**, **B**, and **C** that may be used for illustrating the associativity property (Eq. 2.22). However, one or more constraints is (are) necessary for the orders of the matrices involved in the distributivity property (Eq. 2.23). What is (are) the constraint(s)?

2. If one further wishes to use the same triplet of matrices **A**, **B**, and **C** for illustrating both Eqs. 2.23 and 2.25—together with a fourth matrix **D** for the latter, of course—

one would have to introduce further constraints on the orders. Formulate the minimal set of constraints (i.e., the set that will leave you maximal freedom in choice of orders) for the stated purpose. Explain why each of these constraints is necessary.

3. Here is a set of four matrices that satisfy the minimal constraints referred to in Exercise 2:

$$\mathbf{A} = [-2,\ 1], \qquad \mathbf{B} = [1,\ 2], \qquad \mathbf{C} = \begin{bmatrix} 2 \\ 1 \\ 3 \\ 1 \end{bmatrix}, \qquad \mathbf{D} = [4,\ -1,\ 2]$$

Use the appropriate subsets of these four matrices to illustrate all four of the properties of Kronecker products stated in Eqs. 2.22 to 2.25.

4. If **A** and **B** are nonsingular square matrices of the same order, show that

$$(\mathbf{A} \otimes \mathbf{B})^{-1} = \mathbf{A}^{-1} \otimes \mathbf{B}^{-1} \tag{2.26}$$

Note that, here again (as in Eq. 2.24), the orders of multiplication are the same on both sides of the equation, whereas they were reversed in ordinary matrix multiplication (see Eq. 2.19).

3

MULTIPLE REGRESSION ANALYSIS

It is assumed that the reader is familiar with simple (or bivariate) regression and has some knowledge of multiple regression analysis. We nevertheless start with a brief review of simple regression, and then introduce multiple regression in a way that highlights the analogy with simple regression. Matrix algebra helps to bridge the gap, as it will over and over again in this book. Thus one of the things that we wish to do in this chapter is to illustrate—in a context already somewhat familiar to the reader—the way in which matrix algebra effects a unification of univariate and multivariate statistical techniques.

3.1 SIMPLE REGRESSION

As the reader is aware, the purpose of simple regression is to fit a linear equation of the form

$$\hat{Y} = a + bX \tag{3.1}$$

to a set of data points $\{(X_i, Y_i);\ i = 1, 2, \ldots, N\}$ so that the error will be as small as possible. Specifically, the (quadratic) error is defined as

$$\begin{aligned} Q &= \sum_{i=1}^{N} e_i^2 \\ &= \sum_{i=1}^{N} (\hat{Y}_i - Y_i)^2 \qquad (3.2) \\ &= \sum_{i=1}^{N} (a + bX_i - Y_i)^2 \end{aligned}$$

and our task is to determine the values of a and b that will make Q as small as possible.

(As readers who are familiar with differential calculus will recognize, the equations obtained by setting the partial derivatives equal to zero are, strictly speaking, only the *necessary* conditions for minimizing Q. However, it is intuitively clear that Q has no maximum, for it can be made as large as we please by locating the line far away from the data points. Hence we will assume that the conditions give us the a and b values that minimize Q, without actually proving that the equations in question are *sufficient*.) These equations are

$$\frac{\partial Q}{\partial a} = 2 \sum_{i=1}^{N} (a + bX_i - Y_i) = 0$$

and

$$\frac{\partial Q}{\partial b} = 2 \sum_{i=1}^{N} (a + bX_i - Y_i)(X_i) = 0$$

which reduce, respectively, to

$$\begin{aligned} Na + (\Sigma X_i)b &= \Sigma Y_i \\ (\Sigma X_i)a + (\Sigma X_i^2)b &= \Sigma X_iY_i \end{aligned} \tag{3.3}$$

These are called the *normal equations in raw-score form*. Solving the first of these equations for a yields

$$a = \overline{Y} - b\overline{X} \tag{3.4}$$

Substituting this in the second equation and rearranging terms, we get

$$(\Sigma X_i^2 - \overline{X} \Sigma X_i)b = \Sigma X_iY_i - \overline{Y} \Sigma X_i$$

which may be written in terms of deviation scores $X_i - \overline{X} = x_i$ and $Y_i - \overline{Y} = y_i$ as

$$(\Sigma x_i^2)b = \Sigma x_iy_i \tag{3.5}$$

which is called the *normal equation in deviation-score form*. This may be solved immediately for b to yield

$$b = (\Sigma x_iy_i)/(\Sigma x_i^2) \tag{3.6}$$

This value of b is then substituted in Eq. 3.4 to get the optimal value of a, and the problem of determining the line of best fit to the data points in the sense of minimizing the error (or, more specifically, the quadratic error) Q of Eq. 3.2 is solved. The straight line represented by Eq. 3.1 with the values of b (the regression coefficient) and a (the Y-intercept) just obtained is called the regression line of Y on X. It is also called the *line of least-squares fit* of Y to X, after the method used to determine it, which is called the *principle of least squares;* the "squares" comes from the fact that the expression Q that was minimized consists of the sum of *squared* errors $(\hat{Y}_i - Y_i)^2$.

The Product-Moment Correlation Coefficient

Although the regression line determined is, by construction, the best-fitting straight line in the least-squares sense, the question remains: *How good* is this best fit? A reasonable measure of the goodness (or poorness) of the fit would seem to be provided by the magnitude of the minimum Q that was achieved by taking the optimal a and b values. Substituting the expressions (3.4) and (3.6) in the last expression for Q in Eq. 3.2, we get

$$\begin{aligned} Q_{\min} &= \Sigma\,(\bar{Y} - b\bar{X} + bX_i - Y_i)^2 \\ &= \Sigma\,[b(X_i - \bar{X}) - (Y_i - \bar{Y})]^2 \\ &= \Sigma\,(bx_i - y_i)^2 \\ &= b^2\,\Sigma\,x_i^2 - 2b\,\Sigma\,x_iy_i + \Sigma\,y_i^2 \\ &= \frac{(\Sigma\,x_iy_i)^2}{(\Sigma\,x_i^2)^2}(\Sigma\,x_i^2) - 2\frac{(\Sigma\,x_iy_i)^2}{\Sigma\,x_i^2} + \Sigma\,y_i^2 \\ &= \Sigma\,y_i^2 - \frac{(\Sigma\,x_iy_i)^2}{\Sigma\,x_i^2} \\ &= (\Sigma\,y_i^2)\left[1 - \frac{(\Sigma\,x_iy_i)^2}{(\Sigma\,x_i^2)(\Sigma\,y_i^2)}\right] \end{aligned}$$

In this expression, the factor $\Sigma\,y_i^2$ is a function of the Y's alone and has nothing to do with how the Y's are related to the X's. Moreover, it is dependent on the scale of measurement of Y. Hence $\Sigma\,y_i^2$ should be excluded from a measure of goodness of fit of the regression line to the data points. The expression in brackets, on the other hand, is scale-free. This goes without saying for the first term, 1. The second term is a rather complicated-looking fraction involving the X's and Y's, but a casual examination suffices to show that the numerator and denominator involve each variable in the same power. Hence the fraction is a scale-free (or dimensionless), pure number. The astute reader will recognize that it is the square of the product-moment coefficient of correlation between X and Y,

$$r_{xy} = \frac{\Sigma\,x_iy_i}{\sqrt{(\Sigma\,x_i^2)(\Sigma\,y_i^2)}}$$

Clearly, the larger the absolute value of r_{xy} is, the smaller Q is, and its smallest value, 0, is taken only when $r_{xy} = \pm 1$. That is, perfect fit is achieved only when there is a perfect correlation (either positive or negative) between X and Y.

Thus the minimum quadratic error, $Q_{\min}$, achieved by using the optimal a and b values, may be expressed in two ways, one of which is absolute but scale-dependent whereas the other is relative but scale-free. Both are useful for future reference, so we display them as equations

$$Q_{\min} = \Sigma\, y_i^2 - \frac{(\Sigma\, x_i y_i)^2}{\Sigma\, x_i^2} \tag{3.7a}$$

$$Q_{\min}/\Sigma\, y_i^2 = 1 - r_{xy}^2 \tag{3.7b}$$

3.2 MULTIPLE REGRESSION

When we have several predictor variables $X_1, X_2, \ldots, X_p$ instead of just one, the problem becomes that of determining the $p + 1$ constants $a, b_1, b_2, \ldots, b_p$ in the equation

$$\hat{Y} = a + b_1 X_1 + b_2 X_2 + \cdots + b_p X_p \tag{3.8}$$

so that the quadratic error

$$\begin{aligned} Q &= \Sigma\, e_i^2 \\ &= \Sigma\, (\hat{Y}_i - Y_i)^2 \\ &= \Sigma\, (a + b_1 X_{1i} + b_2 X_{2i} + \cdots + b_p X_{pi} - Y_i)^2 \end{aligned} \tag{3.9}$$

will be as small as possible.

Just as in simple regression analysis, we take the partial derivatives of Q with respect to the unknown constants and set each of them equal to zero. This time there are $p + 1$ such equations,

$$\frac{\partial Q}{\partial a} = 2\,\Sigma\, (a + b_1 X_{1i} + b_2 X_{2i} + \cdots + b_p X_{pi} - Y_i) = 0$$

$$\frac{\partial Q}{\partial b_1} = 2\,\Sigma\, (a + b_1 X_{1i} + b_2 X_{2i} + \cdots + b_p X_{pi} - Y_i)(X_{1i}) = 0$$

$$\cdots$$

$$\frac{\partial Q}{\partial b_p} = 2\,\Sigma\, (a + b_1 X_{1i} + b_2 X_{2i} + \cdots + b_p X_{pi} - Y_i)(X_{pi}) = 0$$

which simplify to the raw-score normal equations

$$\begin{aligned} Na + (\Sigma\, X_{1i})b_1 + (\Sigma\, X_{2i})b_2 + \cdots + (\Sigma\, X_{pi})b_p &= \Sigma\, Y_i \\ (\Sigma\, X_{1i})a + (\Sigma\, X_{1i}^2)b_1 + (\Sigma\, X_{1i}X_{2i})b_2 + \cdots + (\Sigma\, X_{1i}X_{pi})b_p &= \Sigma\, X_{1i}Y_i \\ \cdots \\ (\Sigma\, X_{pi})a + (\Sigma\, X_{pi}X_{1i})b_1 + (\Sigma\, X_{pi}X_{2i})b_2 + \cdots + (\Sigma\, X_{pi}^2)b_p &= \Sigma\, X_{pi}Y_i \end{aligned} \tag{3.10}$$

Again, solving the first of these equations for a and substituting the result,

$$a = \bar{Y} - b_1\bar{X}_1 - b_2\bar{X}_2 - \cdots - b_p\bar{X}_p \qquad (3.11)$$

in the remaining p equations and simplifying, we get the deviation-score normal equations

$$\begin{aligned}
(\Sigma x_{1i}^2)b_1 + (\Sigma x_{1i}x_{2i})b_2 + \cdots + (\Sigma x_{1i}x_{pi})b_p &= \Sigma x_{1i}y_i \\
(\Sigma x_{2i}x_{1i})b_1 + (\Sigma x_{2i}^2)b_2 + \cdots + (\Sigma x_{2i}x_{pi})b_p &= \Sigma x_{2i}y_i \\
\cdots \\
(\Sigma x_{pi}x_{1i})b_1 + (\Sigma x_{pi}x_{2i})b_2 + \cdots + (\Sigma x_{pi}^2)b_p &= \Sigma x_{pi}y_i
\end{aligned} \qquad (3.12)$$

These equations may be written in extended matrix form as

$$\begin{bmatrix} \Sigma x_{ii}^2 & \Sigma x_{1i}x_{2i} & \cdots & \Sigma x_{1i}x_{pi} \\ \Sigma x_{2i}x_{1i} & \Sigma x_{2i}^2 & \cdots & \Sigma \mathbf{x}_{2i}\mathbf{x}_{pi} \\ & \cdots & & \\ \Sigma x_{pi}x_{1i} & \Sigma x_{pi}x_{2i} & \cdots & \mathbf{x}_{pi}^2 \end{bmatrix} \begin{bmatrix} b_1 \\ b_2 \\ \vdots \\ b_p \end{bmatrix} = \begin{bmatrix} \Sigma x_{1i}y_i \\ \Sigma x_{2i}y_i \\ \vdots \\ \Sigma x_{pi}y_i \end{bmatrix} \qquad (3.13)$$

EXAMPLE 3.1 To illustrate the developments up to this point, suppose that, in the ''example of statistical import'' in Section 2.3, we further had another test Y (which might be a mathematical reasoning test given at the end of the school year). Let the scores of the five pupils on this test be as follows:

$$\mathbf{Y} = \begin{bmatrix} 34 \\ 25 \\ 30 \\ 38 \\ 26 \end{bmatrix}$$

The SSCP matrix required for writing the left-hand side of the appropriate instance of Eq. 3.13 was already computed in the example in Section 2.3. However, for the purpose of numerical illustration of the step going from the normal equations in raw-score form (see Eqs. 3.10) to those in deviation-score form (summarized, in matrix notation, in Eq. 3.13, let us start out with the former set of normal equations. Appropriate substitutions in Eqs. 3.10 yield the following set of equations (the normal equations in raw-score form):

$$\begin{aligned}
5a + 137b_1 + 64b_2 &= 153 \\
137a + 3817b_1 + 1770b_2 &= 4273 \\
64a + 1770b_1 + 834b_2 &= 1990
\end{aligned}$$

From the first equation of this set, we obtain

$$a = 30.6 - 27.4b_1 - 12.8b_2$$

When this expression is substituted for a in the other two equations of the set, we obtain

$$137(30.6 - 27.4b_1 - 12.8b_2) + 3817b_1 + 1770b_2 = 4273$$

and

$$64(30.6 - 27.4b_1 - 12.8b_2) + 1770b_1 + 834b_2 = 1990$$

Upon collecting like terms, these equations reduce to

$$63.2b_1 + 16.4b_2 = 80.8$$
$$16.4b_1 + 14.8b_2 = 31.6$$

or, equivalently,

$$\begin{bmatrix} 63.2 & 16.4 \\ 16.4 & 14.8 \end{bmatrix} \begin{bmatrix} b_1 \\ b_2 \end{bmatrix} = \begin{bmatrix} 80.8 \\ 31.6 \end{bmatrix}$$

in matrix notation. Note that the coefficient matrix on the left-hand side is precisely the SSCP matrix computed for the example in Section 2.3; also, the reader should verify that the elements of the vector on the right-hand side are equal, respectively, to $\Sigma\, x_1y$ and to $\Sigma\, x_2y$.

Solving the Normal Equations

To discuss the matrix-algebraic solution of Eq. 3.13, it is convenient to introduce the following abbreviations for the SSCP matrix and the two column vectors involved: Denote the SSCP matrix among the predictor variables by $\mathbf{S}_{pp}$, the vector of regression weights by **b**, and the vector of sums of cross-products between the predictors and the criterion by $\mathbf{S}_{pc}$. That is, we denote the coefficient matrix on the left by **S** and let

$$\begin{bmatrix} b_1 \\ b_2 \\ \vdots \\ b_p \end{bmatrix} = \mathbf{b} \quad \text{and} \quad \begin{bmatrix} \Sigma\, x_{1i}y_i \\ \Sigma\, x_{2i}y_i \\ \vdots \\ \Sigma\, x_{pi}y_i \end{bmatrix} = \mathbf{S}_{pc}$$

We may then write Eq. 3.13 in symbolic matrix form as

$$\mathbf{S}_{pp}\mathbf{b} = \mathbf{S}_{pc} \tag{3.14}$$

which is a straightforward generalization of Eq. 3.5 for the simple regression case. Furthermore, provided that $\mathbf{S}_{pp}$ is nonsingular (as it almost always will be), the solution to this equation is

$$\mathbf{b} = \mathbf{S}_{pp}^{-1}\mathbf{S}_{pc} \tag{3.15}$$

exactly paralleling Eq. 3.6, which may be rewritten as

$$b = (\Sigma\, x_i^2)^{-1}(\Sigma\, x_i y_i)$$

to make the analogy more evident.

Thus Eq. 3.15 gives the desired solution to the normal equations represented by (3.14)—provided that the SSCP matrix (for the predictor variables) is nonsingular. For real data, except when there is some artificial restriction among the predictor variables (such as their sum being constant for all individuals[1]) we may safely assume that the SSCP matrix is nonsingular and hence that $\mathbf{S}_{pp}^{-1}$ exists. Methods for handling cases when $\mathbf{S}_{pp}$ is singular are discussed later in this chapter and in Section 5.6.

Once the b's (the regression weights) have thus been obtained, it is a simple matter to calculate the additive constant a to complete the regression equation (3.8). We simply substitute the values of the b's in Eq. 3.11 for a.

EXAMPLE 3.2 To continue with the example given above in illustrating the process of constructing the normal equations in deviation score form, let us compute the inverse of the SSCP matrix obtained there. We find

$$\mathbf{S}_{pp}^{-1} = \begin{bmatrix} 63.2 & 16.4 \\ 16.4 & 14.8 \end{bmatrix}^{-1} = (1/666.40)\begin{bmatrix} 14.8 & -16.4 \\ -16.4 & 63.2 \end{bmatrix}$$

$$= \begin{bmatrix} .0222 & -.0246 \\ -.0246 & .0948 \end{bmatrix}$$

For the vector $\mathbf{b}$, we therefore obtain

$$\mathbf{b} = \begin{bmatrix} .0222 & -.0246 \\ -.0246 & .0948 \end{bmatrix}\begin{bmatrix} 80.8 \\ 31.6 \end{bmatrix} = \begin{bmatrix} 1.016 \\ 1.008 \end{bmatrix}$$

Then, substituting in the expression for a, we get

$$a = 30.6 - (27.4)(1.016) - (12.8)(1.008) = -10.141$$

(The reader should substitute the values for a, b_1, and b_2 obtained above in the

[1]Variables with this property are called *ipsative measures*. For example, the percentages of income spent for various purposes and the percentages of time spent on various activities are ipsative measures. By definition, the scores on each such set of variables add up to a constant (100 in the above examples) for every individual.

original raw-score normal equations for this example to see that they are satisfied within rounding error.) Thus the desired multiple regression equation for this example is

$$\hat{Y} = -10.141 + 1.016X_1 + 1.008X_2$$

Derivation of Normal Equation by Symbolic Vector Differentiation

We now give an alternative derivation of the normal equations to familiarize the reader with the process of *symbolic vector differentiation* described in Appendix C. As explained there, to form a symbolic derivative of a quantity with respect to a vector (or a matrix), we simply get the partial derivative of the quantity with respect to each element of the vector (matrix) and arrange the results in a vector (matrix) of the same form as the "differentiator." For example, if

$$\mathbf{x}' = [x_1, x_2] \qquad \text{and} \qquad f = x_1^2 + x_1x_2 + x_2^2$$

then

$$\frac{\partial f}{\partial \mathbf{x}'} = \left[\frac{\partial f}{\partial x_1}, \frac{\partial f}{\partial x_2}\right] = [2x_1 + x_2, \quad x_1 + 2x_2]$$

while

$$\frac{\partial f}{\partial \mathbf{x}} = \begin{bmatrix} 2x_1 + x_2 \\ x_1 + 2x_2 \end{bmatrix}$$

because in the first expression the symbolic derivative is taken with respect to the row vector $\mathbf{x}'$ and in the second, the differentiator is the column vector $\mathbf{x}$.

To apply this principle in getting the symbolic derivative of the quadratic error Q with respect to $a, b_1, b_2, \ldots, b_p$, it is convenient first to rewrite Q in the form

$$\begin{aligned} Q &= \sum_{i=1}^{N} (\hat{Y}_i - Y_i)^2 \\ &= (\hat{\mathbf{Y}} - \mathbf{Y})'(\hat{\mathbf{Y}} - \mathbf{Y}) \\ &= \hat{\mathbf{Y}}'\hat{\mathbf{Y}} - 2\hat{\mathbf{Y}}'\mathbf{Y} + \mathbf{Y}'\mathbf{Y} \end{aligned}$$

where

$$\mathbf{Y} = [Y_1, Y_2, \ldots, Y_N]'$$

and

$$\hat{\mathbf{Y}} = \mathbf{X}^+\mathbf{b}^+$$

with

$$\mathbf{b}^{+} = [a, b_1, b_2, \ldots, b_p]'$$

and

$$\mathbf{X}^{+} = \begin{bmatrix} 1 & X_{11} & X_{21} & \cdots & X_{p1} \\ 1 & X_{12} & X_{22} & \cdots & X_{p2} \\ 1 & X_{13} & X_{23} & \cdots & X_{p3} \\ \vdots & \vdots & \vdots & & \vdots \\ 1 & X_{1N} & X_{2N} & \cdots & X_{pN} \end{bmatrix}$$

Hence

$$Q = \mathbf{b}^{+\prime}(\mathbf{X}^{+\prime}\mathbf{X}^{+})\mathbf{b}^{+} - 2\mathbf{b}^{+\prime}(\mathbf{X}^{+\prime}\mathbf{Y}) + \mathbf{Y}'\mathbf{Y}$$

The symbolic vector derivative of this quantity with respect to $\mathbf{b}^{+}$ is, in accordance with Eq. C.1s,

$$\frac{\partial Q}{\partial \mathbf{b}^{+}} = 2(\mathbf{X}^{+\prime}\mathbf{X}^{+})\mathbf{b}^{+} - 2(\mathbf{X}^{+\prime}\mathbf{Y})$$

Setting this equal to the $(p + 1)$-dimensional null vector $\mathbf{O}$ and dividing both sides by 2 yields

$$(\mathbf{X}^{+\prime}\mathbf{X}^{+})\mathbf{b}^{+} = \mathbf{X}^{+\prime}\mathbf{Y}$$

The reader should verify, by carrying out the multiplication, that this equation is equivalent to the normal equations in raw-score form given in Eq. 3.10.

The Multiple Correlation Coefficient

When the multiple regression equation has been determined by solving for the regression coefficients from Eqs. 3.10 and 3.11, we can be assured that we have got the best linear equation in the sense of minimizing the quadratic error, $Q = \Sigma\,(\hat{Y}_i - Y_i)^2$. But a question remains as to how good (or how poor) this best equation is, just as in the simple-regression case. A reasonable index for providing an answer to this question would seem to be the correlation $r_{\hat{Y}Y}$ between the predicted and actual Y scores in the sample at hand. By definition, this correlation is

$$r_{\hat{Y}Y} = \frac{\Sigma\,\hat{y}y}{\sqrt{\Sigma\,\hat{y}^2}\,\sqrt{\Sigma\,y^2}}$$

From Eqs. 3.8 and 3.11, it is readily found that

$$\hat{y}_i = \hat{Y}_i - \overline{\hat{Y}} = \hat{Y}_i - \overline{Y}$$
$$= b_1 x_{1i} + b_2 x_{2i} + \cdots + b_p x_{pi}$$

Taking the case of $p = 2$ (two predictor variables) for simplicity, let us compute the quantities $\Sigma\, \hat{y}_i y_i$ and $\Sigma\, \hat{y}_i^2$ in the expression for $r_{\hat{Y}Y}$, in forms that involve x_{1i} and x_{2i} in place of $\hat{y}_i$. We find

$$\Sigma\, y_i \hat{y}_i = \Sigma\, y_i(b_1 x_{1i} + b_2 x_{2i})$$
$$= b_1\, \Sigma\, x_{1i} y_i + b_2\, \Sigma\, x_{2i} y_i$$

or in symbolic matrix form,

$$\sum_{i=1}^{N} \hat{y}_i y_i = \mathbf{b}'\mathbf{S}_{pc} \tag{3.16}$$

which holds for any p, not just for $p = 2$.

Similarly,

$$\Sigma\, \hat{y}_i^2 = \sum_{i=1}^{N} (b_1 x_{1i} + b_2 x_{2i})^2$$
$$= \sum_{i=1}^{N} (b_1^2 x_{1i}^2 + b_2 x_{2i}^2 + 2b_1 b_2 x_{1i} x_{2i})$$
$$= b_1\, \Sigma\, x_{1i}^2 + b_2\, \Sigma\, x_{2i}^2 + 2b_1 b_2\, \Sigma\, x_{1i} x_{2i}$$

The reader should verify that the last expression is the expansion of the quadratic form

$$[b_1,\, b_2]\begin{bmatrix} \Sigma\, x_{1i}^2 & \Sigma\, x_{1i} x_{2i} \\ \Sigma\, x_{2i} x_{1i} & \Sigma\, x_{2i}^2 \end{bmatrix}\begin{bmatrix} b_1 \\ b_2 \end{bmatrix}$$

Thus, for any p, we may write, using symbolic matrix form,

$$\Sigma\, \hat{y}_i^2 = \mathbf{b}'\mathbf{S}_{pp}\mathbf{b} \tag{3.17}$$

Substituting Eqs. (3.16) and (3.17) in the expression for $r_{\hat{Y}Y}$, we obtain

$$r_{\hat{Y}Y} = \frac{\mathbf{b}'\mathbf{S}_{pc}}{\sqrt{\Sigma\, y_i^2}\,\sqrt{\mathbf{b}'\mathbf{S}_{pp}\mathbf{b}}} \tag{3.18}$$

Up to this point, we have not used the fact that the **b** in this equation is a special **b** that satisfies Eq. 3.14. Therefore, Eq. 3.18 is applicable for calculating the correlation between Y and *any* linear combination of the X's, with an arbitrary set of combining weights. This fact will come in handy later. Now, however, we utilize the special property of the **b** in Eq. 3.18, which allows us to invoke Eq. 3.14 and write

$$\mathbf{S}_{pp}\mathbf{b} = \mathbf{S}_{pc}$$

Upon premultiplying both sides of this equation by $\mathbf{b}'$, we get

$$\mathbf{b}'\mathbf{S}_{pp}\mathbf{b} = \mathbf{b}'\mathbf{S}_{pc}$$

This means that the expression in the numerator on the right-hand side of Eq. 3.18 is equal to the expression that stands under the second radical in the denominator. Hence by writing $\mathbf{b}'\mathbf{S}_{pc}$ in both numerator and denominator, and carrying out the indicated division, we arrive at

$$r_{\hat{Y}Y} = \frac{\sqrt{\mathbf{b}'\mathbf{S}_{pc}}}{\sqrt{\Sigma\, y_i^2}}$$

This correlation between Y and the special (i.e., optimal) linear combination $\hat{Y}$ of the X's—using the special weights provided by Eq. 3.15—is called the coefficient of multiple correlation between Y and the set of predictors $X_1, X_2, \ldots, X_p$. (For brevity it may also be called the *multiple-R* between Y and the X's.) Using the customary notation, $R_{Y\cdot123\cdots p}$ for the multiple correlation coefficient, we write

$$R_{Y\cdot123\cdots p} = \frac{\sqrt{\mathbf{b}'\mathbf{S}_{pc}}}{\sqrt{\Sigma\, y_i^2}} \tag{3.19}$$

This particular expression for the multiple correlation coefficient may not look familiar even to readers who have studied the multiple-R before. The most widely encountered formula for multiple-R's uses *standard-score* regression weights, often called beta weights, and denoted β_j. Here we denote them by b_j^*, and the expression for the *squared* multiple-R becomes

$$R^2 = b_1^* r_{1y} + b_2^* r_{2y} + \cdots + b_p^* r_{py}$$

This may look entirely different from the square of expression (3.19), but their equivalence may be roughly inferred as follows. From Eq. 3.16 and the preceding, unnumbered equation extended to a general p, we see that the numerator of Eq. 3.19, when written in scalar form, becomes

$$\sqrt{b_1\, \Sigma\, x_{1i}y_i + b_2\, \Sigma\, x_{2i}y_i + \cdots + b_p\, \Sigma\, x_{pi}y_i}$$

If X_j and Y were in standardized form, then $\Sigma_{i=1}^{N} x_{ji}y_i/(N - 1)$ would be equal to r_{jy}. Similarly, the denominator, $\sqrt{\Sigma\, y_i^2}$, of Eq. 3.19, when divided by $\sqrt{N - 1}$, would equal 1. Finally, the b_j's would become b_j^*'s. Thus it can be inferred that the most common expression for R^2 is what the square of expression (3.19) would reduce to if the variables were all in standardized form.

Another expression for R—whose relevance will come to be appreciated when canonical correlation is discussed later in the book—may be derived as follows. Squaring both sides of Eq. 3.19 and noting that

$$\mathbf{b}' = \mathbf{S}_{pc}'\mathbf{S}_{pp}^{-1}$$

we get

$$R^2 = (\mathbf{S}_{cp}\mathbf{S}_{pp}^{-1}\mathbf{S}_{pc})(1/\Sigma\, y_i^2) \tag{3.20}$$

This form for the multiple-R has an advantage besides its relation with the canonical correlation coefficient to be discussed later. Namely, if we write the SSCP matrix for the entire set of $p + 1$ variables including both the predictors and the criterion and then partition it into the four sectors $\mathbf{S}_{pp}$, $\mathbf{S}_{pc}$, $\mathbf{S}_{cc}$ $(= \Sigma\, y_i^2)$, and $\mathbf{S}_{cp}$ $(= \mathbf{S}_{pc'})$, we get

$$\left[\begin{array}{c|c} \mathbf{S}_{pp} & \mathbf{S}_{pc} \\ \hline \mathbf{S}_{cp} & \mathbf{S}_{cc} \end{array}\right]$$

where $\mathbf{S}_{pp}$ is the $p \times p$ SSCP matrix of the predictor variables alone, and $\mathbf{S}_{pc}$ and $\mathbf{S}_{cp}$ (transposes of each other) are vectors of the sum of cross-products between the criterion and the p predictor variables. $\mathbf{S}_{cc}$, as noted above, is simply another symbol for the sum of squares $\Sigma\, y_i^2$ of the criterion variable. (Note that while p stands for the number of predictors and hence the number of rows or columns of the sector matrix, c merely stands for "criterion" and does not denote a number. Of course, it may be regarded as standing for "1," since there is just one criterion. But writing c serves as a mnenomic, so we would not want to replace it by 1 even when in specific contexts we may replace p by the actual number of predictors, like 5, 10 or whatever.) Using these symbols for the four sectors of the $(p + 1) \times (p + 1)$ SSCP matrix, expression (3.20) may be written in a neat, "balanced" form that serves as a good mnemonic, as follows:

$$R^2 = \mathbf{S}_{cp}\mathbf{S}_{pp}^{-1}\mathbf{S}_{pc}\mathbf{S}_{cc}^{-1}$$

That is, R^2 is a quadruple product of the two nonsquare sectors and the inverses of the two square sectors of the SSCP matrix, the four factors being taken in clockwise order starting from the lower-left sector.

Cross-Validation

It was stated earlier that the multiple correlation coefficient "would seem to be" a reasonable measure of how good a fit to the data points has been achieved by the multiple regression equation. A little reflection should show, however, that this measure has the limitation that it is based solely on the set of data for which the fit was optimized. Hence the multiple-R is biased upward as a measure of how well the regression equation might be expected to fit the data in a future sample from the same population, or in the population itself. Since one purpose of having a multiple regression equation in the first place is to make predictions in future samples, this upward bias may sometimes be quite misleading. What is needed, then, is a measure that reflects how well the equation will hold up for future data. Ideally, we should wait for some length of time, gather new data from what we assume to be the same population, and examine the predictive power of the equation we developed in the earlier sample.

For example, if the X's are high school grades and college entrance exam scores and Y is the first-year GPA in college, we could construct the multiple regression equation of Y on the X's from data on one year's entering class. Then, after waiting for a year, we could use the equation that was constructed the year before to predict the freshman GPA for each member of the new entering class. Finally, we correlate the predicted GPAs with the actual GPAs earned by the new class of students.

This process is called *cross-validation,* and the correlation coefficient thus obtained is called the cross-validation correlation (or "cross-validation R"). Note that this is *not* a multiple correlation coefficient, since the prediction equation was not optimized for the new sample but obtained from an earlier one.

Sometimes, we cannot afford to wait for a year (or any other suitable length of time) before carrying out a cross-validation. In such cases we may resort to cross-validation on a "holdout sample." By this is meant that we collect the data on an available sample but set aside a relatively small random subset of this sample and construct our multiple regression equation using only the data from the remaining (larger) subsample. Then the sample that was set aside (i.e., the holdout sample) is used for doing the cross-validation. Obviously, this procedure is not as satisfactory as cross-validating in a genuinely new sample, so it should be used only as a last resort—although in practice it is used quite frequently.

Although the cross-validation R is, by definition, the correlation between predicted criterion scores obtained from the regression equation and the actual criterion scores observed in the new (or holdout) sample, we do not actually need to go through the process of computing the $\hat{Y}$ for each member of the second sample and correlate them with the actual Y's. It is here that the equation we got just before getting our first formula for the multiple-R (i.e., Eq. 3.18) comes in handy. Recall that the $\mathbf{b}$ in Eq. 3.18 could be any set of combining weights; it did not have to be the optimal weights obtained from Eq. 3.15. Thus all we have to do is to substitute the $\mathbf{b}$ obtained from the old sample (called the *derivation* or the *normative sample*) together with the $\mathbf{S}_{pc}$, $\mathbf{S}_{pp}$, and $\Sigma\, y_i^2$ from the new sample (the *validation sample*) in Eq. 3.18. Using (1) and (2) as subscripts to indicate the old and new samples, respectively, we get the following formula for computing the square of the cross-validation R:

$$R_{cv}^2 = \frac{(\mathbf{b}'_{(1)}\mathbf{S}_{pc(2)})^2}{(\Sigma\, y_{i(2)}^2)(\mathbf{b}'_{(1)}\mathbf{S}_{pp(2)}\mathbf{b}_{(1)})} \qquad (3.21)$$

The square root, R_{cv}, of this quantity will give a better indication of how well the multiple regression equation derived from the original sample may be expected to hold up in a new sample drawn from the same population.

To further enhance our confidence that we are getting a reasonable measure of the extent to which we may expect our multiple regression equation to hold up in future samples, we may do a two-way cross-validation, provided that we have a large enough sample. This is done by splitting our sample into roughly equal halves at random, constructing one multiple regression equation based on one of the subsamples and cross-validating it in the other, and vice versa. When this is done, we have carried out a double cross-validation.

Significance Tests Related to Multiple Correlation Coefficients

We now present various types of significance tests that are useful in connection with multiple correlation coefficients in one situation or another. First there is the basic test of the significance of an observed multiple-R itself—that is, whether it is significantly different from zero. This is a straightforward generalization of the t-test of the significance of a simple product-moment correlation, which is to calculate

$$t = \frac{r/1}{\sqrt{(1 - r^2)/(N - 2)}}$$

(where N is the sample size) and compare this value with the appropriate percent point of the t-distribution with $N - 2$ degrees of freedom. The reader will recall that the square of a t-variate follows an F-distribution with 1 d.f. in the numerator and the same d.f. as the t in the denominator. Squaring the above expression, we get

$$F = \frac{r^2/1}{(1 - r^2)/(N - 2)}$$

which follows an F-distribution with 1 and $N - 2$ degrees of freedom. The generalization of this F-statistic to the F for testing the significance of a multiple correlation coefficient with p predictor variables consists of replacing the 1 in the numerator by p and the 2 in the denominator by $p + 1$. We then get

$$F = \frac{R^2/p}{(1 - R^2)/(N - p - 1)} \tag{3.22}$$

which is compared with the appropriate percent point of an F-distribution with p and $N - p - 1$ degrees of freedom.

Although such a test of significance of a multiple-R is theoretically necessary, it is almost a foregone conclusion in practical situations that R will be significant, for we would not be constructing a multiple regression equation in the first place unless we have at least one predictor variable that has a significant correlation with the criterion. The only way, under these circumstances, that we could get a nonsignificant R is if we were to use so many predictors as to have a substantial decrease in the denominator degrees of freedom without a commensurate increase in the value of R compared to the largest r_{jy}. Thus it becomes important to have a test of whether the inclusion of additional predictors leads to a significant increase in R. Such a test is known as a test of the significance of the incremental R (or R^2, strictly speaking). It consists of calculating

$$F = \frac{(R^2_{p+q} - R^2_p)/q}{(1 - R^2_{p+q})/(N - p - q - 1)} \tag{3.23}$$

where R_p is an abbreviation for $R_{Y \cdot 123 \cdots p}$, the multiple-R of Y with the first p predictors (which, of course, reduces to a simple correlation coefficient when $p = 1$), and R_{p+q} stands for the multiple-R when q extra predictors are added. This F is tested against the F-distribution with q and $N - p - q - 1$ degrees of freedom.

The special case of Eq. 3.23 when $q = 1$, applicable when just one new predictor variable is added to the original p predictors, serves as a stopping rule in *stepwise multiple regression* programs. These are computer programs in which, at each step, the predictor that leads to the largest increase in multiple-R is added to the current set until the addition no longer leads to a *significant* increase at a prescribed level. More accurately, the method just described is known as the forward incremental method. There is also the backward deletion method, which consists of starting out with a complete set of possible predictor variables and deleting, at each step, the variable that results in the smallest decrease in multiple-R until a *significant* drop is encountered. Most stepwise programs actually combine both the forward incremental and backward deletion procedures. That is, each time a new predictor is added, the program also checks whether a predictor that was added earlier can now be deleted without leading to a significant decrease in the muliple-R. It is possible for this to happen because a predictor that made a significant contribution at one point may no longer continue to do so when some other predictors have been added to the multiple regression equation.

A word of caution is in order here concerning stepwise multiple regression, since it is often not stated explicitly in published computer packages. This is that the stepwise method does not necessarily (in fact, usually does not) yield the best possible multiple regression equation for the number of predictor variables used. An alternative method is available that does look for the "best" two-predictor model, the "best" three-predictor model, and so on. Although this method—which was developed by Mallows (1973)—is generally considered superior to the stepwise technique, it too is not free of shortcomings. In the first place, what is meant by the "best" m-predictor model here is the best only among the set of m-predictor equations that are looked at by the particular technique, and it is not the best among *all possible* m-predictor equations. To find the best m-variable model (i.e., the one yielding the largest possible multiple-R) given p potential/predictors in all would require a program that computes and compares $\binom{p}{m}$ multiple regression equations—which can become quite a time-consuming task even for today's high-speed computers. (For example, to identify the best seven-variable equation given a total of 15 predictors would require computing 6435 seven-predictor equations.) Moreover, if we wanted to ascertain whether the best m-predictor model is significantly better than the best $(m - 1)$-predictor model, we would need to have $\binom{p}{m-1} + \binom{p}{m}$ multiple regression equations computed. (With $p = 15$ and $m = 7$, this comes to a total of 11,440 six- and seven-predictor equations.) Another shortcoming is that the decision of what m should be becomes a subjective one unless, at each step of going to the "best" model with one more predictor a significance test is conducted. Thus there is really no flawless technique that enables us to select an optimal subset of predictors out of even a moderately large set of potential predictors. In fact, some authors regard the entire idea of a "mechanical" selection procedure for this purpose as misguided and argue that a substantive-theory-based choice of predictors is the only justifiable approach.

Finally, there is a significance test that is useful in connection with cross-validation. A multiple regression equation that was constructed on the basis of a sample from a certain population will clearly become "inoperative" in a subsequent sample if, in the interim, some societal or other environmental event has taken place that in effect *changes* the population. For example, suppose that a multiple regression equation for predicting college success is constructed for one year's entering class, but a severe depression hits during the ensuing year. It is then quite likely that the population of potential candidates for admission to that college the following year will have changed substantially. The applicability of the multiple regression equation to this new crop of applicants would then be questionable, even though it *might* continue to hold up. How can we tell whether we may use the equation without too great a detriment? One way to answer this question would be to check to see whether the drop in multiple-R, upon tentatively using the equation, is significantly greater than the drop that would be expected even if the population had not changed. For this we need to have a standard by which to judge the extent of drop that is to be expected in a new sample from the same population. That is, how much smaller would a cross-validation R be expected to be than the original multiple-R when the two samples are drawn from the same population? If the correlation $r_{Y\hat{Y}}$ between the actual first-year GPA, Y, and that predicted from the regression equation developed before the intervening depression is significantly smaller than the expected cross-validation R in a new sample from the same population, we would have to suspect that a population change had indeed taken place. Use of the old multiple regression equation would then be unjustified; otherwise, we can probably continue to use the equation with impunity.

A standard for the expected drop (or "shrinkage") of validity of a multiple regression equation upon cross validation is provided by Stein's (1960) shrinkage formula. Actually, what it represents is the cross-validity of a multiple regression equation with p predictor variables, constructed in a sample of size N, when that equation is (hypothetically) applied to the entire population from which the derivation sample was drawn. The formula is

$$R_s^2 = 1 - \frac{N-1}{N-p-1} \cdot \frac{N-2}{N-p-2} \frac{N+1}{N} (1 - R^2) \tag{3.24}$$

where R is the observed multiple-R and R_s is the multiple-R as "corrected for shrinkage" by Stein's formula.

Returning to the issue at hand, we may compare the $r_{Y\hat{Y}}$ obtained the following year (or the year when we are questioning the continued applicability of the old multiple regression equation) with R_s. If the difference is not significant, we may with some assurance attribute the drop in validity to normal, expected shrinkage. But if the difference is significant, we should conclude that there was a population change and discontinue use of our old regression equation and construct a new one. In fact, it would be prudent to carry out this checking on a routine basis even when an obvious, potential disrupting event has not taken place, for there can always be unsuspected population changes occurring.

Concretely, the procedure would be to treat R_s (i.e., the square root of the value computed from Eq. 3.24) as the hypothesized population correlation coefficient and to test whether the cross-validation R computed from Eq. 3.21 is significantly smaller than this value. Since no multiple-R as such is involved, we have only to use a test for a simple correlation coefficient. Fisher's well-known Z (i.e., the hyperbolic arctangent) transformation leads to the appropriate test in this case, because the hypothesized value of the population correlation, ρ, is not zero. Thus we use either a table for the Z-transform or the formula

$$Z = (\tfrac{1}{2}) \ln [(1 + r)/(1 - r)]$$

(or $Z = \tanh^{-1} r$ if we are using a computer or an electronic hand-held calculator with the hyperbolic tangent and inverse functions built in) to transform both the R_s from Eq. 3.24 and the R_{cv} from Eq. 3.21 into Z's. Denoting the Z-transform of the "shrunken" R from Eq. 3.24 by Z_s and the Z-transform of the $r_{Y\hat{Y}}$ computed from Eq. 3.21 by $Z_{(2)}$, the appropriate test statistic becomes the unit normal deviate

$$z = (Z_{(2)} - Z_s) \sqrt{N_{(2)} - 3} \tag{3.25}$$

where $N_{(2)}$ is the size of the new sample in which the old multiple regression equation has been used.

EXAMPLE 3.3 Suppose that we constructed a multiple regression equation with $p = 5$ predictors in a sample of $N = 60$ cases, and that the multiple-R was .70. The following year, this multiple regression equation was tried out in a new sample of $N_{(2)} = 50$, and it was found that R_{cv} (i.e., the $r_{Y\hat{Y}}$ for the new sample) was .53. Are there grounds to suspect that some population change had taken place in the intervening year?

We first use Eq. 3.24 to calculate

$$R_s^2 = 1 - (59/54)(58/53)(61/60)(1 - .49) = .3800$$

Hence $R_s = .6165$. We thus get

$$Z_s = (1/2) \ln [(1 + .6165)/(1 - .6165)] = .7193$$

and

$$Z_{(2)} = (1/2) \ln [(1 + .53)/(1 - .53)] = .5901$$

Substituting these values in Eq. 3.25, we obtain

$$z = (.5901 - .7193) \sqrt{50 - 3} = -.89$$

which, of course, is far from being significantly smaller than 0 as a unit normal deviate. That is, a shrinkage from an original multiple-R of .70 to a cross-validation R of .53 in a new sample from the same population is not at all unusual. So there are no grounds for suspecting that a population change had taken place in the intervening 1-year period.

3.3 THE EQUAL-WEIGHTS PREDICTOR COMPOSITE

The fact that a multiple regression equation derived in one sample can suffer a marked deterioration of predictive power when applied in another sample from the same population, as just illustrated, can be stated in another way, as follows: The optimal weights for the several predictors can differ widely from one sample to another. This unhappy fact has been known since the earliest days of multiple regression analysis, and is often referred to by the colorful term "the phenomenon of bouncing betas." Several authors have therefore suggested that it is not worth the trouble to determine the optimal combining weights for the sample at hand, since they will often be far from optimal in the new sample for which we want to use the regression equation. "Why not just use an unweighted sum (i.e., an equal-weights composite) of the predictors?" is the usual conclusion of these authors.

Wainer (1976, p. 214) supported the advocacy of equal weights by proving that when p uncorrelated standardized predictors are used for predicting a standardized criterion, the expected loss of variance explained using equal (.5) weights will be less than p/96 if the βs are uniformly distributed on the interval [.25, .75].

The use of equal weights, however, is not universally supported. For example, Laughlin (1978) and Pruzek and Frederick (1978) find fault with Wainer's arguments and make alternative suggestions. Perhaps the best stance to take in this matter is to concede that when a rough-and-ready prediction equation is sufficient and a large sample is not available, constructing an equal-weights composite is adequate, but when a very large sample *is* available, determining the least-squares regression weights may still be worth the trouble.

It should be noted that even an equal-weights composite has to be "constructed" in that the equal weights are applied to the predictors in standardized form. Hence their means and standard deviations must be computed from some sample in order to obtain a raw-score prediction equation. The reason a standard-score prediction equation is usually unusable for practical purposes is that we may often wish to make predictions for a very small number of subsequent cases (or even just one individual), and hence it is impractical or infeasible to do a standardization in the new sample. Thus the standard-score prediction equation with equal (unit) weights,

$$\tilde{z}_y = z_1 + z_2 + \cdots + z_p$$

has to be transformed into the raw-score prediction equation

$$\frac{\tilde{Y} - \overline{Y}}{s_y} = \frac{X_1 - \overline{X}_1}{s_1} + \frac{X_2 - \overline{X}_2}{s_2} + \cdots + \frac{X_p - \overline{X}_p}{s_p}$$

or

$$\begin{aligned}\tilde{Y} = {} & [\overline{Y} - (s_y/s_1)\overline{X}_1 - (s_y/s_2)\overline{X}_2 - (s_y/s_3)\overline{X}_3 \cdots (s_y/s_p)\overline{X}_p] \\ & + (s_y/s_1)X_1 + (s_y/s_2)X_2 + (s_y/s_3)X_3 + \cdots + (s_y/s_p)X_p\end{aligned} \qquad (3.26)$$

EXAMPLE 3.4 To illustrate the calculations for getting the raw-score version of an equal-weights standard-score predictor composite, let us use only the first two predictors in Exercise 1 at the end of the chapter. Since each standard-deviation ratio, s_y/s_j, in Eq. 3.26 may equivalently be taken as the square root of the ratio $\Sigma\, y^2/\Sigma\, x_j^2$, of the corresponding *SS*'s we obtain

$$s_y/s_1 = (\Sigma\, y^2/\Sigma x_1^2)^{1/2} = (9146.08/4550.67)^{1/2} = 1.418$$

and

$$s_y/s_2 = (9146.08/9656.78)^{1/2} = .973$$

These values, together with

$$\overline{Y} = 80.17 \qquad \overline{X}_1 = 91.96 \qquad \overline{X}_2 = 92.28$$

yield

$$a = 80.17 - (1.418)(91.96) - (.973)(92.28) = -140.02$$

Hence the raw-score form of the equal-weights predictor composite is

$$\tilde{Y} = -140.02 + 1.418X_1 + .973X_2$$

Since the purpose of using an equal-weights composite instead of the least-squares regression equation for prediction is (hopefully) to decrease the shrinkage of predictive efficiency in new samples, we must check to see how we fare. For this we need to compute $r_{\tilde{y}y(1)}$ in the first sample (from which the means and standard deviations of the several predictors and the criterion variable were obtained) and compare it with the $r_{\tilde{y}y(2)}$ in a new sample to see how much they (or their squares) differ. At the same time we have to determine the least-squares regression equation and the multiple-R from the first sample, compute the cross-validation-R in the same new sample, and find their (or their squares') difference. If the first difference is substantially smaller than the second, we will have found evidence in favor of the equal-weights composite. To compute all except the multiple-R among the four correlations just mentioned, we use Eq. 3.18, or a variant of it that applies when all variables are standardized. The latter equation is

$$r^2_{\tilde{z}_y z_y} = \frac{(\mathbf{1}'\mathbf{R}_{pc})^2}{(1)(\mathbf{1}'\mathbf{R}_{pp}\mathbf{1})} \tag{3.27}$$

where $\mathbf{R}_{pc}$ and $\mathbf{R}_{pp}$ are the appropriate vector of validity coefficients and the predictor intercorrelation matrix, respectively, and $\mathbf{1}'$s is a p-dimensional row vector of 1's.

Thus, for calculating $r_{\tilde{y}y(1)}$, we get $\mathbf{R}_{pc(1)}$ and $\mathbf{R}_{pp(1)}$ as the appropriate sub-matrices of the 3×3 correlation matrix computed from the SSCP matrix of Exercise 1 and substitute the results in Eq. 3.27, as follows:

$$\mathbf{R} = D(\mathbf{S})^{-1/2}\mathbf{S}D(\mathbf{S})^{-1/2}$$

$$= \begin{bmatrix} (4550.67)^{-1/2} & 0 & 0 \\ 0 & (9656.78)^{-1/2} & 0 \\ 0 & 0 & (9146.08)^{-1/2} \end{bmatrix} \times$$

$$\begin{bmatrix} 4550.67 & 682.13 & 2013.49 \\ 682.13 & 9656.78 & 3976.28 \\ 2013.49 & 3976.28 & 9146.08 \end{bmatrix} D(\mathbf{S})^{-1/2}$$

$$= \begin{bmatrix} 1 & .1029 & .3121 \\ .1029 & 1 & .4231 \\ .3121 & .4231 & 1 \end{bmatrix}$$

Hence

$$\mathbf{1}'\mathbf{R}_{pc} = [1,\ 1]\begin{bmatrix} .3121 \\ .4231 \end{bmatrix} = .7352$$

and

$$\mathbf{1}'\mathbf{R}_{pp}\mathbf{1} = [1,\ 1]\begin{bmatrix} 1 & .1029 \\ .1029 & 1 \end{bmatrix}\begin{bmatrix} 1 \\ 1 \end{bmatrix} = 2.2058$$

(Note that 2.2058 is simply the sum of all the elements of $\mathbf{R}_{pp}$. This will always be true for the special quadratic form $\mathbf{1}'\mathbf{A}\mathbf{1}$, where $\mathbf{A}$ is a $p \times p$ matrix and $\mathbf{1}$ is a vector of p 1's.)

Substituting these results in Eq. 3.27, we get

$$r^2_{\tilde{y}y(1)} = (.7352)^2/(1)(2.2058) = .2450$$

and

$$r_{\tilde{y}y} = .4950$$

This is the correlation coefficient that was denoted $r_{\tilde{z}_y z_y(1)}$ above. We now need to get the correlation, $r_{\tilde{y}y(2)}$, between the raw scores estimated from the foregoing predictor composite for a new sample with the scores actually observed in that sample. For the new sample we use the sample of 10 cases given in Exercise 4 and first compute its SSCP matrix, getting

$$\mathbf{S} = \begin{bmatrix} 894.9 & 142.1 & 246.6 \\ 142.1 & 1372.9 & 39.4 \\ 246.6 & 39.4 & 624.4 \end{bmatrix}$$

Hence from Eq. 3.18 and the coefficients of the raw-score predictor composite found above, we calculate

$$r^2_{\tilde{y}y(2)} = \frac{\left([1.418,\ .973]\begin{bmatrix} 246.6 \\ 39.4 \end{bmatrix}\right)^2}{(624.4)[1.418,\ .973]\begin{bmatrix} 894.9 & 142.1 \\ 142.1 & 1372.9 \end{bmatrix}\begin{bmatrix} 1.418 \\ .973 \end{bmatrix}}$$

$$= \frac{(388.015)^2}{(624.4)(3491.28)} = .0691$$

or

$$r_{\tilde{y}y(2)} = .2628$$

Thus the shrinkage in $r_{\tilde{y}y}$ between the first and second samples is

$$.4950 - .2628 = .2322$$

Let us now compare this shrinkage with that of the multiple-R in the first sample to the cross-validation R in the second. Going back to the SSCP matrix given in Exercise 1, we first compute the vector **b** of multiple regression weights from Eq. 3.14 and the multiple-R from Eq. 3.19. The results are

$$b = \begin{bmatrix} 4550.67 & 682.13 \\ 682.13 & 9656.78 \end{bmatrix}^{-1} \begin{bmatrix} 2013.49 \\ 3976.28 \end{bmatrix} = \begin{bmatrix} .38481 \\ .38458 \end{bmatrix}$$

and

$$R^2_{y.12} = [.38481, .38458] \begin{bmatrix} 2013.49 \\ 3976.28 \end{bmatrix} \div 9146.08 = .2519$$

or

$$R_{y.12} = .5019$$

On the other hand, the cross-validation R in the second sample is, from Eq. 3.21, the square root of

$$R^2_{cv} = \frac{\left([.38481, .38458] \begin{bmatrix} 246.6 \\ 39.4 \end{bmatrix} \right)^2}{(624.4)[.38481, .38458] \begin{bmatrix} 894.9 & 142.1 \\ 142.1 & 1372.9 \end{bmatrix} \begin{bmatrix} .38481 \\ .38458 \end{bmatrix}} = .05136$$

Hence $R_{cv} = .2266$, and the shrinkage is $.5019 - .2266 = .2753$. Thus the shrinkage of the correlation of the equal-weights predictor composite from the first sample to the second is slightly smaller (by about .04) than the shrinkage of the multiple-R to the cross-validation R between the same two samples. This difference is not much to speak of, but in some situations the advantage of the equal-weights composite over the least-squares multiple regression equation may be greater. Unfortunately, it is difficult if not impossible to tell just when it might be better to use an equal-weights composite—except for the general rule that when the predictor variables are highly intercorrelated, the least-squares regression weights become so much the more unstable. In such cases it may well be that the equal-weights composite will show considerably greater sampling stability in predictive power.

Another technique for coping with cases of high predictor intercorrelation—which are often referred to as *multicollinearity* cases—is the method of *ridge regression*. Essentially, what this technique does is to add a suitable positive number less than 1 to the diagonal elements of the predictor intercorrelation matrix so as to avoid the near-singularity of the matrix that results from the high correlations among the predictor variables. The interested reader is referred to articles by Hoerl and Kennard (1970) and by Marquardt and Snee (1975), but should also be aware that (just as in the case of the equal-weights composite) not all statisticians agree that this technique is a good one. A critical view is expressed by Smith and Campbell (1980), and a number of comments—both pro and con—are made by several authors in the same issue of the journal.

3.4 RESEARCH EXAMPLES

Although multiple regression has been popular in educational research for over half a century, for most of that time it was used primarily to demonstrate the existence of regressions and to develop prediction equations. Recently there has been a great deal of enthusiasm for the idea of testing *causal network theories* by means of multiple regression. Such a theory explicates a series of causal hypotheses linking several variables. Each specific causal linkage between two variables is expressed as a *path* from the independent variable to the dependent variable for that relationship. Usually there will be two or more paths into a particular dependent variable (called an *endogenous* variable), and the path coefficients for the causal influences along these paths will be estimated as multiple regression weights. A network arises because a variable which is the dependent variable for one causal hypothesis is allowed to be one of the independent variables for another causal hypothesis. Thus an endogenous variable may have paths out of it as well as paths into it. Variables in the network which serve only as independent variables, thus having no paths into them, are termed *exogenous* variables.

Path models explicating causal network theories may be presented in diagrams in which the path from each causal variable to an endogenous variable is drawn as an arrow on which the regression coefficient is written. However, the fundamental expression of the network theory is a set of linear equations, called *structural equations,* in which there is one equation for each endogenous variable. Each of these equations is a multiple regression model equation. These are *model* equations, not prediction equations, because the residual term is included in each equation, so that each equation explains a complete observation on a particular endogenous variable. In the parlance of path analysis, the residual from a multiple regression is called a *disturbance*.

A successful estimation of a path model can contribute to the plausibility of the causal network theory expressed by the model. It cannot prove the theory.

The theory is always subject to *specification error*. This error occurs when a theory fails to involve an important causal variable which is correlated with one or more of the causal variables invoked by the theory. It is well known that adding another regressor to a multiple regression, when the added regressor is correlated with the previous regressors, will change all the regression coefficients. If those weights are given a causal interpretation, as they are in estimation of path models, the hazard posed by an unspecified regressor is serious indeed. Because of this hazard, the best use of path modeling occurs when two or more competitive models, similar except for the addition or deletion of particular exogenous variables, are compared for their fits to the data. This strategy of causal analysis can be studied in Blalock (1964, 1971, 1979), Duncan (1966, 1975), Goldberger (1970), Heise (1975), and Kenny (1979).

Biniaminov and Glasman (1983) reported a path model for selected determinants of school achievement in Israel. They said explicitly that their theory "assumes a causal relationship among four factors" (p. 252). They acknowledged that the theory was necessarily an oversimplification of reality, but argued: "Because of the difficulties in actually demonstrating causal relationships, the notions of cause and effect are confined to models of the real world, which consist of only finite numbers of explicitly defined variables" (p. 252). With school as the unit of analysis, they measured their four variables on the sample of 32 secondary schools. The names of the variables were:

X_1: Level of disadvantaged students
X_2: Level of fiscal resources
X_3: Teaching experience in the same school
X_4: Certificated graduates as percentage of 12th graders

Three structural equations were given:	(correlations among variables)		
$X_2 = p_{21}X_1 + d_1$	.441		
$X_3 = p_{31}X_1 + p_{32}X_2 + d_2$	−.060	.140	
$X_4 = p_{41}X_1 + p_{42}X_2 + p_{43}X_3 + d_4$	−.356	−.456	.285

The authors gave means, standard deviations, and correlations as Table I (p. 260). They asserted that the path coefficients could be computed as ordinary least-squares (OLS) "beta" coefficients. The values reported were $p_{21} = .441$, $p_{31} = -.151$*, $p_{32} = .206$*, $p_{41} = -.141$*, $p_{42} = -.441$, $p_{43} = .339$ (* indicated judged insignificant). The finding was that level of disadvantaged students affects school achievement only indirectly through its effect upon level of fiscal resources. This placed an interpretation of spuriousness upon the correlation between X_1 and X_4 which would not have been available without the modeling analysis. This is an excellent example of structural modeling by means of multiple regression.

Boli, Allen, and Payne (1985) employed the statistical modeling procedure to develop a path model for causal influences of gender and other variables on performance in college mathematics and chemistry courses. OLS estimation of their model led to the conclusion that gender had zero direct effect on performance, but that it had significant indirect effect mediated by other variables. They presented the path model as a summary analysis following a series of other statistical analyses, in which role it seems to have been appropriate and useful.

Keith and Page (1985) undertook to demonstrate that a specification error invalidated the finding of Greeley (1982), which was that Catholic schools did a substantially better job of educating minority students than did public schools. Greeley had regressed achievement on family background and type of school. Keith and Page claimed that when student ability was added to the specification of causes of achievement, the apparent causal influence of type of school was diminished considerably, although they conceded that it remained above zero. Their work represents an exemplification of the strategy of a priori factor analysis to establish the measurement model for the data, followed by OLS estimation of the structural model for the factors in a conventional path analysis way. They factor analyzed six verbal and nonverbal ability tests, and scored an ability factor. Their family-background and achievement variables were also composite scores. School type was a dummy 0,1 variate. Their data consisted of two samples, both from the 1980 High School and Beyond survey. One sample contained 3,922 black seniors (3,552 from public schools and 370 from Catholic schools). The second sample contained 3,146 Hispanic seniors (2,661 from public schools and 485 from Catholic schools).

Regressing achievements only on family background and school type yielded, for the blacks, path coefficients of .172 from family and .277 from school type, with a path of .371 from family to type. For the Hispanics, the path to achievement from family was .217 and from school type was .278, with a path of .292 from family to type. However, when achievement was regressed on family, ability, and school type, the path coefficients were, for blacks (Hispanic values in parentheses), .094 (.106) from family, .145 (.161) from school type, and .552 (.575) from ability. The path from family to type was .321 (.242), from ability to type was .218 (.199), and from family to ability was .230 (.252). Keith and Page would seem to have made their point. However, a difficulty is that the six ability tests were taken on the same occasion as the achievement tests, so it is not clear that they can be interpreted as aptitude measures. They may have siphoned off some of the achievement variance which properly belongs to the school type variable.

In a further development of their theory for the data, Keith and Page added educational aspiration and academic courses variables, building the number of variables in the path diagram up to six (Fig. 3, p. 344), and were able to show that most of the apparent remaining type of school effect can be moved over to the academic courses variable. That is, minority students in Catholic high schools take more academic courses than do their counterparts in public high schools, and that factor seems to make most of the difference.

A feature of the Keith and Page paper which is highly commendable is that they presented the correlations, means, and standard deviations for the six factors, in Table I for the blacks (p. 341) and Table II for the Hispanics (p. 342), making it possible for others to estimate different models for the data.

Carter (1984) used a set of four path models for four sample cohorts to report the findings of the largest study ever made of elementary education. Starting in 1975, the Sustaining Effects Study collected data on about 120,000 students in a national sample of more than 300 elementary schools. Four cohorts, one each starting in grades K, 1, 2, and 3, were followed for 3 years. Although the mission of the study was to evaluate the effects of Title I interventions, the data provided an extraordinary opportunity to evaluate the effects of education generally. The importance of the mission is indicated by the fact that over $40 billion has been expended on Title I programs since 1965, yet most evaluations have been highly discouraging. The broader importance of this study is that its longitudinal data base contained measures of input aptitude, family background, classroom instructional processes, and achievement in reading and mathematics. The successful utilization of OLS path models has to represent one of the finest hours of this methodology in the field of educational studies.

A major strength of the study was its use of the individual student as the unit of analysis. Carter said, "We are convinced that it is important to follow the progress of individual students rather than groups of students" (p. 12). Cooley and Lohnes (1976, p. 151–4) presented the arguments for the student as unit in educational evaluations in detail.

To dispel the suspense, the finding in regard to the mission was that "it appears that Title I was effective for students who were only moderately disadvantaged, but it did not improve the relative achievement of the most disadvantaged part of the school population" (p. 7).

It is noteworthy that Carter chose to describe the data analysis in the rubrics of modeling. "Using the techniques of causal analysis, the fit was determined between a rational model of the educational process and the data" (p. 8). However, that fit was not good, and it was necessary to modify the theory for the data to account for the facts. "It was found that whereas the rational model formed the basis for a reasonable model of the educational process, the actual process was considerably more complicated than the rational model had postulated" (p. 8). The problem was that a priori reasoning had hypothesized direct relationships of causes with achievement, but the data required heavy use of indirect relationships mediated by an "opportunity to learn" (p. 9) variable (see Cooley and Lohnes, 1976, p. 191), operationalized as a Learning Experience Composite.

As usual, aptitude (A) turned out to have the largest path coefficient to Achievement Composite (Y) in all four cohorts. Aptitude also had the largest path coefficient to Learning Experience (L) in all cohorts. The direct path from Family Background (B) to Achievement had a small coefficient in all cohorts, but the influence of Background on Learning Experience was moderate in all cases. The path from Learning Experience to Achievement had a small coefficient in the three older cohorts. However, for the Kindergarden-to-2nd Grade cohort the path from Learning Experience to Achievement had a moderate coefficient, which Carter took as showing that the opportunity variable is quite important, at least for beginners in the game of schooling. The third exogenous variable in the path diagrams was School Characteristics (S). The two standardized structural equations for each cohort were, with nonsignificant effects and disturbance effects omitted as in the original:

K-2 cohort: $L = .41A + .18S + .28B \quad Y = .44A + .35L + .08B$

1-3 cohort: $L = .41A + .23S + .27B \quad Y = .64A + .06L + .13B + .08S$

2-4 cohort: $L = .44A + .21S + .29B \quad Y = .65A + .18L + .11B$

3-5 cohort: $L = .48A + .07S + .28B \quad Y = .72A + .06L + .14B$

The steadily increasing influence of Aptitude on Achievement as cohorts get older is noteworthy. Especially noteworthy, and not discussed in the report, is the constant moderate influence of Family Background on Learning Experience, and the somewhat stronger constant influence of Aptitude on that opportunity variable. Despite the efforts of the Title I programs and other interventions to equalize opportunity, the data testify that it continues to be the case in America's elementary schools that those who have most get most.

Another popular and useful method for analyzing multiple regressions was recommended to the educational research community by Mood (1971), who named it "partitioning variance." This was an unfortunate naming, because the method does not partition the variance accounted for by a regression model among the predictors. Rather, it compares the variance accounted for by each of several regression models, one being the *full model* employing all predictors, and each of the others being a *reduced model* from which one predictor has been removed. The difference between R^2 for the full model and R^2 for the reduced model with a particular predictor removed is taken as the *unique* contribution to explained variance of that predictor. That is, the *uniqueness* of a predictor is the portion of the full-model R^2 which is lost if that predictor is removed. The difference between the full-model R^2 and the sum of the uniqueness for all the predictors is taken as the *commonality,* which is construed as the portion of the explained variance which is not attributable uniquely to any single predictor, but must be viewed as a joint contribution of the predictors based on their mutual confounding (i.e., their intercorrelatedness). (The fact that in some cases *suppressor variable* phenomena can cause the commonality to be negative is ample evidence that this is not a true variance-partitioning algorithm.) This interesting approach to analyzing a multiple regression is perhaps best named the *uniqueness-commonality method*.

Using data from the 1977–78 National Assessment of Educational Progress,

Welch, Anderson, and Harris (1982) were able to show a very substantial unique contribution of a schooling variable, number of semesters of mathematics study, to the explanation of mathematics learning variance, after the effects of eight background variables were entered. The background variables accounted for about 27% of criterion variance, and the additional variance due to semesters of mathematics was about 31%. Very large random samples were involved, making the results important indeed, especially for their challenge to the trend in other studies which found small unique effects of schooling variables. What the report lacked which a path model could have provided is explicit ideas about how background variables influenced the number of semesters of mathematics taken, and estimation of the hypothetical paths involved.

EXERCISES

Exercises 1–5 are based on the SSCP matrix and the means for four variables, X_1, X_2, X_3, and Y. The deviation-scores SSCP matrix on three predictors,

$$X_1 = \text{math test score}$$
$$X_2 = \text{mechanical reasoning score}$$
$$X_3 = \text{creativity test score}$$

and a criterion

$$Y = \text{achievement in physical science scores}$$

for a sample of $N = 50$ college sophomores was as follows:

$$\mathbf{S} = \begin{array}{c} \\ \end{array}\left[\begin{array}{ccc|c} 4550.67 & 682.13 & 373.54 & 2013.49 \\ 682.13 & 9656.78 & 2715.82 & 3976.28 \\ 373.54 & 2715.82 & 6891.93 & 3006.65 \\ \hline 2013.49 & 3976.28 & 3006.65 & 9146.08 \end{array}\right]\begin{array}{c} X_1 \\ X_2 \\ X_3 \\ Y \end{array}$$

(columns: X_1, X_2, X_3, Y)

Given further that the means on the four variables were $\overline{X}_1 = 91.96$, $\overline{X}_2 = 92.28$, $\overline{X}_3 = 86.89$, and $\overline{Y} = 80.17$, do the following problems.

1. Construct the multiple regression equation of Y on X_1, X_2, X_3. Also compute the multiple-R, $R_{y.123}$.

2. Convert the first three elements $\Sigma\, x_1 y$, $\Sigma\, x_2 y$, $\Sigma\, x_3 y$ of the fourth column of the SSCP matrix into validity coefficients r_{1y}, r_{2y}, r_{3y}. Now *delete* the variable with the smallest r_{jy} and construct the multiple regression equation of Y on the remaining two predictors. Also compute the multiple-R for this equation.

3. Is the two-predictor multiple-R found in Exercise 2 significantly smaller than the $R_{y.123}$ found in Exercise 1?

4. Next, suppose that the original sample actually contained 60 individuals but that 10 had been set aside for cross-validation purposes. The scores for these 10 students on the four variables were as follows:

X_1	X_2	X_3	Y
94	93	98	82
100	100	73	95
88	97	79	86
88	82	100	94
94	98	92	71
98	100	89	96
100	97	91	81
81	99	78	77
99	62	93	83
69	81	77	79

(a) Find the cross-validation correlation, R_{cv} for this sample of 10 cases. [It would be instructive to do this in two ways: (1) using Eq. 3.21 for R^2_{cv}; (2) actually computing $\hat{Y}$ values for the 10 cases and correlating them with the Y values given above. The results in (1) and (2) should agree within rounding error.]

(b) Is the R_{cv} found above significantly smaller than what you would expect on the basis of the shrinkage formula given in Eq. 3.24?

5. Now get the predictor composite Y' using "equal weights" and find the correlation between Y' and Y in both the first sample of 50 cases and the second sample of 10 cases. Are the results closer to each other than are $R_{y.123}$ and R_{cv}?

6. Let X_1 = sex, X_2 = SES, X_3 = locus of control, X_4 = self-concept, X_5 = motivation, and Y = writing achievement test in the "High School and Beyond" (HSB) data given in Appendix F.

(a) Construct the raw-score multiple regression equation of Y on X_1 through X_5. Also compute the multiple correlation coefficient $R_{y \cdot 12345}$.

(b) Use the forward-selection method in a computer program for stepwise multiple regression, with $F = 3.8$ as the criterion for entering a variable, and construct another multiple regression equation, also computing the multiple-R.

7. Using the same data as in Exercise 6, test whether including the two demographic variables, sex and SES, in addition to the three affective variables, locus of control, self-concept and motivation, increases the multiple correlation coefficient significantly at the 1% level.

8. Partition the 600 HSB cases into two subsamples: those with odd-numbered IDs and those with even-numbered IDs. Using the first sample of 300 cases as the derivation sample, construct the multiple regression equation of Y on $X_1, X_2, \ldots, X_5$, where the variables are as defined in Exercise 6. Then compute the cross-validation R of this equation in the second sample of 300 cases. Is R_{cv} significantly smaller than the multiple-R of the derivation sample after it has been "shrunken" in accordance with Eq. 3.24?

4

MULTIVARIATE SIGNIFICANCE TESTS OF GROUP DIFFERENCES

The utility of multivariate analysis comes from the fact that many problems in educational and psychological research require the simultaneous use of several dependent variables. Of course, multiple regression analysis is an exception in that it uses only one dependent (i.e., criterion) variable. As we have seen in the previous chapter, this fact does not detract from its position of paramount importance among multivariate techniques. For example, the outcomes of an instructional program in the language arts may be assessed in terms of vocabulary acquisition, reading comprehension, effectiveness of verbal and written communication, ability to discriminate between good and poor literary style, and so on. Suppose that two or more such instructional programs are to be compared in terms of these several criteria. We may, of course, be interested in the differences among the programs on each of these criteria, separately, in its own right. In that case, we might conduct a series of univariate significance tests (t-tests or F-tests)—one test for each criterion. But what if the outcomes (of two programs, for example) differed only slightly on each criterion, so that none of the univariate tests detected a significant difference? Would we conclude that the two programs seem no different in toto even if the differences on the several criteria were slight but consistent? Or do we conclude that the small differences, taken together, point to a real difference? Alternatively, it may be that the difference is significant (at a given α level) for some criteria but not others. Can we take the significant and nonsignificant differences at their face values? Or do we have to concede that given a large number of criterion variables, we would expect to find significant differences on a few of them by pure chance (1 out of 20 if $\alpha = .05$, for example)?

To answer the foregoing questions, we need a method (or several methods) for studying the differences among the groups in terms of many dependent variables *considered simultaneously*. The multivariate significance tests described in this chapter are among those that serve this purpose. If we wish also to know along what dimensions the stable differences occur, we shall need to invoke the more refined technique of *discriminant analysis,* described in Chapter 6.

4.1 THE MULTIVARIATE NORMAL DISTRIBUTION

Throughout the remainder of the book a basic assumption needed for strict validity of the significance tests is that the variables under study follow a *multivariate normal distribution*. For the case of two variables X_1 and X_2, the bivariate normal density function is

$$\phi(X_1, X_2) = \frac{1}{2\pi\sigma_1\sigma_2\sqrt{1-\rho^2}} \exp\left\{\frac{-1}{2(1-\rho^2)}\left[\frac{(X_1-\mu_1)^2}{\sigma_1^2} + \frac{(X_2-\mu_2)^2}{\sigma_2^2} - 2\rho\frac{(X_1-\mu_1)(X_2-\mu_2)}{\sigma_1\sigma_2}\right]\right\} \quad (4.1)$$

where μ_i and σ_i^2 are the mean and variance of X_i ($i = 1, 2$), and ρ is the correlation coefficient. The surface represented by this equation resembles a bell-shaped mound, distorted by being stretched out in one direction and compressed in the direction perpendicular to the first—the degrees of distortion being dependent on the value of ρ and the ratio σ_1/σ_2. Figure 4.1 depicts the general appearance of the bivariate normal surface for $\rho = .60$ and $\sigma_1/\sigma_2 = 1$. Since such a perspective drawing is difficult to construct and conveys little or no quantitative information, it is customary to represent a bivariate normal surface by drawing an arbitrarily selected *isodensity contour*—that is, the cross section of the surface made by a plane parallel to the (X_1, X_2)-plane. The equation for such a contour curve is obtained by setting the expression in brackets in Eq. 4.1 equal to a positive constant C. The smaller this constant, the larger is the altitude (representing density) along the contour, since the altitude is proportional to the negative exponential of $C/2(1-\rho^2)$.

By use of elementary analytic geometry, it can be shown that the equation in question, namely

$$\frac{(X_1-\mu_1)^2}{\sigma_1^2} + \frac{(X_2-\mu_2)^2}{\sigma_2^2} - 2\rho\frac{(X_1-\mu_1)(X_2-\mu_2)}{\sigma_1\sigma_2} = C \quad (4.2)$$

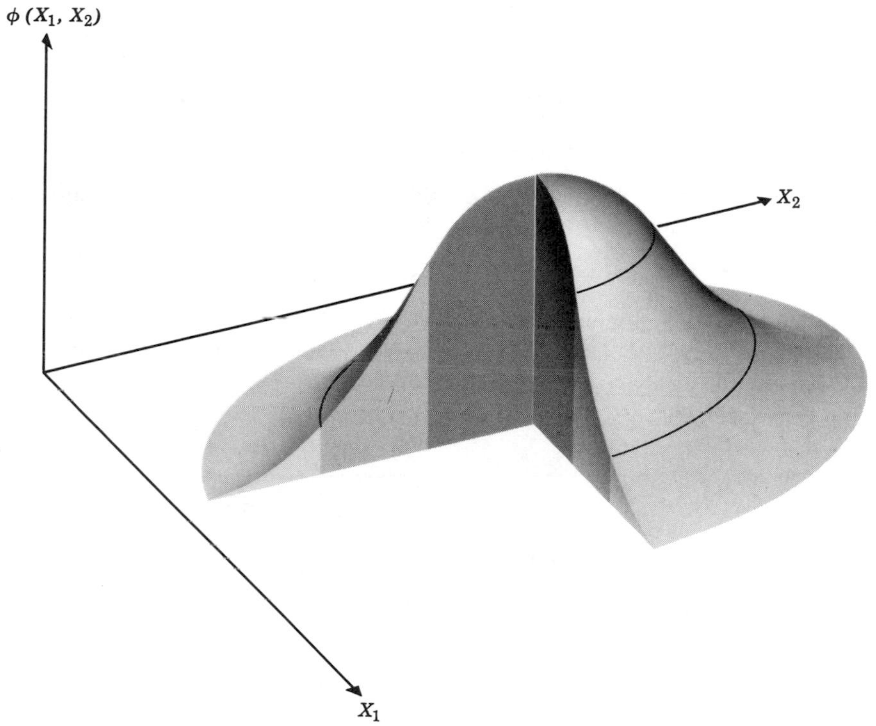

Figure 4.1. *Bivariate normal density surface.*

represents an ellipse with center at the point (μ_1, μ_2), which is called the *centroid* of the bivariate population, and major or minor axis along the line passing through this point and making the following angle with the positive X_1-axis:

$$\theta = \begin{cases} \frac{1}{2} \arctan \dfrac{2\rho\sigma_1\sigma_2}{\sigma_1^2 - \sigma_2^2} & \text{when } \sigma_1 \neq \sigma_2 \\ 45^\circ & \text{when } \sigma_1 = \sigma_2 \end{cases}$$

(This line contains the major axis if $\rho > 0$, the minor axis if $\rho < 0$.) Since the angle θ depends only on σ_1, σ_2, and ρ, and is independent of C, it follows that taking various values of C (in other words, making cross sections of the density surface with planes at various elevations) generates a family of concentric ellipses, all with the same orientation. Thus a representation by a series of contour lines of the bivariate normal surface depicted in Fig. 4.1 would look something like Fig. 4.2. If the reader imagines piling up a large number of elliptic disks of sizes as shown in Fig. 4.2 and thicknesses of about $\frac{1}{8}$ inch each, he or she will have a good idea of what a bivariate normal surface looks like.

EXAMPLE 4.1. Suppose that X_1 and X_2 follow a bivariate normal distribution with $\mu_1 = 15$, $\mu_2 = 20$, $\sigma_1 = \sigma_2 = 5$, and $\rho = .60$. Then the expression in brackets in Eq. 4.1 for the density function becomes

$$(X_1 - 15)^2/25 + (X_2 - 20)^2/25 - 2(.60)(X_1 - 15)(X_2 - 20)/25$$

Setting this expression equal to any positive constant C gives the equation of an isodensity contour ellipse. Thus, treating C as a parameter, the equation

$$(X_1 - 15)^2 + (X_2 - 20)^2 - 1.2(X_1 - 15)(X_2 - 20) = 25C$$

represents a family of concentric ellipses centered at (15, 20), with major axis making a 45° angle with the X_1-axis, because $\sigma_1 = \sigma_2$ and $\rho > 0$. Figure 4.2 shows several members of this family of ellipses for selected values of C, the first three of which are 5.89, 2.95, and 2.06, and the last three are .29, .13, and .01.

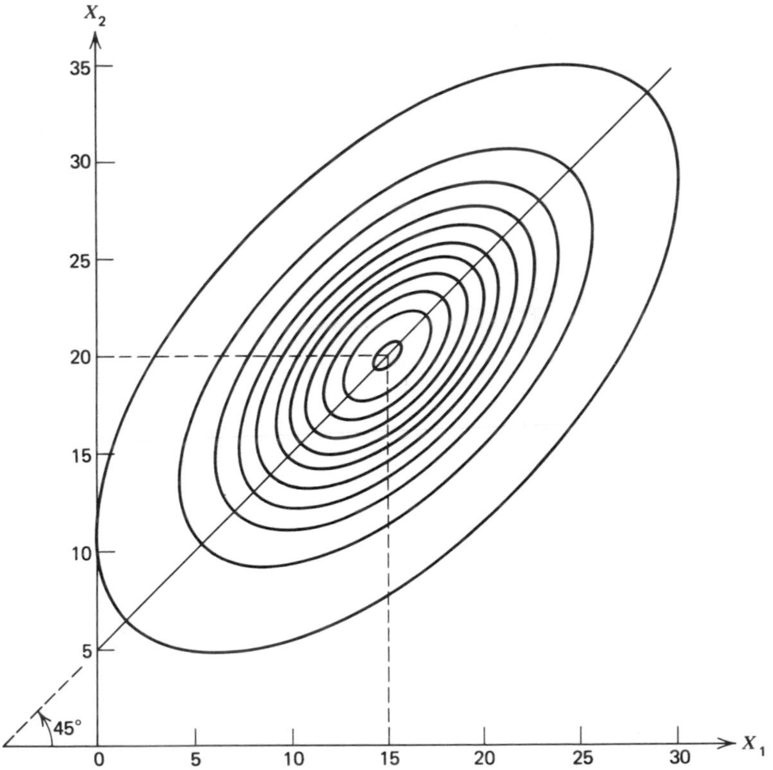

Figure 4.2. *Several members of the family of ellipses $(X_1 - 15)^2 + (x_2 - 20)^2 - 1.2\,(X_1 - 15)(X_2 - 20) = 25\,C$ with selected values for the parameter C.*

Before giving the equation for the multivariate normal density function for more than two variables, it is convenient to rewrite the quantities in Eq. 4.1 in matrix notation. The generalization to p variables will then be almost self-evident. We first define the *covariance matrix* (or *dispersion matrix*) for a bivariate population as follows:

$$\Sigma_2 = \begin{bmatrix} \sigma_1^2 & \rho\sigma_1\sigma_2 \\ \rho\sigma_2\sigma_1 & \sigma_2^2 \end{bmatrix} \tag{4.3}$$

(where the subscript 2 on Σ simply indicates the number of variates.) The diagonal elements of this matrix are the variances of the two variables, and the off-diagonal element is the covariance between the two variables. The determinant of this matrix is

$$|\Sigma_2| = \sigma_1^2\sigma_2^2(1 - \rho^2) \tag{4.4}$$

Hence the inverse of Σ_2 is given, in accordance with Eq. 2.16, by

$$\begin{aligned} \Sigma_2^{-1} &= 1/\sigma_1^2\sigma_2^2(1 - \rho^2)\begin{bmatrix} \sigma_2^2 & -\rho\sigma_1\sigma_2 \\ -\rho\sigma_2\sigma_1 & \sigma_1^2 \end{bmatrix} \\ &= 1/(1 - \rho^2)\begin{bmatrix} 1/\sigma_1^2 & -\rho/\sigma_1\sigma_2 \\ -\rho/\sigma_2\sigma_1 & 1/\sigma_2^2 \end{bmatrix} \end{aligned} \tag{4.5}$$

It is now readily seen that the expression in the exponent of Eq. 4.1, apart from the factor $-\frac{1}{2}$, is equivalent to the quadratic form

$$[X_1 - \mu_1, X_2 - \mu_2]\ \Sigma_2^{-1}\begin{bmatrix} X_1 - \mu_1 \\ X_2 - \mu_2 \end{bmatrix}$$

We let χ^2 symbolize this expression, which, on introducing

$$\mathbf{x}' = [X_1 - \mu_1, X_2 - \mu_2] \tag{4.6}$$

may be written as

$$\chi^2 = \mathbf{x}'\mathbf{\Sigma}_2^{-1}\mathbf{x} \tag{4.7}$$

Next, the constant factor $1/2\pi\sigma_1\sigma_2\sqrt{1 - \rho^2}$ of the expression for $\phi(X_1, X_2)$ may be written as $(2\pi)^{-1}|\mathbf{\Sigma}_2|^{-1/2}$, since $\sigma_1\sigma_2\sqrt{1 - \rho^2}$ is the square root of $|\mathbf{\Sigma}_2|$, as seen from Eq. 4.4.

Thus Eq. 4.1, specifying the bivariate normal density function, may be written compactly as

$$\phi(X_1, X_2) = (2\pi)^{-1}|\mathbf{\Sigma}_2|^{-1/2} \exp(-\chi^2/2) \tag{4.8}$$

with $\mathbf{\Sigma}_2$ and χ^2 defined by Eqs. 4.3 and 4.7, respectively.

The extension to the p-variate case is now almost obvious. If we define the covariance matrix as

$$\boldsymbol{\Sigma} = \begin{bmatrix} \sigma_1^2 & \rho_{12}\sigma_1\sigma_2 & \cdots & \rho_{1p}\sigma_1\sigma_p \\ \rho_{21}\sigma_2\sigma_1 & \sigma_2^2 & \cdots & \rho_{2p}\sigma_2\sigma_p \\ \vdots & & \ddots & \vdots \\ \rho_{p1}\sigma_p\sigma_1 & \rho_{p2}\sigma_p\sigma_2 & \cdots & \sigma_p^2 \end{bmatrix} \tag{4.9}$$

where σ_i^2 is the variance of X_i, and ρ_{ij} $(i \neq j)$ is the coefficient of correlation between X_i and X_j, and let

$$\chi^2 = \mathbf{x}'\boldsymbol{\Sigma}^{-1}\mathbf{x} \tag{4.10}$$

with

$$\mathbf{x}' = [X_1 - \mu_1, X_2 - \mu_2, \ldots, X_p - \mu_p] \tag{4.11}$$

then the p-variate normal density function is given by

$$\phi(X_1, X_2, \ldots, X_p) = K \exp(-\chi^2/2)$$

where only the constant K remains to be determined.

Examination of Eq. 4.8 alone is insufficient to permit our inferring what powers of 2π and of $|\boldsymbol{\Sigma}|$ are involved in K. But a comparison with the univariate normal density function,

$$\phi(X) = \frac{1}{\sqrt{2\pi}\,\sigma} \exp[-(X - \mu)^2/2\sigma^2] \tag{4.12}$$

in which the normalizing constant is

$$(2\pi)^{-1/2}(\sigma^2)^{-1/2}$$

leads to the following conjecture: The power of 2π seems to be $-\frac{1}{2}$ times the number of variables, while the power of $|\boldsymbol{\Sigma}|$ (which reduces to σ^2 in the univariate case) is $-\frac{1}{2}$ regardless of the number of variables. We thus infer that, for the p-variate case,

$$K = (2\pi)^{-p/2}|\boldsymbol{\Sigma}|^{-1/2}$$

This conjecture proves to be correct (see, e.g., Anderson, 1958, p. 17), and we have, as the complete equation for a p-variate normal density function,

$$\phi(X_1, X_2, \ldots, X_p) = (2\pi)^{-p/2}|\boldsymbol{\Sigma}|^{-1/2} \exp(-\chi^2/2) \tag{4.13}$$

with $\boldsymbol{\Sigma}$ and χ^2 defined by Eqs. 4.9 and 4.10, respectively. We shall denote this distribution by the symbol $\mathrm{N}(\boldsymbol{\mu}, \boldsymbol{\Sigma})$, meaning a multivariate normal distribution with centroid $\boldsymbol{\mu} = [\mu_1, \mu_2, \ldots, \mu_p]'$ and covariance matrix $\boldsymbol{\Sigma}$.

The close analogy between the p-variate normal density function (4.13) for $N(\boldsymbol{\mu}, \boldsymbol{\Sigma})$ and the familiar univariate normal density function (4.12) for $N(\mu, \sigma^2)$ should be noted. In particular, we point out that the expression

$$-\chi^2/2 = -(\mathbf{x}'\boldsymbol{\Sigma}^{-1}\mathbf{x})/2$$

occurring in the exponent in Eq. 4.13 is a "natural" generalization of the expression

$$-(X - \mu)^2/2\sigma^2 = -[(X - \mu)(\sigma^2)^{-1}(X - \mu)]/2$$

which occurs in the univariate case.

The reason for denoting the expression $\mathbf{x}'\boldsymbol{\Sigma}^{-1}\mathbf{x}$ as χ^2 may best be understood with reference to the special case in which the p variables are pairwise uncorrelated, that is, when $\rho_{ij} = 0$ for all $i \neq j$. For variables whose joint distribution is multivariate normal, uncorrelatedness implies statistical independence—that is, that their joint density function can be expressed as a product of their respective marginal density functions (see Exercise 1 at the end of this section). In this case, the covariance matrix given by Eq. 4.9 reduces to a diagonal matrix:

$$\boldsymbol{\Sigma} = \begin{bmatrix} \sigma_1^2 & 0 & \cdots & 0 \\ 0 & \sigma_2^2 & \cdots & 0 \\ \vdots & & \ddots & \vdots \\ 0 & 0 & \cdots & \sigma_p^2 \end{bmatrix}$$

and Eq. 4.10 for χ^2 reduces to

$$\chi^2 = [X_1 - \mu_1, X_2 - \mu_2, \ldots, X_p - \mu_p]$$
$$\times \begin{bmatrix} \sigma_1^{-2} & & & \\ & \sigma_2^{-2} & & \\ & & \ddots & \\ 0 & & & \sigma_p^{-2} \end{bmatrix} \begin{bmatrix} X_1 - \mu_1 \\ X_2 - \mu_2 \\ \vdots \\ X_p - \mu_p \end{bmatrix}$$
$$= \sum_{i=1}^{p} (X_i - \mu_i)^2/\sigma_i^2 \qquad (4.14)$$

Each term in this sum is the square of a unit normal variate (a z^2 in customary notation), and is hence a chi-square variate with 1 degree of freedom (d.f.). Then, using the additivity property of independent chi-square variates, we obtain the important conclusion that when the X_i are statistically independent, the sum denoted by χ^2 is in fact a chi-square variate with p d.f.'s. In Chapter 5 it will be shown that when the X_i are not independently distributed, a new set of variables Y_i can be derived that *are* independently distributed, and for which the χ^2 quantity is equivalent to that for the original X_i's. Pending this demonstration, therefore, we have proved the following

Theorem. 1. Given a p-variate normal population $N(\boldsymbol{\mu}, \boldsymbol{\Sigma})$ with density function

$$\phi(X_1, X_2, \ldots, X_p) = (2\pi)^{-p/2}|\boldsymbol{\Sigma}|^{-1/2}\exp(-\chi^2/2)$$

the quantity $\chi^2 = \mathbf{x}'\boldsymbol{\Sigma}^{-1}\mathbf{x}$ in the exponent is a chi-square variate with p d.f.'s.

The importance of this theorem lies in its enabling us to determine the multivariate analogs of the various centile points of a normal distribution, *without* the aid of special tables of multivariate normal integrals. We return to the bivariate case to describe the procedure in detail.

Regions Enclosing Specified Percentages of a Bivariate Normal Population

Recall from Eq. 4.2 that an isodensity contour ellipse of a bivariate normal distribution may be specified by choosing a suitable positive constant C and setting

$$\frac{(X_1 - \mu_1)^2}{\sigma_1^2} + \frac{(X_2 - \mu_2)^2}{\sigma_2^2} - 2\rho\frac{(X_1 - \mu_1)(X_2 - \mu_2)}{\sigma_1\sigma_2} = C$$

In our present notation, the left-hand side of this equation is equal to $(1 - \rho^2)\mathbf{x}'\boldsymbol{\Sigma}^{-1}\mathbf{x}$, or $(1 - \rho^2)\chi^2$. Hence an isodensity ellipse may also be specified by choosing a particular value for χ^2 in the equation

$$\mathbf{x}'\boldsymbol{\Sigma}^{-1}\mathbf{x} = \frac{1}{1 - \rho^2}\left[\frac{(X_1 - \mu_1)^2}{\sigma_1^2} + \frac{(X_2 - \mu_2)^2}{\sigma_2^2} - 2\rho\frac{(X_1 - \mu_1)(X_2 - \mu_2)}{\sigma_1\sigma_2}\right]$$
$$= \chi^2 \qquad (4.15)$$

Note that this corresponds, in the univariate normal case, to specifying two points with equal ordinates (representing density) by fixing the value of z^2 (which is χ_1^2) in

$$\frac{(X - \mu)^2}{\sigma^2} = z^2$$

For any choice of z, say z_0, the abscissas of these points are $X = \mu - z_0\sigma$ and $X = \mu + z_0\sigma$, and the interval along the X-axis between these points is such that the area under the normal curve in this interval is equal to some particular value between 0 and 1, determined by the value z_0. This area represents the probability that a random observation X drawn from $N(\mu, \sigma^2)$ will fall in the interval $[\mu - z_0\sigma, \mu + z_0\sigma]$. For example, with $z^2 = (1.645)^2 = 2.706$, we have

$$p(\mu - 1.645\sigma \leq X \leq \mu + 1.645\sigma) = .90$$

or if we denote the interval $[\mu - 1.645\sigma, \mu + 1.645\sigma]$ by $R_1(2.706)$, where the subscript indicates that we refer to the univariate case, and the argument is the z^2 value, we may write

$$p(X \in R_1(2.706)) = .90$$

where the symbol "$\in$" is read "belongs to" or "is a member of."

In the same way, for the bivariate case, the ellipse specified by a particular choice of χ^2 value defines a region in the (X_1, X_2)-plane such that the point (X_1, X_2), representing a random observation from $N(\boldsymbol{\mu}, \boldsymbol{\Sigma})$, has a specified probability of falling in this region. This probability, according to the theorem stated earlier, is equal to the probability that a chi-square variate with 2 d.f.'s has a value not exceeding the value selected for χ^2. For example, from Table E.3, we find that $p(\chi^2 \leq 4.605) = .90$. Hence, if we denote by $R_2(4.605)$ the region bounded by and including the ellipse

$$\frac{1}{1 - \rho^2}\left[\frac{(X_1 - \mu_1)^2}{\sigma_1^2} + \frac{(X_2 - \mu_2)^2}{\sigma_2^2} - 2\rho\frac{(X_1 - \mu_1)(X_2 - \mu_2)}{\sigma_1\sigma_2}\right] = 4.605$$

we can assert that

$$p[(X_1, X_2) \in R_2(4.605)] = .90$$

Geometrically, this means that the volume of the three-dimensional region $R_2(4.605)$ between the bivariate normal surface and the (X_1, X_2)-plane, bounded laterally by the right elliptic cylinder based on the ellipse $\mathbf{x}'\boldsymbol{\Sigma}^{-1}\mathbf{x} = 4.605$, is equal to .90. The statistical implication is that if we draw many independent observations at random from a bivariate normal population $N(\boldsymbol{\mu}, \boldsymbol{\Sigma})$ and each time determine a point with coordinates (X_1, X_2) given by the paired observation, then 90% of these points will lie in the region $R_2(4.605)$, that is, inside or on the ellipse $\mathbf{x}'\boldsymbol{\Sigma}^{-1}\mathbf{x} = 4.605$.

EXAMPLE 4.2. Consider once again the bivariate normal population of Example 4.1, with $\boldsymbol{\mu}' = [15, 20]$ and

$$\boldsymbol{\Sigma} = \begin{bmatrix} 25 & 15 \\ 15 & 25 \end{bmatrix} \qquad \text{(i.e., } \sigma_1 = \sigma_2 = 5,\ \rho = .60\text{)}$$

We are now interested in the expression $\mathbf{x}'\boldsymbol{\Sigma}^{-1}\mathbf{x}$, which, according to Eq. 4.15, is $1/(1 - \rho^2)$ times the expression used in Example 4.1. Hence

$$\begin{aligned}\mathbf{x}'\boldsymbol{\Sigma}^{-1}\mathbf{x} &= [1/(1 - \rho^2)](1/25)[(X_1 - 15)^2 + (X_2 - 20)^2 \\ &\qquad - 1.2(X_1 - 15)(X_2 - 20)] \\ &= (1/16)[(X_1 - 15)^2 + (X_2 - 20)^2 \\ &\qquad - 1.2(X_1 - 15)(X_2 - 20)]\end{aligned}$$

or, using deviation-score notation ($X_1 - 15 = x_1$, $X_2 - 20 = x_2$),

$$\mathbf{x}'\boldsymbol{\Sigma}^{-1}\mathbf{x} = \tfrac{1}{16}(x_1^2 + x_2^2 - 1.2x_1x_2)$$

Geometrically, this is equivalent to having translated the coordinate axes so that the new origin is at the population centroid. In this new reference system the 90% ellipse (i.e., the ellipse inside or on which 90% of the population lies) is specified by equating the above expression to 4.605, which is the value of a chi-square variate with 2 d.f.'s that is exceeded 10% of the time:

$$x_1^2 + x_2^2 - 1.2x_1x_2 = (16)(4.605) = 73.6890$$

The larger of the two ellipses shown in Fig. 4.3 is the 90% ellipse.

The two points P_1 and P_2 shown inside the larger ellipse have coordinates (8, 5) and (−5, −5), respectively; point P_3, lying on this ellipse, has coordinates (4, −5.565); P_4, which is outside the ellipse, has coordinates (−3, 8). Computing the quantity

$$Q = \mathbf{x}'\boldsymbol{\Sigma}^{-1}\mathbf{x} = \tfrac{1}{16}(x_1^2 + x_2^2 - 1.2x_1x_2)$$

for each of these four points, we find the values to be as follows:

Point	Q-value
P_1 (8, 5)	2.563
P_2 (−5, −5)	1.250
P_3 (4, −5.565)	4.605
P_4 (−3, 8)	6.363

Note that $Q \leq 4.605$ for the three points that are inside or on the 90% ellipse, while $Q > 4.605$ for P_4, which lies outside this ellipse. In general, for any chosen value of χ^2, it is true that $\mathbf{x}'\boldsymbol{\Sigma}^{-1}\mathbf{x} \leq \chi^2$ for any point in $R_2(\chi^2)$, and $\mathbf{x}'\boldsymbol{\Sigma}^{-1}\mathbf{x} > \chi^2$ for any point outside $R_2(\chi^2)$. This fact was tacitly assumed, in addition to the theorem about $\mathbf{x}'\boldsymbol{\Sigma}^{-1}\mathbf{x}$ being a chi-square variate, when it was asserted that the various percent ellipses could be determined by selecting the appropriate values of χ^2 from a table of chi-square distributions.

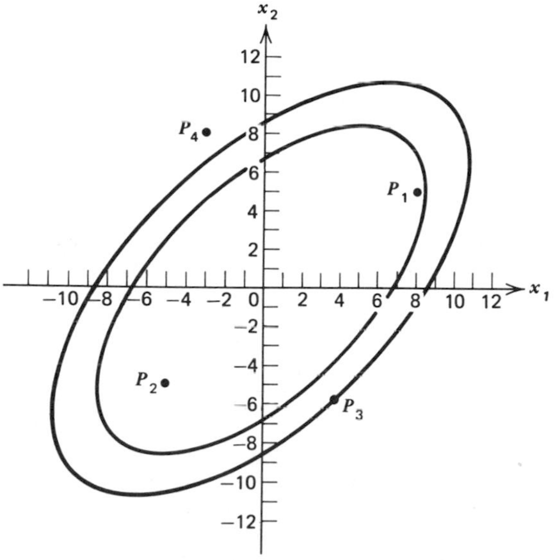

Figure 4.3. *Two isodensity ellipses of the bivariate normal distribution with* $\sigma_1 = \sigma = 5$, $p = .60$.

When the number of variables is three or greater, Eq. 4.7 represents a family of ellipsoids or hyperellipsoids instead of ellipses, but the interpretation exactly parallels that for the bivariate case. For example, when $p = 4$, the 90% hyperellipsoidal region is bounded by the hypersurface

$$\mathbf{x}'\boldsymbol{\Sigma}^{-1}\mathbf{x} = 7.779$$

the value 7.779 being the 90th percentile of the chi-square distribution with n.d.f. $= 4$. This region, $R_4(7.779)$, is such that 90% of the points in four-dimensional space representing observation quadruples (x_1, x_2, x_3, x_4) drawn from the four-variate normal population $\mathrm{N}(\mathbf{0}, \boldsymbol{\Sigma})$ lie inside it. Furthermore, the value of $\mathbf{x}'\boldsymbol{\Sigma}^{-1}\mathbf{x}$ computed for a point with coordinates (x_1, x_2, x_3, x_4) is less than or equal to 7.779 if and only if $P \in R_4(7.779)$.

EXERCISES

1. If p variables $X_1, X_2, \ldots, X_p$ are independently normally distributed with mean μ_i and variance σ_i^2 (for each $i = 1, 2, \ldots, p$), their joint density function is given by the product of their respective univariate density functions. That is

$$\phi(X_1, X_2, \ldots, X_p) = \prod_{i=1}^{p} \left\{ \frac{1}{\sqrt{2\pi}\,\sigma_i} \exp[-(X_i - \mu_i)^2/2\sigma_i^2] \right\}$$

Show that this is a special case of the general multivariate density function (4.13) with

$$\boldsymbol{\Sigma} = \begin{bmatrix} \sigma_1^2 & & & \\ & \sigma_2^2 & & 0 \\ & 0 & \ddots & \\ & & & \sigma_p^2 \end{bmatrix}$$

2. Write the equation for the 75% ellipse of the bivariate normal population considered in Example 4.2. By making a rough plot, verify that the smaller ellipse shown in Fig. 4.3 corresponds to this. Then show that the values of $\mathbf{x}'\boldsymbol{\Sigma}^{-1}\mathbf{x}$ for the points indicated in Fig. 4.3 satisfy the proper magnitude relations.

3. For the same population as in Exercise 2, determine approximately what percent of the population lies in the region bounded by the contour ellipse passing through the point (5, 8).

NOTE: When a region is specified in terms of a point on the boundary ellipse, we say that the points inside the region are "closer" to the centroid than is the given point. Here we are measuring "closeness" by a probabilistic notion of "distance," not by the ordinary Euclidean distance. Thus, if two contour ellipses differing only slightly in size are constructed, there will be many points on the smaller ellipse that are *geometrically* farther from the centroid than are some points on the larger ellipse. Nevertheless, *any* point on the smaller ellipse is said to be "closer" to the centroid in the above sense than is *any* point on the larger ellipse.

4.2 THE SAMPLING DISTRIBUTION OF SAMPLE CENTROIDS

The reader should be familiar with the fact that when independent, random samples of size N are drawn from a univariate normal population $N(\mu, \sigma^2)$, the sample means are distributed as $N(\mu, \sigma^2/N)$. An exact analog to this fact holds in the multivariate case.

If a sample of size N is drawn from a p-variate normal population, there are p sample means $\bar{X}_1, \bar{X}_2, \ldots, \bar{X}_p$. Over successive, independent random samples, these means jointly follow a p-variate normal distribution with the same centroid $\boldsymbol{\mu} = [\mu_1, \mu_2, \ldots, \mu_p]'$ as the parent population, and with covariance matrix equal to $1/N$ times the covariance matrix $\boldsymbol{\Sigma}$ of the parent population. Thus if we define a *sample centroid* as the vector

$$\bar{\mathbf{X}} = \begin{bmatrix} \bar{X}_1 \\ \bar{X}_2 \\ \vdots \\ \bar{X}_p \end{bmatrix}$$

we may summarize the foregoing in the form of a theorem.

Theorem 2. Sample centroids $\overline{\mathbf{X}}$ based on independent, random samples of size N from a population $N(\boldsymbol{\mu}, \boldsymbol{\Sigma})$ have the sampling distribution $N(\boldsymbol{\mu}, \boldsymbol{\Sigma}/N)$.

The density function for $N(\boldsymbol{\mu}, \boldsymbol{\Sigma}/N)$ is found, by making the appropriate substitutions in Eq. 4.13, to be

$$\phi(\overline{X}_1, \overline{X}_2, \ldots, \overline{X}_p) = (2\pi)^{-p/2}N^{p/2}|\boldsymbol{\Sigma}|^{-1/2} \exp\,[-N(\overline{\mathbf{x}}'\boldsymbol{\Sigma}^{-1}\overline{\mathbf{x}})/2] \quad (4.17)$$

where $\overline{\mathbf{x}} = \overline{\mathbf{X}} - \boldsymbol{\mu}$ is the deviation of the sample centroid from the population centroid.

We can apply Theorem 1 in Section 4.1 to the sampling distribution of $\overline{\mathbf{X}}$, and thereby determine elliptical regions within which any specified percentage of the centroids based on random samples of a given size will lie. Or, conversely, we can determine what percent of the samples of a given size will have centroids that lie in a specified elliptical region.

Suppose, for example, that samples of size $N = 10$ are drawn from the bivariate normal population of Example 4.2. Then the region in which 90% of the sample centroids lie is bounded by the ellipse $N\overline{\mathbf{x}}'\boldsymbol{\Sigma}^{-1}\overline{\mathbf{x}} = 4.605$, or

$$10[\overline{X}_1 - 15, \overline{X}_2 - 20]\begin{bmatrix} 25 & 15 \\ 15 & 25 \end{bmatrix}^{-1}\begin{bmatrix} \overline{X}_1 - 15 \\ \overline{X}_2 - 20 \end{bmatrix} = 4.605$$

where the constant on the right is the 90th percentile of the chi-square distribution with 2 d.f.'s, as before. Dividing through by 10 (or, in general, the sample size N), the equation becomes

$$\overline{\mathbf{x}}'\boldsymbol{\Sigma}^{-1}\overline{\mathbf{x}} = .4605$$

in which the left-hand side is formally identical to that for the 90% ellipse of the parent population, but the right hand side is smaller by a factor of $1/N$. Figure 4.4 shows the 90% ellipse of the sampling distribution of $\overline{\mathbf{x}}$ and the corresponding ellipse for the parent population (in dashed lines). Note that the former is about one-third, in linear size, of the latter; more precisely, the ratio of the linear dimensions of the two ellipses is $1\!:\sqrt{10}$, or $1\!:\sqrt{N}$ in general.

What we have just seen is a 90% *probability region* for sample centroids, given the centroid and covariance matrix of a multivariate normal population. It corresponds, in the univariate case, to the symmetric 90% probability interval $[\mu - 1.64\sigma/\sqrt{N},\ \mu + 1.64\sigma/\sqrt{N}]$ for means of samples of size N from $N(\mu, \sigma^2)$. Now the reader is probably familiar with the transition from a probability interval for sample means to a confidence interval for the population mean, effected by algebraic manipulation of the inequalities enclosed by parentheses in a probability statement like

$$p(\mu - 1.64\sigma/\sqrt{N} \le \overline{X} \le \mu + 1.64\sigma/\sqrt{N}) = .90$$

transforming it into

$$p(\overline{X} - 1.64\sigma/\sqrt{N} \le \mu \le \overline{X} + 1.64\sigma/\sqrt{N}) = .90$$

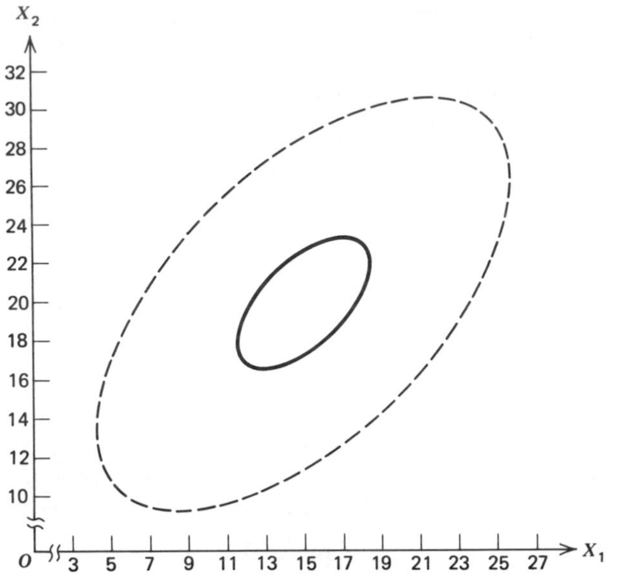

Figure 4.4. *A 90% ellipse of sampling distribution of centroids of samples of size 10 from a bivariate normal population, and a 90% ellipse of the parent population.*

In exactly the same way, the probability statement

$$p[(\overline{\mathbf{X}} - \boldsymbol{\mu})'\boldsymbol{\Sigma}^{-1}(\overline{\mathbf{X}} - \boldsymbol{\mu}) \leq 4.605/N] = .90$$

is readily transformed into

$$p[(\boldsymbol{\mu} - \overline{\mathbf{X}})'\boldsymbol{\Sigma}^{-1}(\boldsymbol{\mu} - \overline{\mathbf{X}}) \leq 4.605/N] = .90$$

and is construed as follows: Imagine that we know the dispersion matrix $\boldsymbol{\Sigma}$, but not the centroid $\boldsymbol{\mu}$, of a bivariate normal population. We draw successive random samples of size N from this population, and for each sample we construct an ellipse

$$(\mathbf{Y} - \overline{\mathbf{X}})'\boldsymbol{\Sigma}^{-1}(\mathbf{Y} - \overline{\mathbf{X}}) = 4.605/N \tag{4.18}$$

centered at the *sample* centroid $\overline{\mathbf{X}} = [\overline{X}_1, \overline{X}_2]$, using a coordinate system (Y_1, Y_2). Then 90% of these ellipses will contain the population centroid $\boldsymbol{\mu}$ as an interior (or boundary) point.

Consequently, for any single sample of size N, the region bounded by the ellipse (4.18), with center at the particular sample centroid $\overline{\mathbf{X}}$, is called a 90% *confidence region* for the unknown population centroid $\boldsymbol{\mu}$ of the bivariate normal population $N(\boldsymbol{\mu}, \boldsymbol{\Sigma})$. More generally, a $100(1 - \alpha)\%$ confidence region for the centroid $\boldsymbol{\mu}$ of a p-variate normal population $N(\boldsymbol{\mu}, \boldsymbol{\Sigma})$ with known $\boldsymbol{\Sigma}$ may be determined by constructing the ellipse, with center at $\overline{\mathbf{X}}$, defined by the equation

$$(\mathbf{Y} - \overline{\mathbf{X}})'\boldsymbol{\Sigma}^{-1}(\mathbf{Y} - \overline{\mathbf{X}}) = \chi^2_{p,1-\alpha}/N \tag{4.19}$$

where $\chi^2_{p,1-\alpha}$ denotes the 100(1 − α) percentile point of the chi-square distribution with p d.f.'s, and N is the sample size.

EXERCISE

1. Suppose that a bivariate normal population has a covariance matrix

$$\begin{bmatrix} 25 & 12 \\ 12 & 16 \end{bmatrix}$$

and that a random sample of 20 observations yielded a centroid [18, 15]. Construct the elliptical 95% confidence region for $\boldsymbol{\mu}$, centered around the sample centroid.

4.3 SIGNIFICANCE TEST: ONE-SAMPLE PROBLEM

Although the purposes differ, conducting a significance test and establishing a confidence region are formally identical in rationale: The test statistic takes a value that leads to rejection, at a given α level, of a hypothesis about a parameter if and only if the 100(1 − α)% confidence region fails to include the hypothesized parameter value.

To illustrate with a familiar univariate example, suppose that the variance of a normal population is known to be equal to 144, and that the mean is unknown but hypothesized to be equal to 100. If a sample of 36 observations from this population yields a mean of 104, the test statistic has the value

$$z = \frac{104 - 100}{12/\sqrt{36}} = 2$$

which leads to a rejection of the hypothesis that $\mu = 100$ at the .05 level (against a two-sided alternative). At the same time, the 95% confidence interval for μ, based on this sample, is

$$[104 - 1.96(12/\sqrt{36}),\ 104 + 1.96(12/\sqrt{36})]$$

or

$$[100.08,\ 107.92]$$

which does not include the hypothesized value, $\mu = 100$.

In the same way, a hypothesis concerning the centroid of a multivariate normal population *with known covariance matrix* $\boldsymbol{\Sigma}$ can be tested by substituting the hypothesized population centroid $\boldsymbol{\mu}_0$ and the observed sample centroid, respectively, for $\mathbf{Y}$ and $\overline{\mathbf{X}}$ in the left-hand side of Eq. 4.19. If the quadratic form thus computed has a value exceeding $\chi^2_{p,1-\alpha}/N$, then the hypothesis $\boldsymbol{\mu} = \boldsymbol{\mu}_0$ is rejected at the α level of significance. The basis for this decision is that from the argument presented in Section 4.1, we know that

$$(\boldsymbol{\mu}_0 - \overline{\mathbf{X}})'\boldsymbol{\Sigma}^{-1}(\boldsymbol{\mu}_0 - \overline{\mathbf{X}}) >, =, \text{or} < \chi^2_{p,1-\alpha}/N$$

according to whether the point $\boldsymbol{\mu}_0$ lies outside, on, or inside the ellipse (4.19) centered at $\overline{\mathbf{X}}$, respectively. In practice, it is more convenient to use

$$Q = N(\overline{\mathbf{X}} - \boldsymbol{\mu}_0)'\boldsymbol{\Sigma}^{-1}(\overline{\mathbf{X}} - \boldsymbol{\mu}_0) \tag{4.20}$$

as the test statistic, and reject the hypothesis that $\boldsymbol{\mu} = \boldsymbol{\mu}_0$ at the α level if and only if $Q > \chi^2_{p,1-\alpha}$.

EXAMPLE 4.3. Suppose that a sample of 25 observations is drawn from a bivariate normal population with unknown centroid $\boldsymbol{\mu}$ and covariance matrix

$$\boldsymbol{\Sigma} = \begin{bmatrix} 16 & 8 \\ 8 & 9 \end{bmatrix}$$

If the sample centroid is found to be $\overline{\mathbf{X}}' = [15.4, 9.9]$, test the hypothesis that $\boldsymbol{\mu}' = [17, 10]$ at the 5% significance level.

Since the inverse of the population covariance matrix is

$$\boldsymbol{\Sigma}^{-1} = \tfrac{1}{80}\begin{bmatrix} 9 & -8 \\ -8 & 16 \end{bmatrix}$$

the test statistic Q of Eq. 4.20 has the value

$$Q = (25)(\tfrac{1}{80})[15.4 - 17, 9.9 - 10]\begin{bmatrix} 9 & -8 \\ -8 & 16 \end{bmatrix}\begin{bmatrix} 15.4 - 17 \\ 9.9 - 10 \end{bmatrix} = 6.45$$

From Table E.3 we find that the 95th percentile of the χ^2 distribution with 2 d.f.'s is 5.991. Thus $Q > \chi^2_{2,.95}$, so the hypothesis that $\boldsymbol{\mu}' = [17, 10]$ is rejected.

The Reason for Possible Differences Between Outcomes of Multivariate and Univariate Tests

We digress a little here to clarify further the difference between the conclusion based on a set of tests of univariate null hypotheses

$$H_{01}: \mu_1 = \mu_{10}, \quad H_{02}: \mu_2 = \mu_{20}, \quad \ldots, \quad H_{0p}: \mu_p = \mu_{p0}$$

and that based on single multivariate test

$$H_0: \boldsymbol{\mu} = \boldsymbol{\mu}_0$$

Taking a bivariate example for simplicity, assume that we know

$$\mathbf{\Sigma} = \begin{bmatrix} 25 & 15 \\ 15 & 25 \end{bmatrix}$$

and wish to test

$$H_0\colon \boldsymbol{\mu}' = [15, 20] \tag{4.21}$$

or, equivalently, that

$$H_{01}\colon \mu_1 = 15 \quad \textit{and} \quad H_{02}\colon \mu_2 = 20 \tag{4.22}$$

[Note that the assertion of the bivariate hypothesis (4.21) and the joint assertion of the two univariate hypotheses (4.22) are logically equivalent. It is the judgment of the truth or falsity of (4.21) and (4.22), based on a bivariate test and a pair of univariate tests, respectively, that may differ.]

Now suppose that a sample of 100 bivariate observations from the population in question yielded $\overline{X}_1 = 14.2$ and $\overline{X}_2 = 20.4$ (i.e., $\overline{\mathbf{X}}' = [14.2, 20.4]$). Let us first test the bivariate null hypothesis 4.21 at the 5% significance level, using the test statistic Q defined in Eq. 4.20. The value of Q for this problem is found to be

$$Q = (100)[-.8, .4]\begin{bmatrix} 25 & 15 \\ 15 & 25 \end{bmatrix}^{-1}\begin{bmatrix} -.8 \\ .4 \end{bmatrix} = 7.4$$

Since $7.4 > 5.99 = \chi^2_{2,.95}$, the null hypothesis is rejected, and we conclude that $\boldsymbol{\mu}' \neq [15, 20]$.

Next, let us test the two univariate null hypotheses stated in (4.22), using the familiar unit-normal statistic z. For H_{01} we get

$$|z| = \frac{|14.2 - 15|}{5/\sqrt{100}} = 1.6 < 1.96 = z_{.975}$$

Hence we cannot reject $H_{01}\colon \mu_2 = 15$ against a two-sided alternative. For H_{02},

$$|z| = \frac{20.4 - 20}{5/\sqrt{100}} = .8 < 1.96$$

so $H_{02}\colon \mu_2 = 20$ cannot be rejected either.

How do we reconcile the contradictory results of the multivariate and univariate tests? To do so, we need to recall the basic "philosophy" underlying significance tests: A null hypothesis is rejected when the sample observed is a "highly unlikely one" under the condition that H_0 is true. So the question becomes: Why is our sample with $\overline{X}_1 = 14.2$ and $\overline{X}_2 = 20.4$ deemed to be "highly unlikely" by the multivariate test but not so unlikely by the univariate tests, given that $\mu_1 = 15$ and $\mu_2 = 20$? Since the multivariate test looks at both variables, X_1 and X_2, simultaneously while each univariate test examines just one variable as though the other were nonexistent, the answer must have something to do with the "tie-in" between the two variables.

From the covariance matrix $\mathbf{\Sigma}$, we see that X_1 and X_2 are positively correlated; specifically, $\rho_{12} = 15/\sqrt{(25)(25)} = .60$. Hence the two sample means should also be positively correlated over all possible samples. That is, a sample that has a "large" $\overline{X}_1$ (i.e., $\overline{X}_1 > \mu_1$) would tend also to have a large $\overline{X}_2$ ($\overline{X}_2 > \mu_2$). Conversely, a sample with $\overline{X}_1 < \mu_1$ would tend to have $\overline{X}_2 < \mu_2$. A sample in which $\overline{X}_1 > \mu_1$ and $\overline{X}_2 < \mu_2$ or one in which $\overline{X}_1 < \mu_1$ and $\overline{X}_2 > \mu_2$ would, therefore, be a rare sample.

Recalling that our particular sample had $\overline{X}_1 = 14.2 < 15 = \mu_1$ while $\overline{X}_2 = 20.4 > 20 = \mu_2$, we see that this would be one of the "unlikely" samples if it were indeed the case that $\boldsymbol{\mu}' = [15, 20]$, as the null hypothesis asserts; specifically, such a sample would be one of less than 5% of the most unlikely samples. Hence the multivariate test renders the judgment that the observation $\overline{\mathbf{X}}' = [14.2, 20.4]$ is a highly unlikely one if H_0 were in fact true; thus H_0 is rejected by the multivariate significance test.

On the other hand, if X_1 were the only variable being considered (i.e., if X_2 simply did not exist), it would not be at all unusual that $\overline{X}_1 = 14.2$, falling short of $\mu_1 = 15$ by 1.6 standard-error units. Similarly, if X_2 alone were being observed, it would be even less unusual to get $\overline{X}_2 = 20.4$, which exceeds $\mu_2 = 20$ by a mere eight-tenths of a standard error. This is why neither of the univariate tests leads to rejection of the relevant null hypothesis.

Population Covariance Matrix Unknown

The foregoing discussions assumed the population covariance matrix to be known. In practice, however, when a hypothesis concerning $\boldsymbol{\mu}$ is to be tested, $\mathbf{\Sigma}$ is usually also unknown and has to be estimated from the sample. In the corresponding univariate case, the appropriate test statistic is the familiar

$$t = \frac{\overline{X} - \mu_0}{s/\sqrt{N}}$$

where s is the square root of the unbiased estimate $\Sigma(X - \overline{X})^2/(N - 1)$ of σ^2. This statistic, as is well known, follows Student's t-distribution with $N -$ d.f.'s, instead of N(0, 1), which is the distribution of

$$z = \frac{\overline{X} - \mu_0}{\sigma/\sqrt{N}}$$

when σ is known.

The multivariate counterpart of t, or rather its square,

$$t^2 = \frac{(\overline{X} - \mu_0)^2}{s^2/N} = N(\overline{X} - \mu_0)(s^2)^{-1}(\overline{X} - \mu_0)$$

is obtained by replacing the scalar quantities $\overline{X} - \mu_0$ and s^2 in the last expression by their matrix analogs $\overline{\mathbf{X}} - \boldsymbol{\mu}_0$ and $\mathbf{S}/(N - 1)$, respectively ($\mathbf{S}$ being the sample SSCP matrix). The result is, of course, very similar to the Q of Eq. 4.20; the only difference is that $\boldsymbol{\Sigma}$ is replaced by its unbiased estimate $\mathbf{S}/(N - 1)$. This statistic was denoted T^2 by Hotelling (1931), who first studied its sampling distribution. We shall use a subscript 1 to indicate that it is the T^2 for a one-sample problem, to distinguish it from the T^2 for two-sample problems, discussed in the next section. Thus

$$T_1^2 = N(N - 1)(\overline{\mathbf{X}} - \boldsymbol{\mu}_0)'\mathbf{S}^{-1}(\overline{\mathbf{X}} - \boldsymbol{\mu}_0) \tag{4.23}$$

What Hotelling showed was that the following multiple of T_1^2 has an F-distribution:

$$\frac{N - p}{(N - 1)p} T_1^2 = F^p_{N-p} \tag{4.24}$$

where the superscript p and the subscript $N - p$ indicate the numerator and denominator d.f.'s of the F-statistic. (Note that for $p = 1$, that is, the univariate case, Eq. 4.24 reduces to the familiar relation $t^2_{N-1} = F^1_{N-1}$.)

EXAMPLE 4.4 The centroid and SSCP matrix for a sample of 22 observations from a bivariate normal population were

$$\overline{X}' = [32.6, 33.5] \qquad \text{and} \qquad \mathbf{S} = \begin{bmatrix} 47.25 & 42.02 \\ 42.02 & 111.09 \end{bmatrix}$$

Test the hypothesis that $\boldsymbol{\mu}_0 = [31, 32]'$ at the 1% level of significance.

The inverse of the given SSCP matrix is found to be

$$\mathbf{S}^{-1} = \begin{bmatrix} 3.1891 & -1.2063 \\ -1.2063 & 1.3564 \end{bmatrix} \times 10^{-2}$$

From Eqs. 4.23 and 4.24, the quantity distributed as an F-variate is

$$\begin{aligned} \frac{N - p}{(N - 1)p} T_1^2 &= \frac{N(N - p)}{p} (\overline{\mathbf{X}} - \boldsymbol{\mu}_0)\mathbf{S}^{-1}(\overline{\mathbf{X}} - \boldsymbol{\mu}_0) \\ &= [(22)(20)/2][1.6, 1.5] \\ &\quad \times \begin{bmatrix} 3.1891 & -1.2063 \\ -1.2063 & 1.3564 \end{bmatrix} \begin{bmatrix} 1.6 \\ 1.5 \end{bmatrix} \times 10^{-2} \\ &= 11.9 \end{aligned}$$

From Table E.4 we find the 99th percentile of the F distribution with $p = 2$ and $N - p = 20$ d.f.'s to be 5.85. Since the observed F value exceeds this, we may reject the hypothesis $\boldsymbol{\mu}_0 = [31, 32]'$ at the 1% level.

The one-sample T_1^2-test finds an important application in testing the significance of the difference between the centroids of matched samples (where each member of one sample is paired with a member of the other sample in some meaningful fashion), or the centroids of the same sample in a test–retest design. Just as in the univariate case, the multivariate matched-samples problem is reduced to a one-sample problem by using the differences between the several paired observations as the dependent variables.

Thus, in the p-variate case, the ith pair of observation vectors yields a difference vector

$$\mathbf{d}_i' = [X_{1i}^{(1)} - X_{1i}^{(2)}, X_{2i}^{(1)} - X_{2i}^{(2)}, \ldots, X_{pi}^{(1)} - X_{pi}^{(2)}] \qquad (4.25)$$

and the mean difference vector, or the centroid of the difference scores, is given by

$$\bar{\mathbf{d}}' = \left(\sum_{i=1}^{N} \mathbf{d}_i'\right)/N = [\bar{X}_1^{(1)} - \bar{X}_1^{(2)}, \bar{X}_2^{(1)} - \bar{X}_2^{(2)}, \ldots, \bar{X}_p^{(1)} - \bar{X}_p^{(2)}] \qquad (4.26)$$

which is the same as the difference between the centroids of the two matched samples.

The hypothesis to be tested is that the population centroid of difference scores,

$$\boldsymbol{\mu}_d' = [\mu_1^{(1)} - \mu_1^{(2)}, \mu_2^{(1)} - \mu_2^{(2)}, \ldots, \mu_p^{(1)} - \mu_p^{(2)}]$$

is equal to the null vector [0, 0, . . . , 0]. Hence the appropriate T_1^2-statistic for the matched-samples problem is given b

$$T_d^2 = N(N - 1)\bar{\mathbf{d}}'\mathbf{S}_d^{-1}\bar{\mathbf{d}} \qquad (4.27)$$

where $\mathbf{S}_d$ is the sample SSCP matrix of the p difference scores. The test is carried out by computing expression (4.24) and comparing the result with the appropriate percentile value of the F_{N-p}^{p}-distribution.

EXAMPLE 4.5. A researcher at a school for the deaf gave several motor tests to the resident students, and also tested a group of hearing children, paired child for child on the basis of sex, age, and height with the deaf children. Scores for 10 deaf girls and their hearing counterparts on a test of grip (X_1) and a test of balance (X_2) were as shown below. Using Hotelling's T_1^2-test, test the significance of the difference between the centroids of the deaf and hearing groups at $\alpha = .01$.

		Pair No.									
		1	**2**	**3**	**4**	**5**	**6**	**7**	**8**	**9**	**10**
X_1	D	25	22	28	35	37	48	49	54	65	57
	H	26	22	29	39	34	51	42	54	77	68
X_2	D	2.0	2.0	2.7	2.7	3.0	1.7	2.0	2.0	2.7	1.0
	H	2.3	1.0	3.7	3.3	10.0	4.3	4.7	7.0	3.3	1.7

The difference scores (H − D) for each pair on each of the two variables are:

d_1	1	0	1	4	−3	3	−7	0	12	11	$\Sigma d_1 = 22$
d_2	0.3	−1.0	1.0	0.6	7.0	2.6	2.7	5.0	0.6	0.7	$\Sigma d_2 = 19.5$

The centroid of the difference scores, the SSCP matrix, and its inverse are (as the reader should verify as an exercise) as follows:

$$\bar{\mathbf{d}} = \begin{bmatrix} 2.20 \\ 1.95 \end{bmatrix} \qquad \mathbf{S}_d = \begin{bmatrix} 301.600 & -56.400 \\ -56.400 & 53.325 \end{bmatrix}$$

and

$$\mathbf{S}_d^{-1} = \begin{bmatrix} 4.133 & 4.371 \\ 4.371 & 23.376 \end{bmatrix} \times 10^{-3}$$

Therefore, from Eq. 4.25 we compute

$$T_d^2 = (10)(9)[2.20, 1.95]\begin{bmatrix} 4.133 & 4.371 \\ 4.371 & 23.376 \end{bmatrix}\begin{bmatrix} 2.20 \\ 1.95 \end{bmatrix} \times 10^{-3}$$

$$= (90)(146.39) \times 10^{-3}$$

and from Eq. 4.24

$$F_8^2 = \frac{8}{(9)(2)}(90)(146.39) \times 10^{-3}$$

$$= (40)(146.39) \times 10^{-3} = 5.86$$

which falls short of the 99th percentile of the F-distribution with 2 and 8 d.f.'s (which is 8.65). Therefore, the null hypothesis that the population centroid of the difference scores is [0, 0] is not rejected.

4.4 SIGNIFICANCE TEST: TWO-SAMPLE PROBLEM

The multivariate analog of the familiar t-ratio for testing the significance of the difference between two independent means (i.e., means based on two unrelated samples) may also be written out by making appropriate replacements of scalars by vectors or matrices in the expression for t^2; namely,

$$t^2 = \frac{(\bar{X}_1 - \bar{X}_2)^2}{s^2(1/n_1 + 1/n_2)} = (\bar{X}_1 - \bar{X}_2)\left(s^2 \frac{n_1 + n_2}{n_1 n_2}\right)^{-1}(\bar{X}_1 - \bar{X}_2)$$

where

$$s^2 = \frac{\Sigma x_1^2 + \Sigma x_2^2}{n_1 + n_2 - 2}$$

is the pooled within-groups estimate of the assumed common variance σ^2 of the two populations.

To obtain the multivariate counterpart of s^2, we divide the within-groups SSCP matrix $\mathbf{W}$ $(= \mathbf{S}_1 + \mathbf{S}_2)$ by its n.d.f., $n_1 + n_2 - 2$. Thus the required analog of s^2 is

$$\frac{\mathbf{W}}{n_1 + n_2 - 2}$$

which is an unbiassed estimate of the assumed common covariance matrix $\boldsymbol{\Sigma}$ of the two populations. The difference $\overline{X}_1 - \overline{X}_2$ between the two means is, of course, replaced by the difference between the two centroid vectors, that is, $\overline{\mathbf{X}}^{(1)} - \overline{\mathbf{X}}^{(2)}$, in the multivariate case. On making these replacements in the expression above for t^2, we obtain Hotelling's T^2-statistic for the two-sample problem, which we here denote by T_2^2:

$$\begin{aligned} T_2^2 &= (\overline{\mathbf{X}}^{(1)} - \overline{\mathbf{X}}^{(2)})'\left(\frac{\mathbf{W}}{n_1 + n_2 - 2}\,\frac{n_1 + n_2}{n_1 n_2}\right)^{-1}(\overline{\mathbf{X}}^{(1)} - \overline{\mathbf{X}}^{(2)}) \\ &= \frac{n_1 n_2 (n_1 + n_2 - 2)}{n_1 + n_2}(\overline{\mathbf{X}}^{(1)} - \overline{\mathbf{X}}^{(2)})'\,\mathbf{W}^{-1}(\overline{\mathbf{X}}^{(1)} - \overline{\mathbf{X}}^{(2)}) \end{aligned} \tag{4.28}$$

Just as in the case of the one-sample T_1^2-statistic, T_2^2 is derivable from a quantity like Q (Eq. 4.20), which in turn derives from considering a confidence region for the difference between the centroids of two multivariate normal populations with the same dispersion matrix $\boldsymbol{\Sigma}$. After obtaining the appropriate form of Q for the two-sample case, one has merely to replace $\boldsymbol{\Sigma}$ by its pooled within-groups estimate, $\mathbf{W}/(n_1 + n_2 - 2)$. We shall not go into the derivation here, but merely point out that the sampling distribution of the difference between pairs of sample centroids based on pairs of independent samples drawn from $\mathrm{N}(\boldsymbol{\mu}_1, \boldsymbol{\Sigma})$ and $\mathrm{N}(\boldsymbol{\mu}_2, \boldsymbol{\Sigma})$, respectively, is given by the multivariate normal distribution

$$\mathrm{N}\left[\boldsymbol{\mu}_1 - \boldsymbol{\mu}_2, \boldsymbol{\Sigma}\left(\frac{1}{n_1} + \frac{1}{n_2}\right)\right]$$

What Hotelling (1931) showed was that the following multiple of the T_2^2, thus derived, is distributed as an F-variate with p and $n_1 + n_2 - p - 1$ d.f.'s:

$$\frac{n_1 + n_2 - p - 1}{(n_1 + n_2 - 2)p} T_2^2 = F^p_{n_1 + n_2 - p - 1} \tag{4.29}$$

(Note, again, that for the univariate case, with $p = 1$, this reduces to the relation $t^2 = F^1_{n_1 + n_2 - 2}$.)

EXAMPLE 4.6. A miniature, fictitious example will suffice to illustrate the computations, and will also indicate the difference between a multivariate significance test and a sequence of univariate significance tests. Suppose that two treatment groups, in an experiment using the randomized-groups design, were measured on two criterion variables X_1 and X_2, and that the scores were as follows:

	Group 1		**Group 2**	
	X_1	X_2	X_1	X_2
	3	6	2	11
	9	6	7	14
	16	8	13	18
	19	13	19	18
	24	12	23	20
Group totals	71	45	64	81
Group means	14.2	9.0	12.8	16.2

If univariate t-tests are conducted for the two criterion variables separately, it will be found (as the reader should verify) that

$$t < 1 \qquad \text{for } X_1$$

$$t = -3.27 \qquad \text{for } X_2\ (.01 < p < .02)$$

Would you conclude that the two groups are significantly different?

To conduct the T_2^2-test, we first compute the SSCP matrices for the two groups:

$$\mathbf{S}_1 = \begin{bmatrix} 274.8 & 96.0 \\ 96.0 & 44.0 \end{bmatrix} \quad \text{and} \quad \mathbf{S}_2 = \begin{bmatrix} 292.8 & 119.2 \\ 119.2 & 52.8 \end{bmatrix}$$

Hence

$$\mathbf{W} = \mathbf{S}_1 + \mathbf{S}_2 = \begin{bmatrix} 567.6 & 215.2 \\ 215.2 & 96.8 \end{bmatrix}$$

and

$$\mathbf{W}^{-1} = \begin{bmatrix} 1.121 & -2.493 \\ -2.493 & 6.575 \end{bmatrix} \times 10^{-2}$$

The vector of centroid difference is

$$\overline{\mathbf{X}}^{(1)'} - \overline{\mathbf{X}}^{(2)'} = [14.2 - 12.8 \quad 9.0 - 16.2]$$

$$= [1.4 \quad -7.2]$$

From the foregoing intermediate results, we compute T_2^2 in accordance with Eq. 4.26, thus:

$$T_2^2 = [(5)(5)(8)/(10)][1.4, \ -7.2]\begin{bmatrix} 1.121 & -2.493 \\ -2.493 & 6.575 \end{bmatrix} \times \begin{bmatrix} 1.4 \\ -7.2 \end{bmatrix} \times 10^{-2} = 78.66$$

Hence the corresponding F-statistic is, by Eq. 4.29, computed as

$$F_7^2 = [7/(8)(2)](78.66) = 34.41$$

which far exceeds the 99th percentile of the F-distribution with 2 and 7 d.f.'s. The two group centroids are therefore significantly different beyond the 1% level—a conclusion that is not between the respective results for the two univariate t-tests, it should be noted.

4.5 SIGNIFICANCE TEST: *K*-SAMPLE PROBLEM

The multivariate significance tests discussed in the foregoing used the covariance matrix $\mathbf{\Sigma}$ as a multivariate analog of the variance σ^2 of a univariate distribution. There is another extension of the concept of variance to multivariate distributions, and this is the *determinant* $|\mathbf{\Sigma}|$ of the covariance matrix, which is called the *generalized variance*.

We digress briefly to give a geometric interpretation of the generalized variance, which emerges when we use the *N-space* (short for N-dimensional person space) representation of multivariate observations—that is, a representation in which the coordinate axes correspond to individuals instead of variates, as they do in test space. Thus, each point (or vector from the origin to the point) in N-space represents a *separate test;* its coordinates (or components of the vector) are the scores earned by N individuals on that test. Each such vector is called a *test vector*. This representation has two important properties. First, when the scores are in deviations from the mean ($X - \bar{X} = x$), the length of a test vector is

$$|\mathbf{x}| = \sqrt{\sum_{i=1}^{N} x_i^2} = \sqrt{N-1}\, s$$

that is, $\sqrt{N-1}$ times the standard deviation of that test. Second, the cosine of the angle between any two test vectors (likewise in deviation form) is equal to the product-moment correlation between these tests.[1]

With this background, let us consider the sample counterpart of $|\mathbf{\Sigma}|$, namely $|\mathbf{S}/(N - 1)|$. For simplicity, we treat the bivariate case, but the result is immediately generalizable to the p-variate case, and is equally applicable to the population covariance matrix.

The sample covariance matrix may be written as

$$\mathbf{S}/(N - 1) = \frac{1}{N - 1}\begin{bmatrix} x_{11} & x_{12} & \cdots & x_{1N} \\ x_{21} & x_{22} & \cdots & x_{2N} \end{bmatrix}\begin{bmatrix} x_{11} & x_{21} \\ x_{12} & x_{22} \\ \vdots & \vdots \\ x_{1N} & x_{2N} \end{bmatrix} = \begin{bmatrix} s_1^2 & rs_1s_2 \\ rs_2s_1 & s_2^2 \end{bmatrix}$$

where x_{ji} ($j = 1, 2, i = 1, 2, \ldots, N$) is the score of the ith individual on the jth test. Hence the sample generalized variance for the bivariate case is

$$|\mathbf{S}/(N - 1)| = s_1^2s_2^2(1 - r^2) = (s_1s_2)^2(1 - \cos^2 \theta) = (s_1s_2 \sin \theta)^2$$

where θ is the angle between the test vectors in deviation-score form. But as shown above, each standard deviation is $1/\sqrt{N - 1}$ times the length of the corresponding test vector. Therefore,

$$s_1s_2 \sin \theta = \frac{|\mathbf{x}_1|}{\sqrt{N - 1}} \times \frac{|\mathbf{x}_2|}{\sqrt{N - 1}} \sin \theta$$

is equal to the area of the parallelogram formed by the "rescaled" test vectors $\mathbf{x}_1/\sqrt{N - 1}$ and $\mathbf{x}_2/\sqrt{N - 1}$ in N-space, as seen from Fig. 4.5. Thus the gen-

[1]This may be seen using a definition of the scalar product of two vectors that is an alternative (but of course equivalent) to the more common and practical one based on Eq. 2.5. The alternative definition is

$$\mathbf{x}'\mathbf{y} = |\mathbf{x}||\mathbf{y}| \cos \theta$$

where θ is the angle between the two vectors. From this it follows that

$$\cos \theta = \frac{\mathbf{x}'\mathbf{y}}{|\mathbf{x}| \cdot |\mathbf{y}|}$$

If $\mathbf{x}$ and $\mathbf{y}$ are deviation-score test vectors, and we use the Eq. -2.5 definition for the scalar product in the numerator and the definition for the scalar product and the first expression for the length of a vector given above, we obtain

$$\cos \theta = \frac{\sum_{i=1}^{N} x_iy_i}{\sqrt{\sum_{i=1}^{N} x_i^2}\sqrt{\sum_{i=1}^{N} y_i^2}}$$

which is precisely the definition of the product-moment correlation coefficient between tests X and Y.

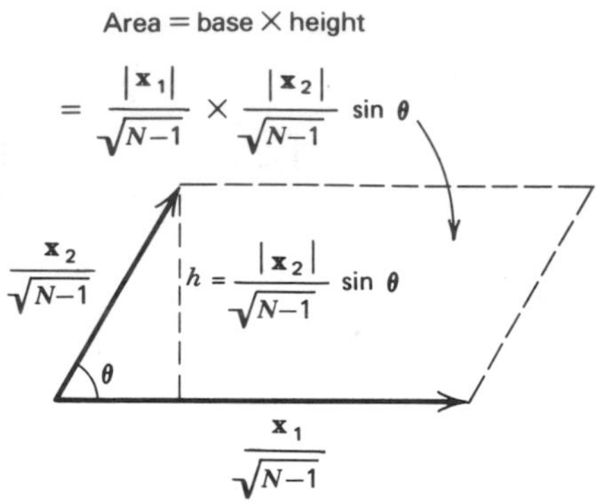

Figure 4.5. *Generalized variance as a squared area (for p = 2).*

eralized variance is the square of this area. In the p-variate case, the generalized variance is the square of the p-dimensional "volume" of the parallelotope formed by the rescaled test vectors

$$\mathbf{x}_i/\sqrt{N-1} \qquad (i = 1, 2, \ldots, p)$$

in the N-dimensional person space. For a more detailed discussion of the generalized variance, see Takeuchi, Yanai and Mukherjee (1982, pp. 76–78).

We thus see that the two multivariate analogs of variance, $\mathbf{\Sigma}$ and $|\mathbf{\Sigma}|$, have their geometric interpretations in the p-dimensional test space (as the matrix of isodensity ellipsoids) and the N-dimensional person space, respectively. It is the latter analog of variance that plays a prominent role in the Λ criterion, due to Wilks (1932), for testing the significance of the overall difference among several sample centroids, constituting a multivariate extension of the F-ratio test in simple analysis of variance.

The rationale for the Λ-criterion stems from the likelihood-ratio principle,[2] but here we shall confine ourselves to showing that Λ reduces to a function of the familiar F-ratio when $p = 1$. Greater insight into the nature of the Λ-criterion will be gained in Chapter 7 after the necessary mathematical background has been developed in Chapter 5. It will then be seen that the Λ-criterion is applicable to many other situations besides that of a multivariate extension of one-way ANOVA.

For the present, we confine our attention to the multivariate significance test for the K-sample problem. In this context—the simplest case for which Wilks' likelihood-ratio criterion is used—Λ is defined as follows:

$$\Lambda = \frac{|\mathbf{W}|}{|\mathbf{T}|} \tag{4.30}$$

[2]See Anderson (1958, pp. 187–191).

where **T** is the total-sample SSCP matrix, and **W** is the within-groups SSCP matrix, which is the sum of the separate SSCP matrices of the K groups; that is,

$$\mathbf{W} = \mathbf{S}_1 + \mathbf{S}_2 + \cdots + \mathbf{S}_K$$

Since $\mathbf{W}/(N - K)$ is the within-group covariance matrix $\mathbf{C}_w$(say) and $\mathbf{T}/(N - 1)$ is the total-sample covariance matrix $\mathbf{C}_t$, the Λ defined by Eq. 4.30 is a multiple of the ratio of two types of sample generalized variances. That is,

$$\Lambda = \frac{|\mathbf{C}_w|}{|\mathbf{C}_t|} \times \left(\frac{N - K}{N - 1}\right)^p \tag{4.31}$$

To see what Λ reduces to when $p = 1$, we note that in this case $|\mathbf{W}| = SS_w$ and $|\mathbf{T}| = SS_t = SS_w + SS_b$. Hence

$$\Lambda_{(p=1)} = \frac{SS_w}{SS_w + SS_b} = \frac{1}{1 + (SS_b/SS_w)}$$

But the customary F-ratio for simple ANOVA is given by

$$F = \frac{SS_b/(K - 1)}{SS_w/(N - K)} = \frac{SS_b}{SS_w}\frac{N - K}{K - 1}$$

where K is the number of groups and $N = n_1 + n_2 + \cdots + n_K$ is the total sample size. Consequently,

$$\frac{SS_b}{SS_w} = \frac{K - 1}{N - K}F$$

which, when substituted in the preceding expression for $\Lambda_{(p=1)}$, yields

$$\Lambda_{(p=1)} = \frac{1}{1 + [(K - 1)/(N - K)]F}$$

One fact immediately evident from this relationship between Λ and F in the univariate case, is that the former is an *inversely* related measure of disparity among the several group means. The larger this disparity is (relative to within-group variability), the larger is F, and hence the smaller is Λ. This observation holds true in the multivariate case as well: The greater the disparity among the several group centroids (relative to the within-groups generalized variance), the smaller the value of Λ.

The sampling distribution of Λ under the null hypothesis (i.e., in the present context, when the K population centroids are equal) was obtained by Schatzoff (1964, 1966a). His results showed that exact numerical computations are feasible only when p is an even number or K is an odd number. For other cases (i.e., p is odd *and* K is even), linear interpolation between tabled values for adjacent p or K was shown to give reasonably accurate approximations.

For reasons of tabular economy Schatzoff's tables were not constructed to give the percentile points of Λ itself, but those of a certain logarithmic function of Λ. This function is

$$\begin{aligned} V &= -[N - 1 - (p + K)/2] \ln \Lambda \\ &= -2.3026[N - 1 - (p + K)/2] \log \Lambda \end{aligned} \tag{4.32}$$

which was shown by Bartlett (1947) to be distributed approximately as a chi-square with $p(K - 1)$ degrees of freedom, provided that $N - 1 - (p + K)/2$ is large. Actually, what Schatzoff's tables explicitly display are "correction factors" for converting selected percentiles of $\chi^2_{p(K-1)}$ to the corresponding percentiles of the statistic V. One multiplies the appropriate chi-square value (shown at the foot of each column) by the proper entry in his tables to obtain the critical value of V for significance at the specified α level. Table 4.1, adapted from one of Schatzoff's sequence of tables, shows the critical values of V at $\alpha = .05$, .025, and .01 for different sample sizes (N) when $p = 10$ and $K = 7$. The values in the bottom row ($N \to \infty$) are exactly equal to the corresponding percentiles of $\chi^2_{p(K-1)}$, because the correction factors (which always decrease monotonically with increasing N) converge to unity as the sample size becomes indefinitely large, no matter what values of p, K, and α are considered.

From Table 4.1 we see that, with $p = 10$ and $K = 7$, the chi-square approximation to V is correct to two significant digits when N is about 100 or

TABLE 4.1 95th, 97.5th, and 99th Percentiles of *V* with *p* = 10, *K* = 7 for Selected Values of *N* (Sample Size)

N \ α	.05	.025	.01
20	91.26	96.71	103.32
22	86.75	91.71	97.75
24	84.45	89.21	94.92
26	83.11	87.71	93.24
28	82.24	86.72	92.09
30	81.61	86.05	91.38
32	81.14	85.47	90.76
34	80.74	85.13	90.41
36	80.50	84.88	90.06
40	80.19	84.47	89.71
46	79.87	84.13	89.26
56	79.56	83.80	88.91
76	79.32	83.55	88.64
100	79.21	83.44	88.53
136	79.16	83.38	88.47
180	79.14	83.34	88.43
∞	79.08	83.30	88.38

more, and correct to three digits when N exceeds 180. The same degree of accuracy is achieved with smaller N's when the number of variables and number of groups are smaller. For example, with $p = K = 5$, a total sample size of 70 (i.e., an average of 14 members per sample group) is sufficient to assure three-digit accuracy. In view of the difficulty of providing extensive tables for the exact distribution of V (since it depends on the three quantities, p, K, and N), the main practical contribution of Schatzoff's theoretically important work may well lie in its having demonstrated, as just exemplified, that the chi-square approximation to Bartlett's V is reasonably good for moderately large sample sizes. Another implication is that, since Schatzoff's correction factors are always greater than 1 for finite N, using the chi-square approximation to V always offers a conservative test: If the value of V as computed from Eq. 4.32 is smaller than the $100(1 - \alpha)$th percentile of $\chi^2_{p(K-1)}$, we can be sure that the exact test will not lead to rejection of the null hypothesis at the $100\alpha\%$ level of significance. It is when the value of V slightly exceeds the specified percentile of $\chi^2_{p(K-1)}$ that one would do well to consult Schatzoff's tables to see whether the V is significant at that level.

There is another function of Λ that seems to offer a better approximate test than does Bartlett's V (without Schatzoff's correction), and is therefore useful when Schatzoff's tables are not available, or when $p(K - 1)$ exceeds 70—which, due to limitations in computer capacity at the time, is the maximum value of $p(K - 1)$ for which Schatzoff's tables are currently in existence. This function, due to Rao (1952), is given by the formula

$$R = \frac{1 - \Lambda^{1/s}}{\Lambda^{1/s}} \frac{ms - p(K - 1)/2 + 1}{p(K - 1)} \tag{4.33}$$

where

$$m = N - 1 - (p + K)/2 \qquad \text{and} \qquad s = \sqrt{\frac{p^2(K - 1)^2 - 4}{p^2 + (K - 1)^2 - 5}}$$

The statistic R is distributed approximately as an F-variate with $\nu_1 = p(K - 1)$ and $\nu_2 = ms - p(K - 1)/2 + 1$ degrees of freedom. When ν_2 is not an integer, we may either interpolate in the F-table or simply use the integer closest to ν_2 as the n.d.f. for the denominator.

It should be pointed out that Bartlett's V and Rao's R each has its advantages. Besides leading to an exact test with the aid of Schatzoff's tables, V has the advantage that being a logarithmic function of Λ, it can be expressed as a sum of several terms—each of which is itself an approximate chi-square variate—when Λ is expressed as a product of several factors, as it will turn out to be the case in more advanced applications of Λ (see Chapter 7). On the other hand, the R-statistic, in addition to permitting a closer approximate test, has the property that it reduces to an *exact* F-variate when $p = 1$ or 2, *or* when $K = 2$

TABLE 4.2 Functions of Λ Distributed as *F*-Ratios for Special Values of *p* and *K*

Any p,	$K = 2$	$\frac{1 - \Lambda}{\Lambda} \frac{N - p - 1}{p} = F^{p}_{N-p-1}$
	$K = 3$	$\frac{1 - \Lambda^{1/2}}{\Lambda^{1/2}} \frac{N - p - 2}{p} = F^{2p}_{2(N-p-2)}$
Any K,	$p = 1$	$\frac{1 - \Lambda}{\Lambda} \frac{N - K}{K - 1} = F^{K-1}_{N-K}$
	$p = 2$	$\frac{1 - \Lambda^{1/2}}{\Lambda^{1/2}} \frac{N - K - 1}{K - 1} = F^{2(K-1)}_{2(N-K-1)}$

or 3. These special cases (which were known long before the R-statistic was developed) are shown in Table 4.2, adapted from Rao (1952, p. 260; 1965, p. 471).

Let us verify, for $K = 3$, the reduction of R to the tabled F-variate. Making the appropriate substitutions in the expressions for m and s, following Eq. 4.30, we obtain

$$m = (2N - p - 5)/2 \qquad s = \sqrt{\frac{p^2(2)^2 - 4}{p^2 + (2)^2 - 5}} = 2$$

and

$$\frac{ms - p(K - 1)/2 + 1}{p(K - 1)} = \frac{2N - p - 5 - p + 1}{2p} = \frac{N - p - 2}{p}$$

Therefore,

$$R = \frac{1 - \Lambda^{1/2}}{\Lambda^{1/2}} \frac{N - p - 2}{p}$$

which is precisely the function of Λ displayed in Table 4.2 for $K = 3$.

It is also of interest to note that for $K = 2$ the function in Table 4.2 has the same distribution as

$$T_2^2 \frac{N - p - 1}{(N - 2)p}$$

In fact, it can be shown (Section 6.6) that

$$\frac{1 - \Lambda}{\Lambda} = \frac{T^2}{N - 2}$$

in the two-group case; that is, Wilks' Λ and Hotelling's T^2 are functionally related in this case.

Another interesting—and perhaps somewhat baffling—point found in Table 4.2 is that for $p = K = 2$, there are *two* functions of Λ that have exact F distributions, namely the first entry with $p = 2$ and the last entry with $K = 2$. These have F^2_{N-3} and $F^2_{2(N-3)}$ distributions, respectively, and one may wonder if they will always yield the same probability of committing a type I error. The answer is that they do, and the outline of a proof is given in an exercise at the end of this chapter. The interested reader may find it profitable to complete the proof.

EXAMPLE 4.7.[3] Employees in three job categories of an airline company were administered an activity preference questionnaire consisting of three bipolar scales: X_1 = outdoor–indoor preferences; X_2 = convivial–solitary preferences; X_3 = conservative–liberal preferences. (A high score on each scale indicates a preponderance of choices of activities of the first-named type over those of the second-named type.) The means of the three groups of employees on the three scales were as follows:

			Means on:		
k	Group	n_k	X_1	X_2	X_3
1.	Passenger agents	85	12.59	24.22	9.02
2.	Mechanics	93	18.54	21.14	10.14
3.	Operations control persons	66	15.58	15.45	13.24

The within-groups SSCP matrix $\mathbf{W}$ (i.e., the sum of the SSCP matrices $\mathbf{S}_1$, $\mathbf{S}_2$, and $\mathbf{S}_3$, for the three groups taken separately), and the total SSCP matrix $\mathbf{T}$ (with the three groups merged into a single conglomerate sample) were found to be as follows:

$$\mathbf{W} = \begin{bmatrix} 3967.8301 & 351.6142 & 76.6342 \\ 351.6142 & 4406.2517 & 235.4365 \\ 76.6342 & 235.4365 & 2683.3164 \end{bmatrix}$$

and

$$\mathbf{T} = \begin{bmatrix} 5540.5742 & -421.4364 & 350.2556 \\ -421.4364 & 7295.5710 & -1170.5590 \\ 350.2556 & -1170.5590 & 3374.9232 \end{bmatrix}$$

Given the data above, we wish to test whether the three group centroids are significantly different from one another; that is, to determine whether the data

[3]Source of contrived data: Rulon, Tiedeman, Langmuir and Tatsuoka (1954).

warrant our rejection of the null hypothesis that the centroids of the three populations, from which our samples were drawn, are all equal.

To find the value of Wilks' Λ, we compute the determinants of **W** and **T**, as required by Eq. 4.30. The results are:

$$|\mathbf{W}| = 46.348 \times 10^9 \qquad \text{and} \qquad |\mathbf{T}| = 127.679 \times 10^9$$

The likelihood ratio Λ is then given by the ratio

$$\Lambda = |\mathbf{W}|/|\mathbf{T}| = 46.348/127.679 = .36301$$

Since the number of groups is three in this example, an exact F-test is possible, based on the Λ just computed. From the appropriate entry of Table 4.2, we see that

$$\frac{1 - \Lambda^{1/2}}{\Lambda^{1/2}} \frac{N - p - 2}{p} = \frac{1 - \sqrt{.36301}}{\sqrt{.36301}} \frac{244 - 3 - 2}{3}$$
$$= 52.57$$

may be referred to the F-distribution with $2p = 6$ d.f.'s in the numerator and $2(N - p - 2) = 478$ (or, for practical purposes, infinite) d.f.'s in the denominator. The observed value 52.57 far exceeds any tabled percentile point of the F^6_∞-distribution, so we reject the null hypothesis that three population centroids are identical.

Purely for illustrative purposes, let us also compute the approximate chi-square statistic V, given in Eq. 4.32, for the Λ-value above:

$$V = -2.3026[244 - 1 - (3 + 3)/2] \log (.36301)$$
$$= (-2.3026)(240)(-.44008) = 243.20$$

which, as a chi-square with $p(K - 1) = 6$ d.f.'s, far exceeds the 99.9th percentile. Since a chi-square value divided by its n.d.f. is an F-variate with the same n.d.f. for the numerator and infinite d.f.'s for the denominator, it is possible to make a check on the degree of approximation offered by the V statistic in this example.

$$V/6 = 243.20/6 = 40.53$$

is to be compared with the exact F-value obtained above. The approximation is seen to be only moderate.

Statistical versus Practical Significance

The reader is probably aware that finding the differences among sample centroids to be statistically significant does not necessarily imply that these differences are significant in the practical sense. Given sufficiently large samples, differences of infinitesimal magnitude can be "statistically significant" even though they make no practical difference whatever. A measure that comes closer to reflecting the *practical* significance of observed differences in one-way univariate analysis of variance is the correlation ratio, defined as

$$\eta^2 = 1 - \frac{SS_w}{SS_t}$$

It can readily be seen, from Eq. 4.30, that the quantity $1 - \Lambda$ is a direct generalization of the correlation ratio to the case when multiple dependent variable are involved, and may hence be called the *multivariate correlation ratio,* and denoted η^2_{mult}. For the data of Example 4.7, which yielded $\Lambda = .363$, the multivariate correlation ratio is $\eta^2_{\text{mult}} = 1 - .363 = .637$. This value may be interpreted (somewhat loosely) as the proportion of generalized variance of the set of dependent variables that is "attributable" to the independent variable(s) indicating membership in one of the three groups (Passenger Agents, Mechanics, and Operations Control Personnel). It therefore gives some indication of the practical significance of differences among the three groups centroids—in the sense of "how important" are the differences. Even more loosely, but perhaps more intuitively, the square-root of this value, slightly less than .80, may be interpreted analogously to a correlation coefficient, and we may say that there is a correlation of just less than .80 between the three Activity Preference variables and group membership.

While the multivariate correlation ratio is quite adequate as a descriptive statistic indicating the strength of association between the independent and dependent variables in the sample at hand, it is highly positive biased; i.e., it grossly overestimates the strength of association in the population. Mindful of this fact, Tatsuoka (1970) proposed a multivariate generalization of Hays' (1963) $\hat{\omega}^2$, as follows:

$$\hat{\omega}^2_{\text{mult}} = 1 - \frac{N}{(N - K)(1 + \lambda_1)(1 + \lambda_2) \ldots (1 + \lambda_r) + 1} \tag{4.34}$$

where $\lambda_1, \lambda_2, \ldots, \lambda_r$ are the non-zero eigenvalues of the matrix $\mathbf{W}^{-1}\mathbf{B}$ (which will be defined in Chapter Five). Equivalently, $\hat{\omega}^2_{\text{mult}}$ may be expressed as

$$\hat{\omega}^2_{\text{mult}} = 1 - \frac{N\Lambda}{(N - K) + \Lambda} \tag{4.35}$$

as was done by Sachdeva (1972). Subsequently, Tatsuoka (1973) did a Monte Carlo of the distributional properties of his $\hat{\omega}^2_{\text{mult}}$ and found that it, too, was positively biased, although less so than was the multivariate correlation ratio η^2_{mult}. He therefore proposed a "corrected" $\hat{\omega}^2_{\text{mult}}$, defined as

$$\hat{\omega}^2_{\text{mult,c}} = \hat{\omega}^2_{\text{mult}} - \frac{p^2 + (K - 1)^2}{3N}(1 - \hat{\omega}^2_{\text{mult}}) \tag{4.36}$$

which was found to be very nearly unbiased when $p(K - 1) \leq 49$ and $75 \leq N \leq 2000$.

For the data of Example 4.7, we find, by substitution in Eq. 4.35,

$$\hat{\omega}^2_{\text{mult}} = 1 - (244)(.363)/241.363 = .633$$

The corrected, or "shrunken" $\hat{\omega}^2_{\text{mult}}$ is, from Eq. 4.36,

$$\hat{\omega}^2_{\text{mult,c}} = .633 - (13)(.367)/(3)(244) = .626$$

In this case, the difference from the value of the multivariate correlation ratio, .637, is not large, but it is evident from Eqs. 4.35 and 4.36 that when N is relatively small, the difference can be considerable.

4.6 TEST OF EQUALITY OF POPULATION COVARIANCE MATRICES

The tests of equality of population centroids that were discussed in the two preceding sections assumed that the population covariance matrices were equal. Although these tests, like their univariate counterparts, are fairly robust with respect to moderate violations of these assumptions, there is of course a limit beyond which departures from homogeneity of covariance matrices will render the tests invalid—that is, will shift the probabilities of type I errors away from their purported (or "nominal") values.

It is therefore desirable to be able to test the null hypothesis of equality of the population covariance matrices as a precondition for the strict validity of Hotelling's T_2^2 test and Wilks' likelihood-ratio test for equality of population centroids. A multivariate generalization of Bartlett's test of homogeneity of K population variances is available for this purpose. A precise statement of the problem and the test procedure are as described below.

Given that $\mathbf{X}_{ki} \sim N(\boldsymbol{\mu}_k, \boldsymbol{\Sigma}_k)$ $(k = 1, 2, \ldots, K;\ \ i = 1, 2, \ldots, n_k)$, we want to test

$$H_0\colon \boldsymbol{\Sigma}_1 = \boldsymbol{\Sigma}_2 = \cdots = \boldsymbol{\Sigma}_K \quad (= \boldsymbol{\Sigma})$$

against the alternative that not all of the population covariance matrices are equal. Let $\mathbf{C}_k = \mathbf{S}_k/(n_k - 1)$ be the unbiased estimate of $\boldsymbol{\Sigma}_k$, based on a sample of size n_k from the kth population, $\mathbf{S}_k$ being the SSCP matrix of that sample. Then if the null hypothesis stated above is true, the pooled within-groups covariance matrix

$$\mathbf{C} = \frac{\sum_{k=1}^{K} (n_k - 1)\mathbf{C}_k}{N - K} = \frac{\sum_{k=1}^{K} \mathbf{S}_k}{N - K} \tag{4.37}$$

would be an unbiased estimate of the common population covariance matrix $\boldsymbol{\Sigma}$.

The hypothesis is tested by a modified log-likelihood-ratio statistic

$$M = (N - K) \ln |\mathbf{C}| - \sum_{k=1}^{K} (n_k - 1) \ln |\mathbf{C}_k| \tag{4.38}$$

Box (1949) showed that when this statistic is multiplied by the scale factor

$$h = 1 - \frac{2p^2 + 3p - 1}{6(p + 1)(K - 1)} \left(\sum \frac{1}{n_k - 1} - \frac{1}{N - K} \right) \tag{4.39}$$

the product Mh is asymptotically distributed as a chi-square with $p(p + 1) \times (K - 1)/2$ degrees of freedom.

When all the n_k are equal (and are hence equal to N/K), the scale factor h reduces to

$$h = 1 - \frac{(2p^2 + 3p - 1)(K + 1)}{6(p + 1)(N - K)} \tag{4.39a}$$

EXAMPLE 4.8 Suppose that two English tests—a long one on paragraph comprehension and a short one on vocabulary—were given to three classes of high school seniors, pursuing three different curricular programs. Prior to testing the significance of differences among the three centroids, we wish to test the tenability of the assumption of equal covariance matrices in the three populations. The SSCP matrices for the three classes were as follows:

$$\mathbf{S}_1 = \begin{bmatrix} 3352.71 & 100.09 \\ 100.09 & 286.97 \end{bmatrix} \quad \mathbf{S}_2 = \begin{bmatrix} 2647.87 & 149.22 \\ 149.22 & 251.57 \end{bmatrix} \quad \mathbf{S}_3 = \begin{bmatrix} 2948.02 & 61.00 \\ 61.00 & 181.83 \end{bmatrix}$$

with

$$n_1 = 33 \qquad n_2 = 37 \qquad n_3 = 24$$

Test the null hypothesis that $\boldsymbol{\Sigma}_1 = \boldsymbol{\Sigma}_2 = \boldsymbol{\Sigma}_3$ against the alternative that not all three of the population covariance matrices are equal, using $\alpha = .10$. Since the three sample SSCP matrices rather than covariance matrices are given, we use the third member of Eqs. 4.37 to calculate the pooled within-groups covariance matrix as

$$\mathbf{C} = (\mathbf{S}_1 + \mathbf{S}_2 + \mathbf{S}_3)/(94 - 3)$$

$$= \begin{bmatrix} 98.336 & 3.410 \\ 3.410 & 7.916 \end{bmatrix} \quad \text{with} \quad |\mathbf{C}| = 766.80$$

The covariance matrices and their determinants for the three classes are

$$\mathbf{C}_1 = \begin{bmatrix} 104.772 & 3.128 \\ 3.128 & 8.968 \end{bmatrix} \quad \mathbf{C}_2 = \begin{bmatrix} 73.552 & 4.145 \\ 4.145 & 6.988 \end{bmatrix} \quad \mathbf{C}_3 = \begin{bmatrix} 128.175 & 2.652 \\ 2.652 & 7.906 \end{bmatrix}$$

with

$$|\mathbf{C}_1| = 929.81 \qquad |\mathbf{C}_2| = 496.80 \qquad |\mathbf{C}_3| = 1006.32$$

Taking the natural logarithms of the four determinants and substituting in Eq. 4.38, we get

$$\begin{aligned} M &= (91)(6.6422) - [(32)(6.8350) + (36)(6.2082) + (23)(6.9141)] \\ &= 3.2007 \end{aligned}$$

Next, from Eq. 4.39 we find

$$h = 1 - [(13)(4)/(6)(3)(2)](1/32 + 1/36 + 1/23 - 1/91) = .9670$$

Hence $Mh = 3.095$, which is smaller than the 90th percent point of the chi-square distribution with $(2)(3)(2)/2 = 6$ degrees of freedom, 10.645. Thus there is insufficient evidence to reject H_0, and we conclude that the assumption of equal covariance matrices in the three populations is tenable.

The reader may wonder why we used $\alpha = .10$ instead of the customary .05 in this example. The reason is that, when we test a null hypothesis as a precondition for another test that is of greater concern to us (here, the test for equality of the three centroids), it is desirable to accept a relatively large probability of committing a type I error in order to decrease the probability of a type II error. This will guard us against unwittingly going ahead to carry out the main test, which may not be valid because its precondition is not satisfied. Some authors even recommend using $\alpha = .20$ under these circumstances.

Another point worth mentioning in connection with the Bartlett–Box test of equality of population covariance matrices is that it is also highly sensitive to the departure from multivariate normality of one or more of the population distributions. Hence some authors recommend not using this test, on the grounds that both the Hotelling's T^2 test and the Wilks' likelihood-ratio test are more robust under violation of multivariate normality than they are when the population covariance matrices are unequal. An exception to this is when the sample n's are all equal, in which case inequality of the covariance matrices has little or no effect.

4.7 RESEARCH EXAMPLES

Lohnes (1966) selected a 2% random sample of the Project TALENT data files and scored his Measuring Adolescent Personality (MAP) factors (11 ability factors plus 11 motive factors) for the 3100 subjects in the sample. Since the MAP factors were orthogonal within the four cells of a Gender × Grade (9th, 12th) design, the covariances among the 22 variables were described as covar-

TABLE 4.3 Ability Contrasts

MAP Factor	12 M	9 M	9 F	12 F
Verbal knowledge	.19	.34	−.20	−.20
English language	−1.04	−.88	1.04	.95
Visual reasoning	.64	.55	−.59	−.72
Mathematics	.99	.65	−.63	−.95
Perceptual speed	−.19	−.05	.22	.07
Screening	.28	.21	−.22	−.33
Hunting, fishing	1.19	1.01	−1.04	−1.27
Memory	−.58	−.50	.57	.55
Color, foods	−1.09	−.79	.83	1.11
Etiquette	−.67	−.49	.57	.62
Games	.41	.26	−.19	−.31

iances of cell means. For the test of equality of the four centroids, Wilks' Λ was .101, and the F with n.d.f. 55 and 9200 was 161. The large sample size delivered a very powerful test. The ability contrasts table (Table 4.3) revealed that in general the gender differences were much larger than the grade differences. Females were markedly superior on English language, memory, color and foods, and etiquette. Males were markedly superior on visual reasoning, mathematics, and hunting and fishing. There were essentially no grade differences in the motive contrasts table (Table 4.4). Females had markedly higher conformity needs and cultural interests, whereas males had markedly higher outdoor and shop interests, science interests, and impulsion. We shall return to this example in Chapter 7 to demonstrate how discriminant analysis builds naturally upon these multivariate analysis of variance (MANOVA) test procedures.

Gribbons and Lohnes (1968, 1969) employed the multivariate significance tests to compare the predictive validity of a battery of four career development

TABLE 4.4 Motive Contrasts

MAP Factor	12 M	9 M	9 F	12 F
Conformity needs	−.45	−.44	.49	.41
Business interests	−.21	−.21	.29	.16
Outdoor, shop interests	1.33	1.33	−1.36	−1.26
Scholasticism	−.06	−.17	.25	.22
Cultural interests	−1.15	−1.11	1.03	1.17
Science interests	1.16	1.13	−1.16	−1.10
Activity level	.22	.23	−.33	−.29
Leadership	−.05	.04	−.06	−.02
Impulsion	.39	.31	−.32	−.42
Sociability	.18	.15	−.05	−.03
Introspection	.21	.14	−.11	−.16

TABLE 4.5 Predictors of Career Development Correlations (N = 110)

Variable	Gender	SES	IQ	RCP
Gender (male = 1, female = 2)	1.00	.01	.03	−.18
Socioeconomic status (1 = high, 7 = low)		1.00	−.35	−.14
Otis beta form intelligence			1.00	.31
Readiness for career planning				1.00

predictors collected on a sample of 110 8th graders in 1958 for an aspirations variable collected in 1958 and a similar aspirations variable collected a decade later in 1967. They were able to demonstrate that the overall strength of predictive validity was almost the same for the two aspirations criteria, despite the 10 years of career development separating them in the lives of the subjects. There were four cells for the aspiration variable in each year. For equality of the centroids, the MANOVA F with n.d.f. 12 and 270 was 6.5 for the 1958 criterion, while the MANOVA F witht he same n.d.f. was 6.3 for the 1967 criterion. These F-ratios translated into MANOVA η^2 = .70 for the 1958 criterion and MANOVA η^2 = .69 for the 1967 criterion. For equality of dispersions, the MANOVA F with n.d.f. 30 and ∞ was 1.6 for the 1958 criterion, and 1.2 for the 1967 criterion. Table 4.5, giving the predictors of career development correlations, showed that the measurement variables were only modestly correlated among themselves, with the largest correlation being −.35 between SES and IQ (negative because SES was scaled 1 = high, 7 = low.

While the overall predictive validity of the measurement battery was about the same for both aspiration years, the pattern of variable predictive validities was markedly different, and therein lay an important finding for the research program. As shown in Table 4.6 for the MANOVA for 1958 aspiration groups, gender and RCP were the best predictors of aspirations group when the subjects weie in eighth grade. However, as shown in the MANOVA for 1967 aspiration

TABLE 4.6 MANOVA for 1958 Aspiration Groups

1958 measures	College science (N = 33) means	Noncollege technology (N = 16) means	Noncollege business (N = 24) means	College cultural (N = 37) means	Pooled within groups S.D.'s	F^3_{106}	η
Gender	1.3	1.1	1.8	1.7	.5	11.4	.49
SES	3.5	3.8	4.9	3.8	1.6	4.3	.27
Otis IQ	109.8	103.8	105.5	109.6	9.3	2.4	.20
RCP	37.5	24.8	27.0	34.7	9.7	9.5	.39
	Equality of dispersions: F^{30}_{∞} = 1.6						
	Equality of centroids: F^{12}_{270} = 6.5 and η = .70						

TABLE 4.7 MANOVA for 1967 Aspiration Groups

1958 measures	College science (N = 10) means	Noncollege technology (N = 19) means	Noncollege business (N = 49) means	College cultural (N = 31) means	Pooled within groups S.D.s	F^3_{105}	η
Gender	1.3	1.3	1.8	1.8	.5	6.6	.33
SES	3.7	4.4	4.6	2.8	1.5	9.8	.40
Otis IQ	113.1	101.1	106.3	112.8	8.6	9.2	.38
RCP	32.8	30.2	29.4	37.9	10.2	4.6	.28

Equality of dispersions: $F^{30}_{\infty} = 1.2$
Equality of centroids: $F^{12}_{270} = 6.3$ and $\eta^2 = .69$

groups (Table 4.7), SES and IQ were the best predictors of aspiration group when the subjects were 4 years out of high school. This shift had great meaning for the theory of career development which the authors presented.

Zimmerman and Pons (1986) researched the validity of a set of 14 interview scales in a domain they named "Use of Self-Regulated Learning Strategies" (herafter SRLS) to discriminate between 10th-grade students in a high-achievement track and 10th-grade students in a low-achievement track. They hoped to show that their structured interview schedule applied theorems of social learning theory to provide a practical assessment of the extent to which students self-regulate their learning and the methods of self-regulation they employ. They also hoped to show that degree and type of self-regulation of learning can explain a considerable portion of the variance in assignment of students by the secondary school to achievement tracks. In pursuit of these research goals, the authors chose a method of planned comparison of two populations by Wilks' Λ test on the centroids of random samples in the 14-dimensional SRLS scales space, followed by inspection of a discriminant function to determine the most effective scales for separating the two groups. They also presented 14 univariate *t*-test values, presumably on the assumption that the univariate tests for the individual scales can serve as tests of the significance of discriminant-function coefficients (an issue discussed in Chapter 7). A classification hit-ratio was reported to strengthen the claim that the SRLS assessment can predict achievement track placement (see Chapter 10). An earlier version of this textbook Tatsuoka (1971), was cited as the source of information about the selected methods of analysis.

Random samples of size 40 were drawn from the advanced achievement track and the other (lower) achievement tracks in a single high school "serving a middle-class suburban community of a large metropolitan area" (p. 616). As is usually the case when true random samples are achieved in educational research, the populations were extremely parochial. So be it. The use of small and parochial samples in this research may be justified by the difficulties and costs of doing interview research. Sex ratios turned out to be unbalanced and different for the two samples. Nothing was said about races of the students.

As is customary in such research, the Λ test of the equality of the dispersions for the two populations was not performed, or if performed, was not reported. The χ^2 with n.d.f. 16 for the Λ test of the equality of centroids was 87, permitting a very comfortable rejection of the null hypothesis. Both discriminant-function scoring coefficients and pooled-within group correlations of scales with the discriminant function were reported. The latter would have been sufficient from our perspective (see Chapter 7). It was also reported that 93% of the 80 sample members could be correctly classified for track on the basis of their SRLS assessment profiles. Unfortunately this classification must be subject to considerable shrinkage (i.e., reduction if replicated on new random samples using the estimates from this sample). Incidentally, all 14 scales were dichotomously scored rather than continuously scored, but the authors did not comment on this radical departure from the assumptions of their MANOVA-discriminant analysis. Since their criterion was also dichotomous, it is possible that they had the ideal conditions for a log-linear modeling analysis (see Cooper and Weeks, 1983).

The authors also developed a three-scale scoring procedure for the interview protocol, providing measures of *Strategy Use* (SU, dichotomous), *Strategy Frequency* (SF, integer values 1 through 6), and *Strategy Consistency* (SC, integer values 1 through 4). They performed a similar series of statistical analyses in this 3-space. For the equality of centroids, the Λ-based χ^2 with n.d.f. 3 was 81, and the classification hit rate was 91%. This time, only discriminant-function scoring coefficients were reported, and significance tests for them based on their standard errors were reported in lieu of univariate *t*-test values. This was a strange inconsistency with the reporting of the other discriminant analysis. We believe that the best report for both analyses would have been in terms of the correlations of the scales with the discriminant function and the significance tests for the function weights based on their standard errors.

The authors went on to compare the validities of the variables Gender and SES with those of their interview-scoring systems, concluding that their interview scoring systems added importantly to Gender and SES in the modeling of track variance. Unfortunately, their interview scoring systems also consumed degrees of freedom and there was need for statistical testing of the presumed unique contributions of the scoring systems, especially the one employing 14 categories. The biggest problem with all of this was, however, that no covariate representing general verbal ability, or intelligence, was employed. The fact is that any interview schedule of the type involved in this research is to a considerable extent a verbal ability test, and it is necessary to partial out this latent supertrait before any conclusions can be reached about the usefulness of additional latent trait constructs such as those social learning theory constructs indicated by the SRLS scales. In contemporary parlance, there was a crippling *specification error* in the hypothesized model for the data, in that it did not specify an operationalization of a latent factor of general intellectual development as a principal exogenous explanatory variable. Interestingly, the found discriminant function of the 14 strategy scales correlated positively with every one of those scales, and had only one negative scoring weight (a nearly negligible $-.11$), so it can be viewed as a general ability factor.

The Λ test for the equality of population dispersions has not been popular in the educational research literature. The reason it is seldom seen is that the conventional multivariate analysis of variance strategy which subsumes it depends on the assumption of equal dispersions as a justification for modeling on the means, whether this modeling is done in the space of the research variates or in the space of discriminant functions. Yet when desirable sample sizes are obtained the test for the equality of dispersions is a powerful one and is likely to yield a rejection, which simply embarrasses the modeling strategy, if not the researcher. Such a rejection implies that the MANOVA model is incorrect, but does not suggest the calculus for a parsimonious and effective alternative modeling procedure. No wonder we have usually swept the issue of equality of dispersions under the rug.

With the recent emergence of LISREL (an excellent description of which can be found in Jörskog and Sörbom, 1983) and similar computer algorithms for maximum likelihood estimation (MLE) of linear structural equations, a new set of conventions has entered the educational research marketplace. MLE theory provides a χ^2 measure of the goodness-of-fit of the estimated model with the data, and also a χ^2 significance test for the difference in goodness-of-fit between two competitive models for the same data. MLE theory also clarifies the calculus for fitting a model that employs a separate dispersion for each population in multiple-population designs. Doing so continues to be hurtful from the standpoint of parsimony, but at least the algorithmic machinery for doing so is readily at hand in computer packages such as LISREL or EQS (Bentler, 1985). In the MLE approach the equality of dispersions is *not* tested directly as it is in the MANOVA approach, but instead the fit of a model fixing the dispersions to be equal is compared with the fit of a model freeing the dispersions to differ. Of course the fit for the latter model has to be better, but it is achieved at the expense of a larger set of free parameters to be estimated, and the issue for the analyst is whether the consumption of degrees of freedom and the elaboration of the theory for the data are justified by the improvement in fit. This is not necessarily solely, or even primarily, a statistical-inference issue, even if the strong assumptions driving the MLE inference theory are met.

This new approach to the equality of dispersions question is demonstrated by Entwisle, Alexander, Cadigan and Pallas (1986). They had two samples of data on first-grade pupils obtained a decade apart in Baltimore schools. They had in mind an elaborate theory of linear structural relationships among variables in several domains, including achievement, expectation, family influence, peer influence, and intelligence. They were convinced that it was necessary to provide separate parameter estimates for three populations of children, namely (1) children attending white middle-class elementary schools, (2) children attending integrated working-class schools, and (3) children attending black working-class schools. Interestingly, they did *not* test this assumption on the data, but simply installed it in the analysis design on a priori grounds. What they hoped to avoid was the need for different parameter estimates for the two decade-separated samples available for each of the three populations. That is, they hypothesized that the two decade-separated dispersions for each population were essentially similar, so that the two sets of data could be *pooled* (in our parlance) or *stacked* (in the parlance of the four authors) to estimate a common model. This hypothesis was, in effect, that the process of first-grade education had not changed over the decade in each of the populations, even though it was assumed to be different among the populations.

Two excerpts from their elaborate report of goodness-of-fit statistics (Table II, p. 602) will give the flavor of their analysis. The analysis was further complicated by the decision to model separately the educational process for the first-half of the first-grade year (termed *Cycle 1*) and the educational process for the second-half of the first-grade year (termed *Cycle 2*). For the white middle-class population and Cycle 1, the 1970s sample included 173 pupils and the χ^2 with n.d.f. 18 was 23.6, for which $P = .17$. This outcome indicated an acceptable goodness-of-fit of the model with the data. For the same population and cycle, the 1980s sample included 70 pupils (mind you, there were 19 variates) and the χ^2 with n.d.f. 18 was 20.3, for which $P = .31$, again indicating an acceptable goodness-of-fit of the model with the data, and indeed a better fit than that for the 1970s sample. (It is disturbing that sample size does not impact on the n.d.f. for these chi-squares, although it does on their computed values.) For the *stacked model* the χ^2 was 141 with n.d.f. 76, for which $P = .00$, indicating an unacceptable looseness-of-fit of a common set of estimates with the pooled data. Also, for the "difference in fit between the stacked model and individual models taken jointly" (p. 601) the χ^2 was 97 with n.d.f. 40, for which $P = .00$, indicating that the deterioration between the fit of separate models and the fit of a common model is unacceptable. A second example leading to a decision in favor of the stacked model is provided by their results for the samples of students attending black working-class schools. For the black working-class population and Cycle 2, the 1970s sample included 65 pupils and the χ^2 with n.d.f. 42 was 37.2, for which $P = .68$. This outcome indicated an acceptable goodness-of-fit of the model with the data. For the same population and cycle, the 1980s sample included 137 pupils and the χ^2 with n.d.f. 42 was 67.6, for which $P = .01$, indicating an unacceptable goodness-of-fit of the model with the data, which is a curious outcome indeed. For the *stacked model* the χ^2 was 144 with n.d.f. 125, for which $P = .12$, indicating an acceptable goodness-of-fit of a common set of estimates with the pooled data. Also, for the difference in fit between the stacked model and individual models the χ^2 was 39.5 with n.d.f. 41, for which $P = .50$, an ideal outcome to support the decision that "a common set of parameters fits the data as well as separate parameters" (p. 601). The authors emphasize that "concluding that a common set of parameters does as well as separate parameters (the second test) has no necessary implications regarding the overall fit of model to data (the first test)" (p. 601). In this example, the stacked model passes the first test.

Additional excellent examples of this MLE-based methodology for making decisions about whether or not to pool dispersions from several populations are provided by Wolfle (1985) and Lomax (1985). If you wish to extend your exploration of multivariate statistical procedures beyond the coverage provided by this textbook, your best bet would be to undertake some study of the MLE approach to structural equations. Bentler (1980) may provide the best entry to the literature.

EXERCISES

1. Ten pairs of college freshmen with matching IQs were used in a problem-solving experiment, members of each pair being assigned at random to the experimental and control groups. The task was to solve two sets of riddle-like problems, one set being entirely verbal and the other set involving some numerical reasoning. The problems in the two sets had similar logical structures, so that some practice effect could be anticipated even in the control group. The experimental group was given instructions that should facilitate performance, especially on the second set of problems.

The scores earned by the 10 experimental *S*'s and their matched controls on the two sets of problems were as follows:

Exp.	X_1	19	28	30	31	33	34	41	35	45	53
	X_2	26	33	37	34	41	36	45	40	42	56
Control	X_1	19	27	31	30	34	34	39	36	42	50
	X_2	18	31	36	31	36	37	39	41	43	52

Test the significance of the difference between the experimental and control group centroids at the 5% level of significance.

2. A group of 20 drug addicts and a group of 30 chronic alcoholics were given two psychological tests *A* and *B*, each yielding standardized scores on a 10-point scale. The means of the two groups on the two tests, and the within-groups SSCP matrix **W** were as follows:

	A	*B*	(n_k)
Drug addicts	5.8	2.9	(20)
Alcoholics	6.7	2.3	(30)

$$\mathbf{W} = \begin{bmatrix} 95.4 & 11.6 \\ 11.6 & 29.2 \end{bmatrix}$$

Test the significance of the difference between the drug-addict and alcoholic group centroids, using $\alpha = .01$.

3. Sixty subjects were assigned at random to three experimental conditions, 20 S's per group, in a learning experiment. The criterion performance was scored in two ways: speed (X_1) and accuracy (X_2). The means of the three groups on the two criterion measures, the within-groups SSCP matrix **W** and the total SSCP matrix **T**, were as shown below.

	$\bar{X}_1$	$\bar{X}_2$
Group 1	38.5	19.6
Group 2	34.0	22.4
Group 3	30.1	16.2

$$\mathbf{W} = \begin{bmatrix} 982.8 & 205.2 \\ 205.2 & 643.5 \end{bmatrix} \qquad \mathbf{T} = \begin{bmatrix} 1689.6 & 472.8 \\ 472.8 & 1029.1 \end{bmatrix}$$

Test the significance of the overall difference among the three group centroids, using $\alpha = .01$.

[NOTE: If the reader is puzzled to find that knowledge of the groups' centroids is apparently not needed in carrying out the significance test, he or she should observe that this problem could have been solved with only the centroids and the **W** matrix given. Matrix **T** can then be computed without its being given. Although the procedure anticipates material discussed in Chapter 6, the reader may find it informative and challenging to try it at this point. The first step is to compute the between-groups SSCP matrix **B** in accordance with Eqs. 6.3 and 6.4. The total SSCP matrix is then obtained as $\mathbf{W} + \mathbf{B} = \mathbf{T}$, which is the multivariate counterpart of the relation $SS_w + SS_b = SS_t$ in univariate, one-way analysis of variance.]

4. Given that a bivariate normal population has the covariance matrix

$$\mathbf{\Sigma} = \begin{bmatrix} 25 & 15 \\ 15 & 25 \end{bmatrix}$$

we wish to test the following null hypotheses about the means of the two variables:

$$H_{01}: \mu_1 = 15 \qquad \text{and} \qquad H_{02}: \mu_2 = 20$$

Suppose that a sample of 100 bivariate observations from the population yielded $\bar{X}_1 = 16.1$ and $\bar{X}_2 = 21.0$. Test the null hypotheses about μ_1 and μ_2 against two-sided alternatives at the 5% significance level in two ways: **(a)** using two univariate tests; and **(b)** using a single multivariate test. Explain the discrepancy that you should have found between the conclusions in (a) and (b).

5. (Theoretical) It was stated on page 95 that when $p = K = 2$, two different functions of Wilks' Λ (the first and last expressions listed in Table 4.2) have exact F distributions. To show that using either of these F variates as test statistics will always lead to the same conclusion, we proceed as follows: First, solve for $1/\Lambda$ in terms of F_n^2 from the first equation, and in terms of F_{2n}^2 from the last equation, in Table 4.2. We must then show that the two resulting expressions are equal for every percent point of the two F variates they respectively involve. (Here n stands for $N - 3$.) To prove this mathematically, we start from the fact that the density function of the $F_{\nu_2}^{\nu_1}$-distribution is

$$g(F) = KF^{(\nu_1/2)-1}(1 + \nu_1 F/\nu_2)^{-(\nu_1+\nu_2)/2}$$

with

$$K = (\nu_1/\nu_2)^{\nu_1/2}/B(\nu_1/2, \nu_2/2)$$

where

$$B\left(\frac{\nu_1}{2}, \frac{\nu_2}{2}\right) = \int_0^1 (1 - x)^{(\nu_1/2)-1}x^{(\nu_2/2)-1}\,dx$$

is the beta function with parameters $\nu_1/2$ and $\nu_2/2$. From this it follows that when $\nu_1 = 2$ and $\nu_2 = n$,

$$g_1(F) = (1 + 2F/n)^{-(n+2)/2}$$

and that when $\nu_1 = 2$ and $\nu_2 = 2n$

$$g_2(F) = (1 + F/n)^{-n-1}$$

Integrating the first density function from 0 to F_n^2 and the second from 0 to F_{2n}^2, we obtain the respective (cumulative) distribution functions

$$G_1(F_n^2) = 1 - (1 + (2/n)F_n^2)^{-n/2}$$

and

$$G_2(F_{2n}^2) = 1 - (1 + (1/n)F_{2n}^2)^{-n}$$

Upon equating the expressions for $G_1(F_n^2)$ and $G_2(F_{2n}^2)$ we get the relation that must hold between each percentile point of F_n^2 and the corresponding percentile point of F_{2n}^2. This relation turns out to be precisely the relation that was previously obtained by equating the expressions for $1/\Lambda$ that were yielded by the first and last entries of Table 4.2.

Readers with the appropriate mathematical background may find it profitable to prove each statement starting from the assertion, "From this it follows that . . ." above. Others may prefer simply to get the relation by equating the two expressions for $1/\Lambda$ and then verifying numerically for selected values of α that this relation does hold, by referring to Table E.4 in Appendix E.

6. Test the significance of the difference between the boys' centroid and the girls' centroid of the five cognitive variables in the HSB data, using $\alpha = .01$. Next, conduct a univariate t-test for the significance of the two means on each of the five variables. Compare the results of the two analyses.

7. Test the significance of the difference among the centroids of the three different high school program groups (HSP) on the five cognitive variables, using $\alpha = .01$. (In this as well as Exercise 6, use a matrix-operations program instead of a canned multivariate significance test program if possible.)

8. Test the null hypothesis that $\mathbf{\Sigma}_1 = \mathbf{\Sigma}_2 = \mathbf{\Sigma}_3$, where $\mathbf{\Sigma}_k$ is the covariance matrix of the five cognitive variables in the population corresponding to the kth HSP group ($k = 1, 2, 3$). Use $\alpha = .05$. Also compute the generalized variances of the three groups. Does the rank order of the magnitudes of the three generalized variances accord with the commonsense expectation of how the HSP groups should rank from most homogeneous to least homogeneous in terms of academic achievement as a whole?

5

MORE MATRIX ALGEBRA: LINEAR TRANSFORMATIONS, AXIS ROTATION, AND EIGENVALUE PROBLEMS

Reference was made in Chapter 4 to the fact (Theorem 1 on p. 72) that the quantity $\chi^2 = \mathbf{x}'\mathbf{\Sigma}^{-1}\mathbf{x}$ in the exponent of the p-variate normal density function is a chi-square variate with p degrees of freedom. This fact was proved only for the special case when the p random variables X_i are independently distributed, and hence uncorrelated. The proof for the general case of p-variate normal populations with nonzero correlation coefficients hinges on the fact that a set of p independently normally distributed variables can be constructed from linear combinations of the original p variables. In this chapter we develop matrix methods that enable us, among other things, to determine these linear combinations. To understand how this works, it is helpful to relate the algebraic process of forming linear combinations to the geometric operation of rotating the axes of a coordinate system.

5.1 LINEAR TRANSFORMATIONS AND AXIS ROTATION

In discussing the p-variate normal distribution, we associated a p-dimensional vector $[X_1, X_2, \ldots, X_p]$ with a point in p-dimensional space having the coordinates $(X_1, X_2, \ldots, X_p)$. Let us go back to the simplest case of $p = 2$ to study the relationship between linear combinations of variables and axis rotation.

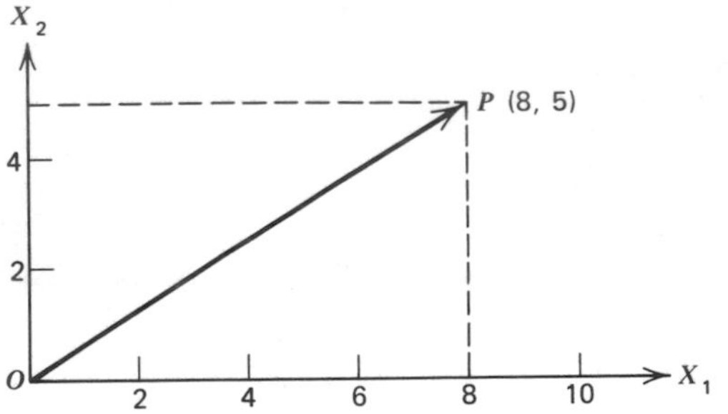

Figure 5.1. *Vector representing a person who scores 8 and 5 on Tests* X_1 *and* X_2.

A person with scores 8 and 5, respectively, on two tests X_1 and X_2 may be represented by a vector [8, 5]. This vector, in turn, can be geometrically represented by an arrow from the origin O of a Cartesian coordinate system in the (X_1, X_2)-plane to the point $P(8, 5)$, as in Fig. 5.1.

Now let us rotate the axes 20° in the counterclockwise direction, and call the resulting coordinate system (OY_1, OY_2), as shown in Fig. 5.2. The coordinates of point P in the new system may be found by letting $X_1 = 8$, $X_2 = 5$, and $\theta = 20°$ in the axis-rotation formulas

$$\begin{aligned} Y_1 &= (\cos\theta)X_1 + (\sin\theta)X_2 \\ Y_2 &= (-\sin\theta)X_1 + (\cos\theta)X_2 \end{aligned} \tag{5.1}$$

Substituting the appropriate trigonometric function values for $\theta = 20°$, the results are

$$\begin{aligned} Y_1 &= (.940)(8) + (.342)(5) = 9.230 \\ Y_2 &= (-.342)(8) + (.940)(5) = 1.964 \end{aligned}$$

Thus, in the new coordinate system, the vector OP has the representation [9.230, 1.964], as may be graphically verified by referring to Fig. 5.2.

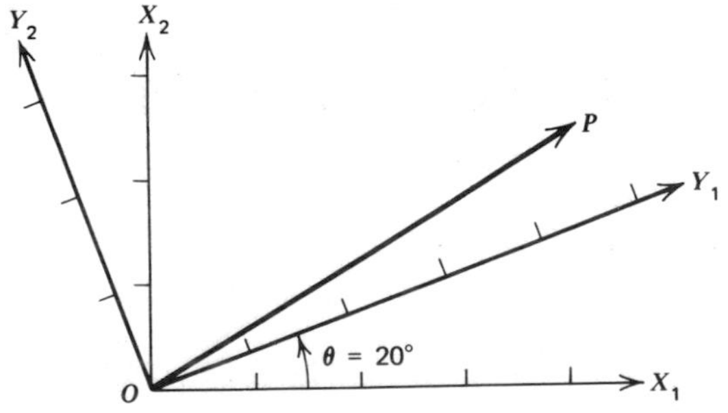

Figure 5.2. *Vector* OP *shown with reference to the original coordinate system* (OX_1, OX_2) *and a rotated system* (OY_1, OY_2).

What is more to the point for our present purposes is to note that each of the transformation equations (5.1) expresses one of the new coordinates of point P as a linear combination of the original coordinates X_1 and X_2. Alternatively, we may regard these equations as defining two new *variables* Y_1 and Y_2 in terms of linear combinations of the original variables X_1 and X_2. We thus see that when we perform a rotation of axes, we are in effect defining a new set of variables, each of which is a linear combination of the original variables.

Is it the case, conversely, that *any* linear combination of a given set of variables expresses the result of rotating one of the axes of a coordinate system through some angle? Not quite so. The coefficients of the linear combination must satisfy a certain condition before that linear combination qualifies as a "rotated axis." This condition may be inferred by examining the coefficients in the expressions for Y_1 and Y_2 in Eqs. 5.1. Note that, from the trigonometric identities

$$(\cos\theta)^2 + (\sin\theta)^2 = 1$$

and

$$(-\sin\theta)^2 + (\cos\theta)^2 = 1$$

the *squares* of the coefficients add up to unity in each case. That this is a sufficient condition for a linear combination to represent a rotated axis may be established by the following observation:

If

$$Y = aX_1 + bX_2 \qquad \text{and} \qquad a^2 + b^2 = 1$$

then an angle θ *can always be found such that cos* $\theta = a$ *and sin* $\theta = b$

Thus a linear combination of two variables, X_1 and X_2, formed by using coefficients whose squares sum to unity can always be represented by an axis accessible by some rotation in the (X_1, X_2)-plane from the X_1-axis.

EXAMPLE 5.1. Let $Y = .6X_1 + .8X_2$. Then, since $(.6)^2 + (.8)^2 = 1$, an angle θ may be found such that $\cos\theta = .6$ and $\sin\theta = .8$. Determine such an angle and draw an axis OY superimposed on a facsimile of Fig. 5.1. Also, verify that the foot of the perpendicular dropped from point P to line OY is $(.6)(8) + (.8)(5) = 8.8$ units away from the origin O. (A more concise way of expressing this fact is to say that *the projection of vector OP onto axis OY has length* 8.8.)

From a table of trigonometric functions, we find $\sin 53°8' = .80003$. So we may take $\theta = 53°8'$. Alternatively, since $\tan\theta = \sin\theta/\cos\theta = .8/.6 = 4/3$, we may determine a point Q whose ordinate and abscissa stand in the ratio 4:3. The line through the origin O and such a point Q gives the desired Y-axis, without our actually having to find angle θ numerically.

Next, given two linear combinations of X_1 and X_2 that are both representable as rotated axes, what further condition among the coefficients must hold in order that the two new axes be perpendicular to each other? The answer is again exemplified in Eqs. 5.1:

$$(\cos\theta)(-\sin\theta) + (\sin\theta)(\cos\theta) = 0$$

To generalize, if the coefficients of two linear combinations

$$Y = aX_1 + bX_2 \quad \text{and} \quad Y' = a'X_1 + b'X_2$$

are such that

$$a^2 + b^2 = a'^2 + b'^2 = 1 \tag{5.2}$$

and the further condition,

$$aa' + bb' = 0 \tag{5.3}$$

holds, then the axes OY and OY', representing the two linear combinations, are perpendicular, or *orthogonal,* to each other. In other words, if both Eqs. 5.2 and 5.3 are satisfied by the coefficients of two linear combinations of a pair of variables X_1 and X_2, then the axes OY and OY' together constitute a coordinate system resulting from a *rigid* (or angle-preserving) rotation of the original rectangular coordinate system (OX_1, OX_2). (A slight qualification of this statement, which we do not wish to introduce at this point, will subsequently be made.)

EXAMPLE 5.2. If $Y = .6X_1 + .8X_2$ and $Y' = -.8X_1 + .6X_2$, we have

$$(.6)^2 + (.8)^2 = (-.8)^2 + (.6)^2 = 1$$

and

$$(.6)(-.8) + (.8)(.6) = 0$$

Therefore, axes OY and OY' are mutually orthogonal. On the other hand, let

$$Y'' = .707X_1 - .707X_2$$

Since

$$(.707)^2 + (-707)^2 = 1$$

Y'' is also representable by an axis in the (OX_1, OX_2)-plane. However,

$$(.6)(.707) + (.8)(-.707) \neq 0$$

so axes OY and OY'' are *not* mutually orthogonal. The foregoing statements may be verified by determining angles θ, θ', and θ'' such that $\cos\theta = .6$, $\cos\theta' = -.8$, and $\cos\theta'' = .707$, and drawing axes OY, OY', and OY'', making these angles, respectively, with axis OX_1 of the original system.

---

mensionality. Thus, for

$\theta_{31})X_3$

$\theta_{32})X_3$ (5.5″)

$\theta_{33})X_3$

iginal system to the jth

mpactly by introducing

, respectively, become,

(5.4′)

(5.5′)

s. 5.4′ and 5.5′ are not
as immediately clear in
separate conditions on
cognizable as those for
plane. The conditions,

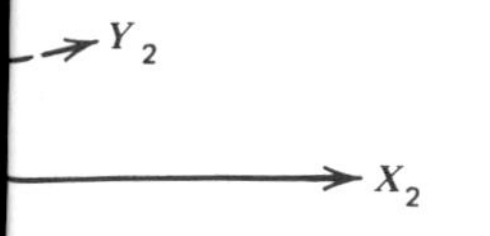

elated angles θ_{ij}.

is rotation in a plane
by introducing a re-
long various pairs of
θ between axes OX_1
$_{21}$, and θ_{22}, as shown
X_i-axis to the (new)

ious θ is as follows:

ons of complementary

) = $\cos\theta_{12}$

may rewrite Eqs. 5.1

(5.4″)

itates generalization to
nts of the form $\cos\theta_{ij}$,
of which $\cos\theta_{ij}$ is the
whose expression the

te set of equations cor-

ted angles θ_{ij}.

responding to rigid rotations of axes in spaces of any d
three dimensions, we have

$$Y_1 = (\cos \theta_{11})X_1 + (\cos \theta_{21})X_2 + (\text{cc}$$

$$Y_2 = (\cos \theta_{12})X_1 + (\cos \theta_{22})X_2 + (\text{cc}$$

$$Y_3 = (\cos \theta_{13})X_1 + (\cos \theta_{23})X_2 + (\text{cc}$$

where θ_{ij} is again the angle from the ith axis of the o
axis of the new system, as depicted in Fig. 5.4.

The transformation equations may be written more c
the abbreviation

$$v_{ij} = \cos \theta_{ij}$$

Equations 5.4″ and 5.5″, for two and three dimensions
in this notation,

$$Y_1 = v_{11}X_1 + v_{21}X_2$$

$$Y_2 = v_{12}X_1 + v_{22}X_2$$

$$Y_1 = v_{11}X_1 + v_{21}X_2 + v_{31}X_3$$

$$Y_2 = v_{12}X_1 + v_{22}X_2 + v_{32}X_3$$

$$Y_3 = v_{13}X_1 + v_{23}X_2 + v_{33}X_3$$

The systematicness and compactness of notation in E
gained without some sacrifice in completeness. What w
Eqs. 5.1, using sin θ and cos θ, must now be stated a
the coefficients v_{ij} before Eqs. 5.4′ are unambiguously r
a rigid rotation of a rectangular coordinate system in a

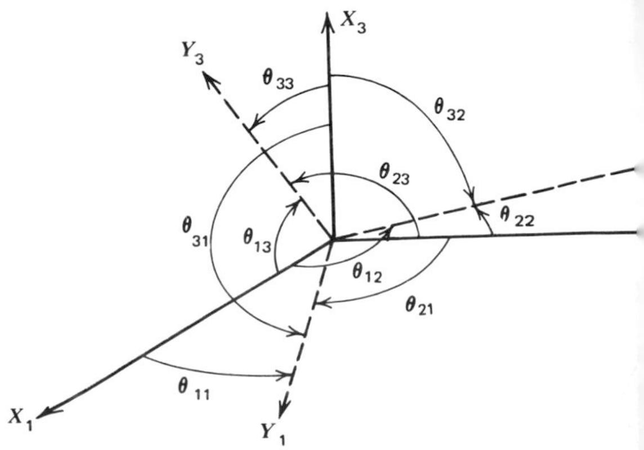

Figure 5.4. *Rigid rotation in space designated by nine*

which were stated in different notation in Eqs. 5.2 and 5.3, now become

$$\begin{aligned} v_{11}^2 + v_{21}^2 &= v_{12}^2 + v_{22}^2 = 1 \\ v_{11}v_{12} + v_{21}v_{22} &= 0 \end{aligned} \tag{5.6}$$

In the three-dimensional case, in order to assure that Eqs. 5.5′ refer to a rigid rotation of axes, the conditions to be stated are six in number:

$$\begin{aligned} v_{11}^2 + v_{21}^2 + v_{31}^2 &= v_{12}^2 + v_{22}^2 + v_{32}^2 \\ &= v_{13}^2 + v_{23}^2 + v_{33}^2 = 1 \\ v_{11}v_{12} + v_{21}v_{22} + v_{31}v_{32} &= v_{11}v_{13} + v_{21}v_{23} + v_{31}v_{33} \\ &= v_{12}v_{13} + v_{22}v_{23} + v_{32}v_{33} = 0 \end{aligned} \tag{5.7}$$

The number of conditions rapidly increases with the dimensionality p of the space. There is one condition requiring that the sum of the squares of the coefficients appearing in each transformation equation be equal to unity (totalling p conditions); and one condition requiring that the sum of cross products between coefficients in each *pair* of transformation equations be equal to zero (totalling $p(p - 1)/2$ conditions). Thus, for a p-dimensional space, the total number of conditions to be satisfied by the p^2 coefficients v_{ij} in the transformation equations is $p(p + 1)/2$. However, a useful notational convention allows us to express all $p(p + 1)/2$ conditions in the form of a single equation type. This is to use the symbol δ_{jk}, known as *Kronecker's delta,* defined as follows:

$$\delta_{jk} = \begin{cases} 1 & \text{if } j = k \\ 0 & \text{if } j \neq k \end{cases} \tag{5.8}$$

It may be readily verified that using this symbol, the sets of Eqs. 5.6 and 5.7 are each embraced by the single equation type

$$\sum_{i=1}^{p} v_{ij}v_{ik} = \delta_{jk} \qquad (j = 1, 2, \ldots, p; \quad k = j, j + 1, \ldots, p) \tag{5.9}$$

An easier way of stating the $p(p + 1)/2$ conditions to be satisfied by the coefficients in the set of p transformation equations representing a rigid rotation of a rectangular coordinate system in p-dimensional space results from writing the set of equations in matrix form. Equations 5.4′ and 5.5′ then become

$$[Y_1, Y_2] = [X_1, X_2] \begin{bmatrix} v_{11} & v_{12} \\ v_{21} & v_{22} \end{bmatrix} \tag{5.4}$$

and

$$[Y_1, Y_2, Y_3] = [X_1, X_2, X_3] \begin{bmatrix} v_{11} & v_{12} & v_{13} \\ v_{21} & v_{22} & v_{23} \\ v_{31} & v_{32} & v_{33} \end{bmatrix} \tag{5.5}$$

respectively. The obvious extension to p-dimensional space is

$$[Y_1, Y_2, \ldots, Y_p] = [X_1, X_2, \ldots, X_p] \begin{bmatrix} v_{11} & v_{12} & \cdots & v_{1p} \\ v_{21} & v_{22} & \cdots & v_{2p} \\ \vdots & \vdots & \ddots & \vdots \\ v_{p1} & v_{p2} & \cdots & v_{pp} \end{bmatrix}$$

or, in symbolic matrix notation,

$$\mathbf{Y}' = \mathbf{X}'\mathbf{V} \tag{5.10}$$

where

$$\mathbf{V} = (v_{ij})$$

We digress briefly to mention that any transformation of the form of Eq. 5.10, with no restriction on the $p \times p$ matrix $\mathbf{V}$, is called a *linear transformation,* since all quantities involved appear only in the first degree. The matrix $\mathbf{V}$ is called the *transformation matrix,* and we say that the vector $\mathbf{X}$ is transformed linearly into vector $\mathbf{Y}$ by matrix $\mathbf{V}$. An important property of linear transformations is that if $\mathbf{X}_1'$ and $\mathbf{X}_2'$ are two p-dimensional row vectors and $\mathbf{V}$ is a $p \times p$ transformation matrix, it follows from the distributive law of matrix multiplication (Eq. 2.10b) that

$$(\mathbf{X}_1' + \mathbf{X}_2')\mathbf{V} = \mathbf{X}_1'\mathbf{V} + \mathbf{X}_2'\mathbf{V}$$

(i.e., the linear transform of the sum of two vectors is equal to the sum of their respective linear transforms).

Returning to our main present concern, we inquire (in the terminology just introduced) what condition the transformation matrix $\mathbf{V}$ must satisfy in order that the linear transformation (5.10) represent a rigid rotation. Since the elements v_{ij} must satisfy Eqs. 5.9, it follows that the matrix $\mathbf{V}$ must satisfy the condition

$$\mathbf{V}'\mathbf{V} = \mathbf{I} \tag{5.11'}$$

as the reader may easily verify for the simple cases when $p = 2$ or 3. (More generally, observe that those members of the set of Eqs. 5.9 for which $j = k$ assert that the diagonal elements of $\mathbf{V}'\mathbf{V}$ are all equal to unity, and those members with $j \neq k$ assert that every off-diagonal element of $\mathbf{V}'\mathbf{V}$ is equal to zero.)

In the foregoing discussions, we have acted as though the stated conditions on the coefficients of linear combinations—presented in various forms, Eqs. 5.2, 5.3, 5.9, and finally 5.11′—were both necessary and sufficient for ensuring that transformation Eq. 5.10 represent a rigid rotation. It is now time to introduce a qualification that we held in abeyance to simplify our discussion. The fact is that condition (5.11′) is necessary, but not quite sufficient by itself to restrict (5.10) to rigid rotations in the true sense of the term.

To see why this is so, we again examine the two-dimensional case. The linear transformation

$$[Y_1, Y_2] = [X_1, X_2] \begin{bmatrix} \cos\theta & \sin\theta \\ \sin\theta & -\cos\theta \end{bmatrix}$$

is an instance of (5.10) with

$$\mathbf{V} = \begin{bmatrix} \cos\theta & \sin\theta \\ \sin\theta & -\cos\theta \end{bmatrix}$$

which satisfies condition (5.11′):

$$\mathbf{V}'\mathbf{V} = \begin{bmatrix} \cos\theta & \sin\theta \\ \sin\theta & -\cos\theta \end{bmatrix} \begin{bmatrix} \cos\theta & \sin\theta \\ \sin\theta & -\cos\theta \end{bmatrix} = \begin{bmatrix} 1 & 0 \\ 0 & 1 \end{bmatrix}$$

However, a comparison with Eq. 5.1, which may be written in the form

$$[Y_1, Y_2] = [X_1, X_2] \begin{bmatrix} \cos\theta & -\sin\theta \\ \sin\theta & \cos\theta \end{bmatrix}$$

reveals that our present transformation matrix **V** differs from that of Eq. 5.1 in the signs of the second-column elements, and only in this respect. That is, the first column of the two matrices are identical, but the second column of **V** is -1 times the second column of the transformation matrix of 5.1—which unquestionably represents a rigid rotation. Yet **V** satisfies condition (5.11′), as we just saw.

What, then, is the nature of the transformation effected by our present **V**, and what further condition besides (5.11′) must we impose on a transformation matrix before (5.10) definitely represents a rigid rotation and nothing else? To distinguish between the linear combinations displayed in Eq. 5.1 and those generated by **V**, let us denote the latter by Y_1' and Y_2'. Thus

$$Y_1' = (\cos\theta)X_1 + (\sin\theta)X_2$$
$$Y_2' = (\sin\theta)X_1 + (-\cos\theta)X_2$$

whereas

$$Y_1 = (\cos\theta)X_1 + (\sin\theta)X_2$$
$$Y_2 = (-\sin\theta)X_1 + (\cos\theta)X_2$$

It is then evident that $Y_1' = Y_1$ and $Y_2' = -Y_2$. Consequently, if Eq. 5.1 represents a rotation by an angle θ, then our present transformation by **V**, leading to Y_1' and Y_2', must represent the same rotation followed by a *reversal* in the positive sense of the Y_2-axis. Such a reversal in direction is called a *reflection*. Thus the transformation (5.1) represents a "pure" rigid rotation, while the transformation by $\mathbf{V}$ represents a rigid rotation combined with a reflection of the Y_2-axis. Figure 5.5 depicts these two transformations.

What property of the transformation matrix makes the difference between pure rotation and rotation plus reflection? From Eq. 5.11′ and the fact that the determinant of the product of two square matrices is equal to the product of their respective determinants (see Eq. A.6, Appendix A), it follows that

$$|\mathbf{V}'\mathbf{V}| = |\mathbf{V}'||\mathbf{V}| = |\mathbf{V}|^2 = |\mathbf{I}| = 1$$

Hence

$$|\mathbf{V}| = \pm 1$$

That is, any matrix satisfying Eq. 5.11′ has the property that its determinant is equal to either 1 or -1. On evaluating the determinant of the transformation matrix of (5.1), we find that

$$\begin{vmatrix} \cos\theta & -\sin\theta \\ \sin\theta & \cos\theta \end{vmatrix} = 1$$

Hence our present $|\mathbf{V}|$, whose second column is the negative of that of the above, must be equal to -1, as the reader may verify directly.

The generality of the foregoing argument for any 2×2 matrix satisfying (5.11′) is obvious. Any transformation matrix representing pure rigid rotation must be of the form of Eq. 5.1,

$$\begin{bmatrix} \cos\theta & -\sin\theta \\ \sin\theta & \cos\theta \end{bmatrix}$$

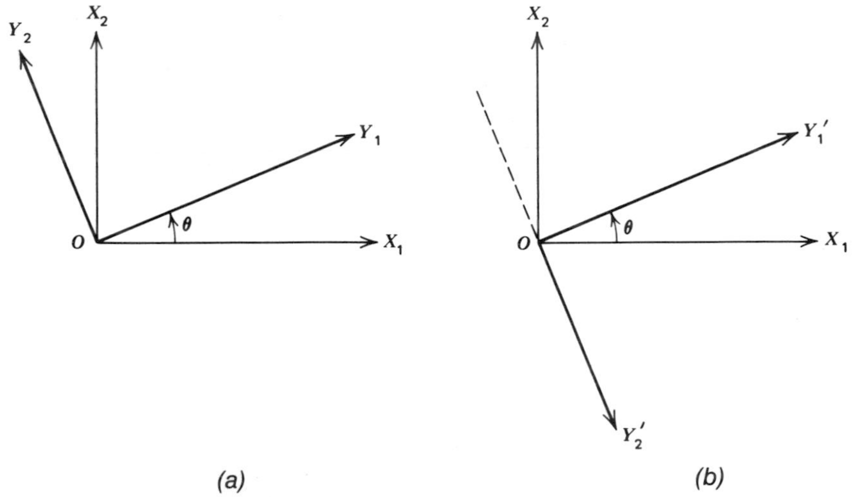

Figure 5.5. *(a) Rigid rotation through an angle θ, and (b) rigid rotation and reflection of second axis.*

for some angle θ. Then any 2×2 matrix

$$\begin{bmatrix} a & a' \\ b & b' \end{bmatrix}$$

whose elements satisfy Eqs. 5.2 and 5.3, that is,

$$a^2 + b^2 = a'^2 + b'^2 = 1$$
$$aa' + bb' = 0$$

(which are together equivalent to Eq. 5.11′), must either be of the above form (as implied in the oversimplified argument on p. 114), *or* differ from it only by a change of signs in one or both columns. The latter cases are possible because replacing a by $-a$ and b by $-b$, exclusively or in addition to replacing a' by $-a'$ and b' by $-b'$, will not affect the conditions (5.2) and (5.3). However, if both columns differed in signs from those of the transformation matrix in (5.1), the new matrix would be equal to

$$\begin{bmatrix} \cos(\theta + 180°) & -\sin(\theta + 180°) \\ \sin(\theta + 180°) & \cos(\theta + 180°) \end{bmatrix}$$

which represents a pure rotation by an angle $\theta + 180°$, and its determinant would equal 1. Thus, only when the transformation matrix involves sign changes in just one of the columns will it represent something other than a pure rotation: namely, rotation plus reflection of one axis, and in this case the determinant is equal to -1. We thus conclude that among 2×2 transformation matrices satisfying (5.11′), those whose determinants are equal to 1 represent pure rigid rotation, while those with determinants equal to -1 represent rigid rotation *and* reflection of one axis.

The foregoing arguments may be extended to three-dimensional space. The transformation matrix

$$\mathbf{V} = \begin{bmatrix} v_{11} & v_{12} & v_{13} \\ v_{21} & v_{22} & v_{23} \\ v_{31} & v_{32} & v_{33} \end{bmatrix}$$

of a rigid rotation may be shown to satisfy $|\mathbf{V}| = 1$ in addition to Eq. 5.11′, while a matrix with sign changes in one or *all three* columns (i.e., an odd number of columns) from the above will have a determinant equal to -1. Such a transformation matrix represents rotation plus a reflection of one axis or of all three axes. Readers should convince themselves—by using the thumb, forefinger, and middle finger spread out at right angles to one another in both hands—that no rotation alone can lead from one coordinate system (the one represented by the left hand, say) to another in which one or all three axes (fingers) are pointed in opposite directions from the first. But if two of the columns of $\mathbf{V}$ above are subjected to sign changes, the resulting matrix will again have a determinant equal to 1. Correspondingly, a pure rotation will suffice to go from one coordinate system to another in which two of the axes are respectively pointed in opposite directions from the former.

For transformations in spaces of dimensionality greater than three, concrete geometric argument no longer avails. Nevertheless, we still speak of rotations followed by an even number of axis reflections as "pure rotations," and those followed by an odd number of axis reflections as "rotation plus reflection." The transformation matrices for pure rotations have determinants equal to 1, and those for rotation plus reflection have determinants equal to -1. We thus arrive at the necessary and sufficient condition for a linear transformation of the form of Eq. 5.10 to represent a rigid rotation by adding the condition $|\mathbf{V}| = 1$ to Eq. 5.11′; that is,

$$\mathbf{V}'\mathbf{V} = \mathbf{I} \quad \text{and} \quad |\mathbf{V}| = 1 \tag{5.11}$$

As a further point of terminology, a square matrix that satisfies conditions (5.11) is called an *orthogonal matrix* (or, more specifically, an *orthonormal* matrix). By contrast, a square matrix that satisfies (5.11′) but not (5.11)—that is, one whose determinant is equal to -1—is sometimes referred to as an *improper orthogonal matrix*. We mention in passing that both "proper" and improper orthogonal matrices have the convenient property that their inverse is simply their transpose:

$$\mathbf{V}' = \mathbf{V}^{-1} \tag{5.12}$$

which follows from postmultiplying both sides of Eq. 5.11′ by $\mathbf{V}^{-1}$.

Using the terminology just introduced, we may summarize the entire development of this section as follows: A linear transformation $\mathbf{Y}' = \mathbf{X}'\mathbf{V}$ represents a rigid rotation of a Cartesian coordinate system if and only if $\mathbf{V}$ is an orthogonal matrix; if $\mathbf{V}$ is an improper orthogonal matrix, the transformation represents a rigid rotation followed by reflections of an odd number of coordinate axes.

EXERCISE

1. Prove that two rigid rotations performed in succession may be replaced by a single rigid rotation. That is, if $\mathbf{Y}' = \mathbf{X}'\mathbf{V}$ and $\mathbf{Z}' = \mathbf{Y}'\mathbf{U}$ are linear transformations representing rigid rotations, then $\mathbf{Z}'$ may be obtained directly from $\mathbf{X}'$ by a rigid rotation.

5.2 THE EFFECT OF A LINEAR TRANSFORMATION ON AN SSCP MATRIX

Having established the relationship between a rotation of axes and a linear transformation, we turn now to the question: How does an SSCP matrix transform when the variables are subjected to a linear transformation? To state the question more operationally, we want to investigate the following problem: Given the SSCP matrix $\mathbf{S}(X)$ for a set of variables $X_1, X_2, \ldots, X_p$, how do we find the SSCP matrix $\mathbf{S}(Y)$ for a set of variables $Y_1, Y_2, \ldots, Y_q$ $(q \leq p)$, each of which is defined as a linear combination of the X's? Note that the number of new (transformed) variables may be equal to or less than the number of original variables. In fact, in some applications we shall be concerned with just one linear combination of the X's, in which case $\mathbf{S}(Y)$ will of course reduce to a single sum of squares.

The first step in determining $\mathbf{S}(Y)$ is to express the means of the Y's in terms of the means of the X's and the transformation matrix. Consider just one linear combination Y_1 of the X's:

$$Y_1 = v_{11}X_1 + v_{21}X_2 + \cdots + v_{p1}X_p$$

If we denote the score of the αth individual on X_i $(i = 1, 2, \ldots, p)$ by $X_{\alpha i}$, then his/her Y_1 score is

$$Y_{\alpha 1} = v_{11}X_{\alpha 1} + v_{21}X_{\alpha 2} + \cdots + v_{p1}X_{\alpha p}$$

For a sample of N individuals, the Y_1 mean is

$$\begin{aligned}\overline{Y}_1 &= \frac{1}{N}\sum_{\alpha=1}^{N} Y_{\alpha 1} = \frac{1}{N}\sum_{\alpha=1}^{N} (v_{11}X_{\alpha 1} + v_{21}X_{\alpha 2} + \cdots + v_{p1}X_{\alpha p}) \\ &= v_{11}\left(\frac{1}{N}\sum_{\alpha=1}^{N} X_{\alpha 1}\right) + v_{21}\left(\frac{1}{N}\sum_{\alpha=1}^{N} X_{\alpha 2}\right) + \cdots \\ &\qquad + v_{p1}\left(\frac{1}{N}\sum_{\alpha=1}^{N} X_{\alpha p}\right) \\ &= v_{11}\overline{X}_1 + v_{21}\overline{X}_2 + \cdots + v_{p1}\overline{X}_p\end{aligned}$$

where $\overline{X}_i$ denotes the mean of X_i for the sample. In other words, the mean of a linear combination of several variables is given by the same linear combination of the means of the original variables.

If several linear combinations $Y_1, Y_2, \ldots, Y_q$ of the X's are considered simultaneously, then each Y mean is expressible as the appropriate linear combination of the X means, and hence the Y centroid $[\overline{Y}_1, \overline{Y}_2, \ldots, \overline{Y}_q]$ is related to the X centroid as follows:

$$[\overline{Y}_1, \overline{Y}_2, \ldots, \overline{Y}_q] = [\overline{X}_1, \overline{X}_2, \ldots, \overline{X}_p]\begin{bmatrix} v_{11} & v_{12} & \cdots & v_{1q} \\ v_{21} & v_{22} & \cdots & v_{2q} \\ \vdots & \vdots & & \vdots \\ v_{p1} & v_{p2} & \cdots & v_{pq}\end{bmatrix}$$

where the elements of the jth column of the rectangular matrix (v_{ij}) are the coefficients for the jth linear combination Y_j. Introducing the abbreviations $\overline{\mathbf{X}}'$ and $\overline{\mathbf{Y}}'$ for the X and Y centroids and $\mathbf{V}$ for the transformation matrix (v_{ij}), the foregoing equation may be written as

$$\overline{\mathbf{Y}}' = \overline{\mathbf{X}}'\mathbf{V} \tag{5.13}$$

which is identical in form to the linear transformation Eq. 5.10. Of course, the $\mathbf{V}$ here need not be an orthogonal matrix (nor even a square matrix unless $q = p$) if the Y's are simply an arbitrary set of linear combinations of the X's. However, the case when the Y's correspond to a set of coordinate axes resulting from a rigid rotation is included in Eq. 5.13, just as in Eq. 5.10. To represent this special case, we need merely to specify that $\mathbf{V}$ is an orthogonal matrix.

Next, let us apply Eq. 5.10 to each of N row vectors

$$\begin{aligned}
\mathbf{X}'_1 &= [X_{11}, X_{12}, \ldots, X_{1p}] \\
\mathbf{X}'_2 &= [X_{21}, X_{22}, \ldots, X_{2p}] \\
&\vdots \\
\mathbf{X}'_N &= [X_{N1}, X_{N2}, \ldots, X_{Np}]
\end{aligned}$$

representing the scores of N individuals on p variables. The resulting transformed row vectors

$$\begin{aligned}
\mathbf{Y}'_1 &= \mathbf{X}'_1\mathbf{V} \\
\mathbf{Y}'_2 &= \mathbf{X}'_2\mathbf{V} \\
&\vdots \\
\mathbf{Y}'_N &= \mathbf{X}'_N\mathbf{V}
\end{aligned}$$

will then represent the same N individuals' scores on p (or, more generally, $q \le p$) linear combinations $Y_1, Y_2, \ldots, Y_p$ (or Y_q) of the X's.

Now suppose that we collect the original N row vectors into a single score matrix, which we denote by $\mathbf{X}$ as in Chapter 2; that is,

$$\mathbf{X} = \begin{bmatrix} X_{11} & X_{12} & \cdots & X_{1p} \\ X_{21} & X_{22} & \cdots & X_{2p} \\ \vdots & \vdots & & \vdots \\ X_{N1} & X_{N2} & \cdots & X_{Np} \end{bmatrix}$$

It then follows (as the reader should verify) that $\mathbf{XV}$ will be the $N \times p$ (or $N \times q$) transformed score matrix consisting of the row vectors $\mathbf{Y}'_1$,

$\mathbf{Y}'_2, \ldots, \mathbf{Y}'_N$. If we denote this matrix by $\mathbf{Y}$, we have

$$\mathbf{Y} = \mathbf{XV} \tag{5.10a}$$

which is formally identical to Eq. 5.10 except for the lack of primes on $\mathbf{X}$ and $\mathbf{Y}$. This difference may seem confusing at first, but it serves to remind us that in Eq. 5.10 a row vector $\mathbf{X}'$ is being transformed, while in Eq. 5.10a an $N \times p$ matrix $\mathbf{X}$ (each row of which is an instance of the earlier $\mathbf{X}'$) is being transformed.[1]

Similarly, by writing N identical rows

$$\overline{\mathbf{X}}' = [\overline{X}_1, \overline{X}_2, \ldots, \overline{X}_p]$$

in the form of an $N \times p$ matrix, we obtain the mean-score matrix of the X's, which was also introduced in Chapter 2 and denoted by $\overline{\mathbf{X}}$ (without a prime). It then follows that

$$\overline{\mathbf{Y}} = \overline{\mathbf{X}}\mathbf{V} \tag{5.13a}$$

is the corresponding mean-score matrix of the Y's, consisting of N identical rows $\overline{\mathbf{Y}}'$, each having as its elements the values of the means $\overline{Y}_1, \overline{Y}_2, \ldots, \overline{Y}_p$ (or $\overline{Y}_q$).

According to Eq. 2.7, an SSCP matrix may be expressed in terms of the score matrix and the mean-score matrix as

$$\mathbf{S}(X) = \mathbf{X}'\mathbf{X} - \overline{\mathbf{X}}'\overline{\mathbf{X}} \tag{5.14}$$

Correspondingly, for the transformed variables, the SSCP matrix is

$$\mathbf{S}(Y) = \mathbf{Y}'\mathbf{Y} - \overline{\mathbf{Y}}'\overline{\mathbf{Y}} \tag{5.15}$$

We now substitute in this equation the expressions for $\mathbf{Y}$ and $\overline{\mathbf{Y}}$ in terms of $\mathbf{X}$, $\overline{\mathbf{X}}$, and $\mathbf{V}$ from Eqs. 5.10a and 5.13a, and obtain

$$\begin{aligned}\mathbf{S}(Y) &= (\mathbf{XV})'(\mathbf{XV}) - (\overline{\mathbf{X}}\mathbf{V})'(\overline{\mathbf{X}}\mathbf{V}) \\ &= (\mathbf{V}'\mathbf{X}')(\mathbf{XV}) - (\mathbf{V}'\overline{\mathbf{X}}')(\overline{\mathbf{X}}\mathbf{V}) \\ &= \mathbf{V}'(\mathbf{X}'\mathbf{X})\mathbf{V} - \mathbf{V}'(\overline{\mathbf{X}}'\overline{\mathbf{X}})\mathbf{V} \\ &= \mathbf{V}'(\mathbf{X}'\mathbf{X} - \overline{\mathbf{X}}'\overline{\mathbf{X}})\mathbf{V}\end{aligned}$$

where we have used the associative law in going from the second to the third expression on the right, and the distributive law (twice) in going from the third to the last expression. Now the quantity in parentheses in this last expression is, according to Eq. 5.14, none other than $\mathbf{S}(X)$, the SSCP matrix for the original

[1]The reason why the row vectors in Eq. 5.10 were not denoted by primed lowercase letters $\mathbf{x}'$ and $\mathbf{y}'$ is, of course, that these symbols are reserved for vectors of deviation scores. We have thus made an exception to the general rule of denoting vectors by lowercase letters.

variables. We thus have the following important relation between the SSCP matrices for the original and the transformed variables:

$$\mathbf{S}(Y) = \mathbf{V}'\mathbf{S}(X)\mathbf{V} \tag{5.16}$$

An alternative method for deriving Eq. 5.16 that dispenses with using the $N \times p$ mean-score matrices $\overline{\mathbf{X}}$ and $\overline{\mathbf{Y}}$ is to utilize Eq. 2.8 for $\mathbf{S}(X)$ and an equivalent formula for $\mathbf{S}(Y)$. We then have

$$\mathbf{S}(Y) = \mathbf{Y}'\mathbf{Y} - (\mathbf{1}'\mathbf{Y})'(\mathbf{1}'\mathbf{Y})/N$$

in which we have only to substitute Eq. 5.10 and simplify the results. Details are left for the reader to complete.

EXAMPLE 5.3. A small numerical example may clarify the procedures developed above. Suppose that a group of 10 college students was given a personal preferences questionnaire, which yields, among other things, scores on two scales: X_1 = outdoor scale and X_2 = gregariousness scale. The score matrix was

$$\mathbf{X} = \begin{bmatrix} 21 & 16 \\ 22 & 23 \\ 18 & 17 \\ 25 & 22 \\ 10 & 21 \\ 19 & 20 \\ 23 & 18 \\ 23 & 17 \\ 16 & 15 \\ 20 & 17 \end{bmatrix}$$

Hence the mean-score matrix is

$$\overline{\mathbf{X}} = \begin{bmatrix} 19.7 & 18.6 \\ 19.7 & 18.6 \\ \vdots & \vdots \\ 19.7 & 18.6 \end{bmatrix} \quad \text{(10 identical rows)}$$

From these, the SSCP matrix is computed, in accordance with Eq. 5.14, as

$$\begin{aligned} \mathbf{S}(X) &= \mathbf{X}'\mathbf{X} - \overline{\mathbf{X}}'\overline{\mathbf{X}} \\ &= \begin{bmatrix} 4049 & 3673 \\ 3673 & 3526 \end{bmatrix} - \begin{bmatrix} 3880.9 & 3664.2 \\ 3664.2 & 3459.6 \end{bmatrix} \\ &= \begin{bmatrix} 168.1 & 8.8 \\ 8.8 & 66.4 \end{bmatrix} \end{aligned}$$

Subsequently, a researcher defines two derived scores that are of special interest to her. They are:

$$Y_1 = X_1 + X_2 \qquad \text{(indicating overall ``strength'' of preferences)}$$

and

$$Y_2 = X_1 - X_2 \qquad \text{(indicating excess of outdoor over gregarious preferences)}$$

She wishes to compute the SSCP matrix for these linearly transformed variables without having to start from scratch; that is, she wants to utilize the fact that the SSCP matrix $\mathbf{S}(X)$ for the original variables is already known.

The transformation matrix here is

$$\mathbf{V} = \begin{bmatrix} 1 & 1 \\ 1 & -1 \end{bmatrix}$$

Note that this matrix satisfies Eq. 5.3 but not Eq. 5.2. Hence $\mathbf{Y}' = \mathbf{X}'\mathbf{V}$ does not represent a rotation. However, we can construct an orthogonal matrix based on $\mathbf{V}$ by normalizing each of its columns to unity—that is, by dividing the elements of its first column by

$$\pm\sqrt{(1)^2 + (1)^2} = \pm\sqrt{2}$$

and those of the second column by

$$\pm\sqrt{(1)^2 + (-1)^2} = \pm\sqrt{2}$$

The signs of the divisors must be chosen so that the resulting matrix, say $\mathbf{V}_0$, has determinant equal to 1. Using $\sqrt{2}$ as the first-column divisor and $-\sqrt{2}$ as the second, we obtain

$$\mathbf{V}_0 = \begin{bmatrix} 1/\sqrt{2} & -1/\sqrt{2} \\ 1/\sqrt{2} & 1/\sqrt{2} \end{bmatrix}$$

which is orthogonal. (What is the angle θ of the rotation represented by the transformation $Y' = X'V_0$?)

Going back to the main problem, we can compute the SSCP matrix for the Y's in accordance with Eq. 5.16, thus:

$$\begin{aligned}\mathbf{S}(Y) &= \mathbf{V}'\mathbf{S}(X)\mathbf{V} \\ &= \begin{bmatrix} 1 & 1 \\ 1 & -1 \end{bmatrix}\begin{bmatrix} 168.1 & 8.8 \\ 8.8 & 66.4 \end{bmatrix}\begin{bmatrix} 1 & 1 \\ 1 & -1 \end{bmatrix} \\ &= \begin{bmatrix} 176.9 & 75.2 \\ 159.3 & -57.6 \end{bmatrix}\begin{bmatrix} 1 & 1 \\ 1 & -1 \end{bmatrix} \\ &= \begin{bmatrix} 252.1 & 101.7 \\ 101.7 & 216.9 \end{bmatrix}\end{aligned}$$

Readers should verify this result by computing $\mathbf{S}(Y)$ from scratch. That is, they should first compute the score matrix and mean-score matrix for Y_1 and Y_2 by using Eqs. 5.10a and 5.13a, and then obtain $\mathbf{S}(Y)$ in accordance with Eq. 5.15.

5.3 VARIANCE-MAXIMIZING ROTATIONS

We have seen how axis rotation is expressed matrix algebraically, and how an SSCP matrix transforms under axis rotation, or under linear transformations more generally. We are now in a position to investigate the problem of how to determine a rotated axis or system of axes possessing certain desirable properties in terms of the transformed SSCP matrix.

One such desirable property is rooted in a general tenet of all scientific endeavor—the canon of parsimony—which seeks to explain as much as possible of observed systematic variation in terms of as few variables as possible. Carried to the logical extreme and translated into multivariate statistical language, this tenet leads one to pose the following task: Given a set of p-variate observations, determine a rotated axis such that the variable thereby defined as a linear combination of the original p variables has maximum variance.

Intuitive Considerations for Multivariate Normal Distributions

The task posed above can obviously be set and carried out irrespective of the nature of the multivariate distribution followed by the observations. However, if the distribution happens to be multivariate normal, there is a simple geometric interpretation of the axis being sought. We describe this in detail with reference to the bivariate normal distribution. In doing so we are, of course, "shifting gears" from considering a sample to considering a population. Hence, we can no longer talk about an SSCP matrix (which is not defined for a population).

Instead, we shall talk about the population covariance matrix and how *it* transforms under axis rotation. The sample SSCP matrix and the population covariance matrix stand in the relation $E[\mathbf{S}/(N - 1)] = \mathbf{\Sigma}$.

The bivariate normal density function, Eq. 4.1, may be written in deviation-score form as

$$\phi(x_1, x_2) = \frac{1}{2\pi\sigma_1\sigma_2\sqrt{1 - \rho^2}} \exp\left[\frac{-1}{2(1 - \rho^2)}\left(\frac{x_1^2}{\sigma_1^2} + \frac{x_2^2}{\sigma_2^2} - 2\rho\frac{x_1 x_2}{\sigma_1\sigma_1}\right)\right] \quad (5.17)$$

where $x_1 = X_1 - \mu_1$ and $x_2 = X_2 - \mu_2$. As described in Chapter 4 the density surface is shaped like a somewhat distorted bell such that the cross sections by planes parallel to the (x_1, x_2)-plane are all ellipses, several of which are shown in Fig. 5.6. Since the density function is in deviation-score form, the center of these isodensity ellipses coincides with the origin of the coordinate system.

Now imagine the cross sections of the density surface made by various planes containing the density axis, that is, planes perpendicular to the (x_1, x_2)-plane and passing through the origin O. It can be shown by analytic geometry that these cross sections all have the shape of univariate normal density curves. Considering that the intersection of each such plane with the (x_1, x_2)-plane is a straight line indicating a possible position of a rotated axis, it should be intuitively plausible that the cross-sectional curve represents the univariate normal density function of the new variable corresponding to the rotated axis. Actually, the height of the curve has to be proportionally adjusted so that the area under it becomes unity before the curve is exactly that of the density function. However, this adjustment does not affect the "width" of the curve, which is a measure of the variance of the transformed variable.

Consequently, the relative magnitudes of the variances of transformed variables defined by different rotated axes may be assessed by comparing the lengths of the line segments (e.g., $A''B''$, $A'B'$, and AB in Fig. 5.6) that are cut off from the several axes by a given isodensity ellipse of the original bivariate normal surface. For example, if the largest ellipse in Fig. 5.6 represents the 90% ellipse of the bivariate distribution, then segment $A''B''$ is the symmetric 90% interval

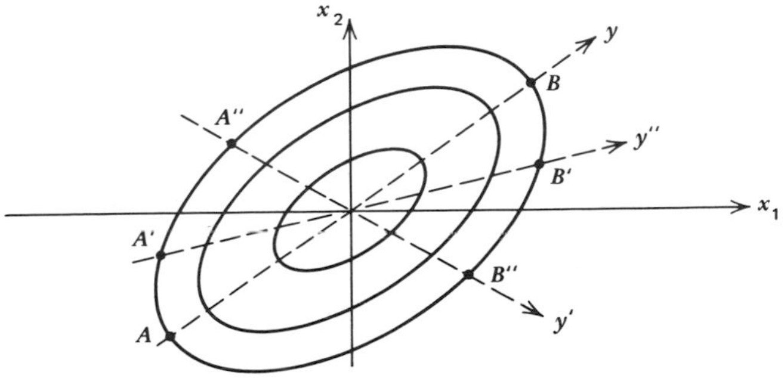

Figure 5.6. *Isodensity ellipses of a bivariate normal distribution.*

of the normal curve which represents the distribution of the transformed variable defined by axis Oy''. Similar remarks hold for segments $A'B'$, AB, and any other diameter of the ellipse in question. Since all the transformed variables have normal distributions, the longer the 90% interval, the larger the variance.

On the basis of the foregoing intuitive reasoning, it is clear that the axis defining the variable with maximum variance must coincide with the major axis of any isodensity ellipse. (In Fig. 5.6, Oy represents this axis.) This, then, is the special geometric interpretation that obtains for the variance-maximizing rotated axis when the original variables follow a bivariate normal distribution. The argument can be extended to higher dimensions, and we have the following theorem.

Theorem 1. Given p variables following a p-variate normal distribution, the axis defining the linearly transformed variable with maximum variance is the major axis of the isodensity hyperellipsoid

$$\mathbf{x}'\mathbf{\Sigma}^{-1}\mathbf{x} = C$$

where $\mathbf{x}' = [X_1 - \mu_1, X_2 - \mu_2, \ldots, X_p - \mu_p]$, $\mathbf{\Sigma}$ is the variance–covariance matrix, and C is an arbitrary positive constant.

Before developing a general method for determining the variance-maximizing axis for any set of multivariate data, let us utilize the formula for locating the major axis of an ellipse, cited in Chapter 4, to determine the desired axis in the case of a bivariate normal distribution. We use Example 5.3 to illustrate the procedure. (In so doing we are temporarily pretending that the variance–covariance matrix obtainable by dividing the SSCP matrix of that example by 10 is that of a bivariate normal distribution; we shall subsequently see that such a pretense can be dispensed with.)

For the stated numerical example, the covariance matrix is

$$\mathbf{\Sigma} = \begin{bmatrix} 16.81 & .88 \\ .88 & 6.64 \end{bmatrix}$$

Any isodensity ellipse of the bivariate normal distribution with this variance–covariance matrix, therefore, has the equation

$$(1/16.81)x_1^2 + (1/6.64)x_2^2 - [(2)(.88)/(16.81)(6.64)]x_1x_2 = C$$

where C is some positive constant. (Note that the coefficient of x_1x_2 here is equivalent to the coefficient $-2\rho/\sigma_1\sigma_2$ given in Eq. 5.17 because $\rho\sigma_1\sigma_2/\sigma_1^2\sigma_2^2 = \rho/\sigma_1\sigma_2$.)

Therefore, the angle θ from the x_1-axis to the major axis (since $\rho > 0$) of the ellipse is the smallest positive angle satisfying the equation

$$\tan 2\theta = \frac{2\rho\sigma_1\sigma_2}{\sigma_1^2 - \sigma_2^2} = (2)(.88)/(16.81 - 6.64) = .17306$$

Consulting a table of trigonometric functions, we find $2\theta = 9°49'$, and hence $\theta = 4°54.5'$. Evidently, the major axis of the isodensity ellipse slopes only very slightly away from the x_1-axis—a fact that could have been anticipated by noting that the correlation coefficient is very small,

$$\rho = .88/\sqrt{(16.81)(6.64)} = .0833$$

To verify that the axis just located is indeed the variance-maximizing axis, let us compute the variance of the transformed variable thereby defined and compare it with the variances of several other transformed variables corresponding to other rotated axes. We use a special case of Eq. 5.16 in which the transformation matrix **V** is a two-dimensional column vector with elements cos θ and sin θ. Again consulting a table of trigonometric functions, we find

$$\cos 4°54.5' = .996332$$

and

$$\sin 4°54.5' = .085572$$

Therefore, the variance of the new variable y defined by the rotation

$$y = .996332x_1 + .085572x_2$$

is given by

$$\sigma_y^2 = [.996332 \quad .085572] \begin{bmatrix} 16.81 & .88 \\ .88 & 6.64 \end{bmatrix} \begin{bmatrix} .996332 \\ .085572 \end{bmatrix}$$

$$= 16.8856$$

Let us take two other variables y' and y'' defined by axes slightly off from and on opposite sides of Oy, by choosing $\theta = 4°$ and $\theta = 6°$ as the angles of rotation. Carrying out the same operations as we just did, but using the transformation matrices

$$\begin{bmatrix} \cos 4° \\ \sin 4° \end{bmatrix} \quad \text{and} \quad \begin{bmatrix} \cos 6° \\ \sin 6° \end{bmatrix}$$

we find

$$\sigma_{y'}^2 = [.99756 \quad .06976] \begin{bmatrix} 16.81 & .88 \\ .88 & 6.64 \end{bmatrix} \begin{bmatrix} .99756 \\ .06976 \end{bmatrix}$$

$$= 16.8830$$

and

$$\sigma_{y''}^2 = [.99452 \quad .10453] \begin{bmatrix} 16.81 & .88 \\ .88 & 6.64 \end{bmatrix} \begin{bmatrix} .99452 \\ .10453 \end{bmatrix}$$

$$= 16.8818$$

It is thus seen that σ_y^2 is greater than either $\sigma_{y'}^2$ or $\sigma_{y''}^2$. It will be instructive for the reader to compute the variances for several other rotated axes, with angles of rotation 45°, 60°, 90°, 120°, 135°, and so on. It should be found that the variance steadily decreases until the angle of rotation is about 90°, and then increases again. The reader may be able to guess just what angle of rotation would yield the minimum variance. More will be said about the significance of the "variance-minimizing" axis when we discuss the variance-maximizing problem more generally.

EXERCISES

The following exercises refer to a bivariate normal distribution with covariance matrix

$$\mathbf{\Sigma} = \begin{bmatrix} 23 & -12 \\ -12 & 16 \end{bmatrix}$$

1. Plot the graph of an arbitrary isodensity ellipse of the bivariate normal distribution specified above, taking the origin at the (unspecified) centroid.
2. Using the tan 2θ equation, locate the major axis of the ellipse you drew above.
3. Compute the variance of the transformed variable represented by the major axis located above. Compare this value with the variances of variables represented by rotated axes 10° away on the two sides of the major axis.
4. Knowing that the minor axis (i.e., shortest diameter) of an ellipse is perpendicular to the major axis, determine (from your result in Exercise 2) the angle of rotation from Ox_1 to the minor axis.
5. Repeat Exercise 3 for the transformed variable represented by the minor axis, including comparisons with variances along two nearby axes.

General Formulation of Variance-Maximizing Rotation Problems

We shall now develop a more general approach to the problem of determining the rotated axis with maximum variance—one that does not depend on the identifying of this axis with the major axis of an isodensity ellipsoid. Conceptually, the approach is quite straightforward, even though its execution requires rather involved techniques. We have, in Eq. 5.16, a means for computing the sum of squares of a linearly transformed variable defined by an arbitrary $p \times 1$ transformation matrix $\mathbf{V}$. Since the latter is a column vector, we will henceforth denote it by $\mathbf{v}$, reserving the uppercase $\mathbf{V}$ for $p \times q$ transformation matrices with $q > 1$. The formula for computing the sum of squares $\Sigma\, y^2$ of a transformed variable y therefore becomes

$$\Sigma\, y^2 = \mathbf{v}'\mathbf{S}(X)\mathbf{v} \tag{5.18}$$

and since we are concerned only with those transformed variables that are defined by axis rotation, we impose the restriction

$$\mathbf{v}'\mathbf{v} = 1 \tag{5.19}$$

which is a special case of Eq. 5.11′, on the vector of coefficients $\mathbf{v}$.

Our problem is thus reduced to the following: Given an SSCP matrix $\mathbf{S}(X)$, determine a vector $\mathbf{v}$ such that maximizes $\mathbf{v}'\mathbf{S}(X)\mathbf{v}$, subject to the condition $\mathbf{v}'\mathbf{v} = 1$. A new interpretation of the restriction Eq. 5.19 emerges in this mathematical formulation of the problem. Although we have hitherto associated this restriction with axis rotation, it is now seen that the maximizing problem itself is meaningless without specifying the norm $(\mathbf{v}'\mathbf{v})^{1/2}$ of the transformation vector $\mathbf{v}$. For the quantity $\mathbf{v}'\mathbf{S}(X)\mathbf{v}$ can be made as large as we please by taking numbers with large absolute values as the elements of $\mathbf{v}$, if the norm of $\mathbf{v}$ were left arbitrary. For example, the two linear combinations $y = .6x_1 + .8x_2$ and $y' = 1.2x_1 + 1.6x_2$ have the same *relative* weights for x_1 and x_2, and are in this sense equivalent. But the variance of y' would be four times as large as that of y for any set of x_1 and x_2 values because

$$y' = 2(.6x_1 + .8x_2) = 2y$$

Similarly, a third linear combination

$$y'' = 1.8x_1 + 2.4x_2 \ (= 3y)$$

would have a variance nine times that of y. Thus the maximizing problem makes sense only if we exclude from consideration various multiples of a vector of coefficients. This is most readily done by specifying that the only vectors to be considered will be those with a fixed length or norm. The fixed length is customarily taken as 1, because then the linear transformation will correspond to a "pure" rotation. Other lengths for the transformation vector would correspond to rotation and *change of scale*.

To return to the problem at hand, our task is to find a set of weights $\mathbf{v}' = [v_1, v_2, \ldots, v_p]$ for constructing a linear combination

$$y = v_1x_1 + v_2x_2 + \cdots + v_px_p$$

such that the quantity

$$\Sigma\, y^2 = \mathbf{v}'\mathbf{S}(X)\mathbf{v}$$

will be as large as possible, under the restriction that

$$\mathbf{v}'\mathbf{v} = \sum_{i=1}^{p} v_i^2 = 1$$

A convenient method for solving such a maximization problem with side conditions is that of *Lagrange multipliers,* discussed in most calculus textbooks.

In general terms, the method states that if a function $f(v_1, v_2, \ldots, v_p)$ of several variables is to be maximized (or minimized) under the side condition $g(v_1, \ldots, v_p) = 0$, this can be accomplished by constructing a new function

$$F = f(v_1, v_2, \ldots, v_p) - \lambda g(v_1, v_2, \ldots, v_p)$$

(where λ is a new unknown, called the Lagrange multiplier), maximizing (or minimizing) this new function without any restriction on the variables, and then later imposing the condition $g(v_1, v_2, \ldots, v_p) = 0$. Applied to our problem, the new function to be maximized without restriction on the elements of $\mathbf{v}$ is

$$F = \mathbf{v}'\mathbf{S}(X)\mathbf{v} - \lambda(\mathbf{v}'\mathbf{v} - 1) \tag{5.20}$$

A necessary condition to be satisfied by $\mathbf{v}$ in order that F be maximized can be obtained by finding the symbolic partial derivative of F with respect to $\mathbf{v}$ and setting this equal to the null vector $\mathbf{0}$. As explained in Appendix C and already exemplified in Chapter 3 (see pp. 43–44), symbolic partial differentiation by a vector is nothing more than a compact way of summarizing the results of partial differentiation by each element of the vector in turn. Thus, applying Eq. C.1s of Appendix C, we find that the symbolic partial derivative of F with respect to $\mathbf{v}$ is given by

$$\frac{\partial F}{\partial \mathbf{v}} = 2\mathbf{S}(X)\mathbf{v} - 2\lambda\mathbf{v}$$

which is a p-dimensional column vector. Setting this expression equal to the p-dimensional null vector $\mathbf{0}$ (a column consisting of p zeros), and then dividing through by 2 and factoring out $\mathbf{v}$, we obtain

$$(\mathbf{S}(X) - \lambda\mathbf{I})\mathbf{v} = \mathbf{0} \tag{5.21}$$

as the necessary condition to be satisfied by $\mathbf{v}$ in order that $\mathbf{v}'\mathbf{S}(X)\mathbf{v}$ be maximized, subject to the condition $\mathbf{v}'\mathbf{v} = 1$.

It will presently be seen that there are, in general, several vectors $\mathbf{v}$ that satisfy Eq. 5.21, but that a readily identifiable one of them will always be the required maximizing vector. It also happens that equations of the general form of Eq. 5.21 frequently turn up in connection with the several multivariate statistical techniques to be discussed in the sequel. The rationale for solving an equation of this type is, therefore, well worth our careful study—even though recourse to a computer solution using, for example, the EIGEN command in the matrix routine, PROC MATRIX, of the SAS package becomes a virtual necessity when the order of the matrix $\mathbf{S}(X)$ is larger than 3 or 4. However, for slightly larger matrices (say of order up to 6 or 7), an iterative method described in Appendix D is practicable when a computer is not available.

5.4 SOLUTION OF EQUATION FOR VARIANCE-MAXIMIZING ROTATIONS: EIGENVALUE PROBLEMS

Since equations of the form of (5.21) arise in other connections besides variance-maximizing rotations, we shall replace the $\mathbf{S}(X)$ there by an arbitrary square matrix $\mathbf{A}$, and discuss the solution of the matrix equation

$$(\mathbf{A} - \lambda\mathbf{I})\mathbf{v} = \mathbf{0} \tag{5.22}$$

or, to indicate the elements of the matrix and vectors,

$$\begin{bmatrix} a_{11} - \lambda & a_{12} & \cdots & a_{1p} \\ a_{21} & a_{22} - \lambda & \cdots & a_{2p} \\ \vdots & \vdots & \ddots & \vdots \\ a_{p1} & a_{p2} & \cdots & a_{pp} - \lambda \end{bmatrix} \begin{bmatrix} v_1 \\ v_2 \\ \vdots \\ v_p \end{bmatrix} = \begin{bmatrix} 0 \\ 0 \\ \vdots \\ 0 \end{bmatrix}$$

In most, but not all, applications $\mathbf{A}$ will be a symmetrical matrix and will have a further property known as positive semidefiniteness, discussed later.

Equation 5.22 differs in two notable respects from the set of normal equations encountered in multiple regression analysis and given, in extended and symbolic forms, respectively, as Eqs. 3.13 and 3.14. First, the vector of constants on the right-hand side is a null vector. Such a set of equations is called a set of *homogeneous equations*. Second, the matrix of coefficients on the left-hand side, $\mathbf{A} - \lambda\mathbf{I}$, itself involves an unknown scalar quantity λ.

An obvious feature of homogeneous equations is that setting all the unknowns equal to zero will always satisfy the equations, because $(\mathbf{A} - \lambda\mathbf{I})\mathbf{0} = \mathbf{0}$, regardless of what $\mathbf{A}$ and λ are. Therefore, $\mathbf{v} = \mathbf{0}$ is called a *trivial solution* of any set of homogeneous equations. The important question is: When does a set of homogeneous equations possess a nontrivial solution, that is, a solution other than the trivial one?

Suppose, for a moment, that an appropriate value for λ had somehow been determined, so that the coefficient matrix $\mathbf{A} - \lambda\mathbf{I}$ is completely known. Now, if $\mathbf{A} - \lambda\mathbf{I}$ is nonsingular, and hence possesses an inverse, then premultiplying both members of Eq. 5.22 by $(\mathbf{A} - \lambda\mathbf{I})^{-1}$ will yield

$$\mathbf{v} = (\mathbf{A} - \lambda\mathbf{I})^{-1}\mathbf{0} = \mathbf{0}$$

that is, the trivial solution, as the *only* solution of the equation. We therefore conclude that in order for a set of homogeneous equations to possess a nontrivial solution, the matrix $\mathbf{A} - \lambda\mathbf{I}$ must *not* have an inverse. In other words, the value of λ must be such that $\mathbf{A} - \lambda\mathbf{I}$ becomes a singular matrix. That is, λ must satisfy the equation

$$|\mathbf{A} - \lambda\mathbf{I}| = 0 \tag{5.23}$$

which is called the *characteristic equation* of the matrix $\mathbf{A}$.

Thus, the first step in solving a matrix equation of the form (5.22) is to determine the roots of the characteristic equation, which are the only possible values of λ for yielding nontrivial solutions for $\mathbf{v}$. These λ values are called the *characteristic roots* or *eigenvalues* of $\mathbf{A}$. The characteristic equation is a polynomial equation in λ, of degree equal to the order of $\mathbf{A}$. Hence, if $\mathbf{A}$ is a $p \times p$ matrix, its characteristic equation will have p roots, although it is possible that not all of them are distinct, and that some of them are zero. (If $\mathbf{A}$ is symmetric, all its eigenvalues are real numbers.)

EXAMPLE 5.4 Find the eigenvalues of the covariance matrix considered in Section 5.3 in connection with the variance-maximizing rotation problem. That matrix was

$$\mathbf{\Sigma} = \begin{bmatrix} 16.81 & .88 \\ .88 & 6.64 \end{bmatrix}$$

Its characteristic equation, $|\mathbf{\Sigma} - \lambda\mathbf{I}| = 0$, is

$$\begin{vmatrix} 16.81 - \lambda & .88 \\ .88 & 6.64 - \lambda \end{vmatrix} = 0$$

which, on expanding the determinant, becomes

$$\lambda^2 - 23.45\lambda + 110.844 = 0$$

Using the formula for the roots of a quadratic, we find

$$\lambda_1 = 16.8856 \qquad \text{and} \qquad \lambda_2 = 6.5644$$

as the eigenvalues of $\mathbf{\Sigma}$.

EXAMPLE 5.5. Find the eigenvalues of the matrix

$$\mathbf{A} = \begin{bmatrix} 136 & 104 & 94 \\ 104 & 106 & 71 \\ 94 & 71 & 65 \end{bmatrix}$$

The characteristic equation of $\mathbf{A}$ is

$$\begin{vmatrix} 136 - \lambda & 104 & 94 \\ 104 & 106 - \lambda & 71 \\ 94 & 71 & 65 - \lambda \end{vmatrix} = 0$$

Expanding the determinant either by Sarrus' rule or by cofactors along any row or column, this equation reduces to

$$-\lambda^3 + 307\lambda^2 - 5453\lambda = 0$$

a cubic with no constant term on the left-hand side. Thus one of its roots is 0, and the other two are those of the quadratic,

$$\lambda^2 - 307\lambda - 5453 = 0$$

which are 288.07 and 18.93. The three eigenvalues of **A**, in descending order of magnitude, are $\lambda_1 = 288.07$, $\lambda_2 = 18.93$, $\lambda_3 = 0$.

EXERCISE

1. Find the eigenvalues of the matrix

$$\mathbf{B} = \begin{bmatrix} 7 & 0 & 1 \\ 0 & 7 & 2 \\ 1 & 2 & 3 \end{bmatrix}$$

Continuing with the solution of Eq. 5.22,

$$(\mathbf{A} - \lambda\mathbf{I})\mathbf{v} = \mathbf{0}$$

we note that there will be one vector solution $\mathbf{v}_i$ corresponding to each eigenvalue λ_i of **A**. The vector $\mathbf{v}_i$ is called the *characteristic vector* or *eigenvector* of **A** associated with the characteristic root or eigenvalue λ_i. Each eigenvector is determined as follows:

Step 1. For the given eigenvalue λ_i, write out the matrix $\mathbf{A} - \lambda_i\mathbf{I}$ by subtracting λ_i from each diagonal element of **A**.

Step 2. Compute the adjoint $\mathbf{adj}(\mathbf{A} - \lambda_i\mathbf{I})$ of the matrix $(\mathbf{A} - \lambda_i\mathbf{I})$ in accordance with the steps indicated in Chapter 2. It should be found that the columns of $\mathbf{adj}(\mathbf{A} - \lambda_i\mathbf{I})$ are all proportional to one another.

Step 3. Divide the elements of any column of $\mathbf{adj}(\mathbf{A} - \lambda_i\mathbf{I})$ by the square root of the sum of the squares of these elements. The resulting numbers are the elements of $\mathbf{v}_i$, with $\mathbf{v}_i'\mathbf{v}_i = 1$.

To see the rationale of the procedure described above, let us introduce the abbreviation

$$\mathbf{A} - \lambda_i\mathbf{I} = \mathbf{H}$$

and denote the cofactor of the (j, k)-element of $\mathbf{H}$ by H_{jk}. Considering, for simplicity, the case when $\mathbf{A}$ is of order 3, the adjoint $\mathbf{adj}(\mathbf{A} - \lambda_i\mathbf{I})$ of $\mathbf{A} - \lambda_i\mathbf{I}$ is then given by

$$\mathbf{adj}(\mathbf{H}) = \begin{bmatrix} H_{11} & H_{21} & H_{31} \\ H_{12} & H_{22} & H_{32} \\ H_{13} & H_{23} & H_{33} \end{bmatrix}$$

In accordance with Step 3, we take

$$\mathbf{v}_i' = c[H_{11}, H_{12}, H_{13}]$$

where

$$c = 1/\sqrt{H_{11}^2 + H_{12}^2 + H_{13}^2}$$

is a scalar multiplier to make $\mathbf{v}_i'\mathbf{v}_i = 1$.

To show that this $\mathbf{v}_i$ satisfies Eq. 5.22, we carry out the multiplication,

$$(\mathbf{A} - \lambda_i\mathbf{I})\mathbf{v}_i = \begin{bmatrix} h_{11} & h_{12} & h_{13} \\ h_{21} & h_{22} & h_{23} \\ h_{31} & h_{32} & h_{33} \end{bmatrix} \begin{bmatrix} H_{11} \\ H_{12} \\ H_{13} \end{bmatrix} c$$

$$= c \begin{bmatrix} h_{11}H_{11} + h_{12}H_{12} + h_{13}H_{13} \\ h_{21}H_{11} + h_{22}H_{12} + h_{23}H_{13} \\ h_{31}H_{11} + h_{32}H_{12} + h_{33}H_{13} \end{bmatrix}$$

The first element of the resulting product vector (apart from the factor c) is none other than the expansion of $|\mathbf{H}|$ along its first row. Hence the value of this element is $|\mathbf{H}| = |\mathbf{A} - \lambda_i\mathbf{I}| = 0$, because λ_i is a root of the characteristic equation. Each of the other two elements of the product vector $\mathbf{H}\mathbf{v}_i$ represents a linear combination of the cofactors H_{11}, H_{12}, H_{13}, using elements of a *non-corresponding* row of $|\mathbf{H}|$ as coefficients. Hence both these elements are zero, by virtue of Property 6 of determinants, stated in Appendix A. Thus all three elements of $\mathbf{H}\mathbf{v}_i$ are 0; that is,

$$\mathbf{H}\mathbf{v}_i = (\mathbf{A} - \lambda_i\mathbf{I})\mathbf{v}_i = \mathbf{0}$$

as we set out to prove.

To illustrate the procedure, we use the example given in the exercise for finding eigenvalues. The reader should have found $\lambda_1 = 8$, $\lambda_2 = 7$, and $\lambda_3 = 2$ as the three eigenvalues of the matrix

$$\mathbf{B} = \begin{bmatrix} 7 & 0 & 1 \\ 0 & 7 & 2 \\ 1 & 2 & 3 \end{bmatrix}$$

Step 1. To find the eigenvector $\mathbf{v}_1$ corresponding to $\lambda_1 = 8$, we first write out the matrix

$$\mathbf{B} - 8\mathbf{I} = \begin{bmatrix} -1 & 0 & 1 \\ 0 & -1 & 2 \\ 1 & 2 & -5 \end{bmatrix}$$

Step 2. Compute the adjoint of the matrix in Step 1. The result is

$$\mathbf{adj}(\mathbf{B} - 8\mathbf{I}) = \begin{bmatrix} 1 & 2 & 1 \\ 2 & 4 & 2 \\ 1 & 2 & 1 \end{bmatrix}$$

Note that all three columns are proportional, which offers a partial check on the calculations. (In this example, the first and third columns are identical, but this will not be true in general.) Before carrying out the last step, the reader should verify that

$$(\mathbf{B} - 8\mathbf{I}) \begin{bmatrix} 1 \\ 2 \\ 1 \end{bmatrix} = \mathbf{0} \qquad \text{or that} \qquad \mathbf{B} \begin{bmatrix} 1 \\ 2 \\ 1 \end{bmatrix} = 8 \begin{bmatrix} 1 \\ 2 \\ 1 \end{bmatrix}$$

Step 3. In order to have $\mathbf{v}_1$ satisfy the unit-norm condition, $\mathbf{v}_1'\mathbf{v}_1 = 1$, we divide each element of [1, 2, 1] by $\sqrt{1^2 + 2^2 + 1^2} = \sqrt{6}$, and obtain

$$\mathbf{v}_1' = [.4082 \quad .8165 \quad .4082]$$

EXERCISES

1. Compute the eigenvectors $\mathbf{v}_2$ and $\mathbf{v}_3$ associated with the other two eigenvalues λ_2 and λ_3 in the example above.
2. (Theoretical) Show that the matrix

$$\begin{bmatrix} 4 & 1 & -1 \\ 1 & 4 & -1 \\ -1 & -1 & 4 \end{bmatrix}$$

has eigenvalues $\lambda_1 = 6$, $\lambda_2 = \lambda_3 = 3$, and that only the first eigenvector, $[1/\sqrt{3}, 1/\sqrt{3}, -1/\sqrt{3}]$, is determinate. The other two may be any vectors that are orthogonal to the first. This is a property common to all matrices with two or more equal eigenvalues.

Next, in order to compare the results obtained by two different methods, we compute the eigenvectors of the variance–covariance matrix

$$\mathbf{\Sigma} = \begin{bmatrix} 16.81 & .88 \\ .88 & 6.64 \end{bmatrix}$$

based on the SSCP matrix used for the rotation problem of Section 5.3. We have, in Example 5.4, already found the eigenvalues of this matrix to be

$$\lambda_1 = 16.8856 \qquad \text{and} \qquad \lambda_2 = 6.5644$$

Hence

$$\mathbf{\Sigma} - \lambda_1\mathbf{I} = \begin{bmatrix} -.0756 & .88 \\ .88 & -10.2456 \end{bmatrix}$$

$$\mathbf{adj}(\mathbf{\Sigma} - \lambda_1\mathbf{I}) = \begin{bmatrix} -10.2456 & -.88 \\ -.88 & -.0756 \end{bmatrix}$$

and

$$\mathbf{v}_1 = \begin{bmatrix} -10.2456 \\ -.88 \end{bmatrix} \div \sqrt{(-10.2456)^2 + (-.88)^2} = \begin{bmatrix} -.99633 \\ -.08558 \end{bmatrix}$$

Similarly,

$$\mathbf{adj}(\mathbf{\Sigma} - \lambda_2\mathbf{I}) = \begin{bmatrix} .0756 & -.88 \\ -.88 & 10.2456 \end{bmatrix}$$

whence

$$\mathbf{v}_2 = \begin{bmatrix} .0756 \\ -.88 \end{bmatrix} \div \sqrt{(.0756)^2 + (-.88)^2} = \begin{bmatrix} .08558 \\ -.99633 \end{bmatrix}$$

Two important points are exemplified in the foregoing results. First, apart from the signs, the elements of $\mathbf{v}_1$ agree (within rounding error) with the values of $\cos\theta$ and $\sin\theta$ obtained previously for the variance-maximizing rotation. The signs depend merely on whether the angle from the old x_1-axis is taken as terminating at Oy^+ (the positive direction of the new axis) or at Oy^- (the negative direction), as shown in Fig. 5.7, and the facts that $\cos(\theta + 180°) = -\cos\theta$ and $\sin(\theta + 180°) = -\sin\theta$.

We thus see that the assumption of a bivariate (in general, p-variate) normal distribution, made in Section 5.3 for determining the variance-maximizing rotation in terms of the major axis of an ellipse, is actually gratuitous. The only difference is that when the distribution is indeed multivariate normal, the ellipse whose major axis is determined in the variance-maximizing process is an iso-

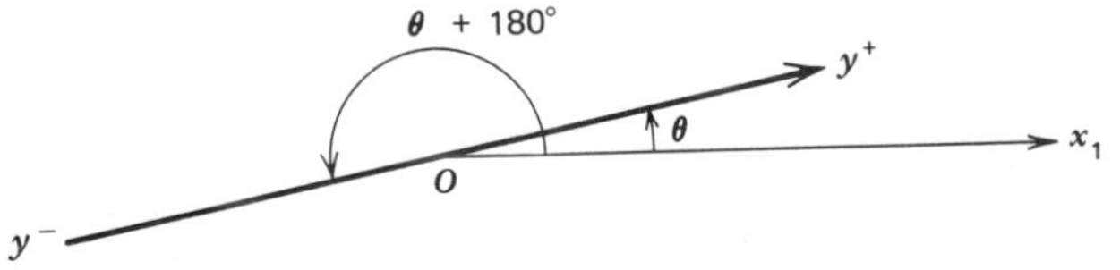

Figure 5.7. *Axis rotations through angles θ and θ + 180°.*

density ellipse of the distribution, and hence the intuitive geometric argument presented earlier makes sense. In other cases, the ellipse is of no particular statistical relevance, and the general approach of the current subsection provides the only adequate rationale.

The second point to be noted in the illustrative example above is that the two eigenvectors $\mathbf{v}_1$ and $\mathbf{v}_2$ satisfy the orthogonality condition, $\mathbf{v}_1'\mathbf{v}_2 = 0$, as may easily be verified. (A general proof is given in the next section.) Since both vectors were determined so as to have unit norms, it follows that the 2×2 transformation matrix $\mathbf{V} = [\mathbf{v}_1 \mid \mathbf{v}_2]$, with $\mathbf{v}_1$ and $\mathbf{v}_2$ as its columns, is an orthogonal matrix—which corresponds to a rigid rotation. Hence the axis defined by $\mathbf{v}_2$ is perpendicular (or orthogonal) to the axis defined by $\mathbf{v}_1$, which we know to be the major axis of an ellipse $\mathbf{x}'\mathbf{\Sigma}^{-1}\mathbf{x} = C$. Consequently, $\mathbf{v}_2$ corresponds to the minor axis of this ellipse, which (following the geometric argument given earlier) is the variance-*minimizing* axis.

It may seem paradoxical that even though we set out to determine the variance-maximizing axis, we got both the maximizing *and* the minimizing axes as solutions of the same matrix equation, $(\mathbf{\Sigma} - \lambda\mathbf{I})\mathbf{v} = \mathbf{0}$. The reason is that this equation is only a *necessary* condition for the variance-maximizing transformation; its solution may yield either the maximum or minimum (or even a stationary point) of the function we set out to maximize.

In the bivariate case, $\mathbf{v}_2$ indeed corresponds to the axis along which the variance is at a minimum. But when there are three or more variables and the variance-covariance matrix has three or more eigenvectors $\mathbf{v}_i$, it turns out that all except the one associated with the smallest eigenvalue define axes along which the variances are *conditionally maximal* in the following sense: If we denote the eigenvectors as $\mathbf{v}_1, \mathbf{v}_2, \mathbf{v}_3, \ldots$, in descending order of magnitude of their associated eigenvalues $\lambda_1, \lambda_2, \lambda_3, \ldots$, and label the corresponding transformed variables $Y_1, Y_2, Y_3, \ldots$, respectively, then Y_1 has the largest possible variance; Y_2 has the largest variance among all linearly transformed variables *that are uncorrelated with* Y_1; Y_3 has the largest variance among all linearly transformed variables *that are uncorrelated with both* Y_1 *and* Y_2; and so on (where it is understood, of course, that by "linearly transformed" we mean so transformed by a vector of unit length).

It may be noted in passing that even the last transformed variable Y_p, although possessing minimum variance, does have the property of "conditional maximizing" of variance. The fact is that once the first $p - 1$ transformed variables are determined, there is only *one* linearly transformed variable that is uncorrelated with all of these, and Y_p is this variable. Hence Y_p vacuously satisfies the statement that it has the largest variance among those variables that are uncorrelated with $Y_1, Y_2, \ldots, Y_{p-1}$. In this respect the bivariate case is no exception: although Y_2 has minimum variance, it is, at the same time, the linearly transformed variable with maximum variance among those uncorrelated with Y_1, since Y_2 is the *only* such variable.

The reader may have anticipated, when we numbered the transformed variables $Y_1, Y_2, Y_3, \ldots$, in descending order of magnitude of the corresponding eigenvalues $\lambda_1, \lambda_2, \lambda_3, \ldots$, that the latter must have something to do with the variances of the former. We now show that for each i, λ_i is precisely equal to the variance of Y_i—even though it was originally introduced artificially as a Lagrange multiplier related to the side condition $\mathbf{v}_i'\mathbf{v}_i = 1$ in our maximizing problem.

Dividing both sides of Eq. 5.18 by N, we see that

$$\text{var}\,(Y_i) = \frac{1}{N}\,\mathbf{v}_i'\mathbf{S}(X)\mathbf{v}_i = \mathbf{v}_i'\boldsymbol{\Sigma}\mathbf{v}_i \tag{5.24}$$

But from Eq. 5.22 with $\mathbf{A}$ replaced by $\boldsymbol{\Sigma}$, we know that for each i,

$$(\boldsymbol{\Sigma} - \lambda_i\mathbf{I})\mathbf{v}_i = \mathbf{0}$$

or, equivalently, that

$$\boldsymbol{\Sigma}\mathbf{v}_i = \lambda_i\mathbf{v}_i \tag{5.25}$$

Substituting the right-hand expression here for $\boldsymbol{\Sigma}\mathbf{v}_i$ in Eq. 5.24, we have

$$\begin{aligned}\text{var}\,(Y_i) &= \mathbf{v}_i'(\lambda_i\mathbf{v}_i)\\ &= \lambda_i(\mathbf{v}_i'\mathbf{v}_i)\end{aligned}$$

(because λ_i is a scalar), or

$$\text{var}\,(Y_i) = \lambda_i \tag{5.26}$$

(because $\mathbf{v}_i'\mathbf{v}_i = 1$, by construction), as we set out to prove.

We conclude this section by introducing the term *principal-axes rotation* to replace our earlier "variance-maximizing rotation." Since by solving the eigenvalue problem $(\boldsymbol{\Sigma} - \lambda\mathbf{I})\mathbf{v} = \mathbf{0}$ we obtain the principal axes of an ellipsoid (or hyperellipsoid), $\mathbf{x}'\boldsymbol{\Sigma}^{-1}\mathbf{x} = C$. The first principal axis is the usual major axis (the longest diameter); the second principal axis is the longest of all the diameters that are orthogonal to the first; the third is the longest of all the diameters that are orthogonal to both the first and the second; and so on.

EXERCISES

Given a set of three variables with covariance matrix

$$\boldsymbol{\Sigma} = \begin{bmatrix} 128.7 & 61.4 & -21.0 \\ 61.4 & 56.9 & -28.3 \\ -21.0 & -28.3 & 63.5 \end{bmatrix}$$

1. Determine the (orthogonal) transformation matrix for the principal-axes rotation.

2. Using Eq. 5.24, verify that the variance of each of the transformed variables is equal to the corresponding eigenvalue of $\mathbf{\Sigma}$, found in the process of determining the transformation matrix above.

5.5 SOME PROPERTIES OF MATRICES RELATED TO EIGENVALUES AND EIGENVECTORS

There are many interesting and useful relationships between the eigenvalues and eigenvectors of a matrix and certain characteristics of the latter, some of which were informally mentioned in Section 5.4. Here we state a number of these properties in a sequence of theorems, some with complete proofs, and others with proof outlines or intuitive supporting arguments (when the proofs are unduly long or too advanced).

Theorem 2. The sum $\Sigma\, \lambda_i$ of the eigenvalues of a matrix $\mathbf{A}$ is equal to the sum $\Sigma\, a_{ii}$ of the latter's diagonal elements. This sum is called the *trace* of $\mathbf{A}$, denoted tr($\mathbf{A}$).

OUTLINE OF PROOF. The eigenvalues λ_i of $\mathbf{A}$ are, by definition, the roots of its characteristic equation,

$$\begin{vmatrix} a_{11} - \lambda & a_{12} & \cdots & a_{1p} \\ a_{21} & a_{22} - \lambda & \cdots & a_{2p} \\ \vdots & \vdots & \ddots & \vdots \\ a_{p1} & a_{p2} & \cdots & a_{pp} - \lambda \end{vmatrix} = 0$$

which is a polynomial equation of degree p in λ. In the theory of equations (see, e.g., Richardson, 1966, p. 363), it is shown that if $\lambda_1, \lambda_2, \ldots, \lambda_p$ are the roots of a polynomial equation

$$c_0\lambda^p + c_1\lambda^{p-1} + c_2\lambda^{p-2} + \cdots + c_p = 0$$

then

$$\Sigma\, \lambda_i = -\frac{c_1}{c_0}$$

For our characteristic equation it can be shown by expanding the determinant that

$$c_0 = (-1)^p \qquad \text{and} \qquad c_1 = (-1)^{p-1}(a_{11} + a_{22} + \cdots + a_{pp})$$

It therefore follows that

$$\sum_{i=1}^{p} \lambda_i = \sum_{i=1}^{p} a_{ii} \qquad (5.27)$$

as was to be proved.

The implication of this theorem for the case in which **A** is a covariance matrix (like the $\mathbf{\Sigma}$ of Example 5.4) is worthy of special note. Then the λ_i is, as shown in Eq. 5.26, the variance of the ith transformed variable Y_i; and a_{ii} is, of course, the variance of the ith original variable X_i. Equation 5.27 therefore implies that

$$\sum_{i=1}^{p} \text{var}\,(Y_i) = \sum_{i=1}^{p} \text{var}\,(X_i) \qquad (5.28)$$

In words, the "aggregate variance" of the entire set of variables remains unchanged from before to after the principal-axes transformation[2] $\mathbf{Y}' = \mathbf{X}'\mathbf{V}$, where the columns of $\mathbf{V}$ are the eigenvectors $\mathbf{v}_i$ of $\mathbf{\Sigma}$. Thus the effect of a principal-axes rotation is to "reshuffle" the aggregate variance so that the largest share is allotted to the first transformed variable, the next largest share to a second transformed variable that is uncorrelated with the first, and so on.

Theorem 3. The product $\lambda_1, \lambda_2, \ldots, \lambda_p$ (which is abbreviated $\prod_{i=1}^{p} \lambda_i$) of the eigenvalues of a matrix **A** is equal to the value of the determinant $|\mathbf{A}|$.

OUTLINE OF PROOF: Borrowing again from the theory of equations, we have the relation

$$\prod_{i=1}^{p} \lambda_i = (-1)^p (c_p/c_0)$$

where c_p is the constant term, and c_0 the coefficient λ^p in the characteristic equation, as in the proof outline for Theorem 2. But the constant term of $|\mathbf{A} - \lambda\mathbf{I}|$ is readily found by setting $\lambda = 0$ in this expression: $c_p = |\mathbf{A} - 0\mathbf{I}| = |\mathbf{A}|$. From this and the fact that $c_0 = (-1)^p$, stated before, it immediately follows that

$$\prod_{i=1}^{p} \lambda_i = |\mathbf{A}| \qquad (5.29)$$

[2]Actually, the property of leaving the aggregate variance unchanged is true not only of the principal-axes rotation, but of *all* rigid rotations. This may be seen from the following intuitive consideration: An arbitrary rigid rotation from the X_i's to a new set of variables $\{Y_i'\}$, say, may be followed by a principal-axes rotation from $\{Y_i'\}$ to $\{Y_i\}$. But this set $\{Y_i\}$ must be the same set as obtainable directly by the principal-axes rotation from $\{X_i\}$ (because two successive rigid rotations can always be replaced by a single rigid rotation). Therefore, Σ var $(Y_i') = \Sigma$ var (Y_i) *and* Σ var $(X_i) = \Sigma$ var (Y_i), from which it follows that Σ var $(Y_i') = \Sigma$ var (X_i). Of course, the rotation vectors must satisfy the scale-preserving condition $\mathbf{v}'\mathbf{v} = 1$.

An important corollary to Theorem 3 is that a matrix has one or more eigenvalues equal to zero if and only if the matrix is singular. This is immediately obvious from Eq. 5.29. Furthermore, the number of zero eigenvalues has an important bearing on the nature of **A**—especially when the latter is a covariance matrix. Even though the presence of just one eigenvalue equal to zero is sufficient to assure that **A** is singular, we somehow feel that it is "more singular" with a greater number of zero eigenvalues. A precise sense in which we can speak of various degrees of singularity will be described subsequently.

Theorem 4. Two eigenvectors $\mathbf{v}_i$ and $\mathbf{v}_j$ associated with two distinct eigenvalues λ_i and λ_j of a *symmetric* matrix are mutually orthogonal; that is, $\mathbf{v}_i'\mathbf{v}_j = 0$.

PROOF: An instance of this theorem was already verified for a numerical example. The general proof is as follows.

By hypothesis, we have

$$\mathbf{A}\mathbf{v}_i = \lambda_i\mathbf{v}_i \tag{a}$$

and

$$\mathbf{A}\mathbf{v}_j = \lambda_j\mathbf{v}_j \qquad \text{with} \qquad \lambda_i \neq \lambda_j \tag{b}$$

Taking the transposes of the two members of Eq. (a), we have

$$\mathbf{v}_i'\mathbf{A} = \lambda_i\mathbf{v}_i' \tag{a$'$}$$

(since $\mathbf{A}' = \mathbf{A}$ by hypothesis). Premultiplying both sides of Eq. (b) by $\mathbf{v}_i'$ yields

$$\mathbf{v}_i'\mathbf{A}\mathbf{v}_j = \lambda_j\mathbf{v}_i'\mathbf{v}_j \tag{c}$$

in the left member of which we may, from Eq. (a′), substitute $\lambda_i\mathbf{v}_i'$ in place of $\mathbf{v}_i'\mathbf{A}$ to obtain

$$\lambda_i\mathbf{v}_i'\mathbf{v}_j = \lambda_j\mathbf{v}_i'\mathbf{v}_j$$

or

$$(\lambda_i - \lambda_j)(\mathbf{v}_i'\mathbf{v}_j) = 0$$

Since $\lambda_i - \lambda_j \neq 0$ by hypothesis, it follows that $\mathbf{v}_i'\mathbf{v}_j = 0$.

Next, instead of considering the eigenvectors in pairs, we treat them all simultaneously by using $\mathbf{v}_1, \mathbf{v}_2, \ldots$ as the successive columns of matrix **V**. If none of the eigenvalues is zero, then **V** will be a square matrix of order equal to that of **A**. Hence when **A** is the covariance matrix $\mathbf{\Sigma}$ of a given set of p variables $X_1, X_2, \ldots, X_p$, **V** is the orthogonal transformation matrix representing a rigid rotation from the X's to the principal axes.

Since $\mathbf{\Sigma v}_i = \lambda_i \mathbf{v}_i$ for each i, it follows that if we define a diagonal matrix

$$\mathbf{\Lambda} = \begin{bmatrix} \lambda_1 & & & \\ & \lambda_2 & & 0 \\ & & \ddots & \\ 0 & & & \lambda_p \end{bmatrix} \tag{5.30}$$

we shall have

$$\mathbf{\Sigma V} = \mathbf{V \Lambda} \tag{5.31}$$

because the columns of $\mathbf{V}$ are the successive $\mathbf{v}_i$'s, and postmultiplication by a diagonal matrix multiplies each column of $\mathbf{V}$ by the corresponding diagonal element.

Now if we premultiply both members of Eq. 5.31 by $\mathbf{V}'$, we obtain

$$\mathbf{V}'\mathbf{\Sigma V} = \mathbf{V}'\mathbf{V \Lambda} = \mathbf{\Lambda} \tag{5.32}$$

(where, in going from the second to the last expression, we utilized the fact that $\mathbf{V}'\mathbf{V} = \mathbf{I}$). But according to a result readily obtainable by dividing both sides of Eq. 5.16 by N, the leftmost member of Eq. 5.32 is none other than the covariance matrix of the transformed variables Y_i corresponding to the principal axes. Thus Eq. 5.32 tells us that the covariances of all pairs (Y_i, Y_j) [with $i \neq j$] of the transformed variables are equal to zero (since they are given by the off-diagonal elements of $\mathbf{\Lambda}$).

This is the basis for the assertion made earlier that Y_2 is uncorrelated with Y_1, Y_3 is uncorrelated with both Y_1 and Y_2, and so on. We summarize the foregoing in a theorem.

Theorem 5. Given a set of variables $X_1, X_2, \ldots, X_p$, with a nonsingular covariance matrix $\mathbf{\Sigma}$, we can always derive a set of uncorrelated variables $Y_1, Y_2, \ldots, Y_p$ by a set of linear transformations corresponding to the principal-axes rotation, that is, the rigid rotation whose transformation matrix $\mathbf{V}$ has, as its columns, the p eigenvectors of $\mathbf{\Sigma}$. The covariance matrix of this new set of variables is the diagonal matrix $\mathbf{\Lambda} = \mathbf{V}'\mathbf{\Sigma V}$, whose diagonal elements are the p eigenvalues of $\mathbf{\Sigma}$.

This theorem may be called the fundamental theorem of *principal-components analysis,* which is the name given to the procedure of performing a principal-axes rotation. The resulting transformed variables $Y_1, Y_2, \ldots, Y_p$ are known as the *principal components,* and determining their covariance matrix $\mathbf{\Lambda}$ as in Eq. 5.32 is referred to as "diagonalizing the matrix $\mathbf{\Sigma}$."

One reason why Theorem 4 is of prime importance is that it enables us to prove the fact, basic to multivariate significance tests, that the quadratic form $\chi^2 = \mathbf{x}'\mathbf{\Sigma}^{-1}\mathbf{x}$ in the exponent of the p-variate normal density function is a chi-square variate with p degrees of freedom. This fact was already proved in Chapter 4 for the special case when $X_1, X_2, \ldots, X_p$ are *independently* normally distributed. In order to extend the proof to the general case of p variables following a multivariate normal distribution with nonzero correlations, we need only show that the quantity $\mathbf{x}'\boldsymbol{\Sigma}^{-1}\mathbf{x}$ remains invariant under a principal-axes rotation whereby we can obtain a set of uncorrelated variables (i.e., the principal components)—because, in the case of normally distributed variables, uncorrelatedness implies independence. Actually, the invariance property just cited holds for *any* rigid rotation, as shown by the following theorem.

Theorem 6. Given a set of variables $X_1, X_2, \ldots, X_p$ with a nonsingular variance–covariance matrix $\boldsymbol{\Sigma}_x$, a new set of variables $Y_1, Y_2, \ldots, Y_p$ is defined by the transformation $\mathbf{Y}' = \mathbf{X}'\mathbf{V}$, where $\mathbf{V}$ is an orthogonal matrix. If the covariance matrix of the Y's is denoted by $\boldsymbol{\Sigma}_y$, the following relation holds between the two covariance matrices:

$$\mathbf{y}'\boldsymbol{\Sigma}_y^{-1}\mathbf{y} = \mathbf{x}'\boldsymbol{\Sigma}_x^{-1}\mathbf{x} \tag{5.33}$$

where $\mathbf{x} = \mathbf{X} - \overline{\mathbf{X}}$ and $\mathbf{y} = \mathbf{Y} - \overline{\mathbf{Y}}$. In words, we say that the quadratic form $\mathbf{x}'\boldsymbol{\Sigma}^{-1}\mathbf{x}$ is invariant under a rigid rotation.

PROOF: The stated transformation holds also for deviation scores.

$$\mathbf{y}' = \mathbf{x}'\mathbf{V} \tag{a}$$

From Eq. 5.16 we know that $\boldsymbol{\Sigma}_y = \mathbf{V}'\boldsymbol{\Sigma}_x\mathbf{V}$, whence $\boldsymbol{\Sigma}_y^{-1} = \mathbf{V}^{-1}\boldsymbol{\Sigma}_x^{-1}(\mathbf{V}')^{-1}$, which, by virtue of the fact that $\mathbf{V}$ is an orthogonal matrix (so that $\mathbf{V}' = \mathbf{V}^{-1}$), reduces to

$$\boldsymbol{\Sigma}_y^{-1} = \mathbf{V}'\boldsymbol{\Sigma}_x^{-1}\mathbf{V} \tag{b}$$

Substituting (a) and (b) in the left member of the equation to be proved, we have

$$\begin{aligned}
\mathbf{y}'\boldsymbol{\Sigma}_y^{-1}\mathbf{y} &= (\mathbf{x}'\mathbf{V})(\mathbf{V}'\boldsymbol{\Sigma}_x^{-1}\mathbf{V})(\mathbf{V}'\mathbf{x}) \\
&= \mathbf{x}'(\mathbf{V}\mathbf{V}')\boldsymbol{\Sigma}_x^{-1}(\mathbf{V}\mathbf{V}')\mathbf{x} \\
&= \mathbf{x}'\boldsymbol{\Sigma}_x^{-1}\mathbf{x} \qquad (\text{because } \mathbf{V}\mathbf{V}' = \mathbf{I})
\end{aligned}$$

which is Eq. 5.33.

When this equation is applied to the case in which $\mathbf{V}$ is the transformation matrix for the principal-axes rotation, we may further invoke Theorem 5 ($\boldsymbol{\Sigma}_y = \boldsymbol{\Lambda}$, a diagonal matrix) and conclude that

$$\mathbf{x}'\boldsymbol{\Sigma}_x^{-1}\mathbf{x} = \mathbf{y}'\boldsymbol{\Lambda}^{-1}\mathbf{y}$$

The right-hand expression is the quadratic form for a set of uncorrelated variables, which was already shown to be a chi-square variate with p degrees of freedom when the Y's are normally distributed. This completes the proof of the assertion made in Chapter 4 and cited at the outset of this chapter.

Degenerate Joint Distributions

In Theorems 5 and 6, it was assumed that the covariance matrix of the initial set of variables was nonsingular. We now investigate the case when this assumption does not hold. In this case we already know, from Theorem 3, that one or more of the eigenvalues of $\mathbf{\Sigma}$ are equal to zero. In practice, we would not expect any eigenvalue of a covariance matrix to be exactly equal to zero, but only approximately so, if at all.

Suppose that just $r(< p)$ of the eigenvalues are *not* zero or negligible. It then follows that we can determine r principal components $Y_1, Y_2, \ldots, Y_r$ (corresponding to the r eigenvectors $\mathbf{v}_1, \mathbf{v}_2, \ldots, \mathbf{v}_r$ associated with the nonzero eigenvalues of $\mathbf{\Sigma}$) such that

$$\sum_{i=1}^{r} \text{var}\,(Y_i) = \sum_{i=1}^{r} \lambda_i = \text{tr}\,(\mathbf{\Sigma})$$

the step from the second to the third expressions following from Theorem 2. In other words, the aggregate variance of the original p variables $X_1, X_2, \ldots, X_p$ is completely accounted for by a smaller number r of transformed variables. The remaining $p - r$ principal components account for no variance at all. Clearly, this means that the swarm of points

$$(X_{\alpha 1}, X_{\alpha 2}, \ldots, X_{\alpha p}) \qquad (\alpha = 1, 2, \ldots, N)$$

representing the N multivariate observations in the original p-dimensional space is actually confined to a subspace of $r(< p)$ dimensions embedded in the total space. Such a "pseudo p-dimensional" swarm of points, or the joint distribution they represent, is said to be *degenerate*.

Ordinarily, the presence of a degeneracy would not be manifest in the original space with the coordinate axes representing $X_1, X_2, \ldots, X_p$ because the confining subspace would not be orthogonal to any of the axes, and hence all p variables would have nonzero variance. But when the principal-axes rotation is performed, the degeneracy (if it exists) is made explicit by the systematic reallocation of variance by the conditional variance-maximizing process described earlier. An example in three-dimensional space will clarify the point.

EXAMPLE 5.6. Show that the joint distribution whose variance–covariance matrix is given below is a degenerate one.

$$\mathbf{\Sigma} = \begin{bmatrix} 4 & -7 & 8 \\ -7 & 13 & -17 \\ 8 & -17 & 28 \end{bmatrix}$$

The characteristic equation $|\mathbf{\Sigma} - \lambda \mathbf{I}| = 0$ reduces to $\lambda(\lambda - 42)(\lambda - 3) = 0$. Hence $\lambda_1 = 42$, $\lambda_2 = 3$, $\lambda_3 = 0$. Therefore, $\lambda_1 + \lambda_2 = 45 = \text{tr}(\mathbf{\Sigma})$, indicating that the first two principal components Y_1 and Y_2 account for all the variance in the purportedly three-dimensional scatter represented by $\mathbf{\Sigma}$. That is, all the observation points lie in a plane parallel to the (Y_1, Y_2)-plane, so there is no spread in the direction perpendicular to this plane.

In the degenerate case, we need only consider the r eigenvectors associated with the nonzero eigenvalues to effect the principal-axes rotation. Thus the transformation matrix $\mathbf{V}$ will be of order $p \times r$, with $\mathbf{v}_1, \mathbf{v}_2, \ldots, \mathbf{v}_r$ as its columns. For the numerical example above, we have

$$\mathbf{V} = \begin{bmatrix} .26726 & .57735 \\ -.53452 & -.57735 \\ .80178 & -.57735 \end{bmatrix}$$

and $\mathbf{\Sigma}$ may be diagonalized into

$$\mathbf{V}'\mathbf{\Sigma}\mathbf{V} = \begin{bmatrix} 42 & 0 \\ 0 & 3 \end{bmatrix}$$

(The reader should verify these results as an exercise.) More generally, the diagonalization of the variance–covariance matrix of a degenerate joint distribution is represented by

$$\mathbf{V}'\mathbf{\Sigma}\mathbf{V} = \mathbf{\Lambda} \tag{5.34}$$

where $\mathbf{V}$ is a $p \times r$ whose columns are of unit norm and are pairwise orthogonal and $\mathbf{\Lambda}$ is an $r \times r$ diagonal matrix with all diagonal elements nonzero.

The Rank of a Matrix

In the foregoing discussion of matrices with one or more zero eigenvalues, we saw that (when the matrix in question is a covariance matrix) the number of nonzero eigenvalues corresponds to the dimensionality of the subspace to which the scatter of observation points is confined. This number, designated r above, is called the *rank* of the (square) matrix.

A more general definition of the rank of a matrix, applicable to rectangular matrices as well, involves the notion of *linear independence* of a set of vectors.

Definition. A set of vectors $\mathbf{v}_1, \mathbf{v}_2, \ldots, \mathbf{v}_k$ is said to be *linearly dependent* if some k scalars $a_1, a_2, \ldots, a_k$, not all of which are zero, can be found such that $a_1\mathbf{v}_1 + a_2\mathbf{v}_2 + \cdots + a_k\mathbf{v}_k = \mathbf{0}$.

If, on the other hand, $\Sigma_{i=1}^{k} a_i\mathbf{v}_i = \mathbf{0}$ only when $a_1 = a_2 = \cdots = a_k = 0$, the set of vectors is said to be *linearly independent*.

The practical import of this definition is that when a set of vectors is linearly dependent, at least one member of the set can be expressed as a linear combination of the remaining members; whereas if a set is linearly independent, no member of the set can be expressed as a linear combination of the others.

EXAMPLE 5.7. The set

$$\mathbf{v}_1' = [1, 2, -1] \qquad \mathbf{v}_2' = [3, -1, 0] \qquad \mathbf{v}_3' = [-3, 15, -6]$$

is linearly dependent because $2\mathbf{v}_1' + (-1)\mathbf{v}_2' + (-\frac{1}{3})\mathbf{v}_3' = [2, 4, -2] + [-3, 1, 0] + [1, -5, 2] = [0, 0, 0]$. Here $a_1 = 2$, $a_2 = -1$, $a_3 = -\frac{1}{3}$, *none* of which is zero, and each member of the set can be expressed as a linear combination of the other two vectors.

EXAMPLE 5.8. The set

$$\mathbf{v}_1' = [2, 4, -1] \qquad \mathbf{v}_2' = [3, 1, 2] \qquad \mathbf{v}_3' = [-1, -2, \tfrac{1}{2}]$$

is linearly dependent because $1.\mathbf{v}_1' + 0.\mathbf{v}_2' + 2\mathbf{v}_3' = \mathbf{0}'$. (The reader should verify this.) Here $a_1 = 1$, $a_2 = 0$, $a_3 = 2$, just two of which are nonzero. In this case only $\mathbf{v}_1$ and $\mathbf{v}_3$ can be expressed as linear combinations of the other two vectors; $\mathbf{v}_2$ cannot be so expressed.

EXAMPLE 5.9. The set

$$\mathbf{v}_1' = [d_1, 0, 0] \qquad \mathbf{v}_2' = [0, d_2, 0] \qquad \mathbf{v}_3' = [0, 0, d_3]$$

where d_1, d_2, and d_3 are arbitrary nonzero numbers, is linearly *in*dependent. This is because $a_1\mathbf{v}_1' + a_2\mathbf{v}_2' + a_3\mathbf{v}_3' = [a_1d_1, a_2d_2, a_3d_3]$, which can equal $[0, 0, 0]$ *only* if $a_1 = a_2 = a_3 = 0$.

EXAMPLE 5.10. Is the set

$$\mathbf{v}_1' = [1, 2, 3] \qquad \mathbf{v}_2' = [1, -1, 1] \qquad \mathbf{v}_3' = [4, 1, -2]$$

linearly dependent or independent?

When a set of vectors is in fact linearly dependent, it suffices to exhibit a set of numbers, not all zero, that makes $\Sigma_{i=1}^{k} a_i\mathbf{v}_i = \mathbf{0}$. But when the set is linearly independent, how can we prove this fact? How can we be sure, for example, that absolutely *no* set of numbers other than $a_1 = a_2 = a_3 = 0$ will satisfy

$$a_1[1, 2, 3] + a_2[1, -1, 1] + a_3[4, 1, -2] = [0, 0, 0] \qquad \text{(a)}$$

in Example 5.10? The answer lies in rearranging the required condition (a) into the form of a set of homogeneous equations, thus:

$$\begin{bmatrix} 1 & 1 & 4 \\ 2 & -1 & 1 \\ 3 & 1 & -2 \end{bmatrix} \begin{bmatrix} a_1 \\ a_2 \\ a_3 \end{bmatrix} = \begin{bmatrix} 0 \\ 0 \\ 0 \end{bmatrix} \tag{b}$$

As we already know, a set of homogeneous linear equations has a nontrivial solution if and only if its coefficient matrix is singular—that is (from Theorem 3), when the matrix has one or more zero eigenvalues. On the other hand, if the matrix is nonsingular (i.e., has no zero eigenvalue), the only solution is the null vector, and we may conclude that the set of vectors under investigation is linearly independent.

The foregoing gives us a systematic method for testing whether a set of vectors is linearly dependent or independent, in the special case when the number of vectors in the set is equal to their dimensionality p. We have merely to put the vectors together as a matrix like that in Eq. (b) and determine the rank r (the number of nonzero eigenvalues) of this $p \times p$ matrix. *If* $r < p$, *the set of vectors is linearly dependent; if* $r = p$, *it is linearly independent.* We now show that an extension of this method is applicable even when the number of vectors in the set is not equal to their dimensionality. Furthermore, when the set is linearly dependent, the extended method will tell us the number of vectors that can form a linearly independent subset of the original set. It was in anticipation of this fact that we stated the criterion for linear dependence in the form $r < p$ rather than mere singularity of the matrix comprising the vectors.

At this point it is convenient to adopt a new, more general definition of the rank of a matrix, applicable to square and rectangular matrices alike. This change should not lead to any confusion, as it will presently be shown that the two definitions actually are equivalent.

Definition. The rank of a $p \times q$ matrix is the largest number r such that there exists at least one set of r columns or r rows which, treated as a set of vectors, is linearly independent.

The following theorem from matrix algebra, which we state without proof, forms the basis on which to establish the said equivalence (which, in the case of rectangular matrices, will necessarily be indirect, since only square matrices have eigenvalues). In what follows we shall understand ''rank'' in the sense of maximal number of linearly independent rows or columns—until its equivalence with the earlier concept has been shown.

Theorem 7. (a) The rank of a product matrix $\mathbf{C} = \mathbf{AB}$ is *less than or equal to* the rank of $\mathbf{A}$ or $\mathbf{B}$, whichever is smaller.

(b) In the special case when **A** and **B** are mutual transposes, the ranks of **A** and **B** are, of course, equal, and the rank of **C** is precisely equal to (rather than "less than or equal to") the common rank of **A** and **B**. In other words, the ranks of both **AA**′ and **A**′**A** are equal to the rank of **A**.

Let **A** be a $p \times q$ matrix, where, for specificity, we assume that $q \le p$. Suppose that the $q \times q$ matrix **A**′**A** has r nonzero eigenvalues. (Of course, $r \le q$.) Then, in accordance with Eq. 5.34, if we take **V** as the $q \times r$ matrix whose columns are the r eigenvectors of **A**′**A** associated with its nonzero eigenvalues, we have $\mathbf{V}'(\mathbf{A}'\mathbf{A})\mathbf{V} = \mathbf{\Lambda}$, the $r \times r$ diagonal matrix with the nonzero eigenvalues of **A**′**A** as its diagonal elements. The rank of $\mathbf{\Lambda}$ is r, since the columns of a diagonal matrix with all diagonal elements nonzero form a linearly independent set, as shown in Example 5.9. Consequently, by Theorem 6(a), the rank of **A**′**A** cannot be *less* than r (since **A**′**A** is one of three factors whose product has rank r). On the other hand, suppose that the rank of **A**′**A** were *greater* than r, say s. Then, by definition, there would exist at least one set of s linearly independent rows and colunns in **A**′**A**. From these, we could construct an $s \times s$ submatrix of **A**′**A** that is nonsingular, and hence has $s > r$ nonzero eigenvalues. This means that a *submatrix* of **A**′**A** can be diagonalized into a larger ($s \times s$) diagonal matrix than can the entire matrix **A**′**A**, which is absurd. (A rigorous demonstration of the absurdity of this conclusion is left as an exercise at the end of this section, with hints for the approach to be taken.) Thus the rank of **A**′**A** can be neither less than nor greater than r, so it must be equal to r. Hence, by Theorem 6(b), the rank of **A** is also equal to r.

We have thus demonstrated the equivalence of the concept of the rank of a matrix **A** as the maximal number of linearly independent rows or columns, and the number of nonzero eigenvalues possessed by either **A** itself (when it is square) or by **A**′**A** (when **A** is rectangular). In doing so, we have also extended the method, stated earlier, for testing whether a set of vectors is linearly dependent or independent, into the following: Given a set of q p-dimensional vectors to be tested for linear independence; and if the set is found to be dependent, to determine the maximal number of linearly independent vectors in the set.

1. Arrange the vectors to form the rows (or columns) of a matrix **A**.
2. Determine the eigenvalues of **A** (if $p = q$), or of **A**′**A** or **AA**′ (whichever has the smaller order[3]) if $p \ne q$. Let the number of nonzero eigenvalues be r.
3. If $r = p$ or q and $q \le p$, the set is linearly independent.
4. If $r = p$ but $q > p$, the set is linearly dependent and p is the largest number of linearly independent vectors in the set.

[3]Actually, it does not matter which one we take; choosing the one with the smaller order merely cuts down on the amount of arithmetic.

5. If r is less than the smaller of p and q, the set is linearly dependent and r is the largest number of linearly independent vectors in the set.

EXAMPLE 5.11 How many linearly independent vectors are there in the following set?

$$\mathbf{v}_1 = \begin{bmatrix} 2 \\ 0 \\ -1 \end{bmatrix} \qquad \mathbf{v}_2 = \begin{bmatrix} 0 \\ -2 \\ 3 \end{bmatrix}$$

$$\mathbf{v}_3 = \begin{bmatrix} 1 \\ -1 \\ 1 \end{bmatrix} \qquad \mathbf{v}_4 = \begin{bmatrix} 1 \\ 1 \\ -2 \end{bmatrix}$$

Arranging the given vectors as the columns of a matrix, we obtain

$$\mathbf{A} = \begin{bmatrix} 2 & 0 & 1 & 1 \\ 0 & -2 & -1 & 1 \\ -1 & 3 & 1 & -2 \end{bmatrix}$$

Since $\mathbf{A}$ is not square, and $\mathbf{AA}'$ has a smaller order than $\mathbf{A}'\mathbf{A}$, we compute

$$\mathbf{AA}' = \begin{bmatrix} 6 & 0 & -3 \\ 0 & 6 & -9 \\ -3 & -9 & 15 \end{bmatrix}$$

and determine its eigenvalues. The characteristic equation of $\mathbf{AA}'$ is

$$\begin{vmatrix} 6-\lambda & 0 & -3 \\ 0 & 6-\lambda & -9 \\ -3 & -9 & 15-\lambda \end{vmatrix} = 0$$

which, on expansion and factorization, yields

$$\lambda(\lambda - 6)(\lambda - 21) = 0$$

Therefore, the eigenvalues of $\mathbf{AA}'$ are $\lambda_1 = 21$, $\lambda_2 = 6$, and $\lambda_3 = 0$, so that $r = 2$. We thus conclude that there are two linearly independent vectors in the given set of four vectors. (Verify that $\mathbf{v}_3$ and $\mathbf{v}_4$ can be expressed as linear combinations of $\mathbf{v}_1$ and $\mathbf{v}_2$.)

EXERCISES

1–4. Redo Examples 5.7 to 5.10 by the method just described.

Gramian Matrices

Throughout the foregoing development, a symmetric matrix of the form $\mathbf{AA}'$ (or $\mathbf{A}'\mathbf{A}$) has played a prominent role in ascertaining the linear dependence or independence of a set of vectors. In fact, the SSCP matrix, which was even earlier found to occupy a central position in multivariate statistical analysis, is also of this form: $\mathbf{S} = \mathbf{xx}'$. Such matrices are called *Gramian matrices,* and they have many special properties, some of which were tacitly assumed in this chapter and the preceding one. We now state several of these properties explicitly, but without proof.

Property 1. Every *principal minor determinant* (defined below) of a Gramian matrix $\mathbf{G}$ is nonnegative (i.e., it is positive or zero in value). In particular, if $\mathbf{G}$ is nonsingular, all of its principal minor determinants are positive.

A principal minor determinant of a matrix $\mathbf{A}$ is the determinant of any submatrix of $\mathbf{A}$ obtained by deleting any number of its *corresponding* row-and-column pairs (e.g., deleting just its first row and first column; or those as well as the fourth row and fourth column). The diagonal elements of $\mathbf{A}$ (which are the "determinants" of submatrices remaining after deletion of all but one corresponding row-and-column pair), as well as $|\mathbf{A}|$ itself, are included among the principal minor determinants of $\mathbf{A}$.

EXAMPLE 5.12. If

$$\mathbf{X} = \begin{bmatrix} 2 & 3 & 3 & 4 \\ 3 & 2 & 4 & 3 \\ 2 & 4 & 4 & 3 \end{bmatrix}$$

$$\mathbf{G} = \mathbf{XX}' = \begin{bmatrix} 38 & 36 & 40 \\ 36 & 38 & 39 \\ 40 & 39 & 45 \end{bmatrix}$$

is a Gramian matrix. Its principal minor determinants are

$$38, 38, 45 \qquad \begin{vmatrix} 38 & 36 \\ 36 & 38 \end{vmatrix} = 148 \qquad \begin{vmatrix} 38 & 40 \\ 40 & 45 \end{vmatrix} = 110$$

$$\begin{vmatrix} 38 & 39 \\ 39 & 45 \end{vmatrix} = 189 \qquad \text{and} \qquad |\mathbf{G}| = 382$$

all of which are positive.

EXAMPLE 5.13

$$\mathbf{H} = \begin{bmatrix} 3 & 1 & 1 & 2 \\ 1 & 3 & 1 & 0 \\ 2 & 2 & 1 & 1 \end{bmatrix} \begin{bmatrix} 3 & 1 & 2 \\ 1 & 3 & 2 \\ 1 & 1 & 1 \\ 2 & 0 & 1 \end{bmatrix} = \begin{bmatrix} 15 & 7 & 11 \\ 7 & 11 & 9 \\ 11 & 9 & 10 \end{bmatrix}$$

is a Gramian matrix. Verify that all of its principal minor determinants are positive except for $|\mathbf{H}|$ itself, which is zero.

EXAMPLE 5.14

$$\mathbf{K} = \begin{bmatrix} 9 & 6 & 3 \\ 6 & 4 & 2 \\ 3 & 2 & 1 \end{bmatrix}$$

has all of its principal minor determinants equal to zero except for its diagonal elements, which are positive. **K** is a Gramian matrix. The reader should try to determine a matrix **M** such that $\mathbf{MM}' = \mathbf{K}$.

Property 2. If **G** is a nonsingular Gramian matrix, $\mathbf{G}^{-1}$ is also a nonsingular Gramian matrix.

Property 3. If **G** is a Gramian matrix, then for all vectors **x**, the quadratic form $Q = \mathbf{x}'\mathbf{G}\mathbf{x}$ has nonnegative values; such as quadratic form is said to be *positive semidefinite*. In particular, if **G** is nonsingular, then $Q > 0$ for all $\mathbf{x} \neq \mathbf{0}$; Q is then said to be *positive definite*.

This property (in combination with Property 2) was implicit in the assertion repeatedly made earlier, that $\mathbf{x}'\mathbf{\Sigma}^{-1}\mathbf{x} = C$, with an arbitrary positive constant C, represents an isodensity ellipsoid of a multivariate normal distribution. Indeed, unless this property were true, the multivariate normal distribution would not have the characteristic, possessed by the univariate normal distribution, that the maximum density occurs at the centroid. For unless $\mathbf{x}'\mathbf{\Sigma}^{-1}\mathbf{x}$ is positive definite, there would be some point **x** at which $\mathbf{x}'\mathbf{\Sigma}^{-1}\mathbf{x} < 0$, and hence

$$[(2\pi)^{-p/2}|\mathbf{\Sigma}|^{-1/2}] \exp(-\tfrac{1}{2}\mathbf{x}'\mathbf{\Sigma}^{-1}\mathbf{x}) > (2\pi)^{-p/2}|\mathbf{\Sigma}|^{-1/2}$$

That is, the density at **x** is greater than the density at the centroid $(0, 0, \ldots, 0)$.

EXAMPLE 5.15. The quadratic form

$$Q = [x, y]\begin{bmatrix} 3 & -1 \\ -1 & 2 \end{bmatrix}\begin{bmatrix} x \\ y \end{bmatrix}$$

whose matrix is Gramian (see Property 5) and nonsingular, is positive definite. [This fact may be verified by expanding Q as $3x^2 - 2xy + 2y^2$ and noting that its discriminant (treating Q as a quadratic in x for fixed y) is $D = (2y)^2 - 4(3)(2y^2) = -20y^2 < 0$ for all $y \neq 0$. Therefore, for each fixed $y \neq 0$, the graph of Q is a parabola open upward and not intersecting the x axis. Hence $Q > 0$ for all values of x and y except $x = y = 0$.]

EXAMPLE 5.16.

$$Q = [x, y, z]\begin{bmatrix} 15 & 7 & 11 \\ 7 & 11 & 9 \\ 11 & 9 & 10 \end{bmatrix}\begin{bmatrix} x \\ y \\ z \end{bmatrix}$$

whose matrix is Gramian but singular (see Example 5.13), is positive semidefinite. See if you can determine values of x, y, z (not all zero) that make $Q = 0$.

Terminology. Although the terms "positive definite" and "positive semidefinite" properly apply to quadratic forms, it is customary to transfer them to the matrices of the quadratic forms as well. Thus a Gramian matrix itself is said to be positive semidefinite or (if it is nonsingular) positive definite.

Property 4. All the eigenvalues of Gramian matrix are nonnegative; hence all the eigenvalues of a nonsingular Gramian matrix are positive.

This fact was implicitly proved when it was shown in Eq. 5.26 that Var $(Y_i) = \lambda_i$ for each principal component Y_i, where λ_i is the ith eigenvalue of the Gramian matrix $\mathbf{\Sigma}$.

Property 5. Any *symmetric* matrix whose principal minor determinants are *all* nonnegative is a Gramian matrix. (Note that symmetry is a prerequisite condition here. A nonsymmetric matrix obviously cannot be Gramian even if all its principal minor determinants are nonnegative.) In other words, given that $\mathbf{A}$ is a symmetric, positive semidefinite (or, of course, positive definite) matrix, there always exists a matrix $\mathbf{B}$ such that $\mathbf{A} = \mathbf{B}'\mathbf{B}$.

This property, which is evidently the converse of Property 1, is of paramount importance as an "existence theorem" for the well-known multivariate technique of *factor analysis*. It assures us that given a correlation matrix [or a "reduced" correlation matrix (see p. 187) which still has the positive semi-definiteness property], we can always factor it as $\mathbf{R} = \mathbf{F}'\mathbf{F}$.

A somewhat restrictive version of this theorem is easy to prove: If $\mathbf{A}$ is a symmetric matrix, all of whose eigenvalues λ_i are positive and distinct, it is Gramian. For we can construct an orthogonal transformation matrix $\mathbf{V}$ (with the associated eigenvectors as its columns) such that (see Eq. 5.32)

$$\mathbf{V}'\mathbf{A}\mathbf{V} = \mathbf{\Lambda}$$

where $\mathbf{\Lambda}$ is the diagonal matrix with the λ_i's as its diagonal elements. Then, by pre- and postmultiplying both members by $\mathbf{V}$ and $\mathbf{V}'$, respectively, we obtain (since $\mathbf{V}$ is an orthogonal matrix)

$$\mathbf{A} = \mathbf{V}\mathbf{\Lambda}\mathbf{V}' \tag{5.35}$$

We have now only to define

$$\mathbf{B} = \mathbf{\Lambda}^{1/2}\mathbf{V}' \tag{5.36}$$

where $\mathbf{\Lambda}^{1/2}$ is the diagonal matrix whose ith diagonal elements is $\sqrt{\lambda_i}$ (which is real, since $\lambda_i > 0$ for each i). It may readily be verified that

$$\mathbf{B}'\mathbf{B} = (\mathbf{\Lambda}^{1/2}\mathbf{V}')'(\mathbf{\Lambda}^{1/2}\mathbf{V}') = \mathbf{A}$$

because of Eq. 5.35.

EXAMPLE 5.17. The following matrix is symmetric and (as the reader should verify) positive definite:

$$\mathbf{A} = \begin{bmatrix} 10 & 3 & -2 \\ 3 & 9 & -3 \\ -2 & -3 & 10 \end{bmatrix}$$

Determine a matrix $\mathbf{B}$ such that $\mathbf{A} - \mathbf{B}'\mathbf{B}$.

The eigenvalues and vectors of $\mathbf{A}$ are found to be $\lambda_1 = 15$, $\lambda_2 = 8$, $\lambda_3 = 6$, and $\mathbf{v}_1' = [1/\sqrt{3}, 1/\sqrt{3}, -1/\sqrt{3}]$, $\mathbf{v}_2' = [1/\sqrt{2}, 0, 1/\sqrt{2}]$, $\mathbf{v}_3' = [-1/\sqrt{6}, 2/\sqrt{6}, 1/\sqrt{6}]$. Therefore, in accordance with Eq. 5.36, we take

$$\mathbf{B} = \begin{bmatrix} \sqrt{15} & & 0 \\ & \sqrt{8} & \\ 0 & & \sqrt{6} \end{bmatrix} \begin{bmatrix} 1/\sqrt{3} & 1/\sqrt{3} & -1/\sqrt{3} \\ 1/\sqrt{2} & 0 & 1/\sqrt{2} \\ -1/\sqrt{6} & 2/\sqrt{6} & 1/\sqrt{6} \end{bmatrix}$$

$$= \begin{bmatrix} \sqrt{5} & \sqrt{5} & -\sqrt{5} \\ 2 & 0 & 2 \\ -1 & 2 & 1 \end{bmatrix}$$

The reader should verify that

$$\mathbf{B}'\mathbf{B} = \begin{bmatrix} 10 & 3 & -2 \\ 3 & 9 & -3 \\ -2 & -3 & 10 \end{bmatrix} = \mathbf{A}$$

Theorems on Eigenvalues of Related Matrices

We have covered most of the important theorems on eigenvalues, eigenvectors, and related matters. All of these have dealt with a single given matrix. We conclude this section with a sequence of theorems concerning the relationship between the eigenvalues and eigenvectors of a given matrix and those of certain related matrices. Some of these theorems play an important role in the sequel; others are useful for certain computational purposes. All of them have the feature in common that they justify the following loose but intuitively appealing characterization of an eigenvector: An eigenvector of a matrix is such a vector that "in its presence" (i.e., when multiplied by it), the matrix acts as though it were a scalar. Not surprisingly, the scalar that the matrix simulates is the associated eigenvalue.

All the theorems we state have a common premise, so we state it once and for all to avoid having to repeat it each time. Given: Matrix $\mathbf{A}$ has an eigenvalue λ and an associated eigenvector $\mathbf{v}$; that is,

$$\mathbf{A}\mathbf{v} = \lambda\mathbf{v}.$$

Then the following conclusions can be drawn:

Theorem 8. The matrix $b\mathbf{A}$ (where b is an arbitrary scalar) has $b\lambda$ as an eigenvalue, with $\mathbf{v}$ as the associated eigenvector.

PROOF. Multiplying both members of $\mathbf{A}\mathbf{v} = \lambda\mathbf{v}$ by b, we have

$$(b\mathbf{A})\mathbf{v} = b(\lambda\mathbf{v}) = (b\lambda)\mathbf{v}$$

Theorem 9. The matrix $\mathbf{A} + c\mathbf{I}$ (where c is an arbitrary scalar) has $(\lambda + c)$ as an eigenvalue, with $\mathbf{v}$ as the associated eigenvector.

PROOF. $(\mathbf{A} + c\mathbf{I})\mathbf{v} = \mathbf{A}\mathbf{v} + c\mathbf{v} = \lambda\mathbf{v} + c\mathbf{v} = (\lambda + c)\mathbf{v}$

Theorem 10. The matrix $\mathbf{A}^m$ (where m is any positive integer) has λ^m as an eigenvalue, with $\mathbf{v}$ as the associated eigenvector.

PROOF. $\mathbf{A}^2\mathbf{v} = \mathbf{A}(\mathbf{A}\mathbf{v}) = \mathbf{A}(\lambda\mathbf{v}) = \lambda(\mathbf{A}\mathbf{v}) = \lambda(\lambda\mathbf{v}) = \lambda^2\mathbf{v}$. The result may be extended to any positive integral power $\mathbf{A}^m$ by induction.

Theorem 11. The matrix $\mathbf{A}^{-1}$ (if it exists) has $1/\lambda$ as an eigenvalue, with $\mathbf{v}$ as the associated eigenvector.

PROOF. Premultiplying both members of $\mathbf{A}\mathbf{v} = \lambda\mathbf{v}$, by $\mathbf{A}^{-1}$ we get

$$\mathbf{v} = \mathbf{A}^{-1}(\lambda\mathbf{v}) = \lambda(\mathbf{A}^{-1}\mathbf{v}) \qquad \text{whence} \qquad \mathbf{A}^{-1}\mathbf{v} = \left(\frac{1}{\lambda}\right)\mathbf{v}$$

As a final, spectacular demonstration of how like its eigenvalues a matrix behaves, we state the celebrated *Cayley–Hamilton theorem*.

Theorem 12. A matrix satisfies its own characteristic equation.

PROOF (UNDER RESTRICTED CONDITIONS). Although a general proof of this theorem requires a much deeper delving into matrix algebra than we have undertaken, a proof for the special case when the matrix is symmetric and nonsingular, and whose eigenvalues are all distinct, is relatively simple.

Let $\mathbf{A}$ be such a matrix, and let its characteristic equation

$$|\mathbf{A} - \lambda\mathbf{I}| = 0$$

be represented by

$$c_0\lambda^p + c_1\lambda^{p-1} + \cdots + c_n = 0$$

Then for each i ($= 1, 2, \ldots, p$), it follows from Theorems 8 and 10 that

$$(c_0\mathbf{A}^p + c_1\mathbf{A}^{p-1} + \cdots + c_n\mathbf{I})\mathbf{v}_i = (c_0\lambda_i^p + c_1\lambda^{p-1} + \cdots + c_n)\mathbf{v}_i$$

$$= 0\mathbf{v}_i = \mathbf{0}$$

Consequently, by collecting $\mathbf{v}_1, \mathbf{v}_2, \ldots, \mathbf{v}_p$ into a matrix $\mathbf{V}$ (which, from Theorem 4 we know to be orthogonal), we have

$$(c_0\mathbf{A}^p + c_1\mathbf{A}^{p-1} + \cdots + c_n\mathbf{I})\mathbf{V} = \mathbf{0}_{p\times p}$$

where $\mathbf{0}_p$ is the $p \times p$ null matrix. Postmultiplying both sides of this equation by $\mathbf{V}'$ ($= \mathbf{V}^{-1}$) yields

$$(c_0\mathbf{A}^p + c_1\mathbf{A}^{p-1} + \cdots + c_n\mathbf{I})\mathbf{I} = \mathbf{0}_{p\times p}$$

or simply

$$c_0\mathbf{A}^p + c_1\mathbf{A}^{p-1} + \cdots + c_n\mathbf{I} = \mathbf{0}_{p\times p} \tag{5.37}$$

as asserted.

An important implication of this theorem is that every positive integral power of a $p \times p$ matrix can be expressed as a linear combination of its powers no higher than the $(p - 1)$th, and the identity matrix.

EXAMPLE 5.18. The characteristic equation $|\mathbf{A} - \lambda\mathbf{I}| = 0$ of the matrix

$$\mathbf{A} = \begin{vmatrix} 7 & 0 & 1 \\ 0 & 7 & 2 \\ 1 & 2 & 3 \end{vmatrix}$$

is

$$-\lambda^3 + 17\lambda^2 - 86\lambda + 112 = 0$$

The square and cube of **A** are

$$\mathbf{A}^2 = \begin{bmatrix} 50 & 2 & 10 \\ 2 & 53 & 20 \\ 10 & 20 & 14 \end{bmatrix} \quad \text{and} \quad \mathbf{A}^3 = \begin{bmatrix} 360 & 34 & 84 \\ 34 & 411 & 168 \\ 84 & 168 & 92 \end{bmatrix}$$

Therefore,

$$\begin{aligned} &-\mathbf{A}^3 + 17\mathbf{A}^2 - 86\mathbf{A} + 112\mathbf{I} \\ &= \begin{bmatrix} -360 & -34 & -84 \\ -34 & -411 & -168 \\ -84 & -168 & -92 \end{bmatrix} + \begin{bmatrix} 850 & 34 & 170 \\ 34 & 90 & 340 \\ 170 & 340 & 238 \end{bmatrix} \\ &+ \begin{bmatrix} -602 & 0 & -86 \\ 0 & -602 & -172 \\ -86 & -172 & -258 \end{bmatrix} + \begin{bmatrix} 112 & 0 & 0 \\ 0 & 112 & 0 \\ 0 & 0 & 112 \end{bmatrix} \\ &= \begin{bmatrix} 0 & 0 & 0 \\ 0 & 0 & 0 \\ 0 & 0 & 0 \end{bmatrix} \end{aligned}$$

which shows that **A** satisfies its own characteristic equation.

EXERCISES

1. Suppose that the following is a covariance matrix:

$$\mathbf{\Sigma} = \begin{bmatrix} 8 & 0 & 1 \\ 0 & 8 & 3 \\ 1 & 3 & 5 \end{bmatrix}$$

Determine the eigenvalues λ_1, λ_2, λ_3 and vectors $\mathbf{v}_1$, $\mathbf{v}_2$, $\mathbf{v}_3$ of $\mathbf{\Sigma}$, and verify Theorems 2 to 6. That is, show that:

(a) $\lambda_1 + \lambda_2 + \lambda_3 = \text{tr}(\mathbf{\Sigma})$.
(b) $\lambda_1\lambda_2\lambda_3 = |\mathbf{\Sigma}|$.
(c) $\mathbf{v}_1'\mathbf{v}_2 = \mathbf{v}_1'\mathbf{v}_3 = \mathbf{v}_2'\mathbf{v}_3 = 0$.
(d) $\mathbf{V}'\mathbf{\Sigma}\mathbf{V} = \mathbf{\Lambda}$.
(e) The quadratic form $\mathbf{x}'\mathbf{\Sigma}^{-1}\mathbf{x}$ is an invariant of the rotation.

2. Assuming that you had not been told that it was a covariance matrix, prove that the matrix $\mathbf{\Sigma}$ given in Exercise 1 is Gramian (see Property 5). Then determine a matrix $\mathbf{B}$ such that $\mathbf{B}'\mathbf{B} = \mathbf{\Sigma}$.

3. The matrix

$$\mathbf{A} = \begin{bmatrix} 6.8 & 2.4 \\ 2.4 & 8.2 \end{bmatrix}$$

has eigenvalues $\lambda_1 = 10$ and $\lambda_2 = 5$, with associated eigenvectors

$$\mathbf{v}_1 = \begin{bmatrix} .6 \\ .8 \end{bmatrix} \quad \text{and} \quad \mathbf{v}_2 = \begin{bmatrix} -.8 \\ .6 \end{bmatrix}$$

respectively. Verify this first, then show that:

(a)
$$5\mathbf{A} = \begin{bmatrix} 34 & 12 \\ 12 & 41 \end{bmatrix}$$

has eigenvalues $5\lambda_1 = 50$ and $5\lambda_2 = 25$, with associated eigenvectors $\mathbf{v}_1$ and $\mathbf{v}_2$ (the same as those of $\mathbf{A}$, given above), thus verifying Theorem 8.

(b)
$$\mathbf{A} + 2\mathbf{I} = \begin{bmatrix} 8.8 & 2.4 \\ 2.4 & 10.2 \end{bmatrix}$$

has eigenvalues $\lambda_1 + 2 = 12$ and $\lambda_2 + 2 = 7$, with associated eigenvectors $\mathbf{v}_1$ and $\mathbf{v}_2$ (as before), thus verifying Theorem 9.

(c)
$$\mathbf{A}^2 = \begin{bmatrix} 52 & 36 \\ 36 & 73 \end{bmatrix}$$

has eigenvalues $\lambda_1^2 = 100$ and $\lambda_2^2 = 25$, with associated eigenvectors $\mathbf{v}_1$ and $\mathbf{v}_2$ (as before), thus verifying Theorem 10.

(d)
$$\mathbf{A}^{-1} = \begin{bmatrix} .164 & -.048 \\ -.048 & .136 \end{bmatrix}$$

has eigenvalues $1/\lambda_1 = .10$ and $1/\lambda_2 = .20$, with associated eigenvectors $\mathbf{v}_1$ and $\mathbf{v}_2$ (as before), thus verifying Theorem 11.

4. (Theoretical) In the course of showing that the rank of a matrix $\mathbf{A}$ is equal to the number of nonzero eigenvalues possessed by $\mathbf{A}'\mathbf{A}$, it was asserted that the rank of a matrix can never exceed the number of nonzero eigenvalues it has. Complete the proof of this assertion, started below.

Suppose that a $p \times p$ matrix $\mathbf{B}$ had r nonzero eigenvalues and $s > r$ linearly independent rows and columns (which we may assume, without loss of generality, to be the first s rows and columns). Then the upper left-hand $s \times s$ submatrix $\mathbf{B}_{ss}$ of $\mathbf{B}$ would be nonsingular and hence have s nonzero eigenvalues. Denote by $\mathbf{V}_{ss}$ the matrix whose columns are the s eigenvectors of $\mathbf{B}_{ss}$ (all corresponding to nonzero eigenvalues, by the supposition above). We now define a $p \times s$ matrix $\mathbf{V}$ formed by augmenting $\mathbf{V}_{ss}$ with $p - s$ rows of zeros (i.e., tacking on a null matrix of order $(p - s) \times s$, $\mathbf{0}_{p-s,s}$), thus:

$$\mathbf{V} = \left[\begin{array}{c} \mathbf{V}_{ss} \\ \hdashline \mathbf{O}_{p-s,s} \end{array}\right]$$

By partitioning the original $p \times p$ matrix $\mathbf{B}$ as

$$\mathbf{B} = \left[\begin{array}{c:c} \mathbf{B}_{ss} & \mathbf{B}_{s,p-s} \\ \hdashline \mathbf{B}_{p-s,s} & \mathbf{B}_{p-s,p-s} \end{array}\right]$$

it can be shown that under the foregoing supposition, we would have

$$\mathbf{V'BV} = \mathbf{\Lambda} \tag{5.38}$$

the $s \times s$ diagonal matrix with the s nonzero eigenvalues of $\mathbf{B}_{ss}$ as its diagonal elements. But this means that $\mathbf{B}$ has $s > r$ nonzero eigenvalues, which is contrary to our hypothesis.

Prove Eq. 5.38. This also illustrates the process of forming the product of two partitioned matrices, which is often useful in multivariate analysis. The general rule for this process is exactly the same as for ordinary matrix multiplication, with elements replaced by submatrices. Thus the partitioning must be done in such a way that the various submatrix products are possible, as exemplified by the following schema:

$$\left[\begin{array}{c:c} \mathbf{A}_{mp} & \mathbf{A}_{mq} \\ \hdashline \mathbf{A}_{np} & \mathbf{A}_{nq} \end{array}\right] \left[\begin{array}{c:c} \mathbf{B}_{pr} & \mathbf{B}_{ps} \\ \hdashline \mathbf{B}_{qr} & \mathbf{B}_{qs} \end{array}\right] = \left[\begin{array}{c:c} \mathbf{C}_{mr} & \mathbf{C}_{ms} \\ \hdashline \mathbf{C}_{nr} & \mathbf{C}_{ns} \end{array}\right]$$

The four submatrices of the product matrix $\mathbf{AB} = \mathbf{C}$ are as follows:

$$\mathbf{C}_{mr} = \mathbf{A}_{mp}\mathbf{B}_{pr} + \mathbf{A}_{mq}\mathbf{B}_{qr} \qquad \mathbf{C}_{ms} = \mathbf{A}_{mp}\mathbf{B}_{ps} + \mathbf{A}_{mq}\mathbf{B}_{qs}$$

$$\mathbf{C}_{nr} = \mathbf{A}_{np}\mathbf{B}_{pr} + \mathbf{A}_{nq}\mathbf{B}_{qr} \qquad \mathbf{C}_{ns} = \mathbf{A}_{np}\mathbf{B}_{ps} + \mathbf{A}_{nq}\mathbf{B}_{qs}$$

5. (Theoretical) If two matrices have the same eigenvalues and eigenvectors, the matrices are identical. Prove this theorem for the special case of symmetric matrices. (HINT: Use Eq. 5.31 to express a symmetric matrix in terms of its eigenvalues and vectors.)

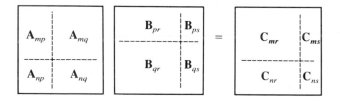

5.6 APPLICATIONS OF PRINCIPAL-COMPONENTS ANALYSIS

In the foregoing discussions, principal-components analysis was described as a purely mathematical technique of principal-axes rotation, without regard to practical applications. We now briefly indicate two such applications.

Principal-Components Analysis as a First-Stage Solution in Factor Analysis

(This subsection is intended for readers who are already familiar with factor analysis; others may skip it without loss of continuity.) The earliest use of principal components analysis for this purpose was made by Hotelling (1933), who developed an iterative computational procedure that is feasible without the aid of electronic computers for moderately large numbers of variables (up to about 10, say).

Actually, Hotelling himself, as well as T. L. Kelley (1935), did not regard principal components analysis as a "first-stage solution" for factor analysis, but as an end in itself—that is, as one approach to factor analysis in its own right. However, most modern factor analysts prefer to take the "first-stage solution" viewpoint, and do not consider a factor analysis complete until further rotations (usually *non*orthogonal, or oblique) are performed, subsequent to the principal-axes rotation, in order to achieve what is known as *simple structure* in the factor matrix. This is discussed further in Chapter 6.

Another difference between Hotelling's original work and the practice of most modern factor analysts is this: Whereas the former used the complete correlation matrix **R** (whose diagonal elements are 1's) as the starting point of analysis, the latter customarily starts with a reduced correlation matrix, whose diagonal elements are estimated *communalities*. The communality of a variable is the proportion of its variance that is accounted for by the *common factors*, that is, factors simultaneously involved in more than one of the set of variables being analyzed. This difference has led many authors to distinguish between component analysis, in which the complete correlation matrix is used as the starting point, and factor analysis, in which the reduced correlation matrix is used.

The reader may well ask what is to be gained—apart from obtaining a set of uncorrelated variables—by a principal-components analysis (or a principal-axes rotation) that is not followed by rotations to achieve simple structure for obtaining "psychologically meaningful" factors. A clue to the answer lies in the observation made earlier (p. 148) that *any degeneracy present in a multivariate distribution is made explicit by a principal-axes rotation.*

Certainly, we do not expect any real data to involve such degeneracy *exactly*—except when artificial constraints are imposed on the variables, as in the case of *ipsative measures,* described later in this section. However, a *near*-degeneracy may often be present (but concealed) in the original variance–covariance (or correlation) matrix. That is, the scatter of observation points in the p-variate space may *almost* be confined to a subspace of smaller dimensionality embedded in the total space. Such a state of affairs would be revealed by principal-components analysis in the following manner: A certain number, say r', of the eigenvalues of the covariance (or correlation) matrix would be substantially positive, but the remaining $p - r'$ eigenvalues would be nearly equal to zero. We may then conclude that the rank of the covariance matrix is "essentially equal to r'," and hence that the scatter of observation points is "practically confined" to an r'-dimensional subspace. Thus, replacing the original p variables by the first r' principal components and discarding the rest will result in very little loss of information. Principal-components analysis, therefore, effects a parsimony of description, which is useful in its own right—provided that we do not insist on "psychological meaningfulness" of the "factors."

The near-degeneracy described above may be expected to hold for the total correlation matrix whenever the original set of variables (battery of tests) consists of several "natural" subsets, the variables in each set measuring more or less the same trait (or "psychological dimension")—that is, whenever the battery of tests is of the sort that one ordinarily uses in factor analytic studies. However, the approximation to degeneracy is usually enhanced in the reduced correlation matrix with communalities instead of unities in the main diagonal because there we are concerned only with the common factor space. But in replacing the unities by estimated communalities, we have to be careful to preserve the Gramian properties (in particular, the positive definiteness) of the matrix. Otherwise, we may end up with one or more negative eigenvalues, and (if these are large in absolute value) the solution would become meaningless—not only psychologically, but logically.

The question remains as to how the "essential dimensionality" r' should be determined. In other words, given the p eigenvalues $\lambda_1 > \lambda_2 > \cdots > \lambda_p > 0$ of a covariance matrix or a correlation matrix (reduced or complete), how do we decide on the smallest of these that will be treated as "truly" nonzero? This is a big question, indeed, for which there is no general consensus on how to go about answering, let alone what constitutes the "correct" answer. We shall not discuss this problem in any detail here, but merely indicate a couple of rules of thumb that should suffice in most situations. One of them involves the proportion of total variance (or total common variance) accounted for by the first r' principal components (or factors), which may be assessed by computing the ratio

$$\frac{\sum_{i=1}^{r'} \lambda_i}{\text{tr}\,(\mathbf{C})}, \qquad \left(\sum_{i=1}^{r'} \lambda_i\right) \Big/ p \qquad \text{or} \qquad \left(\sum_{i=1}^{r'} \lambda_i\right) \Big/ \left(\sum_{i=1}^{p} h_i^2\right)$$

depending on whether the covariance matrix **C**, the complete correlation matrix (unit diagonals) or the reduced correlation matrix with estimated communalities h_i^2 is being analyzed, respectively. If the relevant ratio exceeds .85 or .90, the remaining $p - r'$ components (factors) taken together account for only 10 to 15% of the total relevant variance. Hence the $(r' + 1)$th dimension will probably contribute such a miniscule proportion (perhaps 2 to 5%) to the explained variance that it would be of no practical consequence.

The other, and perhaps the most widely used rule today is that proposed by Kaiser (1960): to retain only those components corresponding to eigenvalues of the (complete) correlation matrix that are greater than unity. This rule accords well with considerations of statistical significance and psychometric reliability among other things, but it, too, is apparently not immune to criticism. Cattell (1966), for example, argues that the 1.0 cut-off value properly applies to eigenvalues of the *population* correlation matrix, not its sample estimate. Also, subsequent rotation may well assign variances greater than 1 to factors beyond those qualifying under the 1.0 cutoff. The scree test, which Cattell proposed instead, will be described in Chapter 6.

Principal-Components Analysis as a Preliminary to Other Multivariate Analyses

It was mentioned in Chapter 3 that when there is an artificial linear constraint on the set of predictor variables (or any subset thereof) in a multiple regression problem, the SSCP matrix becomes singular and hence cannot be inverted to obtain the regression weights in accordance with Eq. 3.15. Such constraints occur most typically with *ipsative measures*—that is, when the scoring system for a set of variables is such that the total score is the same for every individual. Let us first see why the SSCP matrix is singular in such cases.

Letting $X_{\alpha i}$ be the score of the αth individual on the ith variable, we have, by definition,

$$X_{\alpha 1} + X_{\alpha 2} + \cdots + X_{\alpha p} = C \tag{5.39}$$

for each α. Hence the means of the p variables are also subject to the constraint

$$\bar{X}_1 + \bar{X}_2 + \cdots + \bar{X}_p = C$$

and it follows that the deviation scores satisfy the relation

$$x_{\alpha 1} + x_{\alpha 2} + \cdots + x_{\alpha p} = 0 \qquad (\alpha = 1, 2, \ldots, N)$$

Consequently, the columns $\mathbf{x}_i$ of the deviation-score matrix $\mathbf{x}$ are linearly dependent, since

$$\mathbf{x}_1 + \mathbf{x}_2 + \cdots + \mathbf{x}_p = \mathbf{0}$$

Therefore, the rank of $\mathbf{x}$ is less than p. (Most likely it is equal to $p - 1$.) Hence the rank of the $p \times p$ SSCP matrix $\mathbf{S} = \mathbf{xx}'$ is, by Theorem 7(b), also less than p, so that $\mathbf{S}$ is singular. It follows, of course, that the covariance matrix and the correlation matrix are also singular.

When the variables are subject to the constraint (5.39) we could, of course, omit one of them and use only $p - 1$ variables as predictors without any loss of predictive efficiency. However, this procedure would make it impossible to compare the relative importance of all p variables for the prediction. An alternative method, which does not suffer from this defect, is to perform a principal components analysis of the p variables and use the $p - 1$ principal components $Y_1, Y_2, \ldots, Y_{p-1}$ (corresponding to the nonzero eigenvalues of $\mathbf{\Sigma}_x$) as the predictors in lieu of the original set. This approach may be used not only as a preliminary to multiple regression analysis, but also for discriminant and canonical analyses, described in Chapter 7. Here we outline the procedure with respect to multiple regression.

Let $\mathbf{V}$ be the $p \times (p - 1)$ transformation matrix defining the principal components

$$\mathbf{Y}' = \mathbf{X}'\mathbf{V} \tag{a}$$

On carrying out a multiple regression analysis with $Y_1, Y_2, \ldots, Y_{p-1}$ (and usually, other variables as well) as the predictors, and (say) Z as the criterion, we shall have a set of $p - 1$ regression weights to be applied to the Y_i's. Denoting by $\mathbf{d}$ the vector comprising these weights as elements, the regression equation will be of the form

$$\tilde{Z} = a + \mathbf{Y}'\mathbf{d} \tag{b}$$

(plus further terms if there are other predictors). But the Y_i's may be eliminated from Eq. (b) by substituting from (a):

$$\tilde{Z} = a + \mathbf{X}'(\mathbf{Vd}) \tag{5.40}$$

Thus the relative importance of the X's in predicting Z may be assessed from the elements of the p-dimensional vector $\mathbf{Vd}$ after converting them to standard-score form. It also is evident that, in using the regression equation for predictive purposes, it is not necessary to convert a person's X scores into Y's, since Eq. 5.40 is already expressed in terms of the X's.

5.7 THEORETICAL SUPPLEMENT: GENERALIZED INVERSES

When the concept of the inverse of a square, nonsingular matrix was first introduced in Chapter 2, it was mentioned that, under certain conditions, a matrix $\mathbf{X}$ such that $\mathbf{XA} = \mathbf{I}$ can be found even when $\mathbf{A}$ is not square. Discussion of this matter was deferred because it required a knowledge of eigenvalues and eigenvectors. It is now time to describe how such a matrix, called a *left general inverse* of **A**, can be determined when it exists, and to specify the conditions for its existence. This concept is often used in the general linear hypothesis approach to analysis of variance (both univariate and multivariate) for solving the normal equations. Although we use an alternative method for this purpose in Chapter 9, the generalized inverse method also is alluded to there.

Suppose that **A** is an $n \times m$ matrix with rank r, where $n > m \geq r$. Then $\mathbf{AA}'$ is an $n \times n$ matrix and, by Theorem 7 (p. 151) and the ensuing discussions, it will have r nonzero eigenvalues, which may be denoted as $\lambda_1, \lambda_2, \ldots, \lambda_r$. Let **V** be the $n \times r$ matrix whose columns are the r normalized eigenvectors of $\mathbf{AA}'$ associated with the r nonzero eigenvalues. Then, by definition, we have the relation

$$(\mathbf{AA}')\mathbf{V} = \mathbf{V}\mathbf{\Lambda} \tag{5.41}$$

where $\mathbf{\Lambda}$ is the diagonal matrix with diagonal elements λ_i $(i = 1, 2, \ldots, r)$. Similarly, let **U** be the $m \times r$ matrix whose columns are the normalized eigenvectors of the $m \times m$ matrix $\mathbf{A}'\mathbf{A}$ corresponding to its r nonzero eigenvalues. It can be proved, although we shall not do so here, that these eigenvalues are the same as those of $\mathbf{AA}'$, namely, $\lambda_1, \lambda_2, \ldots, \lambda_r$. It therefore follows that

$$(\mathbf{A}'\mathbf{A})\mathbf{U} = \mathbf{U}\mathbf{\Lambda} \tag{5.42}$$

Property 4 (p. 156, of Gramian matrices—which both $\mathbf{AA}'$ and $\mathbf{A}'\mathbf{A}$ are—assures us that $\lambda_1, \lambda_2, \ldots, \lambda_r$ are all positive. Hence we may form the square root of the diagonal matrix $\mathbf{\Lambda}$, which we shall denote by **D**:

$$\mathbf{D} = \mathbf{\Lambda}^{1/2}$$

whose ith diagonal element is $\sqrt{\lambda_i}$ $(i = 1, 2, \ldots, r)$.

An important theorem in matrix algebra, which is an extension of Eq. 5.35 proved in connection with the existence theorem of factor analysis, permits us to express **A** in terms of **D**, **U**, and **V** as follows:

$$\mathbf{A} = \mathbf{VDU}' \tag{5.43}$$

(Readers who wonder why the square root of $\mathbf{\Lambda}$ occurs in this expression whereas Eq. 5.35 involves $\mathbf{\Lambda}$ itself, should recall that the $\mathbf{\Lambda}$ here has the nonzero eigen-values of $\mathbf{AA}'$ as its elements, and hence corresponds to the *square* of the $\mathbf{\Lambda}$ in Eq. 5.35. Thus in the special case when **A** is a square, symmetric matrix, Eq. 5.43 reduces to Eq. 5.35, as it should.) The proof of this theorem is rather involved, and we shall not give it here. The interested reader is referred to Horst (1963, chap. 18), who calls $\mathbf{VDU}'$ the *basic structure* of matrix **A**.

Now let us define an $m \times n$ matrix **X** as follows:

$$\mathbf{X} = \mathbf{UD}^{-1}\mathbf{V}' \tag{5.44}$$

and investigate the product **XA**. Using the right-hand-side expressions in Eqs. 5.44 and 5.43 for **X** and **A**, respectively, and successively transforming, we find

$$\begin{aligned}
\mathbf{XA} &= (\mathbf{UD}^{-1}\mathbf{V}')(\mathbf{VDU}') \\
&= \mathbf{UD}^{-1}(\mathbf{V}'\mathbf{V})\mathbf{DU}' \\
&= \mathbf{UD}^{-1}(\mathbf{I}_r)\mathbf{DU}' \qquad \text{(since the columns of } V \text{ have unit norms and are mutually orthogonal)} \\
&= \mathbf{U}(\mathbf{D}^{-1}\mathbf{D})\mathbf{U}' \\
&= \mathbf{UU}'
\end{aligned} \tag{5.45}$$

Note that although $\mathbf{U}'\mathbf{U} = \mathbf{I}_r$ (since the columns of **U**, like those of **V**, have unit norms and are mutually orthogonal), it is in general *not* true that $\mathbf{UU}' = \mathbf{I}_m$. However, consider the special case when $r = m$, that is, when the columns of **A** are linearly independent, and hence the rank of **A** is equal to its number of columns (which, it will be recalled, is here assumed to be smaller than its number of rows). In this case **U** will be a square matrix of order m, and hence $\mathbf{U}'\mathbf{U} = \mathbf{I}_m$ will also imply $\mathbf{UU}' = \mathbf{I}_m$; that is, **U** is an orthogonal matrix whose transpose is its inverse. Thus the right-hand side of Eq. 5.45 reduces to $\mathbf{I}_m$ in this case, and we arrive at the following conclusion: If a rectangular matrix **A** of order $n \times m$ (with $n > m$) has rank m, it possesses a left general inverse **X**, satisfying the relation

$$\mathbf{XA} = \mathbf{I}_m \tag{5.46}$$

where **X** is as defined in Eq. 5.44.

Similarly, if **A** is an $n \times m$ matrix with $m > n$ (i.e., with more columns than rows), and if its rows are linearly independent so that the rank of **A** is n, then **A** possesses a *right general inverse* **Y** (say) whose order is $m \times n$ and which satisfies

$$\mathbf{AY} = \mathbf{I}_n \tag{5.47}$$

This follows from the fact that in this case, **A′** is a matrix with more rows than columns and whose rank is equal to its number of columns. Hence **A′** has a left general inverse as in Eq. 5.46. If we denote this by **Y′** (an $n \times m$ matrix), we have

$$\mathbf{Y}'\mathbf{A}' = \mathbf{I}_n$$

from which Eq. 5.47 results on taking the transposes of both sides.

In summary, we have shown that a rectangular matrix with maximal rank (i.e., rank equal to its number of rows or its number of columns, whichever is smaller) possesses a left or right general inverse such that pre- or postmultiplication by the latter produces the identity matrix of order equal to the rank of the original matrix. What about a rectangular matrix of nonmaximal rank? It can be shown that such a matrix does not have either a left or right general inverse in the above sense. However, we note that even in this case the matrix **X** defined by Eq. 5.44 satisfies an interesting relation, namely

$$\mathbf{AXA} = \mathbf{A} \tag{5.48}$$

This can easily be proved by premultiplying the first and last members of Eq. 5.45 by **A**, substituting expression (5.43) in the right-hand side of the resulting equation, and utilizing the fact that $\mathbf{U}'\mathbf{U} = \mathbf{I}_r$. Details are left to the reader as an exercise.

In many applications to multivariate analysis, we do not need the full property of an inverse matrix $\mathbf{A}^{-1}$ even when it exists (that is, when **A** is a square, nonsingular matrix). In other words, we often do not utilize the fact that $\mathbf{A}^{-1}\mathbf{A} = \mathbf{AA}^{-1} = \mathbf{I}$ in its full force, but merely require a "weaker" consequence of this fact, namely that $\mathbf{AA}^{-1}\mathbf{A} = \mathbf{A}$. Thus the **X** defined in Eq. 5.44 suffices for this more modest role demanded of the matrix inverse, since it satisfies Eq. 5.48 even when Eq. 5.46 does not hold. The same is true, of course, of a matrix **Y** that would have been a right general inverse of an $n \times m$ matrix **A** (with $m > n$) had the latter been of maximal rank. For this reason, Rao (1965, p. 24) has given the name *generalized inverse* (or "g-inverse" for short) to any matrix **X** that satisfies Eq. 5.48, and denotes such a matrix by the symbol $\mathbf{A}^-$. The g-inverse is a very general concept, since an $\mathbf{A}^-$ exists for any given **A**, regardless of its order and rank—including the case when **A** is a square, singular matrix. In fact, any matrix *except* a square, nonsingular one possesses an infinite number of g-inverses. The regular inverse $\mathbf{A}^{-1}$ of a nonsingular matrix **A** is its only g-inverse as well.

Equation 5.44 (or its counterpart for a matrix with more columns than rows) thus represents just one of many possible ways for constructing a g-inverse, and hence a left (or right) general inverse when the matrix is of maximal rank. We have focused on this method because of its theoretical elegance, deriving from its being based on the important theorem stated in Eq. 5.43—a generalized version of the existence theorem for factor analysis, as was mentioned earlier. Other methods are described by Rao (1965), each method producing a different $\mathbf{A}^-$ for a given **A**. One particularly elementary method, applicable only to rectangular matrices of maximal rank, is given in Exercise 6 at the end of the chapter. Although it may be of limited practical utility, it has the advantage of clearly showing the infinitude of g-inverses.

We may now summarize all our discussions pertaining to matrix inverses, including those in Chapter 2, in the following integrated manner:

1. Any matrix **A** possesses one or more (more precisely, either *just* one or *infinitely* many) generalized inverses (g-inverse) $\mathbf{A}^-$ satisfying Eq. 5.48:

$$\mathbf{AA}^-\mathbf{A} = \mathbf{A}$$

2. If **A** has maximal rank (i.e., rank equal to its number of rows n or number of columns m, whichever is smaller; or rank equal to its order when **A** is square), then its g-inverse is also a left or right general inverse, satisfying Eq. 5.46 or 5.47, as the case may be:

$$\mathbf{A}^{-}\mathbf{A} = \mathbf{I}_m \quad (\text{if } m \leq n) \qquad \text{or} \qquad \mathbf{A}\mathbf{A}^{-} = \mathbf{I}_n \quad (\text{if } n \leq m)$$

3. In particular, if **A** is square and nonsingular (i.e., if $|\mathbf{A}| \neq 0$) its g-inverse is unique, and is denoted as $\mathbf{A}^{-1}$. Both Eqs. 5.46 and 5.47 are satisfied in this case (since $m \leq n$ and $n \leq m$ are both true in this case). Thus

$$\mathbf{A}^{-1}\mathbf{A} = \mathbf{A}\mathbf{A}^{-1} = \mathbf{I}$$

This $\mathbf{A}^{-1}$ is, of course, the inverse of **A** as defined in Chapter 2.

EXAMPLE 5.19 Given the 3×2 matrix

$$\mathbf{A} = \begin{bmatrix} 8.2 & -2.4 \\ 3.0 & 4.0 \\ 2.4 & -6.8 \end{bmatrix}$$

whose columns are obviously linearly independent, determine its left general inverse **X** in accordance with Eq. 5.44. We first compute the products $\mathbf{AA}'$ and $\mathbf{A}'\mathbf{A}$, obtaining

$$\mathbf{AA}' = \begin{bmatrix} 73 & 15 & 36 \\ 15 & 25 & -20 \\ 36 & -20 & 52 \end{bmatrix}$$

and

$$\mathbf{A}'\mathbf{A} = \begin{bmatrix} 82 & -24 \\ -24 & 68 \end{bmatrix}$$

The two nonzero eigenvalues of $\mathbf{AA}'$ (which are also the only two eigenvalues of $\mathbf{A}'\mathbf{A}$) are found to be $\lambda_1 = 100$ and $\lambda_2 = 50$. The normalized eigenvectors of $\mathbf{AA}'$ associated with these eigenvalues are

$$\mathbf{v}_1 = \begin{bmatrix} .8 \\ 0 \\ .6 \end{bmatrix} \quad \text{and} \quad \mathbf{v}_2 = \begin{bmatrix} 3/\sqrt{50} \\ 5/\sqrt{50} \\ -4/\sqrt{50} \end{bmatrix}$$

The normalized eigenvectors of $\mathbf{A}'\mathbf{A}$ are

$$\mathbf{u}_1 = \begin{bmatrix} .8 \\ -.6 \end{bmatrix} \quad \text{and} \quad \mathbf{u}_2 = \begin{bmatrix} .6 \\ .8 \end{bmatrix}$$

(The reader should verify the foregoing results, as well as Eq. 5.43, according to which the triple product

$$\mathbf{VDU}' = \begin{bmatrix} .8 & 3/\sqrt{50} \\ 0 & 5/\sqrt{50} \\ .6 & -4/\sqrt{50} \end{bmatrix} \begin{bmatrix} 10 & 0 \\ 0 & \sqrt{50} \end{bmatrix} \begin{bmatrix} .8 & -.6 \\ .6 & .8 \end{bmatrix}$$

should equal the original matrix **A**.) We then compute **X** in accordance with Eq. 5.44; that is,

$$\begin{aligned} \mathbf{UD}^{-1}\mathbf{V}' &= \begin{bmatrix} .8 & .6 \\ -.6 & .8 \end{bmatrix} \begin{bmatrix} 1/10 & 0 \\ 0 & 1/\sqrt{50} \end{bmatrix} \\ &\quad \times \begin{bmatrix} .8 & 0 & .6 \\ 3/\sqrt{50} & 5/\sqrt{50} & -4/\sqrt{50} \end{bmatrix} \\ &= \begin{bmatrix} .8 & .6 \\ -.6 & .8 \end{bmatrix} \begin{bmatrix} .08 & 0 & .06 \\ .06 & .10 & -.08 \end{bmatrix} \\ &= \begin{bmatrix} .10 & .06 & 0 \\ 0 & .08 & -.10 \end{bmatrix} = \mathbf{X} \end{aligned}$$

We leave it to the reader to verify that $\mathbf{XA} = \mathbf{I}_2$.

In order to emphasize the nonuniqueness of left (or right) general inverses, we point out that the following is another left general inverse of the matrix **A** in this example:

$$\begin{bmatrix} -.650 & 1.310 & 1 \\ -.075 & .205 & 0 \end{bmatrix}$$

as the reader should verify. This was obtained by the method described in Exercise B below.

5.8 RESEARCH EXAMPLES

In his *Journal of Educational Psychology* paper expositing principal components analysis (PCA), Harold Hotelling (1933) acknowledged Truman Lee Kelley's initiation of his inquiry into the viability of a suggestion made by Karl Pearson as early as 1901. Kelley had arranged for the Unitary Traits Committee of the Carnegie Corporation to support Hotelling's research on Pearson's idea. Under the auspices of the Committee, a dialogue among L. L. Thurstone, C. V. Hull, C. Spearman, E. L. Thorndike, and Kelley himself had framed the issues and the viewpoint of the PCA development.What illustrious Godparents this educational research method had! Describing the working assumption that a small set of principal components might be substituted for a larger set of observational variates, Hotelling said that the reduced-rank assumption was "for economy of thought" (p. 420). The hope was that in many educational research situations, a few principal components would operationalize as uncorrelated linear functions of many observational variates the significant latent factors of student performance, thus providing an efficient and parsimonious model of performance phenomena.

Hotelling and his friends were worried about the choice of a metric for each observational variate, and they settled upon disattenuated standardized measurements as their choice. This meant that each score distribution was corrected for the assessed reliability of the scale. Some social scientists still make this choice, but Hotelling was to renounce it in favor of ordinary standardized variates by the time he presented the canonical correlation method in the same journal in 1936.

Basing his example of PCA on disattenuated correlations among four achievement tests, as provided to him by T. L. Kelley, Hotelling discovered what every PCA of a complex of achievement tests has replicated in five decades of such researches, that the major explanatory construct for the battery dispersion is *general ability* (g). As always, the verbal tests were better indicators of g than were the arithmetic tests. Hotelling made the first use of what has become the "eigenvalues larger than one" rule in deciding on the rank of a model for the data. Of course, it wasn't a rule then. It was just an application of his common sense. The *special* (s) second factor was bipolar, as principal components after the first usually are, and it separated students who were better in arithmetic from those who were better in reading. The Hotelling's Original PCA Table (Table 5.1) presents the numerology of an event which was as pioneering in educational research as Lindberg's flight was in aviation.

TABLE 5.1. Hotelling's Original PCA Table

						Structure	
Test	Disattenuated Correlations				Reliability	g	s
Reading speed	1.00	.698	.264	.081	.920	.82	−.44
Reading power	.698	1.00	−.061	.092	.894	.70	−.62
Arithmetic speed	.264	−.061	1.00	.594	.908	.61	.67
Arithmetic power	.081	.092	.594	1.00	.564	.58	.66
	Eigenvalues: (1) 1.85, (2) 1.47, (3) .52, (4) .17						

An issue in terminology which needs to be mentioned is the several meanings of the term *factor* in the educational research literature. In the next chapter of this textbook, a formal mathematical theory called *factor analysis* is exposited. The mathematics of procedures sponsored by this theory project solutions outside the observation space into an imaginary space in order to accommodate estimates of true-score variances of observational variates. Strictly speaking, *only measurement constructs that cannot be measured directly because they incorporate imaginary elements can be factors* in a factor theory for data. In this strict sense, the only method for computing factors which is taught in this textbook is the method of *principal factor analysis* in Chapter 6. The *MLE* method for linear structural equations which was discussed in Chapter 4 as a recommended topic for study beyond the scope of this book is another procedure which produces constructs which are factors in the strict sense. Most of the multivariate analysis procedures taught in this book keep their solutions in the observation space, so that the fitted linear constructs are computable linear functions, or linear composites, of the data. There are those, including the author of this book, who believe that it would be useful to maintain a strict distinction between "components" and "factors." Be that as it may, the reality is that "factor" has come to have a firm meaning in the ordinary language of educational researchers as a generic term for linear constructs fitted to data by any and all methods. In this usage, "factor" is a synonym for "component" which is preferred because it testifies to the intention to relate the data-analytic product to a construct of a theory for the data. In the research examples appended to the chapters of this book, the term *factor* is used in this generic sense. It is the case that almost all of the "factors" which have been presented in the educational research literature in the past 40 years, and there have been ever so many of them, were computed by the components method of PCA, canonical correlation, multiple discriminant analysis, and so on. (Another complication in our language stems from the special meaning of "factor" in the analysis of variance, where it signifies one of the categorical independent variables of a randomized experiment. It's just too lovely a word to be confined to a single denotation.)

Another generic usage followed in these discussions of research examples is that *factor structure* refers to the product-moment correlations between variates and factors based on them. Although factors can be reported in terms of factor-scoring coefficients which make it possible to compute or estimate the factors from the variates on which they are based, or in terms of factor loadings (also called *pattern* coefficients) which show the contributions of the factors to the variates, the structure coefficients are usually considered to provide the best exposition of the factors.

Kendall and Stuart said that Hotelling's components methods "examine a system to see what sort of structure it may have" (1966, p. 306). This is a brilliant characterization of the classical exploratory statistical modeling procedures. Lohnes and Gray (1972) examined the assessment of reading readiness (tested at end of Kindergarden year) and first and second grade achievement (tested at close of those school years) for 3,956 pupils in ten U.S.O.E. Cooperative Reading Studies projects in an effort to demonstrate the adequacy of a theory for the data which postulated that intelligence was the only performance characteristic of importance in the assessment. The readiness battery contained eight tests, the first grade achievement battery contained five tests, and the second grade achievement battery contained six tests, for a total of 19 variates. The first principal component extracted 52.6 percent of the generalized variance for the 19 variates, and all of the 19 loadings were positive and substantial. The only other component of consequence was the second, which extracted 7.7 percent of battery variance, and on which all eight readiness tests loaded negatively, while ten of the 11 achievement tests loaded positively. These results are displayed in the L-G Principal Components Table (Table 5.2). The authors interpreted the first factor as an operationalization of Spearman's *g* and the second factor as "a growth measure that separated those pupils who improved their relative test performances from pre-tests to post-tests from those whose overall level of performance deteriorated, relative to the other pupils" (p. 472). The one achievement test which loaded negatively on the second factor was first grade vocabulary.

Acknowledging that reading skills were measured by some of the 11 achievement tests, Lohnes and Gray argued that their analysis supported the contention that "the best single explanatory principle for observed variance in reading skills was variance in general intelligence," from which they deduced that "reading experts must understand intelligence to understand reading" (p. 475).

TABLE 5.2. Lohnes–Gray Principal Components

		Factor Structure			
	Test	g	s	h^{2a}	R^2
1	Pintner-C	.73	−.38	.68	.60
2	Phonemes	.71	−.19	.54	.51
3	Letter names	.72	−.10	.53	.50
4	Learning rate	.55	−.28	.38	.33
5	Pattern copy	.54	−.33	.40	.34
6	Identical forms	.41	−.39	.32	.22
7	Word meaning	.57	−.45	.53	.40
8	Listening	.49	−.51	.50	.32
9	Words, grade 1	.85	.21	.77	.76
10	Words, grade 2	.84	.22	.76	.76
11	Paragraphs, grade 1	.86	.20	.78	.76
12	Paragraphs, grade 2	.87	.17	.79	.79
13	Vocabulary, grade 1	.76	−.18	.62	.59
14	Spelling, grade 1	.75	.26	.63	.59
15	Spelling, grade 2	.79	.36	.76	.70
16	Study Skills, grade 1	.84	.17	.74	.72
17	Study Skills, grade 2	.81	.18	.68	.66
18	Language, grade 2	.79	.16	.65	.62
19	Arithmetic, grade 2	.64	.08	.41	.39

[a] h^2 stands for *communality,* which is discussed in Chapter 6.

Lohnes and Marshall (1965) undertook to demonstrate that school records are saturated with carriers of general intellectual development and very little else. They collected test scores and course grades in their local junior high school in a New Hampshire college town, and produced a table showing the PCA structure for 13 tests and 8 course grades (Table 5.3), which they considered an eye-popper. Every test and every course grade loaded above .75 on the general factor. (In PCA, a loading represents the path coefficient from the factor to the observational variate, and also represents the correlation between the factor and the variate.) Notice that the almost negligible second factor was bipolar in a fascinating way, with all tests loaded positively and all course grades loaded negatively. This was interpreted as a factor which separated test-pleasers from teacher-pleasers.

TABLE 5.3. PCA Structure for 13 Tests and 8 Course Grades

Variate	g	s	h^2
PGAT verbal	.829	.289	.77
PGAT reasoning	.849	.212	.77
PGAT number	.766	.287	.67
MAT word knowledge	.846	.245	.78
MAT reading	.868	.239	.81
MAT spelling	.835	.025	.70
MAT language	.881	.064	.82
MAT study skills-language	.825	.071	.69
MAT arithmetic computation	.864	.036	.75
MAT arithmetic problems	.852	.153	.75
MAT social studies	.810	.162	.68
MAT study skills-social studies	.822	.166	.70
MAT science	.811	.326	.76
7th-grade English	.877	−.165	.80
8th-grade English	.826	−.397	.84
7th-grade arithmetic	.809	−.195	.69
8th-grade arithmetic	.788	−.174	.65
7th-grade social studies	.855	−.316	.83
8th-grade social studies	.770	−.409	.76
7th-grade science	.789	−.281	.70
8th-grade science	.757	−.406	.74

In Iowa in 1972 Annabel Newman collected test scores on 310 pupils who had been diagnosed as "low reading readiness" as they entered first grade. This meant that their Metropolitan Readiness Battery scores were below the national 60th percentile. Her data were *longitudinal* in that she had 15 test scores in a readiness battery collected at the close of the Kindergarden year and five scores in an achievement battery collected on the same pupils at the close of their first-grade year. She wanted to estimate the regression of the achievement assessment on the readiness assessment in order to test the theory that individual differences in readiness for schooling which exist as children enter the first grade account for a substantial portion of the variance in achievement in first grade. She brought her data to her graduate course in multivariate statistics, where she identified her problem as one of canonical regression. However, her instructor suggested that she not do a canonical regression of five achievement variates on 15 readiness variates in a study providing only 310 subjects, because of the enormous capitalization on chance that must occur. Instead, she and her instructor undertook to *recover degrees of freedom* for the regression analysis by exposing through PCA the structures of the two assessments, then regressing selected principal components of the achievement battery on selected principal components of the readiness battery. Not only would the number of predictors and the number of criteria be reduced, but since both the predictor components and the criterion components would be uncorrelated in their own sets, the only correlations involved would be the cross-correlations of predictor components with criterion components. No matrix inversion would be required, because the product-moment correlation of a predictor with a criterion would provide directly the standardized partial regression weight of a multiple regression equation. This strategy was a direct application of Kendall's famous advice that, confronted with problems of collinearity in multiple regression, ''we might perform a component analysis on both the *y*'s and the *x*'s and then investigate the relationship of the transformed variables'' (1961, p. 75). Years later in their extremely useful text on multiple regression Mosteller and Tukey gave the same advice, asserting that with this strategy ''we can be relatively sure that omitting the small components from further analysis loses us little,'' and ''we are using principal components as a boiling-down process'' (1977, p. 400).

Newman and Lohnes published their analysis of the Iowa data in Cooley and Lohnes (1976, pp. 63-5, 69-76). Table 5.4, showing the structure for three components of five achievement tests illustrates the report of a PCA. The R^2 values in the last column represent the strength of the multiple regression of a variate on all the other variates in the set, and is often taken as a lower bound for the communality, h^2, assigned the variate by the decision to set the rank of the PCA solution at a particular number (in this case, three). Notice the low R^2 value for the vocabulary test, which is the most striking and challenging statistic yielded by the study. Newman and Lohnes chose to use only the first two criterion components in their regression analysis and to use seven components of the readiness battery as the predictors. The result is displayed in Table 5.5 showing the regression of readiness components on achievement components. The major inter-battery relationship emerged as one between general intelligence components, with 36% of the variance in the *g* component of achievement explainable from the seven predictors, but mostly from the variance in the *g* component of readiness. Also, 16% of the variance in the vocabulary component of achievement was explainable from the variance in the seven components of readiness. Thus, to some extent Newman's hypothesis was supported by the data analysis, which also disclosed the structural elements of the continuity in individual differences.

TABLE 5.4. Structure for Three Components of Five Achievement Tests

Stanford achievement battery	Component: Intellectual development	Vocabulary residual	Reading residual	h^2	R^2
Word reading	.86	−.08	−.10	.75	.55
Paragraph meaning	.72	−.24	.65	.99	.33
Vocabulary	.51	.85	.12	.99	.16
Spelling	.85	−.22	−.23	.82	.57
Word study skills	.85	−.01	−.29	.80	.55
Variance extracted	.59	.17	.12		

TABLE 5.5. Regression of Readiness Components on Achievement Components

Readiness component	Achievement Component: Intellectual development	Vocabulary residual
Intelligence	.53	−.22
Alphabet	−.22	−.18
Perceptual speed	−.02	.04
Learning rate	−.04	−.12
Listening	−.01	−.01
Verbal vs. figural	.11	−.18
Word meaning	.09	.19
R^2	**.36**	**.16**

EXERCISES

1. Using Hotelling's iterative method (Appendix D), determine the eigenvalues and eigenvectors of the matrix

$$\mathbf{A} = \begin{bmatrix} .7 & 0 & .1 \\ 0 & .7 & .2 \\ .1 & .2 & .3 \end{bmatrix}$$

2. Determine whether the set of vectors

$$\mathbf{v}_1' = [2, 0, 2] \qquad \mathbf{v}_2' = [1, 3, 4] \qquad \mathbf{v}_3' = [-1, -2, -3]$$

is linearly independent or dependent. If the latter, find a_1, a_2, a_3 (not all zero) such that $a_1\mathbf{v}_1 + a_2\mathbf{v}_2 + a_3\mathbf{v}_3 = \mathbf{0}$.

3. (Theoretical) If $\mathbf{A}$ is a symmetric matrix with eigenvalues $\lambda_1, \lambda_2, \ldots, \lambda_r$ and eigenvectors $\mathbf{v}_1, \mathbf{v}_2, \ldots, \mathbf{v}_r$ (normalized to unity), then

$$\mathbf{A} = \lambda_1\mathbf{v}_1\mathbf{v}_1' + \lambda_2\mathbf{v}_2\mathbf{v}_2' + \cdots + \lambda_r\mathbf{v}_r\mathbf{v}_r'$$

(HINT: Denote the matrix indicated on the right-hand side by $\mathbf{B}$. Then, by successively postmultiplying by $\mathbf{v}_1$, $\mathbf{v}_2$, and so on, show that

$$\mathbf{B}\mathbf{v}_1 = \lambda_1\mathbf{v}_1, \qquad \mathbf{B}\mathbf{v}_2 = \lambda_2\mathbf{v}_2, \qquad \text{etc.}$$

In other words, show that $\mathbf{B}$ has the same eigenvalues and eigenvectors as $\mathbf{A}$. Hence, by Exercise 5 on p. 162, it follows that $\mathbf{B} = \mathbf{A}$.)

4. (Theoretical) If the p variables subject to the linear constraint are the *only* predictors being used, the regression weights $d_1, d_2, \ldots, d_{p-1}$ for the principal components do not have to be computed from scratch, but may be computed on the basis of the SSCP matrix for the original p variables and the criterion. Namely,

$$\mathbf{d} = (\mathbf{V}'\mathbf{S}_{pp}\mathbf{V})^{-1}\mathbf{V}'\mathbf{S}_{pc} = \mathbf{\Lambda}^{-1}\mathbf{V}'\mathbf{S}_{pc}$$

where $\mathbf{V}$ is the $p \times (p - 1)$ transformation matrix, and $\mathbf{S}_{pp}$, $\mathbf{S}_{pc}$ are the appropriate sections of the original SSCP matrix, as defined in Chapter 2.

5. Compute the left general inverse of the following matrix in accordance with Eq. 5.44:

$$\mathbf{A} = \begin{bmatrix} 8 & 3 \\ 0 & 5 \\ 6 & -4 \end{bmatrix}$$

6. (Theoretical) *Alternative Method for Computing a Left General Inverse*. Let $\mathbf{A}$ be an $n \times m$ matrix with rank m ($< n$), partitioned as follows:

$$\mathbf{A} = \begin{bmatrix} \mathbf{A}_{mm} \\ \hline \mathbf{A}_{qm} \end{bmatrix}$$

where the subscripts indicate the orders of the submatrices, and $q = n - m$. Since the rank of $\mathbf{A}$ is m, the submatrix $\mathbf{A}_{mm}$ must be nonsingular. So we can meaningfully define a matrix

$$\mathbf{X}_{mm} = \mathbf{A}_{mm}^{-1} - \mathbf{X}_{mq}\mathbf{A}_{qm}\mathbf{A}_{mm}^{-1}$$

where $\mathbf{X}_{mq}$ is an arbitrary matrix of order $m \times q$. Now form an $m \times n$ matrix $\mathbf{X}$ by adjoining $\mathbf{X}_{mq}$ to the right of $\mathbf{X}_{mm}$, thus:

$$\mathbf{X} = [\mathbf{X}_{mm} \mid \mathbf{X}_{mq}]$$

Prove that this $\mathbf{X}$ is a left general inverse of $\mathbf{A}$—that is, that $\mathbf{X}$ satisfies Eq. 5.46.

(HINT: Review the rule of multiplying two partitioned matrices, given in Exercise 4 on pp. 161–162.)

7. Verify that the matrix displayed at the end of Example 5.19 was obtained by the method described in Exercise 6, using

$$\mathbf{X}_{21} = \begin{bmatrix} 1 \\ 0 \end{bmatrix}$$

as the arbitrary matrix $\mathbf{X}_{mq}$ occurring in the expressions for $\mathbf{X}_{mm}$ and $\mathbf{X}$ above. Also, try using a different $\mathbf{X}_{21}$ to obtain yet another left general inverse of matrix $\mathbf{A}$ of that example.

8. For the HSB data set of Appendix F, do a principal-components analysis of the five cognitive variables (reading achievement through civics achievement). Do this by using a matrix operations program, such as the SAS PROC MATRIX, for getting the covariance matrix and then computing its eigenvectors.

9. Repeat the procedures carried out in Exercise 8, but separately for the boys and the girls this time. Compare the two sets of principal components and comment on their similarity and/or difference.

6

FACTOR ANALYSIS: A BRIEF CONCEPTUAL AND HISTORICAL ACCOUNT[1]

Factor analysis is one of the few multivariate techniques that had its origins in psychology. It is generally acknowledged that Charles Spearman was the originator of this method for studying the structure of human abilities. Spearman formulated what is now known as the two-factor theory of human abilities, which holds that there is, on the one hand, a general factor (g) that pervades all abilities and, on the other, a separate, unique factor specific to each ability, such as verbal, numerical, spatial, mechanical, and so on.

It gradually became evident that Spearman's theory was too restrictive, and alternative theories were proposed by many mathematically inclined psychologists, including Holzinger and Thurstone. Holzinger's modification consisted in changing Spearman's specific factors into group factors that each affected a small group of related abilities instead of just a single ability; this is called the *bi-factor theory*. Thurstone, on the other hand, did away with the general factor and retained only the group factors, establishing what has come to be known either as *multiple-factor* or *common-factor* theory. However, Thurstone's theory allowed also for the existence of *unique* factors, each of which affected just one ability. It can thus be said that Thurstone's model is a union of Spearman's and Holzinger's models *minus* the general factor. Since Thurstone's model seems to be the most widely used today (at least in this country), we confine our discussions to this model. Even at that, the chapter is far from being a self-contained presentation of Thurstone's multiple factor analysis in all its rich detail; for this the reader is referred to Thurstone's original book (1935), Harman (1967), or

[1]Extensive reference was made to Harman (1967, chaps. 4–6) in the preparation of this chapter.

Mulaik (1972). What this chapter seeks to do, then, is to give a simplified conceptual description of factor analysis and to indicate some historical landmarks in the development of factor analysis. It is hoped that this will give those readers who plan to take an advanced course in factor analysis some preparatory background, and will give others a general understanding of what factor analysis is and does, plus a historical perspective of the field.

6.1 THE MODEL EQUATION AND ITS IMPLICATIONS

It was intimated in Chapter 5 that principal-components analysis serves as a first step in factor analysis, or that factor analysis picks up from where principal-components analysis leaves off. We must now point out that these statements were somewhat oversimplified and that there are two or three essential differences between the two types of analyses. The first major difference is that whereas in principal-components analysis the hypothetical variables $Y_1, Y_2, \ldots$ (the principal components) are defined as linear combinations of the observed variables $X_1, X_2, \ldots, X_p$, in factor analysis it is the other way around: The observed variables are conceptualized as being linear composites of the hypothetical constructs, the factors. Second, the starting point of factor analysis is the correlation matrix rather than the covariance matrix. (This, however, is merely a reflection of the fact that in factor analysis, all variables are considered to be standardized. Since a correlation matrix is the covariance matrix of a set of standardized variables, this itself is not an essential difference.) What is essential is that it is not the usual correlation matrix that we start from in factor analysis, but what is called the *reduced* correlation matrix, a matrix in which the 1's in the diagonal of the correlation matrix are replaced by *communalities*—a concept to be explained below.

In line with the remarks above, the basic model equation of Thurstone's multiple factor theory is as follows:

$$z_j = a_{j1}F_1 + a_{j2}F_2 + \cdots + a_{jm}F_m + d_jU_j \qquad (j = 1, 2, \ldots, n) \qquad (6.1)$$

where $z_j = (X_j - \overline{X}_j)/s_j$ = standard score for the jth observed variable

n = number of observed variables

m = number of common factors (to be explained later),

F_k $(k = 1, 2, \ldots, m)$ = kth common factor,

U_j $(j = 1, 2, \ldots, n)$ = jth unique factor,

a_{jk} = loading of z_j on the kth common factor F_k

d_j = loading of z_j on its unique factor (or the uniqueness of the jth observed variable)

The first implication to be deduced from the model equation, (6.1), is based on the fact that z_j is a standardized variable and that hence its variance is unity:

$$\begin{aligned}\text{var}\,(z_j) = 1 &= a_{j1}^2 + a_{j2}^2 + \cdots + a_{jm}^2 + d_j^2 \\ &\quad + 2\sum_{k<l}\sum a_{jk}a_{kl}r_{F_kF_1} + 2\sum_{k=1}^{m} a_{jk}d_j r_{F_kU_j} \\ &= a_{j1}^2 + a_{j2}^2 + \cdots + a_{jm}^2 + d_j^2\end{aligned}$$

the last step holding if and only if all factors are uncorrelated. The correlation between any common factor and any unique factor is zero by definition. Thus the only additional assumption in going from the first to the second right-hand expression in this equation is that the common factors themselves are pairwise uncorrelated. We will make this assumption of orthogonality of the common factors for the time-being, but will later relinquish the assumption. We now define the sum of the squares of the common-factor loadings as the communality of the variable, denoted h_j^2:

$$h_j^2 = a_{j1}^2 + a_{j2}^2 + \cdots + a_{jm}^2 \tag{6.2}$$

From Eq. 6.2 and the immediately preceding equation, we get

$$h_j^2 + d_j^2 = 1 \tag{6.3}$$

In words, this says that the communality and uniqueness of any variable sum to unity; it represents a partition of the variance of each variable into a part unique to itself and a part shared with other variables.

Next, we further partition the uniqueness d_j^2 into what is called specificity, denoted b_j^2, and error of measurement e_j^2. Equation 6.3 then becomes

$$h_j^2 + b_j^2 + e_j^2 = 1$$

or

$$h_j^2 = (1 - e_j^2) - b_j^2 - r_{jj'} - b_j^2 \tag{6.3a}$$

where $r_{jj'}$ is the coefficient of reliability of the jth variable. Hence it follows that

$$h_j^2 < r_{jj'} \tag{6.4}$$

or that the reliability coefficient of a variable is an upper bound of the latter's (unknown) communality.

Returning to Eq. 6.1 itself, let us arrange all the coefficients on the right-hand sides of the n equations represented by it in the form of a matrix that will enable our retrieving the right-hand members by postmultiplying the matrix by an $(m + n)$-dimensional column vector listing the symbols for the common and unique factors. The appropriate matrix is

$$\mathbf{M} = \left[\begin{array}{cccc|cccc} a_{11} & a_{12} & \cdots & a_{1m} & d_1 & & & \\ a_{21} & a_{22} & \cdots & a_{2m} & & d_2 & \text{O} & \\ \vdots & \vdots & & \vdots & \text{O} & & \ddots & \\ a_{n1} & a_{n2} & \cdots & a_{nm} & & & & d_n \end{array}\right] \tag{6.5}$$

This is called the *factor pattern matrix* (or simply the *pattern matrix*), the initial $n \times m$ submatrix being the common-factor pattern matrix and the remaining $n \times n$ diagonal submatrix, the unique-factor pattern matrix. This is the customary end product of the first phase of factor analysis, prior to the next phase of *rotation*. Usually, only the common-factor part is displayed, since the unique-factor pattern is of little or no interest.

Alternatively, we may wish to display the matrix of correlations between the observed variables and the factors instead of the factor loadings that constitute the pattern matrix. The relationship between the pattern matrix and the matrix of variable-factor correlations, which is called the *structure matrix,* may be seen as follows. Imagine writing the jth basic equation represented by (6.1) for each individual in our sample. Denoting the ith individual's score on z_j by z_{ji}, his or her score on F_k by F_{ki}, and that on U_j by U_{ji}, we multiply both sides of the N equations by the F_{pi} with the appropriate i (i.e., the first equation by F_{p1}, the second by F_{p2}, etc.). We then add the N resulting equations to obtain

$$\begin{aligned} r_{z_jF_p} &= \sum_{i=1}^{N} z_{ji}F_{pi}/N \\ &= a_{j1}(\Sigma F_{1i}F_{pi}/N) + a_{j2}(\Sigma F_{2i}F_{pi}/N) + \cdots + a_{jp}(\Sigma F_{pi}^2/N) \\ &\quad + \cdots + a_{jm}(F_{mi}F_{pi}/N) + d_j(\Sigma U_{ji}F_{pi}/N) \\ &= a_{j1}r_{F_1F_p} + a_{j2}r_{F_2F_p} + \cdots + a_{jp} + \cdots + a_{jm}r_{F_mF_p} + d_j \cdot 0 \end{aligned} \tag{6.6}$$

the multiplier of d_j being zero because $\Sigma U_{ji}F_{pi}/N = r_{U_jF_p} = 0$ by definition, and that of a_{jp} being 1 since var $(F_p) = 1$.

We thus see that

$$r_{z_jF_p} = a_{jp} \tag{6.7}$$

if and only if $r_{F_kF_p} = 0$ for all $k \neq p$. That is, the structure coefficient of a variable is equal to the corresponding pattern coefficient (or factor loading) only if the common factors are pairwise uncorrelated; otherwise, it is equal to a linear combination of the variable's loadings on all of the common factors as shown in the last member of Eq. 6.6.

6.2 FOUNDATIONS OF FACTOR ANALYSIS IN MATRIX NOTATION

We now rewrite the basic model equation, (6.1), in matrix form so as to facilitate the derivation of further implications that it entails. First imagine writing the model equation (6.1) for each variable j (= 1, 2, . . . , n) instead of for each individual as we did above in deriving Eq. 6.6. Then, collecting the left-hand members of these n equations into an n-dimensional column vector and writing the right-hand members collectively as the product of an $n \times \overline{m + n}$ matrix times an $(m + n)$-dimensional column vector, we obtain

$$\begin{bmatrix} z_1 \\ z_2 \\ \vdots \\ z_n \end{bmatrix} = \left[\begin{array}{cccc|cccc} a_{11} & a_{12} & \cdots & a_{1m} & d_1 & & & \\ a_{21} & a_{22} & \cdots & a_{2m} & & d_2 & & \\ \vdots & & & & & & \ddots & \\ a_{n1} & a_{n2} & \cdots & a_{nm} & & & & d_n \end{array}\right] \left[\begin{array}{c} F_1 \\ F_2 \\ \vdots \\ F_m \\ \hline U_1 \\ U_2 \\ \vdots \\ U_n \end{array}\right] \begin{array}{l} \left.\vphantom{\begin{array}{c} F_1 \\ F_2 \\ \vdots \\ F_m \end{array}}\right\} \mathbf{f} \\ \left.\vphantom{\begin{array}{c} U_1 \\ U_2 \\ \vdots \\ U_n \end{array}}\right\} \mathbf{u} \end{array} \tag{6.8}$$

In symbolic matrix form, this becomes

$$\begin{aligned} \mathbf{z} &= \mathbf{M}\mathcal{F} \\ &= [\mathbf{A}|\mathbf{D}]\left[\begin{array}{c} \mathbf{f} \\ \hline \mathbf{u} \end{array}\right] = \mathbf{Af} + \mathbf{Du} \end{aligned} \tag{6.9}$$

where $\mathbf{M}$ is called the complete pattern matrix, partitioned into the common-factor part $\mathbf{A}$ and the unique-factor part $\mathbf{D}$. Similarly, $\mathcal{F}$ is the $(m + n)$-dimensional column vector comprising $\mathbf{f}$ and $\mathbf{u}$. (Recall that multiplication of partitioned matrices follows essentially the same rules as ordinary matrix multiplication, as illustrated on page 162.)

We now introduce the *score matrices* corresponding to the "generic" vectors $\mathbf{z}$, $\mathbf{f}$, and $\mathbf{u}$. These consist of N columns of the corresponding vectors, except that each column lists the scores of one individual in the sample on the respective variates. Of course, only $\mathbf{Z}$, the score matrix for $\mathbf{z}$, can list the actual observed scores; the other two, $\mathbf{F}$ and $\mathbf{U}$, are hypothetical score matrices corresponding to $\mathbf{f}$ and $\mathbf{u}$, respectively, since the "scores" are unobservable. That is,

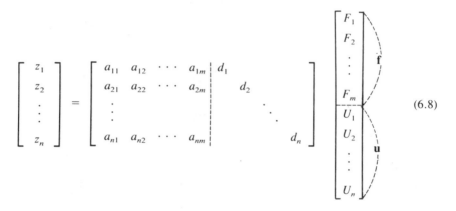

$$\mathbf{Z} = \begin{bmatrix} z_{11} & z_{12} & z_{13} & \cdots & z_{1N} \\ z_{21} & z_{22} & z_{23} & \cdots & z_{2N} \\ \vdots & & & & \\ z_{n1} & z_{n2} & z_{n3} & \cdots & z_{nN} \end{bmatrix}$$

$$\mathbf{F} = \begin{bmatrix} F_{11} & F_{12} & F_{13} & \cdots & F_{1N} \\ F_{21} & F_{22} & F_{23} & \cdots & F_{2N} \\ \vdots & & & & \\ F_{m1} & F_{m2} & F_{m3} & \cdots & F_{mN} \end{bmatrix}$$

and

$$\mathbf{U} = \begin{bmatrix} U_{11} & U_{12} & U_{13} & \cdots & U_{1N} \\ U_{21} & U_{22} & U_{23} & \cdots & U_{2N} \\ \vdots & & & & \\ U_{n1} & U_{n2} & U_{n3} & \cdots & U_{nN} \end{bmatrix}$$

Note that in these score matrices the roles of rows and columns are reversed from what they were in our data matrices up to this point. This may appear confusing at first, but it actually serves to remind us that the space in which the vectors in factor analysis are embedded and that in which the vectors in most other branches of multivariate analysis are embedded have a sort of reversal between themselves, as we shall point out later in this chapter. As a consequence of this reversal of roles between rows and columns in the score matrices, the SSCP matrix of the observed variables is given by $\mathbf{ZZ}'$ instead of $\mathbf{Z}'\mathbf{Z}$ as we might incorrectly expect in analogy with our previous $\mathbf{x}'\mathbf{x}$. Furthermore, since all scores are assumed to be in the standard-score scale, the covariance matrix $\mathbf{ZZ}'/\mathbf{N}$ is actually the correlation matrix. (By tradition, the divisor of the SSCP matrix is taken as N instead of $N - 1$ in factor analysis, presumably because the typical sample size in studies using factor analysis is so large that the difference between N and $N - 1$ is negligible.)

Let us first examine in detail the $n \times (m + n)$ matrix of correlations between the observed variables and the factors, both common and unique. If we denote the $(m + n) \times N$ factor-score matrix by $\mathcal{F}$ (which is the vertical concatenation of the previous $\mathbf{F}$ and $\mathbf{U}$), the model equation applied to individual scores becomes

$$\mathbf{Z} = \mathbf{M}\mathcal{F} \tag{6.10}$$

The matrix of correlations between the observed variables and the $m + n$ factors is, hence, given by

$$\begin{aligned} \mathbf{S} &= \mathbf{ZF}'/\mathrm{N} \\ &= \mathbf{M}(\mathbf{FF}')/\mathrm{N} \end{aligned} \qquad (6.11)$$

(where we have used Eq. 6.10 and the associative law of matrix multiplication in going from the second to the last member). This is called the *complete* structure matrix, and—recalling the definitions of **M** and **F**—it may be rewritten, successively, as

$$\mathbf{S} = [\mathbf{A}|\mathbf{D}]\left[\begin{array}{c} \mathbf{F}\ (m \times N) \\ \hdashline \mathbf{U}\ (n \times N) \end{array}\right]\left[\begin{array}{c:c} \mathbf{F}' & \mathbf{U}' \\ (N \times m) & (N \times n) \end{array}\right] \div N$$

$$= [\mathbf{A}|\mathbf{D}]\left[\begin{array}{c:c} \mathbf{FF}'/N & \mathbf{FU}'/N \\ (m \times m) & (m \times n) \\ \hdashline \mathbf{UF}'/N & \mathbf{UU}'/N \\ (n \times m) & (n \times n) \end{array}\right]$$

$$= [\mathbf{A}|\mathbf{D}]\left[\begin{array}{c:c} \boldsymbol{\phi} & \mathbf{0} \\ (m \times m) & (m \times n) \\ \hdashline \mathbf{0}' & \mathbf{I} \\ (n \times m) & (n \times n) \end{array}\right]$$

where $\boldsymbol{\phi}$ is the matrix of correlations among the m common factors, the two nonsquare sectors are null matrices because the common and unique factors are uncorrelated by definition, and the lower-right $n \times n$ sector is an identity matrix because the different unique factors are pairwise uncorrelated among themselves (making the off-diagonal elements zero) while the diagonal elements are all unities because they each represent the variance of a unique factor, which is in standard-score form.

Carrying out the multiplication of the two partitioned matrices standing in the last expression, and noting that $\mathcal{S}$ may be partitioned into the matrix of correlations between observed variables and *common* factors, followed by the diagonal matrix **D** with the correlations between the observed variables each with its own unique factor as the diagonal elements, we obtain the following result:

$$\mathbf{S} = [\mathbf{S}|\mathbf{D}] = [\mathbf{A}\boldsymbol{\phi}|\mathbf{D}] \qquad (6.12)$$

From this it follows that

$$\mathbf{S} = \mathbf{A}\boldsymbol{\phi}$$

and hence that

$$\mathbf{S} = \mathbf{A} \quad \text{iff} \quad \boldsymbol{\phi} = \mathbf{I} \qquad (6.13)$$

That is, the structure matrix and pattern matrix are equal if and only if the common factors are uncorrelated.

Let us now see how the observed correlation matrix can be expressed when the factor model stated in Eq. 6.10 holds exactly; we then get

$$\begin{aligned}\mathbf{R} &= \frac{\mathbf{ZZ}'}{N} \\ &= \frac{(\mathbf{MF})(\mathbf{MF})'}{N} \\ &= \mathbf{M}\left(\frac{\mathbf{FF}'}{N}\right)\mathbf{M}' \\ &= [\mathbf{A}|\mathbf{D}]\left[\begin{array}{c|c}\boldsymbol{\phi} & \mathbf{0} \\ \hline \mathbf{0}' & \mathbf{I}\end{array}\right]\left[\begin{array}{c}\mathbf{A}' \\ \hline \mathbf{D}\end{array}\right]\end{aligned}$$

Carrying out the multiplication indicated in the last expression (which is analogous to a quadratic form), we get the result that

$$\mathbf{R} = \mathbf{A}\boldsymbol{\phi}\mathbf{A}' + \mathbf{D}^2 \tag{6.14}$$

if and only if the factor model holds exactly. Equating the jth diagonal elements of the two sides of this equation, we find

$$1 = (\mathbf{A}\boldsymbol{\phi}\mathbf{A}')_{jj} + d_j^2$$

Comparing this with Eq. 6.3, it follows that

$$(\mathbf{A}\boldsymbol{\phi}\mathbf{A}')_{jj} = h_j^2 \tag{6.15}$$

when the factor model fits the data perfectly. We therefore define

$$\mathbf{R}^{\dagger} = \mathbf{R} - \mathbf{D}^2 = \mathbf{A}\boldsymbol{\phi}\mathbf{A}' \tag{6.16}$$

and call it the *reduced correlation matrix*, which has communalities instead of unities as the diagonal elements. It further follows that when the common factors are mutually uncorrelated (i.e., when $\boldsymbol{\phi} = \mathbf{I}$),

$$\mathbf{R}^{\dagger} = \mathbf{AA}' \tag{6.17}$$

Hence if the communalities h_j^2 were known, $\mathbf{R}^{\dagger}$ would be completely determined, and we could compute the factor pattern matrix $\mathbf{A}$ as follows: First, compute the $n \times m$ matrix $\mathbf{V}$ whose columns are the eigenvectors of $\mathbf{R}^{\dagger}$ that are associated with its m ($< n$) nonzero eigenvalues. By definition, $\mathbf{V}$ will satisfy the equation

$$\mathbf{R}^{\dagger}\mathbf{V} = \mathbf{V}\boldsymbol{\Lambda}$$

where $\boldsymbol{\Lambda}$ is the $m \times m$ diagonal matrix whose diagonal elements are the nonzero eigenvalues of $\mathbf{R}^{\dagger}$. Consequently, since $\mathbf{V}$ is orthonormal by columns (i.e.,

$\mathbf{VV'} = \mathbf{I}$), we obtain, upon postmultiplying both sides of this equation by $\mathbf{V}'$, the following result:

$$\begin{aligned}\mathbf{R}^{\dagger} &= \mathbf{V\Lambda V'}\\ &= (\mathbf{V\Lambda}^{1/2})(\mathbf{V\Lambda}^{1/2})'\end{aligned}$$

We may therefore define $\mathbf{V\Lambda}^{1/2}$ as our factor common-factor pattern matrix $\mathbf{A}$ satisfying Eq. 6.17, and the factor solution would be complete. Unfortunately, however, the communalities are unknown, so we have to resort to an iterative procedure for estimating the communalities and the pattern matrix simultaneously, such as the one described (in simplified terms) in Section 6.5.

6.3 CONSTRAINTS FOR THE REDUCED CORRELATION MATRIX TO HAVE RANK LESS THAN ITS ORDER

To build up to our discussion of the iterative solution, we first consider the twin problem of estimating the communalities and determining the number of factors, that is, the lowest rank that can be achieved by the reduced correlation matrix $\mathbf{R}^{\dagger}$, defined by Eq. 6.16 as

$$\mathbf{R}^{\dagger} = \begin{bmatrix} h_1^2 & r_{12} & r_{13} & \cdots & r_{1n} \\ r_{21} & h_2^2 & r_{23} & \cdots & r_{2n} \\ \vdots & & \ddots & & \\ r_{n1} & r_{n2} & r_{n3} & \cdots & h_n^2 \end{bmatrix}$$

The problem is to make $\mathbf{R}^{\dagger}$ have a rank smaller than its order by suitable choice of values for its diagonal elements, that is, the communalities $h_1^2, h_2^2, \ldots, h_n^2$— whose "true" values are unknown anyway. Obviously, however, the rank cannot be made arbitrarily small; there is a limit to how small we can make it.

What, then, is the minimum rank that can be achieved by the choice of suitable communalities? To answer this question, we need to invoke a *third* definition of the rank of a matrix, which is applicable only to symmetric matrices and which is related to Property 5 of Gramian matrices. Of course, this new definition is equivalent to each of the two definitions given earlier insofar as they are used for symmetric matrices.

Definition The rank of a symmetric matrix $\mathbf{A}$ is m if and only if (a) there exists a nonzero principal minor $|\mathbf{A}_m|$ of order m, and (b) all principal minors of orders $m + 1$ and $m + 2$ that are obtainable by augmenting $|\mathbf{A}_m|$ with one or two corresponding rows and columns of $\mathbf{A}$ have the value 0.

This definition amounts to saying that "the rank of a symmetric matrix is equal to the order of the largest nonzero principal minor it has," but it is easier to apply in practice than the latter statement. We are, however, interested more in the theoretical implications of this definition than in its practical application.

If the order of $\mathbf{A}$ is n, then there are $n - m$ rows and their corresponding columns that can be annexed to $|\mathbf{A}_m|$ to generate principal minors of order $m + 1$. To require every one of these to be zero is to impose $n - m$ conditions on the elements of $\mathbf{A}$. Similarly, we can choose $\binom{n-m}{2} = (n - m)(n - m - 1)/2$ *pairs* of rows plus their corresponding columns with which to generate principal minors of order $m + 2$. Requiring all of these also to be zero amounts to imposing $(n - m)(n - m - 1)/2$ additional conditions on the elements of $\mathbf{A}$. All in all, therefore, we require

$$c = (n - m) + (n - m)(n - m - 1)/2$$
$$= (n - m)(n - m + 1)/2$$

conditions to be satisfied by the elements of $\mathbf{A}$ in order that the latter's rank will be m.

Let us now apply the foregoing conclusion to a reduced correlation matrix $\mathbf{R}^{\dagger}$. Out of the c conditions to be satisfied by the elements of $\mathbf{R}^{\dagger}$, n of them can be satisfied by assigning suitable values to the n unknown communalities, as mentioned above. Hence, if $c \leq n$, all of the conditions can be met simply by choosing appropriate values for the h_j^2s. The inequality may more conveniently be written as $c - n \leq 0$, or introducing the symbol Δ_{nm} for the difference, as

$$\Delta_{nm} = (n - m)(n - m + 1)/2 - n \leq 0 \tag{6.18}$$

Table 6.1 shows the value of Δ_{nm} for $n = 2, 3, \ldots, 13$ and $m = 1, 2, \ldots, 7$. For each n, the smallest value of m for which the mth row, nth column entry is 0 or negative is the smallest rank to which an $\mathbf{R}^{\dagger}$ of order n can be reduced by choosing suitable values for the communalities. More specifically, if $\Delta_{mn} = 0$, the conditions required for a reduced correlation matrix of order n to have rank m ($<n$) can be "just barely" satisfied by assigning appropriate values to the h_j^2's. This would be the case if, for example, $n = 10$ and $m = 6$, because $\Delta_{10,6} = (10 - 6)(10 - 6 + 1)/2 - 10 = 0$.

If $\Delta_{nm} < 0$, the required conditions cannot only be met by assigning suitable values to the h_j^2's, but we even have the freedom to choose $|\Delta_{nm}|$ of the values at will—within obvious constraints such as $h_j^2 < 1$, of course. For instance, if $n = 9$ and we want to make $m = 6$, we would have $\Delta_{9,6} = -3$ (as the reader should verify); hence we may give arbitrary values to three of the nine communalities, whereupon the remaining six communalities would have to be assigned special values in order to satisfy the conditions.

If $\Delta_{nm} > 0$ (as will most often be the case), the conditions cannot be met

TABLE 6.1 The Number, $\Delta_{nm} = (n - m)(n - m + 1)/2 - n$, of Conditions That Must be Satisfied by the $\binom{n}{2}$ Correlation Coefficients in Order that a Reduced Correlation Matrix $\mathbf{R}^{\dagger}$ of Order n be of Rank m

	n											
m	2	3	4	5	6	7	8	9	10	11	12	13
1	−1	0	2	5	9	14	20	27	35	44	54	65
2	−2	−2	−1	1	4	8	13	19	26	34	43	53
3		−3	−3	−2	0	3	7	12	18	25	33	42
4			−4	−4	−3	−1	2	6	11	17	24	32
5				−5	−5	−4	−2	1	5	10	16	23
6					−6	−6	−5	−3	0	4	9	15
7						−7	−7	−6	−4	−1	3	8

simply by giving suitable values to the h_j^2's, but the off-diagonal elements (i.e., the observed correlation coefficients) must satisfy Δ_{nm} conditions. For example, if $n = 12$ and we want the rank of $\mathbf{R}^{\dagger}$ to be $m = 6$, not only must we assign appropriate values to the h_j^2's but, in addition, the correlation coefficients have to satisfy

$$\Delta_{12,6} = (12 - 6)(12 - 6 + 1)/2 - 12 = 9$$

conditions. But the correlation coefficients [which are $\binom{12}{2} = 66$ in number] are observed quantities that cannot be made to satisfy any conditions at our command! It is nature, so to speak, that has the say as to whether the correlations shall satisfy the required conditions. Of course, we could never expect them to be satisfied exactly, but perhaps they might be approximately satisfied.

We are in a situation that is parallel to the one we encountered in principal-components analysis, when the first few leading eigenvalues of the covariance matrix (say the first three, to be specific) summed to a large percentage (say more than 90%) of the trace of $\mathbf{C}$. We then said that the swarm of points represented by $\mathbf{C}$ was *essentially* confined to three dimensions. In the present case, if $\Delta_{nm} = 9$ (as in the example above), *and* nature has obligingly seen to it that our correlation coefficients approximately satisfied these nine constraints, then we could say that our battery of 12 tests is more or less adequately "explained" by six common factors, assuming that we can determine the appropriate values of the 12 communalities. This, in a nutshell, is the problem we are faced with in factor analysis. We do not know the "appropriate" values of the communalities of our battery of tests, whose correlation matrix we seek to factor-analyze; nor do we know what the "essential" dimensionality of our swarm of points is—that is, how many conditions among the correlation coefficients nature has "seen fit" to impose. (There always must be some constraints holding among them. Otherwise, we would have a random set of correlation coefficients, and factoring their matrix would not yield any result of remote interest.) The dimensionality cannot be determined unless and until we know the communalities; but we cannot know the communality values unless we know the dimensionality. It is this vicious circle that factor analysts have been struggling with and trying, over the decades, to break. With the advent of high-speed computers, many ingenious iterative techniques have been developed that enable the simultaneous solution of this age-old twin problem. However, before we discuss one of these iterative methods, let us look at the admittedly unrealistic but much simpler cases when $\Delta_{nm} \leq 0$. They will provide us with a frame of reference or a backdrop, as it were, that will enable us to see more clearly the nature of factor analysis.

6.4 CASES WHEN COMMUNALITIES AND RANK ARE DETERMINABLE

We now examine the simple, substantively almost trivial, cases when $\Delta_{nm} \leq 0$, and hence the conditions for $\mathbf{R}^{\dagger}$ to have rank less than order can be satisfied just by assigning suitable values to the communalities. Table 6.1 shows the values of Δ_{nm} for $n = 2(1)12$ and $m = 1(1)7$. Any cell with an entry of 0 or a negative number illustrates the cases under consideration. As the simplest possible example of exact determination, let us take the case of $n = 3$, $m = 1$. Since $\Delta_{3,1} = 0$, we know that assigning the appropriate values to h_1^2, h_2^2, and h_3^2 will yield an $\mathbf{R}^{\dagger}$ with exact rank 1. The following example shows how this comes about.

EXAMPLE 6.1. Suppose that we have observed the 3×3 correlation matrix

$$\mathbf{R} = \begin{bmatrix} 1 & .35 & .42 \\ .35 & 1 & .30 \\ .42 & .30 & 1 \end{bmatrix}$$

and wish to get a reduced correlation matrix $\mathbf{R}^{\dagger}$ of rank 1 by replacing the unities in the diagonal by communalities h_1^2, h_2^2, and h_3^2 with suitable values. Let us denote the unknown h_1^2 by x^2 for simplifying the notation. We then define

$$\mathbf{a}_1' = [x,\ .35/x,\ .42/x]$$

(which is obtained by first writing $[x^2, .35, .42]$—i.e., the first row of $\mathbf{R}$ with the 1 replaced by the unknown x^2—and then dividing all three elements by x). Next, we construct the matrix product $\mathbf{a}_1\mathbf{a}_1'$, only the above-diagonal elements of which we write below, since the product is a symmetric 3 × 3 matrix. That is,

$$\mathbf{a}_1\mathbf{a}_1' = \begin{bmatrix} x^2 & .35 & .42 \\ & (.35)^2/x^2 & (.35)(.42)/x^2 \\ & & (.42)^2/x^2 \end{bmatrix}$$

Note that the two off-diagonal elements of the first row (and column) of $\mathbf{a}_1\mathbf{a}_1'$ are equal to those of $\mathbf{R}$, by construction. We now determine x so that the (2,3)-element of $\mathbf{a}_1\mathbf{a}_1'$ will be equal to r_{23}; that is,

$$(.35)(.42)/x^2 = .30$$

or

$$x^2 = (.35)(.42)/.30 = .49$$

Hence

$$h_1^2 = .49 \qquad h_2^2 = (.35)^2/.49 = .25 \qquad h_3^2 = (.42)^2/.49 = .36$$

Thus

$$\mathbf{a}_1' = [.70,\ .50,\ .60]$$

and

$$\mathbf{R}^\dagger = \begin{bmatrix} .49 & .35 & .42 \\ .35 & .25 & .30 \\ .42 & .30 & .36 \end{bmatrix} = \mathbf{a}_1\mathbf{a}_1'$$

which is of rank 1 since it is the cross-product of a single vector.

The example we just saw was a simple case of what is known as *Cholesky decomposition*. Let us now take a slightly more complicated example, with $n = 4$, $m = 2$. The cell entry -1 for $\Delta_{4,2}$ tells us that one of the communalities can be given an arbitrary value (<1), whereupon the other three communalities are uniquely determined in order for $\mathbf{R}^\dagger$ to have a rank of 2. The way these values are determined is illustrated in the following example.

EXAMPLE 6.2. Consider the 4 × 4 reduced correlation matrix

$$\mathbf{R}^\dagger = \begin{bmatrix} h_1^2 & .48 & .32 & -.40 \\ .48 & h_2^2 & .18 & -.12 \\ .32 & .18 & h_3^2 & -.32 \\ -.40 & -.12 & -.32 & h_4^2 \end{bmatrix}$$

Since we may give an arbitrary value less than 1 to one of the communalities, let us take $h_1^2 = (.80)^2 = .64$. As in Example 6.1 (except that here h_1 has the known value .80 instead of the unknown x), we then define

$$\begin{aligned}\mathbf{a}_1' &= [.80, .48/.80, .32/.80, -.40/.80] \\ &= [.80, .60, .40, -.50]\end{aligned}$$

Hence

$$\mathbf{a}_1\mathbf{a}_1' = \begin{bmatrix} .64 & .48 & .32 & -.40 \\ .48 & .36 & .24 & -.30 \\ .32 & .24 & .16 & -.20 \\ -.40 & -.30 & -.20 & .25 \end{bmatrix}$$

Next, we subtract this matrix from the original $\mathbf{R}^\dagger$ (with h_1^2 replaced by .64) and call the result $\mathbf{R}^\dagger_{(1)}$, for "first *residual* reduced correlation matrix." That is,

$$\mathbf{R}^\dagger_{(1)} = \mathbf{R}^\dagger - \mathbf{a}_1\mathbf{a}_1' = \begin{bmatrix} 0 & 0 & 0 & 0 \\ 0 & x - .36 & -.06 & .18 \\ 0 & -.06 & y - .16 & -.12 \\ 0 & .18 & -.12 & z - .25 \end{bmatrix}$$

where we have replaced h_2^2, h_3^2, and h_4^2 by x, y, and z, respectively, to simplify the notation.

We now define

$$\mathbf{a}_2' = [0, \sqrt{x - .36}, -.06/\sqrt{x - .36}, .18/\sqrt{x - .36}]$$

whereupon

$$\mathbf{a}_2\mathbf{a}_2' = \begin{bmatrix} 0 & 0 & 0 & 0 \\ 0 & x - .36 & -.06 & .18 \\ 0 & -.06 & .0036/(x - .36) & -.0108/(x - .36) \\ 0 & .18 & -.0108/(x - .36) & .0324/(x - .36) \end{bmatrix}$$

The second row (and column) of this product are equal to those of $\mathbf{R}^\dagger_{(1)}$ by construction. In order that the three remaining distinct elements of $\mathbf{a}_2\mathbf{a}_2'$ also be equal to the corresponding elements of $\mathbf{R}^\dagger_{(1)}$, we must have

$$-.0108/(x - .36) = -.12$$

or

$$.12x - .0432 = .0108$$

whence

$$x = .45$$

and

$$x - .36 = .09$$

Then

$$.0036/(x - .36) = .0036/.09 = y - .16$$

and

$$y = .20$$

Similarly,

$$.0324/(x - .36) = .0324/.09 = z - .25$$

and

$$z = .61$$

Hence the reduced correlation matrix, with the communalities determined, is

$$\mathbf{R}^{\dagger} = \begin{bmatrix} .64 & .48 & .32 & -.40 \\ .48 & .45 & .18 & -.12 \\ .32 & .18 & .20 & -.32 \\ -.40 & -.12 & -.32 & .61 \end{bmatrix}$$

and since $\sqrt{x - .36} = \sqrt{.09} = .30$, $\mathbf{a}_2'$ becomes

$$\mathbf{a}_2' = [0, .30, -.06/.30, .18/.30]$$
$$= [0, .30, -.20, .60]$$

It may readily be verified that

$$\mathbf{R}^{\dagger} = \mathbf{a}_1\mathbf{a}_1' + \mathbf{a}_2\mathbf{a}_2'$$

In other words,

$$\mathbf{R}^{\dagger} = \mathbf{A}\mathbf{A}'$$

with

$$\mathbf{A} = \begin{bmatrix} .80 & 0 \\ .60 & .30 \\ .40 & -.20 \\ -.50 & .60 \end{bmatrix}$$

EXERCISES

1. In the reduced correlation matrix of the example just discussed, let $h_1^2 = .5625$ instead of .64 and carry out the Cholesky decomposition, thereby determining the remaining three communalities as well as the factor-pattern matrix.

2. Find the Cholesky factors of the reduced correlation matrix

$$\mathbf{R}^{\dagger} = \begin{bmatrix} .81 & -.54 & .27 & .36 \\ -.54 & .61 & .02 & -.34 \\ -.27 & .02 & .25 & .04 \\ .36 & -.34 & .04 & .20 \end{bmatrix}$$

(Here there are no unknown communalities to be determined. Although we have not shown an example of this type, the problem should be simpler than the two examples given above.)

3. Determine h_3^2 in the following reduced correlation matrix so that its rank will be 3.

$$\mathbf{R}^{\dagger} = \begin{bmatrix} .49 & .35 & -.28 & .21 \\ .35 & .61 & -.02 & .03 \\ -.28 & -.02 & .50 & .02 \\ .21 & .03 & .02 & h_3^2 \end{bmatrix}$$

It should be clearly understood that Cholesky decomposition is just a mathematical device for factoring a reduced correlation matrix (or any other symmetrical matrix) with known rank into the product of a triangular matrix and its transpose. Although it can be used as the starting point for factor-analyzing a reduced correlation matrix, with subsequent rotations to simple structure (to be discussed later), its usefulness in real-life factor analysis is quite limited. Its utility lies, rather, in serving as a tool for carrying out what is called the Gram–Schmidt procedure for constructing an orthonormal basis in such contexts as the general linear model.

We now turn to the more customary and realistic case of factor analysis when neither the communalities nor the number of factors (i.e., the dimensionality) is known and we have to use an iterative method. As intimated above, the method we describe is not necessarily an efficient one that would be used in practice. It merely serves to give a simple conceptualization of what is involved in principle.

6.5 A CONCEPTUALLY SIMPLIFIED ITERATIVE PRINCIPAL FACTORING METHOD

It was shown earlier that the reliability of a variable is an upper bound of its communality. A simple argument suffices to show that a *lower* bound to the communality of a variable in the battery to be factor-analyzed is given by the squared multiple-R of that variable with the rest of the variables in the battery, denoted by $R^2_{j.12\ldots)j(\ldots n}$. We have only to realize that this squared multiple-R (or SMR) represents the variance that is shared by z_j and the other $n - 1$ observed variables themselves, which is necessarily "contaminated" by the unique factors of these variables. It should not be difficult to see that this contamination can only act in the direction of decreasing the proportion of variance shared by z_j and all the *common* factors, which is the communality h_j^2 of the jth variable. Thus we now have h_j^2 bounded from both above and below:

$$R^2_{j.12\ldots)j(\ldots n} \leq h_j^2 \leq r_{jj'}$$

With this addition to our repertoire, we are now ready to outline the steps of a simplified iterative procedure.

Step 1. Compute the eigenvalues of the *observed* correlation matrix R (with unities in the diagonal). Let m be the number of eigenvalues that are greater than 1, as required by Kaiser's (1960) criterion.

Step 2. Take $h^2_{j(1)} = R^2_{j.12\ldots)j(\ldots n}$ as the initial estimate of h_j^2.

Step 3. let $\mathbf{R}^{\dagger}_{(1)} = \mathbf{R}_0 + \hat{\mathbf{H}}_{(1)}$, where

$$\mathbf{R}_0 = \begin{bmatrix} 0 & r_{12} & r_{13} & \cdots & r_{1n} \\ r_{21} & 0 & r_{23} & \cdots & r_{2n} \\ \vdots & \cdot & \cdot & \cdot & \vdots \\ r_{n1} & r_{n2} & r_{n3} & \cdots & 0 \end{bmatrix} \quad \text{and} \quad \hat{H}_{(1)} = \begin{bmatrix} \hat{h}^2_{1(1)} & & & \\ & \hat{h}^2_{2(1)} & & 0 \\ 0 & & \ddots & \\ & & & \hat{h}^2_{n(1)} \end{bmatrix}$$

(This is simply a device for separating out the off-diagonal and diagonal parts of the initial trial value of the reduced correlation matrix so that the off-diagonal part—consisting of the observed correlation coefficients—may be kept intact throughout the iteration steps; we want only the diagonal part to change successively.)

Step 4. Compute the m largest eigenvalues $\lambda_{p(1)}$ and their associated eigenvectors $\mathbf{v}_{p(1)}$ ($p = 1, 2, \ldots, m$; where m is the number we get in Step 1).

Step 5. Put the m eigenvectors together in an $n \times m$ matrix, and the square roots of the m eigenvalues in a diagonal matrix $\mathbf{\Lambda}^{1/2}$; then form their product, calling it $\mathbf{A}_{(1)}$:

$$\mathbf{A}_{(1)} = \mathbf{V}_{(1)}\mathbf{\Lambda}^{1/2}_{(1)}$$

Step 6. Compute the diagonal elements of $\mathbf{A}_{(1)}\mathbf{A}'_{(1)}$; let $(\mathbf{A}_{(1)}\,\mathbf{A}'_{(1)})_{jj} = \hat{h}^2_{j(2)}$ and arrange these in a diagonal matrix denoted $\hat{H}_{(2)}$.

Step 7. Define $\mathbf{R}^{\dagger}_{(2)} = \mathbf{R}_0 + \hat{\mathbf{H}}_{(2)}$.

Steps 8–11.
Steps 12–15 etc. Repeat Steps 4–7 several times over, each time increasing all the subscripts in parentheses by 1. Stop when two consecutive sets of communality estimates are approximately equal.

The rule "Stop when two consecutive sets of communality estimates are approximately equal" begs the question, "Approximately equal by what criterion?" Although a test of approximate convergence of communality values is possible, it is more common to examine how close the reproduced correlation matrix as a whole is to the observed correlation matrix. Recalling that we are seeking a factor matrix **A** such that

$$\mathbf{R}^{\dagger} = \mathbf{R} - \mathbf{D}^2 = \mathbf{A}\mathbf{A}'$$

it follows that we need to see how close $\mathbf{A}_k\mathbf{A}_k' + \mathbf{D}_k^2$ is to **R**, where $\mathbf{A}_k$ is the kth iteration value of **A**, and $\mathbf{D}_k^2$ is the kth approximation to $\mathbf{D}^2$, computed as $\mathbf{I} - \hat{\mathbf{H}}_{(k)}$. One test of proximity of the reproduced correlation matrix to the observed one is *Rippe's test*. This calls for computing

$$(N - 1) \ln [|\mathbf{A}_k\mathbf{A}_k' + \mathbf{I} - \hat{\mathbf{H}}_{(k)}|/|\mathbf{R}|] \tag{6.19}$$

and comparing its value to a chi-square with $(n - m)(n - m + 1)/2$ degrees of freedom. If the value exceeds the prescribed $100(1 - \alpha)\%$ point of this distribution, we conclude that the reproduced correlation matrix is not close enough to the observed, and do a few more iterations.

If convergence is not achieved within a reasonable number of iterations, we may suspect that the number of factors extracted (i.e., the m obtained in Step 1)—was too small. We then increase m by 1 and repeat the procedures from Step 4 on.

Alternatively, we may make a direct check to see if the m found in Step 1 is adequate without testing the proximity between the reproduced and observed correlation matrices using Rippe's test. One method for doing this, Cattell's (1966) scree test, may in fact be used instead of the method given as Step 1, although using both would increase our confidence if both indicate the same number of factors. Essentially a refinement of the earliest test in this field, due to Tucker (1938), the scree test consists in plotting the eigenvalues of the observed correlation matrix **R** against their ordinal number and noting the point at which the plot becomes nearly horizontal (i.e., where the eigenvalues start showing little change from one to the next). Figure 6.1 shows such a plot, which starts leveling off at λ_3, so we conclude that the plot from this point on is mere "scree" (which means "rubbish at the foot of a steep seashore"!). Hence we conclude that the number of factors is two. This method is very simple to use, but as the reader has doubtless noticed, it is also rather subjective in that it involves a judgment of "just how level is level."

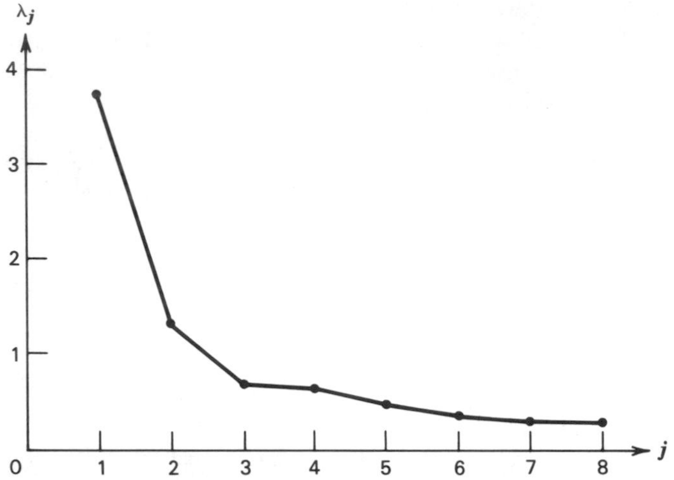

Figure 6.1. *Plot of successive eigenvalues λ_j of an 8 × 8 correlation matrix against their ordinal number,* j.

A method that may be regarded as a sort of more objective scree test is Humphreys' (1969) "random correlation matrix" method. This method does require us first to carry out the iterations described above. To begin, the eigenvalues of the reproduced reduced correlation matrix $\mathbf{A}_k\mathbf{A}_k'$ are plotted against their ordinal number. Next, one generates a correlation matrix with off-diagonal elements based on a Monte Carlo sampling distribution appropriate to the number of variables (n) and the sample size (N) that were used, and with SMRs in the diagonal just as in Step 2. The eigenvalues of this semirandom reduced correlation matrix are then computed and are plotted on the same graph that shows the plot of the eigenvalues of $\mathbf{A}_k\mathbf{A}_k'$. This second plot will be flatter (i.e., it will be close to a horizontal straight line) from the second eigenvalue on. If, however, the scree portion of the first plot coincides approximately with that of the second plot from the mth eigenvalue on, we may conclude that the factoring is complete. If not, the m is increased by 1 as in the preceding methods.

The reader may have wondered why in the foregoing discussions we spoke only of *increasing* the m, never of decreasing it. This is because lack of close agreement between the reproduced and observed correlation matrices is due to *underfactoring* (i.e., extracting too few factors). However, sometimes overfactoring can cause problems too. This is because we would then tend to split a "genuine" factor into a strong factor plus one or more minute or trivial factors. This would show up in the rotation stage (discussed in Section 6.6) by having factors that are very highly correlated. If this happens, we should go back to the iteration steps, decrease the last m by 2, and repeat Steps 4 on.

6.6 ROTATION TO SIMPLE STRUCTURE

The foregoing procedures—which, to repeat, do not constitute an efficient method that would be used in real life—will yield a set of orthogonal, principal factors. These would suffice if the sole purpose were to "account for" the total variability in the variate space and reproduce the intercorrelations by means of a small number of common factors. However, the principal factors would generally not lend themselves easily to interpretation, which is what most factor analysts would insist on doing. For this purpose we need to carry out further processing in the way of rotations to what is called *simple* structure.

Ironically, the concept of simple structure is by no means simple and unambiguous. Introduced by Thurstone (1931) more than 50 years ago, it has been the target of many attempts at clarification and objectification by several authors, with little success. Thus Thurstone's latest version (1947) of the set of criteria for simple structure is perhaps as good as any to cite here. They are that the $n \times m$ factor-pattern matrix should satisfy the following conditions:

1. Each row of the pattern matrix should contain at least one zero.
2. Each column of the pattern matrix should contain at least m zeros.
3. Every pair of columns should contain rows whose loadings are zero in one column but nonzero in the other.
4. If $m > 4$, every pair of columns should contain a large number of rows (variables) whose loadings are zero in *both* columns.
5. For every pair of columns, only a few rows should have nonzero loadings in both columns.

Let us restate these conditions in plain English, allowing some looseness. Condition 1 says that not all of the factors should account for individual differences on any given variable; unless this holds, every variable would be related to every factor—a very "unsimple" situation. Condition 2 is a sort of inverse of (1): It decrees the description of each factor in terms of the observed variables (as in principal-components analysis) should be "simple." Condition 3 says that any two factors should be differentially related to the observed variables. Condition 4 serves to differentiate each pair of factors from other factors. Condition 5 seems to be largely a repetition of (4).

It should be clear that no real-life factor pattern matrix can satisfy conditions 1–5 exactly; we would expect at most an approximate satisfaction of these conditions. Again, how close is close enough is a matter of subjective judgment. With the advent of the computer age, researchers' efforts shifted from trying to clarify Thurstone's conditions; instead, the focus became one of trying to capture in mathematical, programmable form something *resembling* these conditions. Several authors, including Carroll (1953), Neuhaus and Wrigley (1954), and

Saunders (1953), independently developed what is now known as the *quartimax* criterion. Essentially, the computer is programmed to seek rotated axes for the factors that have the property of maximizing the sum, Q, of the fourth powers of the loadings. Although this criterion has a number of intuitively desirable properties, a major drawback is that it does not preclude the establishment of a general factor—which runs counter to Thurstone's multiple-factor model.

For the above reason the quartimax criterion has now been largely superseded by Kaiser's (1956, 1958) *varimax* criterion. What this procedure does is to rotate the axes to orientations that maximize the sum of the variances of squared loadings within the individual columns. The varimax rotation is an orthogonal rotation (as is quartimax). Unfortunately, as mentioned earlier, such orthogonal methods often make for difficulty in arriving at clear, intuitive interpretation of the factors. This in turn gave impetus to the development of oblique rotation criteria such as oblimax, binormamin, biquartimin, and so on, many of which are also due to Kaiser and his associates.

The most recent development in the field of factor rotation is Tucker and Finkbeiner's (1981) automated computer method known as "direct artificial personal probability factor rotation" (DAPPFR). "Personal probability" refers to the judgment of when a test vector (more generally, an attribute vector) has a sufficiently low loading on a factor that it may be regarded as lying in the "plane" perpendicular to the axis representing that factor. The term "personal probability"—and especially the qualifier "artificial"—should not be construed as implying that there is anything particularly subjective about this method compared to other approaches. Rather, it makes explicit the fact that no automated, computer-based method of rotation can eliminate a built-in role for subjective judgment in deciding when the criteria for simple structure have adequately been met. If anything, its role has become even more crucial as a "criterion upon which to judge the quality of results obtained by various procedures."

Thus it is safe to say that although some refinements remain to be made, the state of the art of factor analysis is now such that factor analysts can correctly claim that the model-building phase has essentially been completed and that the main problems remaining are statistical ones relating to the distribution of factor loadings and the extent of variation of factor structures from one population to another. It should also be mentioned that there is an increasing recognition of two rather different functions of factor analysis: *exploratory* and *confirmatory*. The former is the "traditional" function, while the latter is the more recent development starting with the work of Lawley and Maxwell (1963) and brought to fruition by Jöreskog and his coworkers (1969, 1971, 1979). Here the purpose is to confirm (or disconfirm) an a priori theory or theories concerning the factor structure in the domain of interest. Actually, the most recent development, called LISREL (which is also the name of the computer program used for implementing the system), covers both the exploratory and confirmatory factor-analytic models as well as several other models, including path analysis. In short, LISREL (for "linear structural relationships") is a synthesis of virtually all models and techniques that are concerned with the covariance structure of a domain in terms of both observed and latent variables or hypothetical constructs. Since a detailed description of LISREL would take us too far afield, however, the interested reader is referred to an excellent discussion by Kerlinger (1979) as well as to the original writings by Jöreskog and his co-workers just cited.

We close this chapter by illustrating the concept of simple structure by comparing an original and a rotated factor matrix; the example and method of rotation are both due to Kaiser (1976). Table 6.2 shows, for Thurstone's Primary Mental Abilities Battery, both a pattern matrix based on a principal factor analysis without rotation and the pattern matrix after an oblique rotation to simple structure by a program called SEARCH. The original pattern matrix shows a general factor and two bipolar factors. The first bipolar factor (II) is a factor indicating the preponderance of verbal over spatial ability, since the first three tests, which are clearly verbal in nature, have high positive loadings (indicated by X's) and the next two tests, both spatial, have high negative loadings (−X). The second bipolar factor (III) indicates the preponderance of spatial over numerical ability,

TABLE 6.2 Oblique Rotation to Simple Structure by SEARCH

	Original Principal Factor Analysis			Oblique Rotation by SEARCH[a]		
Test	**I**	**II**	**III**	F_1	F_2	F_3
Reading	X	X	0	0	0	X
Vocabulary	X	X	0	0	0	X
Coding	X	X	0	0	0	X
Cubes	X	−X	X	0	X	0
Flags	X	−X	X	0	X	0
Number code	X	0	−X	X	0	0
Addition	X	0	−X	X	0	0
Subtraction	X	0	−X	X	0	0
Multiplication	X	0	−X	X	0	0
Division	X	0	−X	X	0	0
	General Factor	Bipolar Factors		Pure Factors		
		Verbal vs. spatial	Spatial vs. numerical	Numerical	Spatial	Verbal

[a]Rotation by a program called SEARCH
Source: Kaiser (1976); courtesy of the author.

for the two spatial tests have high positive loadings while the last five tests (obviously numerical) have high negative loadings. The factors F_1, F_2, and F_3 resulting from the oblique rotation by SEARCH, on the other hand, are each a unitary factor, clearly representing numerical, spatial, and verbal abilities, respectively. The reader may readily verify that the rotated factor pattern clearly satisfies the five conditions for simple structure stated on page 199.

Although this example is an oversimplified one developed by Kaiser just for illustrating how the SEARCH program works, it should suffice for conveying to the reader what simple structure accomplishes. The increase in interpretability in going from the original to the simple-structure pattern matrix is usually greater than in this example, for the second and subsequent factors (if any) will generally be more complicated than merely bipolar: They may comprise several sets of positive and negative loadings, and not all the tests with loadings of each sign may suggest a single factor.

6.7 RESEARCH EXAMPLES

In 1962, under the auspices of the National Merit Scholarship Corporation, Robert C. Nichols collected responses to 47 self-descriptive bipolar adjective scales from 391 male high school seniors randomly selected from NMSQT takers. He performed a *principal factors analysis* on the correlations among the 47 variates which established that 12 useful factors could be extracted as a solution to the problem of the latent variables implicit in the correlation matrix. He subjected the 12 factors to an orthogonal varimax rotation, and also to an oblique simple structure rotation. Nichols has provided his previously unpublished research as an example for this chapter.

Before computing the principal factors Nichols computed a principal components solution. This produced a squared multiple correlation for the regression of each variate on the other 46 variates, which was then used as the starting estimate of the communality for the variate in the iterative principal factors algorithm. The eigenvalues from the *PCA* were useful in the choice of a rank for the factor model. There were 12 eigenvalues larger than 1.0, thus the choice of rank 12 for the principal factors solution. The largest eigenvalue was 9.1 and the second-largest was only 3.4, indicating that one latent factor would dominate the solution, as indeed it did. Only five of the eigenvalues reached the value of 2. In order to simplify this pedagogical example, only the five major factors will be presented in the tables.

Table 6.3 displays five mysterious factors which it would be difficult to name or to connect with psychological constructs. Clearly rotation was called for. By comparison, Table 6.4 (the structure for selected varimax factors) displays five meaningful factors in a reasonable approximation to simple structure. Names for these factors are withheld in order to provide you with the challenge of

TABLE 6.3. Structure for Selected Principal Factors

Variate	PF_1	PF_2	PF_3	PF_4	PF_5	R^2	Rank 12 h^2
1 Religious–nonreligious	.25	−.09	−.11	−.16	.02	.29	.31
2 Good looking–unattractive	.42	−.04	−.05	.32	−.21	.38	.37
3 Happy–unhappy	.58	−.11	−.22	−.07	.31	.55	.58
4 Satisfied with self–dissatisfied with self	.51	−.11	−.05	.05	.32	.49	.51
5 Considerate–inconsiderate	.39	−.36	−.08	−.08	−.22	.44	.48
6 Well adjusted–maladjusted	.68	−.08	−.11	−.01	.17	.56	.57
7 Dependable–undependable	.42	−.25	.44	−.19	−.11	.58	.68
8 Ambitious–unambitious	.47	.11	.33	−.04	.04	.41	.37
9 Optimistic–pessimistic	.52	−.08	−.18	.06	.30	.47	.43
10 High strung–calm	−.15	.36	.13	−.16	−.11	.29	.26
11 Responsible–irresponsible	.48	−.21	.49	−.19	−.15	.62	.71
12 Lazy–energetic	−.42	.14	−.26	.20	−.05	.42	.47
13 Stubborn–give in easily	.17	.35	.26	.19	−.01	.32	.34
14 Extravert–introvert	.50	.48	−.23	−.26	−.13	.61	.67
15 Critical of others–uncritical	.04	.37	.22	.02	.00	.29	.28
16 Talkative–quiet	.37	.53	−.17	−.25	−.10	.53	.52
17 Like responsibility–avoid responsibility	.58	.16	.16	−.20	−.11	.53	.55
18 Messy–neat	−.33	.27	−.19	.06	.09	.40	.45
19 Easily angered–good natured	−.20	.44	.39	.02	.09	.44	.55
20 Worried–carefree	−.26	.23	.32	−.07	−.38	.37	.43
21 Have many friends–have few friends	.52	.08	−.37	−.09	−.11	.55	.56
22 Conforming–nonconforming	.05	−.17	−.19	−.17	−.13	.30	.34
23 Timid–bold	−.50	−.47	−.05	.02	.00	.57	.56
24 Politically liberal–political conservative	−.06	.05	−.01	.02	.02	.19	.22
25 Careless–careful	−.33	.25	−.36	.13	.13	.44	.55
26 Self confident–lack self confidence	.63	.16	−.01	.09	.40	.68	.70
27 Patient–impatient	.25	−.41	−.16	−.05	−.06	.37	.33
28 Successful–unsuccessful	.52	−.14	.18	−.03	.20	.44	.45
29 Persistent–give up easily	.52	−.05	.30	.07	−.08	.46	.47
30 Friendly–unfriendly	.58	−.12	−.35	−.11	−.25	.53	.57
31 Original–unoriginal	.29	.18	.07	.14	.00	.33	.42
32 Strong–weak	.43	−.04	.07	.59	−.25	.57	.65
33 Popular–unpopular	.62	−.08	−.31	.06	−.19	.65	.61
34 Kind–cruel	.32	−.40	−.20	−.06	−.29	.47	.50
35 Hard worker–take it easy	.48	−.17	.41	−.10	−.04	.52	.59
36 Rugged–delicate	.43	.00	.02	.56	−.16	.58	.62
37 Prefer to work alone–prefer to work with others	−.20	−.06	.29	.27	.09	.28	.25
38 Leader–follower	.57	.31	.12	.00	.13	.52	.48
39 Good sense of humor–poor sense of humor	.41	.06	−.19	−.01	−.19	.35	.35
40 Often tired–rarely tired	−.30	.10	−.01	−.08	−.17	.28	.27

TABLE 6.3. (continued) Structure for Selected Principal Factors

Variate	PF_1	PF_2	PF_3	PF_4	PF_5	R^2	Rank 12 h^2
41 Difficulty getting up–little difficulty getting up	−.22	.22	−.14	.18	−.19	.36	.49
42 Masculine–feminine	.43	.03	−.06	.30	−.20	.39	.42
43 Confident–unsure	.68	.15	.02	.21	.36	.70	.72
44 Practical–impractical	.52	−.09	.11	.01	−.01	.48	.50
45 Shy–outgoing	−.54	−.52	.20	.17	.02	.67	.71
46 Sophisticated–unsophisticated	.23	.22	−.03	−.03	−.01	.24	.19
47 Work best at night–work best in morning	−.12	.22	−.08	.12	−.19	.27	.33
PCA Eigenvalue	9.07	3.40	2.83	2.04	1.99		
PFA Eigenvalue	8.61	2.89	2.35	1.56	1.50		

relating the factors to personality constructs and naming them. The proof of the pudding is in the ease and confidence with which you can accomplish this.

In the Manual for his excellent package of multivariate programs, Nichols (1985) expressed his preference for one of the several procedures for oblique rotations. "The best analytic solution to oblique rotation is the *orthoblique* method developed by Harris and Kaiser (1964)" (p. 13–24). Table 6.5 (the pattern for selected orthoblique factors) displays oblique factor loadings for variates which have their major loading on one of the first six factors. Generally speaking, oblique factors will be less efficient than orthogonal factors in extracting battery variance, thus it will be necessary to retain more factors to reach a given level of explanation of observed variances. Since the factors are correlated among themselves, it is not the case that the pattern and structure coefficients are identical, and the pattern coefficients can be considered the fundamental report of the solution. There were moderate correlations among the orthoblique factors representing obtained degrees of obliquity.

Kaiser's Varimax rotation can be applied to uncorrelated components as well as to uncorrelated principal factors. In fact, there are far more reports of varimaxed principal components in the educational research literature than there are of varimaxed principle factors. Confronted with Project TALENT's massive assessment of the academic abilities of American adolescents in 1960 by means of 60 measurements (58 maximum performance tests plus the variates of gender and grade in high school), and its copious assessment of the motivations of adolescents by means of 38 scales (36 typical performance inventories plus gender and grade), and mindful that there were clusters of high correlations in the correlation matrices for both domains, Lohnes (1966) chose to *replace* the 60 abilities with 11 varimaxed components of ability, and to *replace* the 38 motives with 11 varimaxed components of motive. His confidence in the data analysis devices of Hotelling and Kaiser was sufficient to give him the courage to propose this dramatic reformulation of the data resources of the most expan-

TABLE 6.4. Structure for Selected Varimax Factors

Variate	VF_1	VF_2	VF_3	VF_4	VF_5
14 Extravert–introvert	**.76**	.11	.09	.12	.10
16 Talkative–quiet	**.67**	.09	−.03	.17	.06
38 Leader–follower	**.46**	.17	.25	.10	.25
21 Have many friends–have few friends	**.43**	.28	.18	−.13	.09
46 Sophisticated–unsophisticated	**.27**	.16	.04	.12	−.11
37 Prefer to work alone–prefer to work with others	**−.39**	−.04	.06	.20	.07
23 Timid–bold	**−.62**	−.14	−.18	−.15	−.15
45 Shy-outgoing	**−.81**	−.17	−.10	−.03	−.05
3 Happy–unhappy	.16	**.65**	.04	−.08	.12
43 Confident–unsure	.29	**.65**	.28	.00	.14
4 Satisfied with self–dissatisfied with self	.06	**.64**	.08	.01	.11
26 Self confident–lacking self confidence	.34	**.62**	.14	−.09	.22
6 Well adjusted–maladjusted	.24	**.58**	.21	−.03	.10
9 Optimistic–pessimistic	.14	**.58**	.12	−.11	.04
28 Successful–unsuccessful	.03	**.46**	.02	−.04	.39
20 Worried–carefree	−.02	**−.47**	−.02	.30	.11
32 Strong–weak	.02	.08	**.77**	.01	.15
33 Rugged–delicate	.06	.09	**.74**	.01	.12
42 Masculine–feminine	.21	.13	**.53**	−.05	−.06
2 Good looking–unattractive	.10	.19	**.50**	−.06	.09
33 Popular–unpopular	.32	.32	**.37**	−.18	.14
19 Easily angered–good natured	−.03	−.11	−.06	**.64**	.12
15 Critical of others–uncritical of others	.13	−.01	.03	**.49**	−.04
13 Stubborn–give in easily	.08	.09	.18	**.47**	.15
10 High strung–calm	.14	−.18	−.20	**.28**	.03
5 Considerate–inconsiderate	.01	.18	.12	**−.40**	.11
34 Kind–cruel	.00	.06	.12	**−.48**	.09
27 Patient–impatient	−.03	.13	.09	**−.51**	.16
35 Hard worker–take it easy	.04	.10	.12	−.04	**.69**
29 Persistent–give up easily	.08	.19	.12	.04	**.52**
17 Like responsibility–avoid responsibility	.41	.16	.11	−.01	**.42**
8 Ambitious–unambitious	.21	.19	.13	.10	**.38**
12 Lazy–energetic	−.16	−.07	−.01	.14	**−.48**
Percentage of variance extracted	16.6	16.3	11.2	9.4	9.3

Note: Each of the remaining 14 variates has its highest loading on one of the lesser factors, i.e., factors 6 through 12, and thus has been omitted from this table.

TABLE 6.5 Pattern for Selected Orthoblique Factors

Variate	OF_1	OF_2	OF_3	OF_4	OF_5	OF_6
45 Shy–outgoing	**.81**	−.19	−.03	.02	.11	.05
23 Timid–bold	**.59**	−.21	−.04	−.08	.24	−.03
37 Prefer to work alone–prefer to work with others	**.44**	−.04	.00	.13	.01	.09
38 Leader–follower	**−.35**	.14	−.03	.14	.05	.09
16 Talkative–quiet	**−.62**	−.09	.10	−.15	.18	−.02
14 Extravert–introvert	**−.74**	−.12	.14	−.03	.17	.03
26 Self confident–lacking self confidence	−.12	**.84**	.02	−.04	−.12	.08
43 Confident–unsure	−.05	**.78**	−.16	.12	.12	−.06
21 Have many friends–have few friends	−.17	−.08	**.77**	.00	−.09	.00
33 Popular–unpopular	−.07	−.10	**.67**	.19	.01	.07
4 Satisfied with self–dissatisfied with self	.30	.42	**.43**	−.14	.10	−.03
30 Friendly–unfriendly	−.21	−.26	**.38**	.00	.18	.02
32 Strong–weak	.08	.00	.15	**.80**	−.13	.06
36 Rugged–delicate	.01	−.01	−.06	**.80**	.07	.03
42 Masculine–feminine	−.14	.11	−.12	**.50**	.02	−.25
2 Good looking–unattractive	.00	.05	.06	**.44**	.11	.03
1 Religious–nonreligious	−.05	−.10	−.19	−.01	**.65**	.02
3 Happy–unhappy	.15	.23	.19	−.16	**.57**	.01
6 Well adjusted–maladjusted	.02	.23	.21	.02	**.38**	−.09
9 Optimistic–pessimistic	.12	.33	.14	−.03	**.34**	−.10
24 Politically liberal–political conservative	.12	.12	.47	−.07	**−.53**	−.07
35 Hard worker–take it easy	.04	.01	.00	.04	.02	**.75**
29 Persistent–give up easily	.10	−.07	.23	.13	.07	**.50**
28 Successful–unsuccessful	.25	.26	.31	−.17	.06	**.32**
17 Like responsibility–avoid responsibility	−.31	.11	−.17	−.01	.23	**.32**
8 Ambitious–unambitious	−.15	.23	−.17	.05	.07	**.29**
46 Sophisticated–unsophisticated	−.17	.14	.10	−.06	−.06	**−.27**
12 Lazy–energetic	.18	−.05	.07	.06	.13	**−.53**
Percentage of variance extracted	13.0	11.5	9.9	9.3	8.9	7.9

Note: Each of the remaining 19 variates has its highest loading on one of the lesser factors, i.e., factors 7 through 12, and thus has been omitted from this table.

sive and expensive educational research project in history (to that date). Project TALENT had collected its initial data from 440,000 adolescents in a national 5% probability sample, and intended to follow its subjects for at least a decade beyond high school graduation. Its main purpose was to develop regressions of developmental criteria in early adulthood on the measured abilities and motives of adolescence. Lohnes proposed to develop those regressions on his 22 MAP factors rather than on the 96 observational scales. (MAP is an acronymn for Measuring Adolescent Personality.) Cooley and Lohnes (1968) carried out this program of regression research, and were able to report a series of useful regressions of career development outcomes on the MAP factors. Their 1976 book incorporated these predictive validities of the MAP factors into a comprehensive measurement theory for American education.

Table 6.6 showing the abilities domain factors reports the names of 11 MAP ability factors and the proportion of total battery variance extracted (v.e.) by each factor, as well as by the two control factors of gender and grade. The proportion of battery variance extracted by the 13 factors was .65. This rather modest claim for common factor variance in such an ambitious set of measurements on such a massive sample (N = 16,785 randomly sampled from the 9th and 12th grade date files) was justifiable because the number of items in each of the 58 tests was quite small and these short tests were relatively unreliable. The major factors of the tests would have much greater reliability than the average test reliability. Although 11 common factors were retained, it was clear that the five important factors were verbal knowledges, English language, visual reasoning, mathematics, and perceptual speed and accuracy. Verbal knowledges was clearly a g-type factor, as evidenced by the fact that every one of the 58 tests correlated positively with it. English language and mathematics were *curriculum* factors with substantial regressions on the high school curriculum patterns of the students. Visual reasoning and perceptual speed and accuracy were *differential aptitudes* easily connected to well-known measurement factors in the literature. These five uncorrelated factors provided the constructs for a theory of the structure of adolescent abilities. The remaining six factors were viewed as a sort of a safety net, intended to insure that the analyst's decision about a rank for the model did not entail the risk of discarding useful common factors.

TABLE 6.6 Abilities Domain Factors

Factor Name	v.e.
Verbal knowledges	.187
Grade (9th or 12th)	.078
English language	.066
Gender	.057
Visual reasoning	.053
Mathematics	.041
Perceptual speed and accuracy	.036
Screening (basic literacy)	.033
Hunting–fishing	.022
Memory	.021
Color, foods	.019
Etiquette	.016
Games	.015

The subsequent longitudinal predictive validity studies established that it was the case that the predictive validity of the battery was almost totally captured by the five main factors.

Table 6.7 showing the motives domain factors reports the names of the 11 common factors of the 38 variates in the motives battery, as located by varimaxing the leading principal components, as well as the proportion of total battery variance extracted by each common factor and by the control factors of gender and grade. It should be noted that the control factors were available because the point-biserial correlations of the measurements with the dichotomous variates of gender and grade (9th versus 12th) were included in the correlation matrices which were factored. These control factors were extracted before PCA was performed, so that the PCA was done on a residuals-from-gender-and-grade covariance matrix which was, in effect, the pooled-within matrix for a two-factors-without-interaction MANOVA. The 13 factors accounted for .72 of the total battery variance. Again, the analyst believed he had obtained six really important dimensions of adolescent motivation in conformity needs, scholasticism, and the four interest factors, and that the other five factors represented a safety net. Subsequent predictive validity studies supported this interpretation.

Almost two decades separate the author at this time from the decisions he made in the MAP factors solution for the Project TALENT measurement batteries. In retrospect he can see how some of the details of the solution might have been improved. However, he remains convinced that the longitudinal predictive validity studies which were the raison d'etre for this massive research program could not have been approached intelligently without the imposition of some common factors solution on the 1960 measurements, and the *replacement* of the excessive number of variates with a small number of common factors to serve as regressors. It is interesting that when the research team which conducted

TABLE 6.7 Motives Domain Factors

Factor Name	v.e.
Conformity needs	.111
Gender	.091
Business interests	.087
Outdoors, shop interests	.068
Scholasticism	.066
Cultural interests	.058
Science interests	.043
Grade (9th or 12th)	.042
Activity level	.040
Leadership	.031
Impulsion	.028
Sociability	.028
Introspection	.024

the equally massive Equality of Educational Opportunity study in 1966 critiqued their own work (Mayeske, Okada, Cohen, Beaton, and Wisler, 1973), their major criticism was that the failure to replace the many measurements indicating features of school and neighborhood with a few common factors as regressors had defeated the search for meaning in the data. They undertook new analyses using a common factors strategy. Mosteller and Tukey (1977) strongly endorse the use of common factors of the predictor battery as regressors in researches that collect data on large numbers of substantially intercorrelated variates. They speak of ''using principal components as a boiling-down process'' (p. 400), and suggest that since measurement errors tend to be compacted downward into the less principal components, the leading principal components can be viewed as somewhat disattenuated variables. Kendall (1957, p. 75) also endorsed this strategy for improving multiple regression studies.

The strategy of scaling research variables by factors analyses conducted before the structural equations are estimated, and then estimating the structure by OLS path analysis method on the intercorrelations among the factors is also exemplified by Bean (1985), who scored factors from separate factor analyses of four independent domains and a dependent domain. His research was concerned with factors affecting readiness to drop out of college, and was based on questionnaire and college records information for 1,406 students distributed over three classes (517 freshmen, 466 sophomores, 423 juniors). The sample was a random selection of 5,235 American students at a midwestern university (foreign students excluded). Attrition from the sample was due primarily to non-response to the questionnaire (34% responded), but also to deletion of cases with missing data. Bean had some comparisons among independent variables and among classes which ought be stimulating to people concerned with the college attrition problem. In his judgment, ''the most important finding from this study is that social life has large significant effects on institutional fit for each class. It indicates that students seem to have a much greater effect on the attitudes of other students than do faculty members'' (p. 60).

EXERCISES

1. Do a principal factor analysis of the three affective and five cognitive variables in the sample of 600 students in the HSB data set of Appendix F.

   Rotate the factor matrix obtained above to orthogonal simple structure, using the varimax method (or any other method available in the computer package that you customarily use).

3. Rotate the factor matrix obtained in Exercise 1 to oblique simple structure by the oblimax method or other available method.

7

DISCRIMINANT ANALYSIS AND CANONICAL CORRELATION

In Chapter 4 we discussed multivariate procedures for testing the significance of the overall difference among several group centroids. If a significant difference is found, we may be further interested in studying the directions or dimensions along which the major differences occur. In Chapter 5 we noted, in connection with principal components analysis, that the geometric concept of directions or dimensions is closely related to the algebraic notion of linear combinations. Hence the problem of studying the dimension(s) along which large group differences exist is, equivalently, a problem of finding linear combinations of the original predictor variables that show large differences in group means. Discriminant analysis is a method for determining such linear combinations. For a more detailed mathematical treatment than given in this chapter, the reader is referred to a monograph by McKeon (1964). A more elementary and less mathematical treatment is given in Tatsuoka (1970). Although studying the nature of group differences is the main purpose for which discriminant analysis is now used, it was developed by Fisher (1936) as a tool for solving classification problems (see Chapter 10). A historical survey of this shift in emphasis was made by Tatsuoka and Tiedeman (1954) and, more recently, by Lohnes (1985).

7.1 THE DISCRIMINANT CRITERION

The first step toward determining a linear combination of a set of variables such that several group means on this linear combination will differ widely among themselves is to decide on a criterion for measuring such group-mean differences. Now, once a linear combination has been constructed, we are dealing with a single transformed variable. Hence the familiar F-ratio for testing the significance of the overall difference among several group means on a single variable suggests an appropriate criterion.

When we have K groups with a total of N individuals, the F-ratio is given by

$$F = \frac{SS_b/(K-1)}{SS_w/(N-K)} = \frac{SS_b}{SS_w}\frac{N-K}{K-1}$$

Since the second factor, $(N - K)/(K - 1)$, in the last expression is a constant for any given problem (where N and K are fixed), the first factor, SS_b/SS_w, is the only essential quantity for measuring how widely a set of group means differ among themselves, relative to the amount of variability present within the groups. Let us, therefore, write this ratio explicitly for a linear combination of a set of variables, and investigate what sort of function it is of the combining weights used.

If there are p predictor variables, $X_1, X_2, \ldots, X_p$, and we form a linear combination,

$$Y = v_1X_1 + v_2X_2 + \cdots + v_pX_p \tag{7.1}$$

of these variables, the within-groups and between-groups sums of squares of Y both turn out to be expressible as quadratic forms analogous to that in Eq. 5.18. The formula for $SS_w(Y)$ is a direct consequence of this equation, obtained by applying it to each of the K groups separately and then adding the results. Thus, if we denote the sum of squares of Y for the kth group by $SS_k(Y)$, and let $\mathbf{v}' = [v_1, v_2, \ldots, v_p]$, we have

$$\begin{aligned} SS_w(Y) &= SS_1(Y) + SS_2(Y) + \cdots + SS_K(Y) \\ &= \mathbf{v}'\mathbf{S}_1\mathbf{v} + \mathbf{v}'\mathbf{S}_2\mathbf{v} + \cdots + \mathbf{v}'\mathbf{S}_K\mathbf{v} \\ &= \mathbf{v}'(\mathbf{S}_1 + \mathbf{S}_2 + \cdots + \mathbf{S}_K)\mathbf{v} \end{aligned}$$

or

$$SS_w(Y) = \mathbf{v}'\mathbf{W}\mathbf{v} \tag{7.2}$$

where

$$\mathbf{W} = \sum_{k=1}^{K} \mathbf{S}_k \tag{7.3}$$

is called the within-groups SSCP matrix, being the multivariate extension of the within-groups sum of squares of univariate ANOVA.

To obtain the corresponding formula for the between-groups sum of squares $SS_b(Y)$ involves a little more work. We first define the between-groups SSCP matrix $\mathbf{B}$ for the original p variables as follows. The diagonal elements of $\mathbf{B}$ are the usual between-groups sums of squares for the variables taken one at a time; that is,

$$b_{ii} = \sum_{k=1}^{K} n_k(\overline{X}_{ik} - \overline{X}_i)^2 \tag{7.4}$$

where n_k is the size of the kth group, $\overline{X}_{ik}$ is the kth group mean of X_i, and $\overline{X}_i$ is the grand mean of X_i. The off-diagonal elements of **B** are the between-groups sums-of-products for pairs of variables. Thus the (i, j)-element is

$$b_{ij} = \sum_{k=1}^{K} n_k(\overline{X}_{ik} - \overline{X}_i)(\overline{X}_{jk} - \overline{X}_j) \tag{7.5}$$

Equations 7.4 and 7.5 provide an adequate definition of the between-groups SSCP matrix for computational purposes, but it is convenient to express **B** in terms of more basic matrices in order to relate it to $SS_b(Y)$. This may be done as follows. Define an $N \times p$ matrix $\overline{\mathbf{X}}$ in which the first n_1 rows are each equal to $[\overline{X}_{12}, \overline{X}_{21}, \ldots, \overline{X}_{p1}]$, the next n_2 rows are each equal to $[\overline{X}_{12}, \overline{X}_{22}, \ldots, \overline{X}_{p2}]$, and so on, until the last n_K rows are each equal to $[\overline{X}_{1K}, \overline{X}_{2K}, \ldots, \overline{X}_{pK}]$; that is, $\overline{\mathbf{X}}$ consists of n_1 repeated listings of the group 1 means of the p variables, n_2 repeated listings of the group 2 means of the p variables, and so on. Next, define another $N \times p$ matrix $\overline{\overline{\mathbf{X}}}$, all of whose rows are listings of the grand means of the p variables, $[\overline{X}_1, \overline{X}_2, \ldots, \overline{X}_p]$. (Note that this is just the matrix we earlier denoted by $\overline{\mathbf{X}}$, first in Chapter 2 and then in Chapter 5. But we now have two classes of means—the group means and the grand means; hence the double bar in the symbol for the matrix of grand means.) It can then be shown, as the reader should verify, that the matrix **B** can be expressed as

$$\mathbf{B} = (\overline{\mathbf{X}} - \overline{\overline{\mathbf{X}}})'(\overline{\mathbf{X}} - \overline{\overline{\mathbf{X}}}) \tag{7.6}$$

Now if we pre- and postmultiply both sides of Eq. 7.6 by $\mathbf{v}'$ and $\mathbf{v}$, respectively, we get

$$\begin{aligned} \mathbf{v}'\mathbf{B}\mathbf{v} &= \mathbf{v}'(\overline{\mathbf{X}} - \overline{\overline{\mathbf{X}}})'(\overline{\mathbf{X}} - \overline{\overline{\mathbf{X}}})\mathbf{v} \\ &= (\overline{\mathbf{X}}\mathbf{v} - \overline{\overline{\mathbf{X}}}\mathbf{v})'(\overline{\mathbf{X}}\mathbf{v} - \overline{\overline{\mathbf{X}}}\mathbf{v}) \end{aligned}$$

But $\overline{\mathbf{X}}\mathbf{v}$ and $\overline{\overline{\mathbf{X}}}\mathbf{v}$ are, in accordance with Eq. 5.13a, vectors whose elements consist of the group means of Y and the grand mean of Y, respectively. More specifically, $\overline{\mathbf{X}}\mathbf{v}$ is an N-dimensional column vector whose first n_1 elements are equal to $\overline{Y}_1$, the next n_2 elements are equal to $\overline{Y}_2$, and so on, until the last n_K elements are equal to $\overline{Y}_K$. $\overline{\overline{\mathbf{X}}}\mathbf{v}$ is an N-vector with all elements equal to $\overline{Y}$. Therefore, it is not difficult to see that the product $(\overline{\mathbf{X}}\mathbf{v} - \overline{\overline{\mathbf{X}}}\mathbf{v})'(\overline{\mathbf{X}}\mathbf{v} - \overline{\overline{\mathbf{X}}}\mathbf{v})$ is equal to

$$\sum_{k=1}^{K} n_k(\overline{Y}_k - \overline{Y})^2$$

which is none other than the between-groups sum of squares of the transformed variable Y. We have thus shown that

$$SS_b(Y) = \mathbf{v}'\mathbf{B}\mathbf{v} \tag{7.7}$$

Using Eqs. 7.2 and 7.7 we are now able to write the ratio of the between-groups to within-groups sums of squares of Y as a function of the vector of combining weights $\mathbf{v}$, thus:

$$\frac{SS_b(Y)}{SS_w(Y)} = \frac{\mathbf{v}'\mathbf{B}\mathbf{v}}{\mathbf{v}'\mathbf{W}\mathbf{v}} \equiv \lambda \tag{7.8}$$

Following the argument presented earlier, we take the ratio λ, defined by Eq. 7.8, as a criterion for measuring the group differentiation along the dimension specified by the vector $\mathbf{v}$. Fisher (1936) proposed this criterion in connection with his two-group discriminant function. We shall refer to λ as the *discriminant criterion*.

EXERCISE

1. (Theoretical) An alternative way to derive Eq. 7.2 is to utilize Eq. 2.8, applying it to the data matrix $\mathbf{X}_k$ (of order $n_k \times p$) of each group to get the individual-group SSCP matrix $\mathbf{S}_k$ for each $k = 1, 2, \ldots, K$, and then to transform the X's to the Y by writing Eq. 7.1 in matrix notation (i.e., $Y = \mathbf{X}'\mathbf{v}$). We start the procedure below and leave it to the reader to complete as an exercise.

From Eq. 2.8 and the definition of t' as applied to the kth group, we may write

$$\begin{aligned} \mathbf{S}_k &= \mathbf{X}_k'\mathbf{X}_k - (\mathbf{X}_k'\mathbf{1})(\mathbf{1}'\mathbf{X}_k)/n_k \\ &= \mathbf{X}_k'(\mathbf{I}_{nk} - \mathbf{1}\mathbf{1}'/n_k)\mathbf{X}_k \end{aligned}$$

Premultiply both sides of this equation by $\mathbf{v}'$, postmultiply them by $\mathbf{v}$, and note that the transformation equation $Y = \mathbf{X}'\mathbf{v}$ when applied to an $n \times p$ data matrix $\mathbf{X}$ to produce an $n \times 1$ data vector $\mathbf{Y}$ reads $\mathbf{Y} = \mathbf{X}\mathbf{v}$. It should then be evident that the new right-hand member of the equation is the SS of Y for the kth group. In the last step, we utilize the univariate counterpart of Eq. 7.3.

7.2 MAXIMIZING THE DISCRIMINANT CRITERION

Once we have decided on a criterion for group differentiation, our task reduces to that of determining a set of weights, $[v_1, v_2, \ldots, v_p]$, which maximizes the discriminant criterion. This is accomplished by taking the partial derivative of λ with respect to each component v_i of $\mathbf{v}$ and setting the result equal to zero. Symbolically, we may find the derivative of λ with respect to the column vector $\mathbf{v}$ and equate the result to the $p \times 1$ null vector. (For details, see Appendix C.)

The vector equation thus obtained is as follows:

$$\frac{\partial \lambda}{\partial \mathbf{v}} = \frac{2[(\mathbf{Bv})(\mathbf{v}'\mathbf{Wv}) - (\mathbf{v}'\mathbf{Bv})(\mathbf{Wv})]}{(\mathbf{v}'\mathbf{Wv})^2} = \mathbf{0}$$

Dividing both numerator and denominator of the middle member by $\mathbf{v}'\mathbf{Wv}$ and using the definition (7.8) of λ, this equation reduces to

$$\frac{2[\mathbf{Bv} - \lambda\mathbf{Wv}]}{\mathbf{v}'\mathbf{Wv}} = \mathbf{0}$$

which is equivalent to

$$(\mathbf{B} - \lambda\mathbf{W})\mathbf{v} = \mathbf{0} \tag{7.9}$$

It may safely be assumed that $\mathbf{W}$ is nonsingular, and hence possesses an inverse $\mathbf{W}^{-1}$, except when a linear restriction of the form

$$a_iX_i + a_jX_j + \cdots + a_qX_q = 0$$

holds, by design, among some subset of the p predictor variables, as would be the case if ipsative measures are included among the predictors. In such a case, one of the variables in this subset may be eliminated or, alternatively, the principal components of this subset may be computed and used in place of these variables in the analysis, as described in connection with multiple regression analysis in Chapter 5.[1] Thus we may, without loss of generality, assume that $\mathbf{W}^{-1}$ exists, and premultiply both sides of Eq. 7.9 by it to obtain

$$(\mathbf{W}^{-1}\mathbf{B} - \lambda\mathbf{I})\mathbf{v} = \mathbf{0} \tag{7.10}$$

This equation is of the form

$$(\mathbf{A} - \lambda\mathbf{I})\mathbf{v} = \mathbf{0}$$

procedures for solving which were fully described in Chapter 5. Its solution, yielding the eigenvalues λ_m and associated eigenvectors $\mathbf{v}_m$ of the matrix $\mathbf{A}$, is therefore well known by now, and we have thus solved the problem of maximizing the discriminant criterion. [Strictly speaking, Eq. 7.10 represents only a necessary condition for maximizing the discriminant criterion, and we should examine the second-order derivative of λ with respect to $\mathbf{v}$ in order to verify that the solutions of this equation yield maxima rather than minima or points of

[1]A third alternative is simply to solve Eq. 7.9 as it stands, without attempting to reduce it to the form of (7.10). It is then called a generalized eigenvalue problem, and the characteristic equaton is

$$|\mathbf{B} - \lambda\mathbf{W}| = 0$$

which is more tedious to formulate with a hand-held calculator than is

$$|\mathbf{A} - \lambda\mathbf{I}| = 0$$

However, with electronic computers, it makes little difference, and most discriminant analysis programs use this approach.

inflection. We shall not carry out this extra step here, but take it as a proven fact that (7.10) is a sufficient as well as necessary condition for maximizing λ.]

Computational Note

It should be noted that the matrix $\mathbf{W}^{-1}\mathbf{B}$ in Eq. 7.10 is, in general, nonsymmetric. Although the procedures described in Chapter 5 for computing the eigenvalues and eigenvectors of a square matrix are equally applicable to symmetric and nonsymmetric matrices, the reader should be alerted to the fact that the eigenroutines given in computer packages such as the SAS PROC MATRIX and the SAS/IML utilize the Jacobi method, which is applicable only to symmetric matrices. Fortunately, there is a simple way to transform a nonsymmetric square matrix into a symmetric one that has the same eigenvalues and eigenvectors that are linearly related to those of the given nonsymmetric matrix. The transformation is based on a more general theorem in matrix algebra, which is as follows:

> ***Theorem 7.1.*** If $\mathbf{F}$ is a square matrix of order p and $\mathbf{S}$ is any *nonsingular* matrix of the same order, then the matrix $\mathbf{G} = \mathbf{S}^{-1}\mathbf{F}\mathbf{S}$ has the same eigenvalues as $\mathbf{F}$ and eigenvectors that are equal to those of $\mathbf{F}$ premultiplied by $\mathbf{S}^{-1}$. (Two matrices that are related to each other in this way are called *similar matrices*.)

In discriminant analysis, we need the eigenvalues and eigenvectors of the nonsymmetric matrix $\mathbf{F} = \mathbf{W}^{-1}\mathbf{B}$. A matrix that is similar to $\mathbf{F}$ but symmetric may be constructed as follows: Let $\mathbf{H}$ be the matrix whose columns are the eigenvectors of $\mathbf{W}$, and $\mathbf{D}$ be the diagonal matrix whose diagonal matrices are the (nonzero) eigenvalues of $\mathbf{W}$. Then, by taking $\mathbf{S} = \mathbf{H}\mathbf{D}^{-1/2}$ as the transformation matrix, we get a matrix $\mathbf{G} = (\mathbf{H}\mathbf{D}^{-1/2})^{-1}\mathbf{F}(\mathbf{H}\mathbf{D}^{-1/2}) = (\mathbf{H}\mathbf{D}^{-1/2})^{-1}(\mathbf{W}^{-1}\mathbf{B})(\mathbf{H}\mathbf{D}^{-1/2})$, which is symmetric (as the reader should verify by noting that, by definition, $\mathbf{W}\mathbf{H} = \mathbf{H}\mathbf{D}$, and that $\mathbf{H}'\mathbf{H} = \mathbf{I}$ because $\mathbf{W}$ is symmetric). Hence, the eigenvalues and vectors of $\mathbf{G}$ can be computed by the Jacobi method. If these are denoted by λ_i and $\mathbf{u}_i$, it follows from the above theorem that the corresponding eigenvalues and vectors of $\mathbf{F} = \mathbf{W}^{-1}\mathbf{B}$ are λ_i and $(\mathbf{H}\mathbf{D}^{-1/2})\mathbf{u}_i$. This fact, long known by mathematicians, was recorded and utilized by Bryan (1950) in his dissertation and was first documented in a book addressed to behavioral scientists by Cooley and Lohnes (1962).

7.3 DISCRIMINANT FUNCTIONS

It was shown in Chapter 6 that the number of nonzero eigenvalues of a square matrix $\mathbf{A}$ is equal to the rank of $\mathbf{A}$. In the present context, with $\mathbf{W}^{-1}\mathbf{B}$ playing the role of $\mathbf{A}$, the number of nonzero eigenvalues depends on the rank of $\mathbf{B}$, since the rank of the product of two matrices cannot exceed the smaller of the two factor matrices' ranks, and $\mathbf{W}^{-1}$ (being nonsingular) must be of full rank p, while the rank of $\mathbf{B}$ is usually smaller than p.

The last-mentioned fact may be seen by referring to Eq. 7.6, which shows that the rank of $\mathbf{B}$ is equal to that of $\overline{\mathbf{X}} - \overline{\overline{\mathbf{X}}}$. Now $\overline{\mathbf{X}} - \overline{\overline{\mathbf{X}}}$ is an $N \times p$ matrix with only K different rows. (Recall that the first n_1 rows of $\overline{\mathbf{X}}$ are all equal, the next n_2 rows are different from the first n_1 but equal among themselves, and so on, until the last n_K rows are different from all preceding rows but equal among themselves, and that $\overline{\overline{\mathbf{X}}}$ has all N rows identical.) Moreover, these K different rows, which may be denoted $\mathbf{b}'_1, \mathbf{b}'_2, \ldots, \mathbf{b}'_K$ in their order of occurrence, are themselves subject to a linear restriction; namely,

$$n_1\mathbf{b}'_1 + n_2\mathbf{b}'_2 + \cdots + n_K\mathbf{b}'_K = \mathbf{0}'$$

because, for each variable X_i, the well-known identity

$$\sum_{k=1}^{K} n_k(\overline{X}_{ik} - \overline{X}_i) = 0$$

holds. Therefore, there are only $K - 1$ linearly independent rows in $\overline{\mathbf{X}} - \overline{\overline{\mathbf{X}}}$. Thus the rank of $\overline{\mathbf{X}} - \overline{\overline{\mathbf{X}}}$, and hence that of $\mathbf{B}$, must be equal to $K - 1$ or p, whichever is smaller. Usually, $K - 1$ will be smaller than p, because we would rarely perform a discriminant analysis using fewer variables than the number of groups being studied. For generality, however, let us denote the rank of $\mathbf{B}$ by $r = \min(K - 1, p)$. [The function $\min(a, b)$ stands for "the smaller of the two numbers a and b."]

Thus, when Eq. 7.10 is solved, we get r nonzero eigenvalues, which will be denoted as $\lambda_1, \lambda_2, \ldots, \lambda_r$ in descending order of magnitude, and r associated eigenvectors $\mathbf{v}_1, \mathbf{v}_2, \ldots, \mathbf{v}_r$. As usual, the eigenvectors are determined only up to an arbitrary multiplier, because if $\mathbf{v}$ satisfies (7.10) for some λ, it is clear that $c\mathbf{v}$ also satisfies the equation for the same λ (where c is an arbitrary constant). It is customary to choose the multiplier for each eigenvector in one of two ways: (1) so that its norm will be unity (i.e., $\mathbf{v}'_m\mathbf{v}_m = 1$, for each m), or (2) so that its largest element will be unity.

Now, from the fact that the eigenvalues λ_m are, by definition, the values assumed by the discriminant criterion for linear combinations using the elements of the corresponding eigenvectors $\mathbf{v}_m$ as combining weights, it is clear that the eigenvector $\mathbf{v}'_1 = [v_{11}, v_{12}, \ldots, v_{1p}]$ provides a set of weights such that the transformed variable

$$Y_1 = v_{11}X_1 + v_{12}X_2 + \cdots + v_{1p}X_p$$

has the largest discriminant criterion, λ_1, achievable by any linear combination of the p predictor variables. What are the properties of the remaining eigenvectors, $\mathbf{v}_2, \mathbf{v}_3, \ldots, \mathbf{v}_r$?

It can be shown (see Exercise 4 at the end of the chapter) that if the elements of $\mathbf{v}_2$ are used as combining weights to form a second linear combination,

$$Y_2 = v_{21}X_1 + v_{22}X_2 + \cdots + v_{2p}X_p$$

then Y_2 has this property: Its discriminant-criterion value, λ_2, is the largest achievable by any linear combination of the X's that is uncorrelated (in the total sample) with Y_1. Similarly,

$$Y_3 = v_{31}X_1 + v_{32}X_2 + \cdots + v_{3p}X_p$$

has the largest possible discriminant-criterion value (λ_3) among all linear combinations of the X's that are uncorrelated with both Y_1 and Y_2; and so on until Y_r, using the elements of $\mathbf{v}_r$ as weights, has the largest possible discriminant-criterion value among linear combinations that are uncorrelated with all the preceding linear combinations $Y_1, Y_2, \ldots, Y_{r-1}$. The linear combinations $Y_1, Y_2, \ldots, Y_r$ are called the first, second, . . . , rth (linear) *discriminant functions* for optimally differentiating among the K given groups.

We thus see that although we started out by seeking to maximize the discriminant criterion, we obtain several discriminant functions, the first of which has the largest possible discriminant-criterion value, and each of the others has a conditionally maximal discriminant-criterion value. It is in this sense, then, that discriminant analysis reveals the "dimensions" of group differences. By this technique we find, simultaneously, the dimension along which maximum group differentiation occurs; the dimension along which is observed the largest group differences that are not accounted for by the first dimension; and so on.

The situation here is reminiscent of principal-components analysis. There, the dimension corresponding to the first component had maximum variance; the second-component dimension had maximum variance among those uncorrelated with the first; and so on. In discriminant analysis, the ratio of between- to within-groups sums of squares takes the place of variance as the criterion in determining the successive dimensions. However, an important difference between the dimensions identified in discriminant analysis and those in component analysis is that the former are generally not mutually orthogonal in test space, even though they are uncorrelated. That is, the axes representing the discriminant functions are not a subset of axes obtainable by rigid rotation of the original system of p axes; the discriminant rotation is an *oblique* rotation.

Just as in component analysis, the dimensions represented by the discriminant functions *may* be susceptible to meaningful interpretations. Even if they are not, we shall still have achieved parsimony by having reduced the dimensionality of the space in which to describe group differences. In seeking to interpret the discriminant functions, we would want to know which of the original p variables contribute most to each function. For this purpose, comparison of the relative magnitudes of the combining weights as given by the elements of each eigenvector of $\mathbf{W}^{-1}\mathbf{B}$ is inappropriate because these are weights to be applied to the predictors in raw-score scales, and hence are affected by the particular unit used for each variable. To eliminate the spurious effects of units on the magnitudes of combining weights, we must compare the weights that would be applied to the predictors in standardized form. The relative magnitudes of these standardized weights may be assessed by multiplying each raw-score weight by the standard deviation of the corresponding variable as computed from the within-groups SSCP matrix. This amounts to multiplying each element of a given eigenvector $\mathbf{v}_m$ by the square root of the corresponding diagonal element of $\mathbf{W}$. Thus, for each m, we define

$$v_{mi}^* = \sqrt{w_{ii}}\, v_{mi} \qquad i = 1, 2, \ldots, p \tag{7.11}$$

as the standardized discriminant weights. The relative contribution of the ith predictor to the mth discriminant function may then be gauged by the magnitude of v_{mi}^* in comparison with the other weights v_{mj}^*.

Up to this point, we have taken the dimensionality of the discriminant space to be equal to the number of nonzero eigenvalues of $\mathbf{W}^{-1}\mathbf{B}$, which is the smaller of the two numbers, $K - 1$ and p. It may often happen, however, that the number of *significant* discriminant dimensions may be even smaller. That is, not all of the discriminant functions may represent dimensions along which statistically significant group differences occur. We now turn to a test for determining the number of discriminant functions that are significant at a prescribed level.

7.4 SIGNIFICANCE TESTS IN DISCRIMINANT ANALYSIS

It was seen in Chapter 4 that a basic statistic for testing the significance of the overall difference among several group centroids is the ratio of the determinants of the within-groups and the total SSCP matrices, known as Wilks' Λ criterion. We shall now see that this criterion figures prominently not only in the overall test, but also in the fine-grained test to determine how many of the discriminant dimensions contribute significantly to group differentiation. The relevance of Λ for this purpose hinges on an interesting algebraic relation existing between it and the discriminant-criterion values for the successive discriminant functions.

By equating the reciprocals of the two members of Eq. 4.30 and successively transforming, it is seen that

$$\begin{aligned} 1/\Lambda &= |\mathbf{T}|/|\mathbf{W}| = |\mathbf{W}^{-1}\mathbf{T}| && (\text{because } |\mathbf{AB}| = |\mathbf{A}||\mathbf{B}|) \\ &= |\mathbf{W}^{-1}(\mathbf{W} + \mathbf{B})| && (\text{because } \mathbf{T} = \mathbf{W} + \mathbf{B}) \\ &= |\mathbf{I} + \mathbf{W}^{-1}\mathbf{B}| \end{aligned}$$

Hence, by use of Theorems 3 and 9 of Chapter 5, it follows that

$$1/\Lambda = (1 + \lambda_1)(1 + \lambda_2) \cdots (1 + \lambda_r) \tag{7.12}$$

where $\lambda_1, \lambda_2, \ldots, \lambda_r$ are the nonzero eigenvalues of $\mathbf{W}^{-1}\mathbf{B}$. Consequently, Bartlett's V statistic for testing the significance of an observed Λ value, given in Eq. 4.32 can be expressed as

$$\begin{aligned} V &= -[N - 1 - (p + K)/2] \ln \Lambda \\ &= [N - 1 - (p + K)/2] \ln [(1 + \lambda_1)(1 + \lambda_2) \cdots (1 + \lambda_r)] \\ &= [N - 1 - (p + K)/2] \sum_{m=1}^{r} \ln (1 + \lambda_m) \\ &= \sum_{m=1}^{r} V_m \end{aligned} \tag{7.13}$$

where

$$V_m = [N - 1 - (p + K)/2] \ln (1 + \lambda_m) \tag{7.14}$$

This statistic, it will be recalled, is distributed approximately as a chi-square with $p(K - 1)$ degrees of freedom.

What we are testing by use of Bartlett's V statistic is the null hypothesis that all the population eigenvalues corresponding to $\lambda_1, \lambda_2, \ldots, \lambda_r$ are zero.[2] We designate this by

$$H_0^{(0)}\colon \lambda_1^* = \lambda_2^* = \cdots \lambda_r^* = 0$$

This is equivalent to the test of

$$H_0\colon \boldsymbol{\mu}_1 = \boldsymbol{\mu}_2 = \cdots = \boldsymbol{\mu}_K$$

discussed in Chapter 4 in the context of the K-sample problem, because if H_0 is true, then the $\boldsymbol{\Sigma}_b$ defined in footnote 2 is a null matrix, so $H_0^{(0)}$ must be true, and vice versa.

However, as intimated at the beginning of this section, the new formulation of V given in Eq. 7.13 enables us to carry out more than just this overall or "global" test if $H_0^{(0)}$ is rejected—which means that not all of the λ_m^* are zero. It then follows that *at least* λ_1^* (the largest eigenvalue of $\boldsymbol{\Sigma}^{-1}\boldsymbol{\Sigma}_b$) is not zero. Bartlett (1947) showed that in this case we may set up a new null hypothesis,

[2]More precisely, the relevant population eigenvalues are those of $\boldsymbol{\Sigma}^{-1}\boldsymbol{\Sigma}_b$ where $\boldsymbol{\Sigma}$ is the common covariance matrix of the K populations, and $\boldsymbol{\Sigma}_b$ is the covariance matrix of the K population centroids. (This is because there are no population counterparts of **W** and **B** themselves, populations are infinite in size, so any population "SSCP matrix" would be infinite.) But since $\boldsymbol{\Sigma}^{-1}\boldsymbol{\Sigma}_b$ corresponds to $[\mathbf{W}/(N - K)]^{-1}[\mathbf{B}/(K - 1)] = \mathbf{W}^{-1}\mathbf{B}(N - K)/(K - 1)$ in the sample, the relevant population eigenvalues correspond to $\lambda_m(N - K)/(K - 1)$. But if the eigenvalues of $\boldsymbol{\Sigma}^{-1}\boldsymbol{\Sigma}_b$ are zero, then each λ_m as well as $\lambda_m(N - K)/(K - 1)$ would differ from zero only by sampling fluctuation.

$$H_0^{(1)}: \lambda_1^* \neq 0 \text{ and } \lambda_2^* = \lambda_3^* = \cdots = \lambda_r^* = 0$$

and test it by means of the *first residual* V, namely,

$$V - V_1 = \sum_{m=2}^{r} V_m \tag{7.15}$$

which is distributed approximately as a chi-square with $(p - 1)(K - 2)$ degrees of freedom. If this is not significant at the α level being used, we conclude that only $\lambda_1^* \neq 0$ while $\lambda_2^*, \lambda_3^*, \ldots, \lambda_r^*$ are all zero, and conduct no further testing; only the first discriminant function, Y_1, is said to be significant in this case. If, on the other hand, $V - V_1$ *is* significant, we conclude that at least λ_2^* is nonzero in addition to λ_1^*, and set up another new null hypothesis,

$$H_0^{(2)}: \lambda_1^* \lambda_2^* \neq 0 \quad \text{and} \quad \lambda_3^* = \lambda_4^* = \cdots = \lambda_r^* = 0$$

and test it using the second residual V,

$$V - V_1 - V_2 = \sum_{m=3}^{r} V_m$$

as an approximate chi-square with $(p - 2)(K - 3)$ degrees of freedom.

We proceed in this manner as long as the successive null hypotheses, $H_0^{(3)}$, $H_0^{(4)}, \ldots$, continue to be rejected. The null hypothesis at the qth stage is

$$H_0^{(q)}: \lambda_1^* \lambda_2^* \cdots \lambda_q^* \neq 0 \quad \text{and} \quad \lambda_{q+1}^* = \lambda_{q+2}^* = \cdots = \lambda_r^* = 0 \tag{7.17}$$

and the statistic for testing it is

$$V - V_1 - V_2 - \cdots - V_q = \sum_{m=q+1}^{r} V_m \tag{7.18}$$

which is an approximate chi-square variate with $(p - q)(K - q - 1)$ degrees of freedom. If $H_0^{(q)}$ is *not* rejected, we conclude that only $\lambda_1^*, \lambda_2^*, \ldots, \lambda_q^*$ are nonzero while the remaining eigenvalues, $\lambda_{q+1}^*, \lambda_{q+2}^*, \ldots, \lambda_r^*$ are zero, and no further testing is carried out.

7.5 NUMERICAL EXAMPLE

Let us follow up the example given in Chapter 4 (pp. 95–96) in connection with the use of Wilks' Λ criterion in multivariate significance testing, and compute the discriminant functions for those data. The within-groups SSCP matrix, already computed there, was as follows:

$$\mathbf{W} = \begin{bmatrix} 3967.8301 & 351.6142 & 76.6342 \\ 351.6142 & 4406.2517 & 235.4365 \\ 76.6342 & 235.4365 & 2683.3164 \end{bmatrix}$$

To illustrate the computations for the between-groups SSCP matrix **B**, we reproduce below the means on the three variables in each group and in the total sample.

			Means on:		
k	Group	n_k	X_1	X_2	X_3
1.	Passenger agents	85	12.59	24.22	9.02
2.	Mechanics	93	18.54	21.14	10.14
3.	Operations control persons	66	15.58	15.45	13.24
	Total sample	N = 244	15.66	20.68	10.59

Equations 7.3 and 7.4 may now be used to compute the diagonal and off-diagonal elements of **B**, respectively. For example,

$$b_{11} = 85(12.59 - 15.66)^2 + 93(18.54 - 15.66)^2 + 66(15.58 - 15.66)^2$$
$$= 1572.92$$

and

$$b_{12} = 85(12.59 - 15.66)(24.22 - 20.68)$$
$$+ 93(18.54 - 15.66)(21.14 - 20.68)$$
$$+ 66(15.58 - 15.66)(15.45 - 20.68)$$
$$= -772.94$$

Proceeding in this manner, the between-groups SSCP matrix is found to be

$$\mathbf{B} = \begin{bmatrix} 1572.7441 & -773.0506 & 273.6214 \\ -773.0506 & 2889.3193 & -1405.9955 \\ 273.6214 & -1405.9955 & 691.6068 \end{bmatrix}$$

[The discrepancies between the values of b_{11} and b_{12} shown in the matrix and those computed above are due to the fact that the raw-score forms of Eqs. 7.4 and 7.5 were used for greater accuracy in the actual calculations yielding the matrix; that is,

$$b_{ij} = \sum_{k=1}^{K} \frac{T_{ik}T_{jk}}{n_k} - \frac{T_iT_j}{N}$$

where T_{ik} and T_{jk} are the totals, in group k, of variables X_i and X_j, and T_i and T_j are their grand totals. The reader may verify the more accurate results by reconstructing the group totals for each variable by multiplying the group means given above by the group sizes and rounding to the nearest integer.]

Before continuing, it is advisable, as a computational check, to verify that $\mathbf{W} + \mathbf{B} = \mathbf{T}$ (where $\mathbf{T}$ is the total SSCP matrix, given on p. 95). We leave this verification for the reader to carry out. Next, we compute the matrix $\mathbf{W}^{-1}\mathbf{B}$. The result is

$$\mathbf{W}^{-1}\mathbf{B} = \begin{bmatrix} .413255 & -.246239 & .093744 \\ -.214244 & .706315 & -.341803 \\ .108967 & -.578917 & .285056 \end{bmatrix}$$

which is the matrix whose eigenvalues and eigenvectors we need to compute.

The characteristic equation, $|\mathbf{W}^{-1}\mathbf{B} - \lambda\mathbf{I}| = 0$, is found to be

$$\lambda^3 - 1.404626\lambda^2 + .350182\lambda = 0$$

(If the reader carries out the computations, he or she will find a constant term of the order of .00002. But we know that the rank of $\mathbf{B}$, and hence that of $\mathbf{W}^{-1}\mathbf{B}$, must be 2, which is the smaller of the two numbers $K - 1$ and p for this example. Therefore, the nonzero constant term is due to rounding error, and we simply omit it from the characteristic equation.) The roots of this equation are, in descending order of magnitude,

$$\lambda_1 = 1.080548 \qquad \lambda_2 = .324078 \qquad \lambda_3 = 0$$

Next, we compute the eigenvectors $\mathbf{v}_1$ and $\mathbf{v}_2$ corresponding to the two nonzero eigenvalues λ_1 and λ_2, following the steps described on pages 137–139. The results (after normalizing each vector) are as follows:

$$\mathbf{v}_1 = \begin{bmatrix} .3524 \\ -.7331 \\ .5818 \end{bmatrix} \quad \text{and} \quad \mathbf{v}_2 = \begin{bmatrix} .9145 \\ .1960 \\ -.3540 \end{bmatrix}$$

The elements of these vectors may be interpreted geometrically as discussed in Chapter 5: namely, as the cosines of the angles between the original axes (representing the variables X_1, X_2, and X_3) and the new axes representing the two discriminant functions.

Our next task is to ascertain whether both or only one of these discriminant functions is statistically significant. (In general, the possibility exists, of course, that none of the functions is significant; but this event is already excluded in the present example, because the F-ratio based on Wilks' Λ-criterion was found to be significant in Example 4.8 at the end of Chapter 4.) We first compute the approximate chi-square statistic V for the overall group centroid differences in accordance with Eq. 7.13:

$$V = \left(N - 1 - \frac{p + K}{2}\right) [\ln (1 + \lambda_1) + \ln (1 + \lambda_2)]$$

$$= \left(244 - 1 - \frac{3 + 3}{2}\right)(2.3026)(\log 2.0805 + \log 1.3241)$$

$$= (240)(2.3026)(.31817 + .12192)$$

$$= 175.82 + 67.37 = 243.19$$

which agrees with the value found in Example 4.8 using Λ, thus verifying the fact that $1/\Lambda = (1 + \lambda_1)(1 + \lambda_2)$, stated in Eq. 7.12. The two numbers given just before the final result, 243.19, for V are the terms V_1 (= 175.82) and V_2 (= 67.37) as computed from Eq. 7.14. Thus the quantity $V - V_1$ to be used for testing the significance of the "residual discrimination" after removing the first discriminant function is simply equal to V_2. [In general, $V - V_1 - V_2 - \cdots - V_s = V_r + V_{r-1} + \cdots + V_{s+1}$; that is, the statistic for testing the residual after the first s discriminants may be obtained as the cumulative sum from V_r (corresponding to the last, or smallest, nonzero eigenvalue λ_r) up through V_{s+1}, based on the next smaller eigenvalue after the sth.] Now the quantity $V - V_1 = V_2 = 67.37$ is an approximate chi-square variate with $(p - 1)(K - 2) = (2)(1) = 2$ degrees of freedom, under the null hypothesis $H_0^{(1)}$ that $\lambda_1^* \neq 0$ and $\lambda_2^* = 0$. This value far exceeds the 99.9th percentile of the chi-square distribution with 2 d.f.'s, which is 13.82. We therefore reject the hypothesis that the group differences after removal of the first discriminant function are due to sampling error, and conclude that both discriminant functions are statistically significant beyond the .001 level.

Having decided to retain both of the discriminant functions as significant, we may next try to see if some intuitively meaningful interpretations can be given to the two dimensions along which our three groups were found to differ. In this attempt it is useful, besides determining the standardized discriminant weights in accordance with Eq. 7.11, to examine the three group means on each of the discriminant functions. These means may be computed by forming the linear combinations of the group means on X_1, X_2, and X_3, using, in turn, the elements of $\mathbf{v}_1$ and those of $\mathbf{v}_2$ as the weights in Eq. 7.1, or, what amounts to the same thing, by using the matrix equation (5.13):

$$\bar{\mathbf{Y}} = \bar{\mathbf{X}}\mathbf{V} = \begin{bmatrix} 12.59 & 24.22 & 9.02 \\ 18.54 & 21.14 & 10.14 \\ 15.58 & 15.45 & 13.24 \end{bmatrix} \begin{bmatrix} .3524 & .9145 \\ -.7331 & .1960 \\ .5818 & -.3540 \end{bmatrix}$$

$$= \begin{bmatrix} -8.07 & 13.07 \\ -3.06 & 17.51 \\ 1.87 & 12.59 \end{bmatrix}$$

The group centroids $(\bar{Y}_{1k}, \bar{Y}_{2k})$ may be represented graphically as in Fig. 7.1. It should be noted, however, that our use of a rectangular coordinate system is

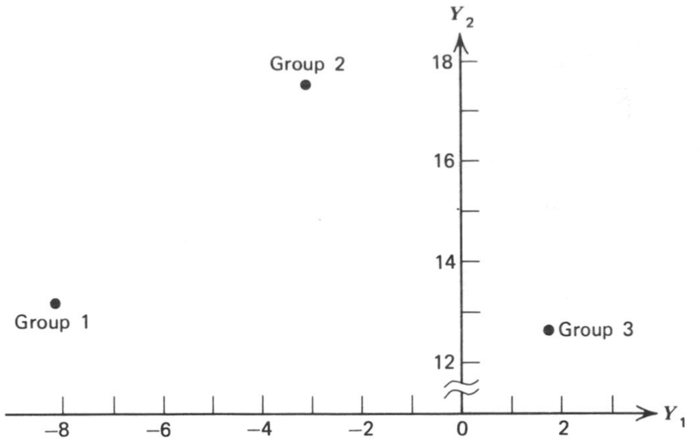

Figure 7.1. *Discriminant function centroids of the three groups.*

merely a matter of convenience and does not imply that the axes representing the discriminant functions in the original three-dimensional test space are mutually orthogonal. [The actual angle between the discriminant axes is such that its cosine is equal to

$$\mathbf{v}_1'\mathbf{v}_2 = (.3524)(.9145) + (-.7331)(.1960) + (.5818)(-.3540) = -.0274$$

Consulting a table of trigonometric functions, the angle is found to be 91°34′.] The plot of discriminant function centroids in Fig. 7.1 shows at a glance that the first discriminant dimension (Y_1) separates the three groups in roughly equal steps: the group 2 mean is about as far from the group 1 mean as the group 3 mean is from group 2 along this dimension. In terms of the second dimension, groups 1 and 3 are practically indistinguishable, but group 2 is separated from them by a considerable distance.

Let us now determine the relative contributions of the three original variables to group differentiation along each of the two dimensions, by computing the standardized discriminant weights in accordance with Eq. 7.11. The results are

$$\mathbf{v}_1^* = \begin{bmatrix} 22.2 \\ -48.7 \\ 30.1 \end{bmatrix} \quad \text{and} \quad \mathbf{v}_2^* = \begin{bmatrix} 57.6 \\ 13.0 \\ -18.3 \end{bmatrix}$$

Actually, we are interested only in rough indications of the relative importance of each variable for each discriminant dimension. So we may take small integers roughly proportional to the elements of $\mathbf{v}_1^*$ and $\mathbf{v}_2^*$ to assess the relative contributions of X_1, X_2, and X_3. We thus find these to be in the proportion 2:−5:3

for the first dimension, and 6:1:−2 in the second dimension. (In this example, the rank order of the standardized weights and that of the raw-score weights are identical, because the diagonal elements of **W** are of the same order of magnitude; but this will not always be the case.) Recalling that X_1, X_2, and X_3 stand for the outdoor, convivial, and conservative scales, respectively, of an activity preference questionnaire, let us try to attach some meaningful labels to the two discriminant dimensions in the light of the relative weights of the three variables.

The Y_1 dimension is most highly weighted, in the negative direction, with the convivial scale; that is, persons with few convivial activity preferences (or, equivalently, with many solitary activity preferences) will score high on Y_1. But the conservative scale also contributes fairly heavily to the first discriminant, with a weight of 3 compared to the −5 for X_2. Therefore, we might say that Y_1 represents a "solitary-conservative" syndrome. Examination of Fig. 7.1, which shows the group means on Y_1 to be ranked in the order (from high to low) $\bar{Y}_{13}$, $\bar{Y}_{12}$, $\bar{Y}_{11}$, corroborates this interpretation. We would expect operations control personnel, as a group, to be less gregarious (either by nature or by occupational habit) and perhaps more conservative in outlook than the other two groups. Passenger agents (group 1) would certainly be expected to be the most convivial and sociable lot, while mechanics would fall in between the other two groups in this respect.

In the Y_2 dimension, X_1 (outdoor activity preference) far outweighs the other two variables in importance. Without doubt, it is the person with many outdoor interests who will score high on the second discriminant. Figure 7.1 shows the mechanics (group 2) to be a class by itself, scoring much higher than the other two groups on Y_2, which is what we would expect.

7.6 SPECIAL CASE OF TWO GROUPS

When there are only two groups under study, the computations for determining the single ($K - 1 = 1$) discriminant function can be simplified a great deal. To see this simplification, we first show that discriminant functions (not only for the two-group case, but in general) can be determined just as well from the equation

$$(\mathbf{T}^{-1}\mathbf{B} - \mu\mathbf{I})\mathbf{v} = \mathbf{0} \tag{7.19}$$

as from Eq. 7.10. We rewrite the latter equation in the form

$$\mathbf{Bv} = \lambda\mathbf{Wv}$$

and add $\lambda\mathbf{Bv}$ to both members, obtaining

$$(1 + \lambda)\mathbf{Bv} = \lambda(\mathbf{W} + \mathbf{B})\mathbf{v}$$

or

$$\left(\mathbf{T}^{-1}\mathbf{B} - \frac{\lambda}{1+\lambda}\mathbf{I}\right)\mathbf{v} = \mathbf{0}$$

Hence, if $\mathbf{v}_0$ is the eigenvector of $\mathbf{W}^{-1}\mathbf{B}$ associated with the eigenvalue λ_0, it is also an eigenvector of $\mathbf{T}^{-1}\mathbf{B}$, and its associated eigenvalue is $\lambda_0/(1 + \lambda_0)$. In other words, the vectors $\mathbf{v}$ satisfying Eq. 7.19 are identical to those satisfying Eq. 7.10, and the eigenvalues stand in the relation

$$\mu = \frac{\lambda}{1+\lambda}$$

Now, in the two-group case, the between-groups SSCP matrix $\mathbf{B}$ assumes a particularly simple form, as follows. First, the expression (7.5) for its general element b_{ij} reduces to

$$b_{ij} = n_1(\overline{X}_{i1} - \overline{X}_i)(\overline{X}_{j1} - \overline{X}_j) + n_2(\overline{X}_{i2} - \overline{X}_i)(\overline{X}_{j2} - \overline{X}_j)$$

But the grand mean $\overline{X}_i$ of the ith variable is, in this case, equal to

$$\frac{n_1\overline{X}_{i1} + n_2\overline{X}_{i2}}{n_1 + n_2}$$

Consequently,

$$\overline{X}_{i1} - \overline{X}_i = \overline{X}_{i1} - \frac{n_1\overline{X}_{i1} + n_2\overline{X}_{i2}}{n_1 + n_2} = \frac{n_2}{n_1 + n_2}(\overline{X}_{i1} - \overline{X}_{i2})$$

and similarly,

$$\overline{X}_{i2} - \overline{X}_i = \frac{n_1}{n_1 + n_2}(\overline{X}_{i2} - \overline{X}_{i1}) = \frac{-n_1}{n_1 + n_2}(\overline{X}_{i1} - \overline{X}_{i2})$$

Substituting these expressions (and corresponding ones for $\overline{X}_{j1} - \overline{X}_j$ and $\overline{X}_{j2} - \overline{X}_j$) in the formula for b_{ij}, we obtain

$$\begin{aligned} b_{ij} &= n_1\left[\frac{n_2}{n_1 + n_2}(\overline{X}_{i1} - \overline{X}_{i2})\right]\left[\frac{n_2}{n_1 + n_2}(\overline{X}_{j1} - \overline{X}_{j2})\right] \\ &\quad + n_2\left[\frac{-n_1}{n_1 + n_2}(\overline{X}_{i1} - \overline{X}_{i2})\right]\left[\frac{-n_1}{n_1 + n_2}(\overline{X}_{j1} - \overline{X}_{j2})\right] \\ &= \left[\frac{n_1n_2^2}{(n_1 + n_2)^2} + \frac{n_2n_1^2}{(n_1 + n_2)^2}\right](\overline{X}_{i1} - \overline{X}_{i2})(\overline{X}_{j1} - \overline{X}_{j2}) \\ &= \frac{n_1n_2}{n_1 + n_2}(\overline{X}_{i1} - \overline{X}_{i2})(\overline{X}_{j1} - \overline{X}_{j2}) \end{aligned}$$

Therefore, if we define a p-dimensional row vector

$$\mathbf{d}' = \left[\overline{X}_{11} - \overline{X}_{12}, \overline{X}_{21} - \overline{X}_{22}, \ldots, \overline{X}_{p1} - \overline{X}_{p2}\right] \tag{7.20}$$

of the differences between the two group means on the p variables, we see that

$$\mathbf{B} = \frac{n_1 n_2}{n_1 + n_2} \mathbf{dd}' \tag{7.21}$$

When this special form of **B** for the two-group case is substituted in Eq. 7.19, we get

$$(c\mathbf{T}^{-1}(\mathbf{dd}') - \mu\mathbf{I})\mathbf{v} = \mathbf{0}$$

where $c = n_1 n_2/(n_1 + n_2)$. Distributing the postmultiplication by $\mathbf{v}$ and transposing, we obtain

$$\begin{aligned}\mu\mathbf{v} &= c\mathbf{T}^{-1}(\mathbf{dd}')\mathbf{v} \\ &= c(\mathbf{T}^{-1}\mathbf{d})(\mathbf{d}'\mathbf{v})\end{aligned}$$

In the last expression, $\mathbf{d}'\mathbf{v}$, although unknown, is a scalar quantity, since d′ is $1 \times p$ and $\mathbf{v}$ is $p \times 1$. We may collect all the scalar quantities into a single multiplier and write

$$\begin{aligned}\mathbf{v} &= c(\mathbf{d}'\mathbf{v})/\mu\mathbf{T}^{-1}\mathbf{d} \\ &= m\mathbf{T}^{-1}\mathbf{d}\end{aligned} \tag{7.22}$$

where m is an unknown scalar.

Although Eq. 7.22 involves an unknown multiplier, it actually gives a solution for the eigenvector $\mathbf{v}$ of $\mathbf{T}^{-1}\mathbf{B}$ in the two-group case, because $\mathbf{v}$ is in any case determined only up to an arbitrary proportionality constant. (See the discussion on page 216; the constant may be determined by normalizing $\mathbf{v}$ to unity or by setting its leading element equal to unity.) We thus arrive at the interesting result that *in the two-group case,* the single discriminant function can be obtained without solving an eigenvalue problem. We need only postmultiply the inverse of the total SSCP matrix by the column of mean differences on the p predictors in order to get the vector of discriminant-function coefficients. Let us now examine the nature of the coefficients thus obtained.

The total SSCP matrix **T** is simply the SSCP matrix of the p predictors, computed for the two groups thrown together as a single sample; it is identical to what we denoted by $\mathbf{S}_{pp}$ in Section 3.2. Comparing Eq. 7.22 with Eq. 3.15, we see a striking resemblance, except for the arbitrary multipler m. We may, in fact, rewrite Eq. 7.22 as

$$\mathbf{v} = \mathbf{S}_{pp}^{-1}(m\mathbf{d})$$

which suggests that the elements of $\mathbf{v}$ might be regarded as some sort of multiple regression weights—provided that the elements of $m\mathbf{d}$ are interpretable as sums of products between predictors and criterion. But what criterion variable do we have in a two-group discriminant analysis? The answer is: a variable indicating each subject's membership in one or the other of the two groups. This would be a dichotomous variable taking the values 1 and 0 (for example) for members of groups 1 and 2, respectively.

Now, an appropriate coefficient of correlation between a continuous predictor X and a dichotomous criterion Y is the point-biserial coefficient, given in most statistics texts as

$$r_{pb} = \frac{\overline{X}_1 - \overline{X}_2}{s_x} \frac{\sqrt{n_1 n_2}}{N}$$

where $\overline{X}_1$ and $\overline{X}_2$ are the two group means on the predictor, s_x is the predictor standard deviation in the total sample of N (= $n_1 + n_2$) cases, and n_1 and n_2 are the numbers of cases with $Y = 1$ and $Y = 0$ (i.e., the two group sizes), respectively. From this coefficient, we may work backward to infer that the sum of products is given by

$$\Sigma\, xy = \frac{n_1 n_2}{n_1 + n_2} (\overline{X}_1 - \overline{X}_2)$$

We thus see that the elements of

$$m\mathbf{d}' = m[\overline{X}_{11} - \overline{X}_{12}, \overline{X}_{21} - \overline{X}_{22}, \ldots, \overline{X}_{p1} - \overline{X}_{p2}]$$

are indeed proportional to the predictor-criterion sum of products, where the criterion is a dichotomous variable indicating group membership.

Our conclusion, then, is that *in the two-group case* the discriminant weights are proportional to the weights for a multiple regression equation of a dichotomous group-membership variable on the p predictors. That is, discriminant analysis reduces, in this case, to a multiple regression analysis in which all group 1 members are assigned to score 1, and all group 2 members the score 0, on a "dummy" criterion variable Y. This fact led many early writers (e.g., Garrett, 1943; Wherry, 1947) to state that discriminant analysis *in general* was nothing more than a special case of multiple regression analysis. It cannot be overemphasized that this reduction holds only in the two-group case. When there are more than two groups under study, discriminant analysis reduces, not to multiple regression, but to canonical correlation analysis, as we shall see in Section 7.8.

EXAMPLE 7.1. In this miniature numerical example we illustrate the computation of a two-group discriminant function as a multiple regression equation, and compare it with the result obtained by the general method of discriminant analysis. Table 7.1 shows the scores of members of two groups on two predictor variables X_1 and X_2, and on a "dummy" criterion variable Y. All members of group 1 are assigned $Y = 1$, and all members of group 2 are given $Y = 0$.

TABLE 7.1 Roster for Two Groups on Two Predictors and a "Dummy" Criterion Variable Y

	Group 1 (n_1 = 8)			Group 2 (n_2 = 10)		
	X_1	X_2	Y	X_1	X_2	Y
	20	6	1	17	12	0
	21	10	1	11	11	0
	15	12	1	15	14	0
	15	8	1	20	16	0
	11	11	1	14	16	0
	24	17	1	13	18	0
	18	13	1	16	13	0
	14	4	1	12	6	0
				17	19	0
				10	16	0
ΣX_i	138	81		145	141	
ΣX_i^2	2508	939		2189	2119	
ΣX_1X_2	1449			2076		

The matrix $\mathbf{S}_{pp}$, whose inverse occurs in Eq. 3.15, is the upper left-hand 2 × 2 submatrix of an augmented SSCP matrix of the form shown on page 47. To compute this matrix, we merge the two groups into a single sample of $n_1 + n_2$ (= 18) cases. In other words, the values of ΣX_i, ΣX_i^2, and ΣX_1X_2, listed separately for each group at the foot of Table 7.1, are to be added across groups before computing the deviation-score sums of squares and sums of products. Further, it should be clear that the various sums involving Y for the total sample are given by

$$\sum Y = \sum Y^2 = n_1(= 8) \qquad \text{and} \qquad \sum X_iY = \sum_{\text{Gr.}1} X_i$$

Using these intermediate results, the augmented SSCP matrix is found to be

$$\left[\begin{array}{cc|c} 247.61 & 34.67 & 12.22 \\ 34.67 & 320.00 & -17.67 \\ \hline 12.22 & -17.67 & 4.44 \end{array}\right]$$

Therefore, according to Eqs. 2.23 and 6.17, the multiple regression weights, to which the two-group discriminant function weights are proportional, are given by the elements of the vector

$$\mathbf{b} = \begin{bmatrix} 247.61 & 34.67 \\ 34.67 & 320.00 \end{bmatrix}^{-1} \begin{bmatrix} 12.22 \\ -17.67 \end{bmatrix}$$

$$= (10^{-2})\begin{bmatrix} .4101 & -.0444 \\ -.0444 & .3173 \end{bmatrix}\begin{bmatrix} 12.22 \\ -17.67 \end{bmatrix} = \begin{bmatrix} .0580 \\ -.0615 \end{bmatrix}$$

Let us now compute the discriminant function by the general method of this chapter. The within-groups and between-groups SSCP matrices of the predictor variables are

$$\mathbf{W} = \begin{bmatrix} 214.00 & 83.25 \\ 83.25 & 249.78 \end{bmatrix}$$

and

$$\mathbf{B} = \begin{bmatrix} 33.61 & -48.58 \\ -48.58 & 70.22 \end{bmatrix}$$

We then obtain

$$\mathbf{W}^{-1}\mathbf{B} = \begin{bmatrix} .2674 & -.3864 \\ -.2836 & .4099 \end{bmatrix}$$

as the matrix to be substituted in Eq. 6.9. The characteristic equation, $|\mathbf{W}^{-1}\mathbf{B} - \lambda\mathbf{I}| = 0$, is found to be

$$\lambda^2 - .6773\lambda = 0$$

whose single nonzero root is $\lambda_1 = .6773$. The adjoint of $\mathbf{W}^{-1}\mathbf{B} - \lambda_1\mathbf{I}$ is

$$\begin{bmatrix} -.2674 & .3864 \\ .2836 & -.4099 \end{bmatrix}$$

and hence the eigenvector $\mathbf{v}_1$ of $\mathbf{W}^{-1}\mathbf{B}$, with the larger element set equal to unity, is

$$\mathbf{v}_1 = \begin{bmatrix} -.2674/.2836 \\ 1 \end{bmatrix} = \begin{bmatrix} -.9429 \\ 1 \end{bmatrix}$$

When the vector of regression weights obtained earlier is similarly rescaled, we find

$$\mathbf{b}^* = \begin{bmatrix} .0580/(-.0615) \\ 1 \end{bmatrix} = \begin{bmatrix} -.9431 \\ 1 \end{bmatrix}$$

which agrees, within rounding error, with the $\mathbf{v}_1$ just obtained by the general method of discriminant analysis.

Theoretical Supplement

As another matter pertaining to the special case of two groups, we recall the relationship between Wilks' Λ and Hotelling's T^2, mentioned without proof at the end of Chapter 4. We are now in a position to prove this relationship:

$$\frac{1 - \Lambda}{\Lambda} = \frac{T^2}{N - 2} \qquad \text{(where } N = n_1 + n_2\text{)}$$

We saw, in Eq. 7.12, that

$$1/\Lambda = (1 + \lambda_1)(1 + \lambda_2) \cdots (1 + \lambda_r)$$

where $\lambda_1, \lambda_2, \ldots, \lambda_r$ are the nonzero eigenvalues of $\mathbf{W}^{-1}\mathbf{B}$. However, in the two-group case $\mathbf{W}^{-1}\mathbf{B}$ has only one nonzero eigenvalue λ_1, since $\mathbf{B}$ is of rank 1. It therefore follows, in this case, that

$$1/\Lambda = 1 + \lambda_1$$

or

$$\frac{1 - \Lambda}{\Lambda} = \lambda_1 \qquad \text{(the only nonzero eigenvalue of } \mathbf{W}^{-1}\mathbf{B}\text{)}$$

On the other hand, from Eq. 4.28 it follows that

$$\frac{T^2}{N - 2} = c\mathbf{d}'\mathbf{W}^{-1}\mathbf{d}$$

where $c = n_1 n_2/(n_1 + n_2)$ and $\mathbf{d}$ is as defined in Eq. 7.20. Premultiplying both sides of this equation by $\mathbf{d}$, we get

$$\mathbf{d}[T^2/(N - 2)] = \mathbf{d}(c\mathbf{d}'\mathbf{W}^{-1}\mathbf{d})$$

or, on regrouping,

$$[T^2/(N - 2)]\mathbf{d} = [(c\mathbf{d}\mathbf{d}')\mathbf{W}^{-1}]\mathbf{d}$$

which becomes, by virtue of Eq. 7.21,

$$[T^2/(N - 2)]\mathbf{d} = (\mathbf{B}\mathbf{W}^{-1})\mathbf{d}$$

This means that $T^2/(N - 2)$ is the sole nonzero eigenvalue of $\mathbf{B}\mathbf{W}^{-1}$ (with $\mathbf{d}$ as the associated eigenvector). But since both $\mathbf{B}$ and $\mathbf{W}^{-1}$ are symmetric matrices, $\mathbf{B}\mathbf{W}^{-1} = (\mathbf{W}^{-1}\mathbf{B})'$. Furthermore, any matrix and its transpose have the same eigenvalue(s). (Prove this.) Therefore, $T^2/(N - 2)$ is also the only nonzero eigenvalue of $\mathbf{W}^{-1}\mathbf{B}$. The relation to be proved follows immediately.

7.7 STRUCTURE MATRICES IN DISCRIMINANT ANALYSIS

Before going on to the alternative method for computing discriminant functions—via canonical analysis—let us revisit the problem of interpreting a discriminant function and discuss an approach proposed by Cooley and Lohnes (1971) as an alternative to that presented in Section 7.3 (which was to compare the relative magnitudes of the standardized discriminant weights defined by Eq. 7.11). This is to compute the structure matrix for discriminant analysis, defined analogously to the structure matrix in factor analysis. It will be recalled (see page 183) that the (i, j)-element of the structure matrix is the correlation between the ith observed variable and the jth factor. Analogously, we define the (i, j)-element a_{ij} of the structure matrix in discriminant analysis as the correlation between the ith original variable X_i and the jth discriminant function Y_j. The computation in this case is more straightforward than in the case of factor analysis, however, because each discriminant function is explicitly defined as a linear combination of the p original variables, $X_1, X_2, \ldots, X_p$. In fact, the procedure is somewhat similar to that leading to Eq. 3.18 as an intermediate step in deriving the formula (3.19) for the multiple correlation coefficient in Chapter 3.

Since the jth discriminant function is

$$Y_j = X_1 v_{1j} + X_2 v_{2j} + \cdots + X_p v_{pj} = \mathbf{X}\mathbf{v}_j$$

it follows that the deviation score $Y_j - \overline{Y}_j = y_j$ is given by

$$y_j = \mathbf{x}'\mathbf{v}_j \tag{7.23}$$

where $\mathbf{x}'$ is the deviation of the vector $\mathbf{X}'$ from the centroid $\overline{\mathbf{X}}'$. Thus if we let $\mathbf{y}_j$ be the $N \times 1$ vector of deviation scores on the jth discriminant function in the total sample and $\mathbf{x}_i$ be the $N \times 1$ vector of deviation scores on X_i (i.e., the ith column of the $N \times p$ deviation-score matrix $\mathbf{x}$), then the correlation between X_i and Y_j is

$$\begin{aligned} r_{x_i y_j} &= \mathbf{x}_i'\mathbf{y}_j/\sqrt{(\mathbf{x}_i'\mathbf{x}_i)(\mathbf{y}_j'\mathbf{y}_j)} \\ &= \mathbf{x}_i'(\mathbf{x}\mathbf{v}_j)/\sqrt{(\mathbf{x}_i'\mathbf{x}_i)(\mathbf{x}\mathbf{v}_j)'(\mathbf{x}\mathbf{v}_j)} \\ &= [(\mathbf{x}'\mathbf{x})\mathbf{V}]_{ij}/[(\mathbf{x}'\mathbf{x})_{ii}]^{1/2}[\mathbf{V}'(\mathbf{x}'\mathbf{x})\mathbf{V}]_{jj}^{1/2} \end{aligned} \tag{7.24}$$

The $\mathbf{x}'$ in the last member of Eq. 7.24 is the transpose of the $N \times p$ deviation-score data matrix $\mathbf{x}$ rather than the "generic" row vector of deviation scores $\mathbf{X}' - \overline{\mathbf{X}}'$ as in Eq. 7.23, and $\mathbf{V}$ is the $p \times r$ matrix whose columns are the successive eigenvectors $\mathbf{v}_1, \mathbf{v}_2, \ldots, \mathbf{v}_r$ of $\mathbf{W}^{-1}\mathbf{B}$.

The matrix $\mathbf{x}'\mathbf{x}$ is simply the SSCP matrix in the total sample, which, in the context of discriminant analysis, we designate by $\mathbf{T}$ instead of $\mathbf{S}$. Then denoting by $\mathbf{D}(\cdot)$ the diagonal matrix consisting of the diagonal elements of the matrix standing inside the parentheses (which of course must be a square matrix), it can be seen that the two quantities whose square roots occur in the denominator of the last member of Eq. 7.24 are the ith and jth elements of $\mathbf{D}(\mathbf{T})$ and $\mathbf{D}(\mathbf{V}'\mathbf{T}\mathbf{V})$, respectively. Hence it can be shown that the entire structure matrix $\mathbf{A}$, whose (i, j)-element is given by expression (7.24) can be written as

$$\mathbf{A} = [\mathbf{D}(\mathbf{T})]^{-1/2}(\mathbf{T}\mathbf{V})[\mathbf{D}(\mathbf{V}'\mathbf{T}\mathbf{V})]^{-1/2} \tag{7.25}$$

Here $[\mathbf{D}(\mathbf{T})]^{-1/2}$ is the inverse of the square-root matrix of $\mathbf{D}(\mathbf{T})$ [or, equivalently, the square root of the inverse $[\mathbf{D}(\mathbf{T})]^{-1}$ of $\mathbf{D}(\mathbf{T})$], and similarly for $[\mathbf{D}(\mathbf{V}'\mathbf{T}\mathbf{V})]^{-1/2}$.

EXAMPLE 7.2 Let us compute the structure matrix for the example given in Section 7.5. Adding the $\mathbf{W}$ and $\mathbf{B}$ given on pp. 220–221, we get

$$\mathbf{T} = \begin{bmatrix} 5540.5742 & -421.4364 & 350.2556 \\ -421.4364 & 7295.5710 & -1170.5590 \\ 350.2556 & -1170.5590 & 3374.9232 \end{bmatrix}$$

Next, collecting the $\mathbf{v}_1$ and $\mathbf{v}_2$ given on page 222 into a single 3×2 matrix $\mathbf{V}$, we get

$$\mathbf{V} = \begin{bmatrix} .3524 & .9145 \\ -.7331 & .1960 \\ .5818 & -.3540 \end{bmatrix}$$

Premultiplying this by $\mathbf{T}$, we obtain

$$\mathbf{T}\mathbf{V} = \begin{bmatrix} 2465.23 & 4860.26 \\ -6177.93 & 1458.91 \\ 2945.10 & -1103.84 \end{bmatrix}$$

and further premultiplying this by $\mathbf{V}'$, we get

$$\mathbf{V}'\mathbf{T}\mathbf{V} = \begin{bmatrix} 7111.25 & 1.01 \\ 1.01 & 5121.41 \end{bmatrix}$$

Hence

$$[\mathbf{D}(\mathbf{V}'\mathbf{T}\mathbf{V})]^{-1/2} = \begin{bmatrix} 1/\sqrt{7111.25} & 0 \\ 0 & 1/\sqrt{5121.41} \end{bmatrix} = \begin{bmatrix} 1/84.33 & 0 \\ 0 & 1/71.56 \end{bmatrix}$$

Similarly,

$$[\mathbf{D}(\mathbf{T})]^{-1/2} = \begin{bmatrix} 1/74.44 & 0 & 0 \\ 0 & 1/85.41 & 0 \\ 0 & 0 & 1/58.09 \end{bmatrix}$$

Substituting in Eq. 7.25, we obtain the structure matrix

$$\mathbf{A} = [\mathbf{D}(\mathbf{T})]^{-1/2}(\mathbf{TV})[\mathbf{D}(\mathbf{V}'\mathbf{TV})]^{-1/2}$$

$$= \begin{bmatrix} 1/74.44 & 0 & 0 \\ 0 & 1/85.41 & 0 \\ 0 & 0 & 1/58.09 \end{bmatrix} \begin{bmatrix} 2465.23 & 4860.26 \\ -6177.93 & 1458.91 \\ 2945.10 & -1103.84 \end{bmatrix}$$

$$\times \begin{bmatrix} 1/84.33 & 0 \\ 0 & 1/71.56 \end{bmatrix}$$

$$= \begin{bmatrix} .3927 & .9124 \\ -.8577 & .2387 \\ .6012 & -.2655 \end{bmatrix}$$

The first column of **A** lists the correlations of the three original variables X_1, X_2, and X_3 with the first discriminant function Y_1, and the second column, their correlations with Y_2.

Interpretation of Discriminant Functions Revisited

In Section 7.3 we mentioned the possibility of interpreting a discriminant function by comparing the standardized discriminant weights accruing to each of the original variables. In the structure matrix we have another possible means for interpreting discriminant functions—just as for factor analysis.

Most authors actually seem to adopt one or the other of the two possible means for interpretation to the exclusion of the other. The present author's standpoint is that each method provides a slightly different type of interpretation. The set of standardized discriminant weights $\sqrt{w_{ii}}v_{ji}$ ($i = 1, 2, \ldots, p$ for each Y_j) enables us to answer the question: What sort of profile on the p variables gives rise to a high (or a low) score on the jth discriminant function? Thus, noting the natures of those variables on which high scores are conducive to a high Y_j score and those on which low scores are so conducive, we may have a means for characterizing Y_j in a somewhat indirect way. On the other hand, recalling that the correlation between two variables is the cosine of the angle between the two test vectors in *person space,* and hence the larger the correlation, the smaller the angle between the two vectors, the elements $r_{x_i y_j}$ of the structure matrix provide an answer to the following question: To which of the

p test vectors of the original variables does the vector of the jth discriminant function (a composite variable) lie closest in person space?'' The answer to this question enables us to characterize Y_j as being ''similar'' to one or more of the original variables.

It should be noted that the statement on page 222—that the elements of the discriminant-weight vectors (i.e., the eigenvectors of $\mathbf{W}^{-1}\mathbf{B}$) are ''the cosines of the angles between the original axes . . . and the new axes representing the . . . discriminant functions''—does not imply that these elements and those of the structure matrix **A** are the same things. It is true that in both cases we have cosines of the angles between entities representing the original variables and those representing the discriminant functions. However, in the case of the eigenvectors, these entities are axes in test space, while in the structure matrix, they are test vectors in person space. As mentioned in Section 4.5, it is only in the latter case (and then, only when the test scores are expressed as deviations from their respective means) that the cosine of the angle between two vectors represents the correlation between the two tests (or, more generally, between a test and a linear composite of several tests or between two linear composites). Thus, even though the elements of the rotation matrix **V** are the cosines of angles between axes representing the original variables and those representing the discriminant functions, they do not stand for the correlations between these variates.

7.8 CANONICAL CORRELATIONS APPROACH TO DISCRIMINANT ANALYSIS

A natural extension of the multiple-regression formulation of two-group discriminant analysis to discriminant analysis for any number of groups was made by Brown (1947) prior to the developments described in Sections 7.1 to 7.3. In this formulation, a set of ''dummy criterion variables'' one fewer than the number of groups is used, and the predictor and criterion sets are treated by the method of canonical correlation analysis.[3] A mathematical proof that the discriminant criterion and canonical correlation approaches yield identical results was given by Tatsuoka (1953).

Suppose that $K = 5$ groups are under study. Then, for carrying out a discriminant analysis via the canonical correlation approach, $K - 1 = 4$ dummy criterion variables Y_1, Y_2, Y_3, Y_4 are used, and ''scores'' on these are assigned to members of the five groups in the following manner:

[3]Although a canonical-correlation approach had been suggested earlier by Bartlett (1938), he proposed using differences among the group means on the predictor variables rather than dummy group-membership variables as the criterion set.

	Y_1	Y_2	Y_3	Y_4
All group 1 members get	1	0	0	0
All group 2 members get	0	1	0	0
All group 3 members get	0	0	1	0
All group 4 members get	0	0	0	1
All group 5 members get	0	0	0	0

The generalization to the case of K groups should be obvious: For each k between 1 and $K - 1$ inclusive, all members of the kth group get a 1 on Y_k and 0's on the other Ys; while members of the last (Kth) group get 0's on all $K - 1$ dummy criterion variables $Y_1, Y_2, \ldots, Y_{K-1}$.

Upon thus assigning "scores" on the $K - 1$ dummy variables to each individual in the entire sample, we shall have $p + (K - 1)$ "observations" on $N (= n_1 + n_2 + \cdots + n_k)$ individuals; namely, p predictor-variable scores and $K - 1$ "criterion" scores. Whereas in the two-group case we had only $2 - 1 = 1$ "criterion" variable, and hence multiple regression analysis enabled us to determine a linear combination of the p predictors having maximum correlation with the criterion, we now have $q = K - 1$ criterion variables. Something more elaborate than multiple regression analysis must now be invoked.

Canonical correlation analysis is a technique by which we determine a linear combination of p predictors on the one hand, and a linear combination of q (= $K - 1$ in the present context) criterion variables on the other, such that the correlation between these linear combinations in the total sample is as large as possible. Formulated mathematically, the problem is as follows. We want to determine one set of weights

$$\mathbf{u}' = [u_1, u_2, \ldots, u_p]$$

for the predictor variables, and another set of weights

$$\mathbf{v}' = [v_1, v_2, \ldots, v_p]$$

for the criterion variables, in such a way that the correlation r_{zw} between

$$Z = u_1X_1 + u_2X_2 + \cdots + u_pX$$

and

$$W = v_1Y_1 + v_2Y + \cdots + v_pY_q$$

is the largest achievable for our particular sample.

Postponing the discussion of the rationale until Section 7.9, here we describe the mechanics of the solution to this problem. The first step is to compute the SSCP matrix of all $p + q$ variables (in the present context, $p + K - 1$ variables) for the total sample of N individuals. This may be done in the usual manner regardless of whether the criterion variables are genuine variables such as test scores, or are "dummy" variables such as the group-membership variables in the discriminant-analysis context. In the latter case, however, some obvious simplifications of the formulas for the raw-score sums, sums of squares, and sums of products involving the Y's may be noted. Namely,

$$\Sigma Y_k = \Sigma Y_k^2 = n_k \qquad (k = 1, 2, \ldots, K-1)$$

$$\Sigma Y_h Y_k = 0 \qquad (h, k = 1, 2, \ldots, K-1; h \neq k)$$

$$\Sigma X_i Y_k = \sum_{\text{Gr.}k} X_i \qquad \text{(i.e., the sum of } X_i \text{ in group } k\text{)}$$

The SSCP matrix is then partitioned into four parts, similar to what was done for the multiple regression problem (see p. 47), except that the part referring to the "criterion" variables is now a $q \times q$ matrix instead of a single scalar, and the parts interrelating predictor and "criterion" variables are $p \times q$ and $q \times p$ matrices instead of column and row vectors, respectively. The partitioning is indicated as follows:

$$\mathbf{S} = \left[\begin{array}{c|c} \mathbf{S}_{pp} & \mathbf{S}_{pc} \\ \hline \mathbf{S}_{cp} & \mathbf{S}_{cc} \end{array}\right] \begin{array}{l} p \text{ rows} \\ q \text{ rows} \end{array}$$

p columns $\quad q$ columns

Next, the quadruple matrix product

$$\mathbf{A} = \mathbf{S}_{pp}^{-1}\mathbf{S}_{pc}\mathbf{S}_{cc}^{-1}\mathbf{S}_{cp} \tag{7.26}$$

is computed.

Finally, the eigenvalues μ_i^2 and the eigenvectors $\mathbf{u}_i$ of the matrix $\mathbf{A}$ are computed. The largest eigenvalue μ_1^2 is the *square* of the maximum r_{zw}, called the first *canonical correlation* between the predictor and criterion sets. The elements of the corresponding eigenvector $\mathbf{u}_1$ are the weights to be used in combining the predictor variables to obtain the optimal linear combination Z_1. (The weights $\mathbf{v}_1$ to be applied to the criterion set to get W_1 are of no interest in the present context, where the "criteria" are simply dummy variables indicating group membership. An equation for computing $\mathbf{v}_i$ is given in the next section, where canonical correlation analysis is considered a greater generality.)

As the reader may have anticipated, the elements of the eigenvector $\mathbf{u}_2$, associated with the second largest eigenvalue μ_2^2, give the weights to be applied to the predictor set to construct a linear combination Z_2 which, among all predictor-set linear combinations that are uncorrelated with Z_1, has the largest possible correlation with any criterion-set linear combination that is uncorrelated with W_1. Similar interpretations hold for the other eigenvectors, $\mathbf{u}_3, \mathbf{u}_4, \ldots, \mathbf{u}_{K-1}$.

The set of linear combinations $Z_1, Z_2, \ldots, Z_{K-1}$ thus obtained is identical (within proportionality) to the discriminant functions obtainable by the approach, described earlier, of maximizing the discriminant criterion. Furthermore, each discriminant-criterion value λ_i is related to the corresponding squared canonical-correlation value μ_i^2 by the equation (proved in Tatsuoka, 1953)

$$\lambda_i = \mu_i^2/(1 - \mu_i^2) \tag{7.27}$$

The following numerical example illustrates these facts.

EXAMPLE 7.3. For this example we add a third group of $n_3 = 12$ individuals to the two-group example of Section 7.7. Table 7.2 shows the rosters of the three groups, including "scores" on the two "dummy" criterion variables Y_1 and Y_2. The SSCP matrix of the two predictors and two "criterion" variables for the total sample of $N = 8 + 10 + 12 = 30$ individuals, partitioned into four submatrices as indicated above, is found to be as follows:

$$\mathbf{S} = \left[\begin{array}{cc|cc} 768.80 & 166.60 & 32.40 & 13.00 \\ 166.60 & 471.87 & -7.53 & 30.33 \\ \hline 32.40 & -7.53 & 5.87 & -2.67 \\ 13.00 & 30.33 & -2.67 & 6.67 \end{array}\right]$$

The computation of the upper left-hand submatrix $\mathbf{S}_{pp}$ should require no comment. The other submatrices $\mathbf{S}_{pc}$ and $\mathbf{S}_{cc}$, too, really involve nothing new, since their elements are of the forms

$$\Sigma\, x_i y_j = \Sigma\, X_i Y_j - (\Sigma\, X_i)(\Sigma Y_j)/N$$

and

$$\Sigma\, y_i y_j = \Sigma\, Y_i Y_j - (\Sigma\, Y_i)(\Sigma\, Y_j)/N$$

However, to illustrate use of the simplified formulas displayed earlier for the various sums involving Y_1 and Y_2, we show the calculations for a few of these elements below. (The reader should compare the figures below with the various entries of Table 7.2 and trace the substitutions made in the formulas.)

$$\begin{aligned} (\mathbf{S}_{pc})_{11} &= \Sigma\, X_1 Y_1 - (\Sigma\, X_1)(\Sigma\, Y_1)/N \\ &= 138 - (138 + 145 + 113)(8)/30 = 32.40 \end{aligned}$$

$$\begin{aligned} (\mathbf{S}_{pc})_{12} &= \Sigma\, X_1 Y_2 - (\Sigma\, X_1)(\Sigma\, Y_2)/N \\ &= 145 - (396)(10)/30 = 13.00 \end{aligned}$$

$$\begin{aligned} (\mathbf{S}_{cc})_{11} &= \Sigma\, Y_1^2 - (\Sigma\, Y_1)^2/N \\ &= 8 - (8)^2/30 = 5.87 \end{aligned}$$

$$(\mathbf{S}_{cc})_{12} = \Sigma Y_1 Y_2 - (\Sigma Y_1)(\Sigma Y_2)/N$$
$$= 0 - (8)(10)/30 = -2.67$$

After constructing the SSCP matrix **S** in this manner, we compute the quadruple matrix product $\mathbf{S}_{pp}^{-1}\mathbf{S}_{pc}\mathbf{S}_{cc}^{-1}\mathbf{S}_{cp}$, whose eigenvalues and eigenvectors we need to determine. The result is

$$\mathbf{A} = \mathbf{S}_{pp}^{-1}\mathbf{S}_{pc}\mathbf{S}_{cc}^{-1}\mathbf{S}_{cp}$$
$$= \begin{bmatrix} 768.80 & 166.60 \\ 166.60 & 471.87 \end{bmatrix}^{-1} \begin{bmatrix} 32.40 & 13.00 \\ -7.53 & 30.33 \end{bmatrix}$$
$$\times \begin{bmatrix} 5.87 & -2.67 \\ -2.67 & 6.67 \end{bmatrix}^{-1} \begin{bmatrix} 32.40 & -7.53 \\ 13.00 & 30.33 \end{bmatrix}$$
$$= \begin{bmatrix} .403219 & .063240 \\ .059355 & .279504 \end{bmatrix}$$

The characteristic equation, $|\mathbf{A} - \mu^2\mathbf{I}| = 0$, is

$$\mu^4 - .6827\mu^2 + .1089 = 0$$

whose roots are

$$\mu_1^2 = .4286 \qquad \text{and} \qquad \mu_2^2 = .2541$$

The associated eigenvectors, each with the largest element set equal to unity, are:

$$\mathbf{u}_1' = [1, .3986] \qquad \text{and} \qquad \mathbf{u}_2' = [-.4246, 1]$$

Hence the two discriminant functions may be written as

$$Z_1 = X_1 + .3986X_2$$

and

$$Z_2 = -.4246X_1 + X_2$$

We now carry out the discriminant analysis by the discriminant-criterion approach. The within-groups and between-groups SSCP matrices are found to be

$$\mathbf{W} = \begin{bmatrix} 448.9167 & 71.4167 \\ 71.4167 & 329.4417 \end{bmatrix}$$

and

$$\mathbf{B} = \begin{bmatrix} 319.8833 & 95.1833 \\ 95.1833 & 142.4250 \end{bmatrix}$$

From these, we compute

$$\mathbf{W}^{-1}\mathbf{B} = \begin{bmatrix} .690413 & .148368 \\ .139253 & .400158 \end{bmatrix}$$

The characteristic equation, $|\mathbf{W}^{-1}\mathbf{B} - \lambda\mathbf{I}| = 0$, is

$$\lambda^2 - 1.0906\lambda + .2556 = 0$$

with roots

$$\lambda_1 = .7496 \qquad \text{and} \qquad \lambda_2 = .3410$$

The associated eigenvectors are

$$\mathbf{v}_1' = [1, .3986] \qquad \text{and} \qquad \mathbf{v}_2' = [-.4247, 1]$$

which agree, within rounding error, with the results just obtained by the canonical correlation approach. Note also that

$$\mu_1^2/(1 - \mu_1^2) = .4286/.5714 = .7501 \simeq \lambda_1$$

and

$$\mu_2^2/(1 - \mu_2^2) = .2541/.7459 = .3407 \simeq \lambda_2$$

asserted in Eq. 7.27.

TABLE 7.2 Roster for Three Groups on Two Predictors and Two "Dummy" Criterion Variables Y_1 and Y_2

	Group 1 ($n_1 = 8$)				Group 2 ($n_2 = 10$)				Group 3 ($n_3 = 12$)			
	X_1	X_2	Y_1	Y_2	X_1	X_2	Y_1	Y_2	X_1	X_2	Y_1	Y_2
	20	6	1	0	17	12	0	1	6	11	0	0
	21	10	1	0	11	11	0	1	10	8	0	0
	15	12	1	0	15	14	0	1	4	6	0	0
	15	8	1	0	20	16	0	1	14	7	0	0
	11	11	1	0	14	16	0	1	13	11	0	0
	24	17	1	0	13	18	0	1	18	11	0	0
	18	13	1	0	16	13	0	1	12	9	0	0
	14	4	1	0	12	6	0	1	12	9	0	0
					17	19	0	1	5	7	0	0
					10	16	0	1	10	5	0	0
									3	14	0	0
									6	12	0	0
ΣX_i	138	81			145	141			113	110		
ΣX_i^2	2508	939			2189	2119			1299	1088		
ΣX_1X_2	1449				2076				1024			

7.9 CANONICAL CORRELATION ANALYSIS

In Section 7.8 we introduced canonical correlation analysis in a backhanded sort of way, as an alternative computational technique for multigroup discriminant functions. We now turn to the method of canonical analysis in its more usual and general application: that of seeking relationships between two sets of variables. The method was developed by Hotelling (1935).

The type of problem for which canonical analysis is useful may be illustrated by the following. An investigator wants to explore the possible relationship between personality variables on the one hand, and academic achievement on the other, among high school students. He or she may administer a personality inventory such as Cattell and Eber's (1967–1968) Sixteen Personality-Factors Questionnaire, the 16 PF (which yields scores on 16 factor-analytically determined personality scales), and also give a battery of achievement tests in various subject-matter areas, to a group of high school students. The essential question is: What sort of personality profile tends to be associated with what sort of parttern of academic achievement? Canonical analysis helps answer this question by determining linear combinations of the personality scales that are most highly correlated with linear combinations of the achievement tests. The technique may therefore be loosely characterized as a sort of "double-barreled principal components analysis." It identifies the "components" of one set of variables that are most highly related (linearly) to the "components" of the other set of variables.

Let the variables in the first set be denoted $X_1, X_2, \ldots, X_p$, and those in the second set, $Y_1, Y_2, \ldots, Y_q$. We construct a linear combination

$$Z = u_1X_1 + u_2X_2 + \cdots + u_pX_p$$

of the first set of variables, and a linear combination

$$W = v_1Y_1 + v_2Y_2 + \cdots + v_qY_q$$

of the second set. Since we want to determine the two sets of coefficients, $\mathbf{u}' = [u_1, u_2, \ldots, u_p]$ and $\mathbf{v}' = [v_1, v_2, \ldots, v_q]$, so as to maximize the correlation between the two linear combinations, our first task is to express r_{zw} as a function of $\mathbf{u}$ and $\mathbf{v}$.

The quantities Σz^2 and Σw^2, which occur in the denominator of the expression $\Sigma zw/\sqrt{(\Sigma z^2)(\Sigma w^2)}$ for r_{zw}, are expressible as quadratic forms, as we already saw in Eq. 5.18. That is,

$$\Sigma z^2 = \mathbf{u}'\mathbf{S}_{xx}\mathbf{u} \tag{7.28}$$

and

$$\Sigma w^2 = \mathbf{v}'\mathbf{S}_{yy}\mathbf{v} \tag{7.29}$$

where $\mathbf{S}_{xx}$ and $\mathbf{S}_{yy}$ are the SSCP matrices of the two sets of variables, respectively. By a similar line of reasoning, it can be shown that

$$\Sigma\, zw = \mathbf{u}'\mathbf{S}_{xy}\mathbf{v} \tag{7.30}$$

where $\mathbf{S}_{xy}$ is the $p \times q$ matrix of sums of products between the X's and the Y's.

Bringing together Eqs. 7.28 to 7.30, the formula for r_{zw} may be written as

$$r_{zw} = \frac{\mathbf{u}'\mathbf{S}_{xy}\mathbf{v}}{\sqrt{(\mathbf{u}'\mathbf{S}_{xx}\mathbf{u})(\mathbf{v}'\mathbf{S}_{yy}\mathbf{v})}} \tag{7.31}$$

As was the case with discriminant analysis, the weights **u** and **v**, which maximize this expression, are determined only up to proportionality constants. For if a and b are two arbitrary constants of the same sign, the value of r_{zw} obtained by using the elements of $a\mathbf{u}$ and $b\mathbf{v}$ as the combining weights is readily seen to be equal to the value resulting from the use of u_i and v_i as weights. It is most convenient, in the present situation, to choose the proportionality constants so that

$$\mathbf{u}'\mathbf{S}_{xx}\mathbf{u} = \mathbf{v}'\mathbf{S}_{yy}\mathbf{v} = 1 \tag{7.32}$$

and hence the entire denominator of expression (7.31) becomes unity. Then, introducing Lagrange multipliers $\lambda/2$ and $\mu/2$ (where the factors $\frac{1}{2}$ are for numerical convenience), the function to be maximized is

$$F(\mathbf{u}, \mathbf{v}) = \mathbf{u}'\mathbf{S}_{xy}\mathbf{v} - (\lambda/2)(\mathbf{u}'\mathbf{S}_{xx}\mathbf{u} - 1) - (\mu/2)(\mathbf{v}'\mathbf{S}_{yy}\mathbf{v} - 1)$$

Taking the symbolic partial derivatives (see Appendix C) of $F(\mathbf{u}, \mathbf{v})$ with respect to **u** and **v**′, and setting each of these equal to the null vector, we have

$$\frac{\partial F}{\partial \mathbf{u}} = \mathbf{S}_{xy}\mathbf{v} - \lambda\mathbf{S}_{xx}\mathbf{u} = \mathbf{0} \tag{7.33}$$

and

$$\frac{\partial F}{\partial \mathbf{v}'} = \mathbf{u}'\mathbf{S}_{xy} - \mu\mathbf{v}'\mathbf{S}_{yy} = \mathbf{0}' \tag{7.34}$$

as the equations to be satisfied by **u** and **v** in order to maximize r_{zw} under the side conditions 7.32. (It can also be shown that Eqs. 7.33 and 7.34 constitute sufficient conditions for the desired maximization.)

Next, we premultiply all members of Eq. 7.33 by **u**′, and postmultiply all members of Eq. 7.34 by **v**. These operations result in the relations

$$\mathbf{u}'\mathbf{S}_{xy}\mathbf{v} = \lambda(\mathbf{u}'\mathbf{S}_{xx}\mathbf{u}) = \mu(\mathbf{v}'\mathbf{S}_{yy}\mathbf{v})$$

which, by virtue of conditions (7.32), reduce to

$$\mathbf{u}'\mathbf{S}_{xy}\mathbf{v} = \lambda = \mu$$

In other words, both λ and μ are equal to the maximum value achievable by the correlation coefficient r_{zw}, which results from using as combining weights the elements of $\mathbf{u}$ and $\mathbf{v}$ that satisfy Eqs. 7.33 and 7.34. Such being the case, the λ in Eq. 7.33 may be replaced by μ. Making this replacement, taking the transpose of both members of Eq. 7.34, and writing $\mathbf{S}_{yx}$ for $\mathbf{S}'_{xy}$, the pair of Eqs. 7.33 and 7.34 may be rewritten as

$$\mathbf{S}_{xy}\mathbf{v} = \mu\mathbf{S}_{xx}\mathbf{u}$$

$$\mathbf{S}_{yx}\mathbf{u} = \mu\mathbf{S}_{yy}\mathbf{v}$$

Assuming $\mathbf{S}_{yy}$ to be nonsingular, we may solve the second of these equations to express $\mathbf{v}$ in terms of $\mathbf{u}$, as follows:

$$\mathbf{v} = (1/\mu)\mathbf{S}_{yy}^{-1}\mathbf{S}_{yx}\mathbf{u} \qquad (7.35)$$

This expression may then be substituted for $\mathbf{v}$ in the first of the above pair of equations to yield

$$\mathbf{S}_{xy}\left(\frac{1}{\mu}\mathbf{S}_{yy}^{-1}\mathbf{S}_{yx}\mathbf{u}\right) = \mu\mathbf{S}_{xx}\mathbf{u}$$

or, premultiplying both members by $\mu\mathbf{S}_{xx}^{-1}$ and rearranging the terms,

$$(\mathbf{S}_{xx}^{-1}\mathbf{S}_{xy}\mathbf{S}_{yy}^{-1}\mathbf{S}_{yx} - \mu^2\mathbf{I})\mathbf{u} = \mathbf{0} \qquad (7.36)$$

It is thus seen that the largest eigenvalue μ_1^2 of the quadruple matrix product $\mathbf{S}_{xx}^{-1}\mathbf{S}_{xy}\mathbf{S}_{yy}^{-1}\mathbf{S}_{yx}$ gives the square of the maximum r_{zw}, and that the elements of the associated eigenvector $\mathbf{u}_1$ provide the weights by which one set of variables (the X's) should be linearly combined in order to achieve this maximum correlation. The vector of combining weights $\mathbf{v}_1$ for the other set of variables (the Y's) may be obtained by substituting μ_1 and $\mathbf{u}_1$ in Eq. 7.35, that is,

$$\mathbf{v}_1 = (1/\mu_1)\mathbf{S}_{yy}^{-1}\mathbf{S}_{yx}\mathbf{u}_1$$

Equation 7.36 will, in general, yield other eigenvalues and vectors besides μ_1^2 and $\mathbf{u}_1$, and their meaning should be familiar to the reader by now. The situation is exactly parallel to those holding in principal-components analysis and in multigroup discriminant analysis. In each case, we start out by seeking a set of combining weights (or two such sets, in canonical correlation analysis) that will maximize a specified criterion for the resulting linear combination (or pair of linear combinations). In each case, the elements of the vector associated with the largest eigenvalue of a certain matrix constitute the weights leading to the desired absolute maximum. In addition, the elements of the vectors associated with the second, third, and subsequent eigenvalues (in descending order of magnitude) yield conditional maxima in the following sense: The linear combination formed by the elements of the second vector has the largest value on the relevant criterion among those that are uncorrelated with the first linear combination; the third linear combination has the maximum criterion value among linear combinations uncorrelated with the first two linear combinations; and so on. In the case of canonical correlation analysis, the linear combinations occur in two sequences, one for each of the two sets of variables. The uncorrelatedness holds both within each sequence and between unmatched pairs of linear combinations across the two sequences. Thus not only is Z_2 uncorrelated with Z_1 and W_2 with W_1, but Z_2 is also uncorrelated with W_1, and W_2 with Z_1. In short, the only nonzero correlations are those between the corresponding members, Z_1 and W_1, Z_2 and W_2, and so on, of the two sequences. These pairs of linear combinations are called the *canonical variates*, and the correlations between them are called the *canonical correlations*. The number of canonical-variate pairs will be equal to the number of variables in the smaller set, that is, the smaller of the two numbers p and q. This is because the rank of the quadruple product matrix $\mathbf{S}_{xx}^{-1}\mathbf{S}_{xy}\mathbf{S}_{yy}^{-1}\mathbf{S}_{yx}$, whose eigenvalues and eigenvectors determine the canonical variates, is equal to p or q, whichever is smaller.

EXAMPLE 7.4[4] A study was conducted to investigate, among other things, whether specific instruction on how to relax when taking tests and how to increase motivation would affect performance on standardized achievement tests in school subjects. Here we use a small part of the data from this study to illustrate canonical correlation analysis.

A group of 65 third- and fourth-grade pupils were rated by their teachers (on 10-point scales), after the instruction and immediately prior to taking the California Tests of Basic Skills (CTBS), with respect to how relaxed they were (X_1) and how motivated they were (X_2). Their percentile-rank scores on the CTBS reading, language, and mathematics scales are denoted by Y_1, Y_2, and Y_3, respectively. The SSCP matrix for the five variables, partitioned into the four sectors $\mathbf{S}_{xx}$, $\mathbf{S}_{xy}$, $\mathbf{S}_{yy}$, and $\mathbf{S}_{yx}$, was as follows:

$$\mathbf{S} = \left[\begin{array}{cc|ccc} 96.1538 & 15.8462 & -41.5365 & 249.231 & 63.6154 \\ 15.8642 & 30.1538 & 418.538 & 558.769 & 452.385 \\ \hline -41.5385 & 418.538 & 40{,}545.4 & 31{,}339.7 & 32{,}959.8 \\ 249.231 & 558.769 & 31{,}339.7 & 36{,}059.8 & 25{,}809.8 \\ 63.6154 & 452.385 & 32{,}959.8 & 25{,}809.9 & 45{,}302.9 \end{array}\right]$$

Although we need the quadruple product $\mathbf{S}_{xx}^{-1}\mathbf{S}_{xy}\mathbf{S}_{yy}^{-1}\mathbf{S}_{yx}$ in the eigenequation 7.36, we obtain this in two steps, because we also need $\mathbf{S}_{yy}^{-1}\mathbf{S}_{yx}$ in Eq. 7.35 in

[4]I wish to thank Professor Kennedy T. Hill of the Institute of Child Behavior and Development, University of Illinois at Urbana-Champaign, for making available to me the data used in this example. Thanks are due also to Ms. Ratna Nandakumar, his research assistant, for carrying out the analysis for me, using the SAS matrix algebra procedures.

order to get the vector **v** of weights for the Y set of variables after determining **u** from Eq. 7.36. The results are

$$\mathbf{S}_{yy}^{-1}\mathbf{S}_{yx} = \begin{bmatrix} -.232232 & -.095539 \\ .236341 & .197163 \\ .048353 & .057038 \end{bmatrix} \times 10^{-1}$$

and

$$\mathbf{S}_{xx}^{-1}\mathbf{S}_{xy} = \begin{bmatrix} -2.977298 & -.505646 & -1.982531 \\ 15.444716 & 18.796356 & 16.044432 \end{bmatrix}$$

From these we get the quadruple product

$$\mathbf{S}_{xx}^{-1}\mathbf{S}_{xy}\mathbf{S}_{yy}^{-1}\mathbf{S}_{yx} = \begin{bmatrix} .047606 & .007167 \\ .163139 & .314552 \end{bmatrix}$$

Substituting this expression in Eq. 7.36 and computing the characteristic equation

$$|\mathbf{S}_{xx}^{-1}\mathbf{S}_{xy}\mathbf{S}_{yy}^{-1}\mathbf{S}_{yx} - \mu^2\mathbf{I}| = 0$$

we get

$$\mu^4 - .362159\mu^2 + .013805 = 0$$

whose roots are

$$\mu_1^2 = .318863 \qquad \text{and} \qquad \mu_2^2 = .043295$$

Substituting the first eigenvalue in the coefficient matrix in Eq. 7.36 and computing its adjoint, we get

$$\mathbf{Adj}(\mathbf{S}_{xx}^{-1}\mathbf{S}_{xy}\mathbf{S}_{yy}^{-1}\mathbf{S}_{yx} - .318863\mathbf{I}) = \begin{bmatrix} -.004310 & -.007167 \\ -.163138 & -.271257 \end{bmatrix}$$

Normalizing either of the two columns of this matrix to unity, we obtain the eigenvector $\mathbf{u}_1$ associated with the eigenvalue μ_1^2:

$$\mathbf{u}_1 = \begin{bmatrix} .02641 \\ .99965 \end{bmatrix}$$

Substituting this vector, the matrix $\mathbf{S}_{yy}^{-1}\mathbf{S}_{yx}$ obtained above and the square root $\sqrt{.318863} = .564680$ of the eigenvalue μ_1^2 in Eq. 7.35, we get

$$(1/.564680)(\mathbf{S}_{yy}^{-1}\mathbf{S}_{yx})\mathbf{u}_1 = \begin{bmatrix} -.101639 \\ .203336 \\ .058295 \end{bmatrix}(1/.564680)$$

Actually, we need not carry out the indicated division by .564680, because we will normalize the resulting vector to unity anyway to get $\mathbf{v}_1$:

$$\mathbf{v}_1 = \begin{bmatrix} -.43310 \\ .86644 \\ .24840 \end{bmatrix}$$

Thus the first pair of canonical variates, with the coefficients rounded to three decimal places, is

$$Z_1 = .026X_1 + 1.000X_2$$

and

$$W_1 = -.433Y_1 + .866Y_2 + .248Y_3$$

Determination of the second pair of eigenvectors $\mathbf{u}_2$ and $\mathbf{v}_2$, associated with the second eigenvalue $\mu_2^2 = .043295$, is left as an exercise.

7.10 SIGNIFICANCE TESTS OF CANONICAL VARIATES

Two types of significance tests are of interest in canonical correlation analysis. The first is an overall test to decide whether there is *any* significant linear relationship between the two sets of variables. If overall significance is found, we would then want to know how many of the canonical-variate pairs are significantly related. As might be expected (since we have seen that discriminant analysis may be regarded as a special case of canonical analysis), the significance tests here are closely related to those described earlier for discriminant analysis.

To see this relationship, we recall that Eq. 7.27 holds between each eigenvalue λ_i from the basic equation,

$$(\mathbf{W}^{-1}\mathbf{B} - \lambda\mathbf{I})\mathbf{v} = \mathbf{0}$$

for discriminant analysis via the discriminant criterion approach and the corresponding eigenvalue μ_i^2 from the basic equation,

$$(\mathbf{S}_{xx}^{-1}\mathbf{S}_{xy}\mathbf{S}_{yy}^{-1}\mathbf{S}_{yx} - \mu^2\mathbf{I})\mathbf{u} = \mathbf{0}$$

in the canonical correlation approach. Namely, the corresponding eigenvalues resulting from these two basic equations are related by the equality

$$\lambda_i = \mu_i^2/(1 - \mu_i^2)$$

As a consequence, Wilks' Λ-criterion, which was shown in Eq. 7.12 to be expressible as

$$\Lambda = 1/\prod_{i=1}^{r} (1 + \lambda_i)$$

may also be expressed in terms of the μ_i^2 as follows:

$$\Lambda = 1/\Pi[1 + \mu_i^2/(1 - \mu_i^2)]$$
$$= 1/\Pi[1/(1 - \mu_i^2)]$$

or

$$\Lambda = \prod_{i=1}^{r} (1 - \mu_i^2) \tag{7.37}$$

The demonstration above, of course, shows only that this alternative expression for Λ holds when the μ_i^2 results from canonical analysis as applied to the problem of discriminant analysis. However, it is quite plausible that Eq. 7.37 will continue to hold for canonical analysis in general. For each μ_i^2, it will be recalled, is a conditionally maximal value of r_{zw}^2, the squared correlation between corresponding pairs of canonical variates constructed from the two sets of variables. Thus, each factor of the product

$$(1 - \mu_i^2)(1 - \mu_2^2) \cdots (1 - \mu_r^2)$$

is in fact the coefficient of alienation between a particular pair of canonical variates. This is consistent with the fact that Λ is a statistic that is *inversely* related to the magnitude of differences or strength of relationship: The smaller the value of Λ, the greater the difference or relationship in question.

There is a problem, however, in that the definition of Λ as the ratio $|\mathbf{W}|/|\mathbf{T}|$ (first given in Chapter 4 and used again in connection with discriminant analysis in this chapter) does not make sense in the context of canonical analysis. The sample in this situation is not composed of several subgroups, and hence there is no such thing as a within-groups SSCP matrix **W**. The resolution of this difficulty lies in introducing a more general concept of which the **W** matrix is a special instance applicable to multigroup significance tests and discriminant analysis. The general concept is the *error SSCP matrix*, which we will denote by $\mathbf{S}_e$. Thus a definition of Λ more general than that given in Eq. 4.30 is as follows:

$$\Lambda = \frac{|\mathbf{S}_e|}{|\mathbf{T}|} \tag{7.38}$$

In the applications of Λ encountered up to now, the within-groups SSCP matrix was the appropriate error SSCP matrix. (This was referred to as "the simplest case" when we introduced Eq. 4.30.) In the context of canonical analysis, the error SSCP matrix is the residual SSCP matrix after the effects of the correlations between the canonical-variate pairs have been removed. Of course, this matrix need not actually be computed in order to determine the value of Λ, since Λ may be obtained from Eq. 7.37 once the required eigenvalues are found. It is necessary only to realize that, in computing Λ from Eq. 7.37, we are indirectly calculating the determinantal ratio $|\mathbf{S}_e|/|\mathbf{T}|$ appearing in Eq. 7.38. A more exact proof of Eq. 7.38 is outlined in an exercise at the end of the chapter.

After Λ has been computed from Eq. 7.37, the overall significance test may be carried out by either the chi-square approximation or the F-ratio approximation described in Chapter 4, with K (the number of groups) replaced by $q + 1$. This is consistent with the fact that in using the canonical correlation approach to discriminant analysis, $K - 1$ "dummy criterion variables" were employed; that is, the number of groups is *one more* than the number of variables in the second set. Thus Bartlett's chi-square approximation becomes

$$V = -[N - 3/2 - (p + q)/2] \ln \Lambda \tag{7.39}$$
$$= -[N - 3/2 - (p + q)/2] \sum_{j=1}^{r} \ln (1 - \mu_j^2)$$

with pq degrees of freedom.

Similarly, Rao's F-ratio approximation is now written as

$$R = \frac{1 - \Lambda^{1/s}}{\Lambda^{1/s}} \frac{ms - pq/2 + 1}{pq} \tag{7.40}$$

where

$$m = N - 3/2 - (p + q)/2 \qquad \text{and} \qquad s = \sqrt{\frac{p^2q^2 - 4}{p^2 + q^2 - 5}}$$

and R is to be referred to the F-distribution with pq degrees of freedom in the numerator and $[ms - pq/2 + 1]$ in the denominator. It should be noted that the special cases (shown in Table 4.1) in which the statistic R becomes an exact F-variate hold just as well in the present context. In fact, the conditions characterizing these special cases are now simpler to state than before. Noting that $K = 2$ and $K = 3$ correspond, respectively, to $q = 1$ and $q = 2$, we see that the conditions stated in Table 4.1 now reduce to this simple rule: If either of the two sets of variables contains just *one* or *two* variables, then (no matter how many variables there are in the other set), the R-statistic defined by Eq. 7.40 becomes an exact F-variate. (The case when at least one set consists of a single variable is trivial in the sense that the canonical correlation then reduces to, at most, a multiple correlation.)

Besides the overall significance test described above, there are tests for deciding how many of the canonical correlations can be regarded as significant. The procedure is exactly analogous to that discussed in connection with discriminant analysis on pages 219–220. That is, the V-statistic of Eq. 7.39 is broken down into its additive components as

$$V = \sum_{j=1}^{r} V_j \tag{7.41}$$

where

$$V_j = -[N - 3/2 - (p + q)/2] \ln (1 - \mu_j^2) \tag{7.42}$$

If the test using V itself, which tests

$$H_0^{(0)}: \mu_1^* = \mu_2^* = \cdots = \mu_r^* = 0$$

is rejected (μ_j^* being the population canonical correlation corresponding to μ_j), we set up a new null hypothesis,

$$H_0^{(1)}: \mu_1^* \neq 0 \quad \text{and} \quad \mu_2^* = \cdots = \mu_r^* = 0$$

and test it by means of the first residual V, namely

$$V - V_1 = \sum_{j=2}^{r} V_j$$

which is distributed approximately as a chi-square with $(p - 1)(q - 1)$ degrees of freedom. If this is not significant at the α level being used, we conclude that only $\mu_1^* \neq 0$ while $\mu_2^* = \mu_3^* = \cdots = \mu_r^* = 0$, and conduct no further testing; only the first canonical-variate pair is said to be significantly correlated. If, on the other hand, $V - V_1$ is significant, we conclude that at least μ_2^* is nonzero in addition to μ_1^* and set up another new null hypothesis,

$$H_0^{(2)}: \mu_1^* \mu_2^* \neq 0 \quad \text{and} \quad \mu_3^* = \mu_4^* = \cdots = \mu_r^* = 0$$

This is tested by using the second residual V,

$$V - V_1 - V_2 = \sum_{j=3}^{r} V_j \tag{7.43}$$

as an approximate chi-square with $(p - 2)(q - 2)$ degrees of freedom. We proceed in this manner until the residual after removing the effects of the first s canonical-variate pairs becomes smaller than the prescribed percentile point of the appropriate chi-square distribution. We then conclude that only the first s canonical correlations are significant.

7.11 INTERPRETATION OF THE CANONICAL VARIATES

As was mentioned earlier, canonical correlations analysis may be regarded as a sort of "double-barreled principal components analysis." Therefore, the rules (or perhaps the art) of interpreting the canonical variates are exactly the same as those for interpreting principal components as well as discriminant functions:

One examines the relative magnitudes and the signs of the several combining weights defining each canonical variate, and sees if a meaningful psychological interpretation can be given.

For example, returning to the hypothetical study of relationships between personality attributes and academic achievement alluded to earlier, suppose that the variables with the largest weights for the first two pairs of canonical variates were as shown below. (The scales on the 16 PF Questionnaire are bipolar. A low score on each scale corresponds to the pole characterized by the first adjective; a high score, to the pole described by the second adjective. Hence a large positive weight for a scale implies that a person close to the *second*-adjective pole of that personality dimension will tend to score high on the canonical variate in question; a large negative weight implies that a person close to the *first*-adjective pole will tend to score high on the said canonical variate.)

First Canonical Variate

Personality-Variable Set	**Achievement-Variable Set**
High Positive Weights	*High Positive Weights*
Practical versus Imaginative	Art
Submissive versus Dominant	Literature
Conservative versus Experimenting	
High Negative Weights	*High Negative Weight*
Emotional versus Calm	Algebra
Impulsive versus Controlled	

Second Canonical Variate

Personality-Variable Set	**Achievement-Variable Set**
High Positive Weights	*High Positive Weights*
Conservative versus Experimenting	Physics
Group-dependent versus Self-sufficient	Algebra
High Negative Weights	*High Negative Weight*
Submissive versus Dominant	Literature
Serious versus Gay	

Given these results, the interpretation is self-evident. (Of course, this is a hypothetical example.) It appears that a personality syndrome characterized as "highly imaginative, dominant (self-asserting), experimenting (nonconforming), emotional (easily upset), and impulsive (following own urges)" tends to go with high achievement in the artistic-literary areas, while the "experimenting, self-sufficient (not very sociable), submissive, and serious (or taciturn)" student tends to excel in the physical science and mathematical areas.

With real data, one would seldom expect to find such clear-cut (and stereotype-confirming) results. But this does not detract from the potential value of canonical analysis. It would simply mean that the dimensions of one domain (such a personality) that are strongly associated with those of another domain (such as academic achievement) are not necessarily susceptible to "meaningful" verbal descriptions within the framework of our intuitive, everyday concepts. It may be that subsequent research will show that precisely these "nonintuitive" dimensions represented by the canonical variates are of greater scientific import.

Although we spoke of "large weights" without qualification in the foregoing discussion, the reader should be aware by now that the relative magnitudes of the weights must be compared in the standardized scale. Thus it is not the relative magnitudes of the elements of vectors **u** and **v** as defined by Eqs. 7.36 and 7.35 that are relevant to the interpretation of canonical variates, but the relative magnitudes of the weights rescaled to be applicable to the respective variables in standardized form. These weights are obtained from the raw-score weights **u** and **v** by multiplying each element by the standard deviation of the corresponding variable. That is,

$$u^*_{mi} = s_{x_i} u_{mi} \qquad \text{and} \qquad v^*_{mi} = s_{y_i} v_{mi}$$

are the weights whose relative magnitudes we should compare.

7.12 CANONICAL REDUNDANCY COEFFICIENTS

An alternative measure of association between two sets of variables, called *redundancy coefficients*, was developed by Stewart and Love (1968) and, independently, by Miller (1969). Unlike the canonical correlation coefficient, this measure is asymmetrical with respect to the two sets. That is, there is one coefficient, denoted $R^2_{x.y}$, representing the redundancy of set X given set Y, and another, $R^2_{y.x}$, for the redundancy of set Y given set X, and the two coefficients are, in general, unequal. Each of these is a weighted sum of the squared canonical correlations, the weights being the proportion of the aggregate variance of the variables in the set indicated by the first subscript that is accounted for by the successive canonical variates of that set. Thus the formula for $R^2_{x.y}$ is

$$R^2_{x.y} = \sum_{j=1}^{r} P_{xj} \mu_j^2 \tag{7.44}$$

where

$$P_{xj} = \left(\sum_{i=1}^{p} r^2_{x_i z_j} \right) / p \tag{7.45}$$

in which Z_j is the jth canonical variate of the X set, containing p variables. Similarly, $R^2_{y.x}$ is given by

$$R^2_{y.x} = \sum_{j=1}^{r} p_{yj}\mu_j^2 \qquad (7.44a)$$

with

$$P_{yj} = \left(\sum_{i=1}^{q} r^2_{y_iw_j}\right)/q \qquad (7.45a)$$

The rationale for weighting each squared canonical correlation by the proportion of variance of the variables in the relevant set that is accounted for (or "absorbed" by) the corresponding canonical variate is this: The square of the jth canonical correlation itself represents the proportion of variance common to the jth pair of canonical variates (Z_j and W_j for the X set and Y set, respectively); but we want a measure of the extent of relationship between the two sets of variables as wholes rather than the degree of relationship between the pairs of canonical variates derived from the two sets of variables.

The only explanation that needs to be made about the computation of $R^2_{x.y}$ and $R^2_{y.x}$ is probably how to calculate the correlations $r_{x_iz_j}$ and $r_{y_iw_j}$ in getting the weights P_{xj} and P_{yj}. The reader will probably recognize that these correlations are analogous to the elements of the structure matrix in discriminant analysis, discussed in Section 7.7. Hence the calculation of each set of correlations may be done in one fell swoop by a formula that is a slight modification of Eq. 7.25. That is, the structure matrix whose elements are the correlations of $X_1, X_2, \ldots, X_p$ each with the successive canonical variates Z_j of that set is computed as

$$\mathbf{A}_x = [\mathbf{D}(\mathbf{S}_x)]^{-1/2}(\mathbf{S}_x\mathbf{U})[\mathbf{D}(\mathbf{U}'\mathbf{S}_x\mathbf{U})]^{-1/2} \qquad (7.46)$$

where S_x is the SSCP matrix of the X set of variables and $\mathbf{U}$ is the $p \times r$ matrix whose columns are the successive eigenvectors calculated from Eq. 7.36 and each $\mathbf{D}(\cdot)$ is a diagonal matrix like those described just before Eq. 7.25 and the meaning of the exponent $-\frac{1}{2}$ is as explained just after that equation. Similarly, the correlations whose squares are involved in calculating the weights P_{yj} are the elements of

$$\mathbf{A}_y = [\mathbf{D}(\mathbf{S}_y)]^{-1/2}(\mathbf{S}_y\mathbf{V})[\mathbf{D}(\mathbf{V}'\mathbf{S}_y\mathbf{V})]^{-1/2} \qquad (7.46a)$$

where $\mathbf{V}$ is the $q \times r$ matrix whose columns are the successive $\mathbf{v}_j$'s given by Eq. 7.35.

A word of explanation may be in order about why we have two different redundancy coefficients $R^2_{x.y}$ and $R^2_{y.x}$. Following Cooley and Lohnes (1971, p. 172), suppose that the eight Differential Aptitude Tests constitute set X, while set Y consists of tests that tap only verbal and reasoning abilities. Then the redundancy of set Y given a student's scores on set X would be quite large because we should be able to infer how he/she did on the Y tests given his/her scores on the X tests. On the other hand, we expect the redundancy of set X given scores on set Y to be fairly small, since set X contains information that is not contained in set Y, which covers only a subset of the domains included in the DAT battery. Hence we need two redundancy coefficients to reflect this asymmetry.

7.13 RESEARCH EXAMPLES

Discriminant Examples

In Chapter 4 some work by Lohnes (1966) was discussed as an example of MANOVA significance tests. A 2% random sample of the Project TALENT data files had provided 3100 subjects who were distributed in the four cells of a gender × grade (9th, 12th) design. Each subject was described by a vector of 22 MAP factor scores. The MANOVA tests supported the assumption of equal dispersions for the populations formed by the four cells of the design, and warranted a rejection of the assumption of equal centroids for the four populations. Contrasts tables for the four cells on the 11 ability factors and 11 motives factors were displayed. The dominance of gender differences over grade differences could also be seen in the fact that a single discriminant function absorbed 99% of the trace of the $\mathbf{W}^{-1}\mathbf{A}$ matrix, and as shown in Table 7.3 giving discriminant function cell means, the primary separation on the function was of the gender groups. (The standard deviation for the discriminant function was 1.00.) The correlations between the 22 MAP factors and the discriminant function are shown in Table 7.4 presenting the structure of the discriminant function. This table also establishes the acronyms for the 22 factors, which will be used in tables further on in this section. The factors with strong correlations with the discriminant function were those which had strong point-biserial correlations with gender (negative sign indicates females superior). Curriculum theorists could be quite discouraged by the small separation of the same-gender grade populations on the discriminant function. The canonical correlation between the discriminant function and a best function of dummy variates for cell membership was a remarkably high .94.

TABLE 7.3 Discriminant Function Cell Means

Group	Mean
12th grade males	1.01
9th grade males	.89
9th grade females	−.91
12th grade females	−.99

TABLE 7.4. Structure of Discriminant Function

MAP factor	Structure coefficient
Abilities	
VKN Verbal knowledge	.24
ENG English language	−.75
VIS Visual reasoning	.54
MAT Mathematics	.66
PSA Perceptual speed, accuracy	−.14
SCR Screening	.27
H-F Hunting, fishing	.79
MEM Memory	−.50
COL Color, foods	−.73
ETI Etiquette	−.54
GAM Games	.30
Motives	
CON Conformity needs	−.44
BUS Business interests	−.23
OUT Outdoor, shop interests	.84
SCH Scholasticism	−.08
CUL Cultural interests	−.79
SCI Science interests	.81
ACT Activity level	.28
LEA Leadership	.02
IMP Impulsion	.35
SOC Sociability	.11
INT Introspection	.17

Cooley and Lohnes remark that the best discriminant plane is very often an adequate model for the data (1971, p.244). Perhaps researchers often find it adequate because it is so convenient to map centroids of groups onto a plane. The opportunity that usually is afforded to reduce the rank of the measurement model to one, two, or three is a most attractive aspect of discriminant research strategy, particularly when the number of variates under measurement in the data matrix is large enough to tend to be mind-boggling, and/or when the sample size is not a large multiple of the number of variates. Although the sample sizes for Project TALENT studies were generally comfortably large, the number of independent variables was uncomfortably large, even after 22 MAP factor scores had been substituted for 98 original test and inventory scores. In their research on the predictive validities of MAP factors for career development criteria, Cooley and Lohnes discovered that approximately the same discriminant plane served to separate criterion groups for a series of 14 studies using different samples and different taxonomic criterion variables. The major discriminant function, named *Science-oriented Scholasticism,* was oriented toward mathematics, verbal ability, and interest in the scholastic pursuits. The minor discriminant function, termed *Technical versus Sociocultural,* was bipolar such that men with technical interests and abilities were separated from men with business, social, and cultural interests and abilities. [Unfortunately, in 1967 when they designed this series of studies, the authors chose to concentrate solely on male career development. They and others believed that at that time the experience of career was quite different for American women than it was for men, and that separate studies should be conducted. Fortunately, H. S. Astin and T. Myint (1971) reported an excellent set of parallel studies of career development of the Project TALENT females.] One of their 14 studies can provide an example for this text. The reference for it is Cooley and Lohnes (1968, pp. 4-68 to 4-75).

The career placements of 9,322 young men was collected by questionnaire 5 years after they left high school. These placements were classified into 12 categories of a careers taxonomy. Table 7.5, giving the career placement groups, lists these categories, assigns them acronyms, and reports the sample sizes for them. Six years earlier, when they were high school seniors, these men had taken the tests and inventories on the basis of which they were assigned 22 MAP factor scores. Two discriminant functions in that 22-space were sufficient to separate the 12 taxonomic groups adequately. Table 7.6 showing the structure of two discriminant functions reports the correlations of the MAP factors with the two functions, and the canonical correlations of the functions with a dummy coding of the taxonomy. Table 7.7 shows the centroids of career groups, that is the means of the groups on the two discriminant functions (for each of which the total sample mean is 0.0 and standard deviation is 1.0).

A Project TALENT study in which a single discriminant function in the 23-space of the 22 MAP factors plus a socioeconomic status variable (SEE) sufficed to separate four criterion groups involved: ninth-grade males classified into four high-school curriculum groups (Cooley and Lohnes, 1976, pp. 112–3). Table 7.8 names the curriculum groups and reports the sample sizes. It also reports the mean of the group on the single discriminant function (for which the standard

TABLE 7.5 Career Placement Groups

Acronym	Group name	Sample size
MED	M.D. medicine and Ph.D. biology	279
BIO	Medicine and biology below M.D. & Ph.D.	438
RES	Physical science and mathematics Ph.D.	221
ENG	Physical science, engineering M.S. & B.S.	939
TEC	Technical worker	1297
LBR	Laborer with no post-h.s. training	706
CLK	Office worker with no post-h.s. training	530
ACT	Accountant or other trained nontechnical	1430
BUS	Business B.A. & B.S.	1214
MGT	Management post-baccalaureate training	270
WEL	Sociocultural M.A. & B.A.	1183
PRF	Sociocultural research degree	815

TABLE 7.6 Structure of Two Discriminant Functions

MAP factor	Science-oriented scholasticism	Technical vs. sociocultural
Abilities		
VKN	.62	−.20
PSA	.02	−.10
MAT	.73	.49
H-F	−.10	.26
ENG	.28	−.23
VIS	−.01	.43
COL	.08	−.10
ETI	.05	−.07
MEM	.00	−.01
SCR	−.33	.25
GAM	.10	.05
Motives		
BUS	−.04	−.31
CON	.21	−.12
SCH	.78	.19
OUT	−.41	.42
CUL	.25	−.47
ACT	−.22	.10
IMP	−.01	−.08
SCI	.54	.36
SOC	−.19	−.47
LEA	.28	−.22
INT	−.06	.03
Canonical *R*	.69	.37

TABLE 7.7 Centroids of Career Groups

Career group	DF_1	DF_2
MED	1.4	.2
BIO	.5	.3
RES	1.2	.9
ENG	.4	.6
TEC	−.7	.4
LBR	−1.2	.2
CLK	−.9	−.2
ACT	−.5	−.3
BUS	.2	−.2
MGT	.9	−.1
WEL	.3	−.4
PRF	.9	−.4

TABLE 7.8 Curriculum Groups

Group name	Sample size	*DF* mean
College preparatory	3407	.45
Business	268	−.84
General	1417	−.73
Vocational	225	−1.05

deviation is 1.0). Table 7.9 shows the structure of the discriminant function, that is, the correlations of the MAP factors with the discriminant function. The high coefficient for VKN was expected, since it is a general intellectual development measure, but the high coefficients for the motives of SCH (Scholasticism) and SCI (Science Interest) were interesting findings. Particularly, the

TABLE 7.9 Structure of Discriminant Function

MAP Factor	DF_1
Abilities	
VKN	.66
MAT	.33
ENG	.35
VIS	.10
PSA	.07
MEM	−.03
H-F	−.05
COL	−.06
ETI	.01
SCR	.01
GAM	.10
Motives	
SCH	.60
SCI	.54
BUS	.09
OUT	−.29
CUL	.10
CON	.36
ACT	−.23
LEA	.06
INT	.02
IMP	−.18
SOC	−.05
Status	
SEE	.56
Canonical *R*	.61

coefficient for SCH tended to validate this as a measure of interest in academic pursuits.

Lyson and Falk (1984) used data from the National Longitudinal Study of the High School Class of 1972 to follow up on the actual career placements of students who planned to be teachers, and also to follow back to the high school plans of people who became teachers. The problem addressed was that almost half of the people who were teachers in 1979 had not been planning teaching careers in 1972, and only one-quarter of the students who were planning to be teachers in 1972 were actually employed as teachers in 1979. On the basis of previous research, the authors knew that academic ability, sex, race, social class origin, and residence place were ''important determinants of teacher recruitment'' (p. 183). Thus they had a weak causal theory to build upon. They sorted 4,112 subjects into six groups representing various career patterns, including in-migration to teaching and out-migration from teaching plans. They stated their intention to ''use these variables to probe for similarities and differences among the various . . . groups'' (p. 183). Their use of the term ''probe'' nicely acknowledged the exploratory nature of the research.

Data analysis revealed that two discriminant functions were useful. The first loaded .86 on SAT and .39 on mother's education, with negligible loadings on the other six variates. Clearly this constructed variable is an operationalization of an academic ability construct. The mapping of the group centroids on this constructed variable indicated that teachers who had planned to be such or who had climbed up into teaching from less ambitious plans were low on academic ability, whereas teachers who dropped down into teaching from more ambitious plans were higher on academic ability, although they were also low on it in comparison to people who obtained other professions. The lesson seemed to be that more able teachers were obtainable from a pool of people who aspired to other lofty careers when they were in high school than from the pool of teacher aspirants. The second discriminant function loaded .62 on high school grades, .40 on sex, −.39 on race, and .37 on SAT, with negligible loadings for residence, mother's education, and father's and mother's occupations. Given the mapping of the groups on it, it might be said to represent a construct of a teacher stereotype as a white female who had good grades in high school but not a good SAT.

Canonical Examples

Harold Hotelling invented canonical correlation and revealed it to the world in 1935 in a *Journal of Educational Psychology* paper titled ''The Most Predictable Criterion.'' He used the differential calculus to define a pair of linear functions, one a best prediction function of the predictor measurements, and the other the most-predictable criterion function of the criterion measurements. The product-moment correlation between these *canonical components* is the canonical correlation coefficients, R_c. For his 1936 narration of his invention in *Biometrika,* Hotelling used the interesting title "Relations Between Two Sets of Variates." He seems to have felt that *variables* in statistical analysis usually must be created as *composites* of observational *variates*. This is a usage with much to recommend it.

Hotelling's example of his new method was based on the same four-variate data set he had borrowed from Truman Kelley in 1933 to illustrate his presentation of principal components analysis. This time he rejected the reliability-disattenuated version of the correlations in favor of the actual product-moment correlations, terming the former "unsatisfactory." His correlation matrix and the pair of canonical components are reproduced in Table 7.10. You should compare this analysis with Hotelling's 1933 principal components analysis of these data, as given in Chapter 5. The canonical variables given by Hotelling align with his second principal component as a contrast between the two speed tests and the two power tests. If you compute a second pair of canonical variables, will they align with the general ability factor located by the first principal component? It is worthwhile for you to compute a full canonical correlation analysis of Hotelling's matrix by a modern algorithm. You will get the *structure* coefficients which represent the actual correlations of the composites with the measurement variates, and you will get the *redundancy* measurement for the analysis. Also, you will get a significance test for the canonical relationship. These statistics represent extensions of Hotelling's original vision.

Lohnes and Marshall (1965) analyzed records of 230 students in one junior high school in an attempt to demonstrate two theses: (1) that the standardized test scores of students contain much the same assessment of student achievement as the teachers' course grades; (2) that both assessments are practically unidimensional, providing an excellent (i.e., highly reliable and valid) assessment of general academic development, and practically nothing else. In Chapter 5 you saw the report of the first two principal components of 13 test scores and 8 course grades, which were taken as verification of these theses. A canonical correlation analysis of the same data, reported in Table 7.11 giving the structure and redundancy in school records, accomplishes the same objective in somewhat different fashion. The first pair of canonical components are highly correlated,

TABLE 7.10 Hotelling's Example

Test	Correlations				CC Coefficients	
Reading speed	1.00	.633	.241	.059	−2.777	
Reading power		1.00	−.055	.066	2.266	
Arithmetic speed			1.00	.425	−2.440	
Arithmetic power				1.00	1.000	
			R_c = .395			

TABLE 7.11 Structure and Redundancy in School Records

Test	CC_1	CC_2	Course grade	CC_1	CC_2
PGAT verbal	.79	−.06	7th-grade English	.85	.32
PGAT reasoning	.83	.16	8th-grade English	.80	.45
PGAT number	.71	.46	7th-grade arithmetic	.95	−.14
MAT word knowledge	.80	.03	8th-grade arithmetic	.88	−.24
MAT reading	.82	−.06	7th grade social studies	.90	−.13
MAT spelling	.89	−.19	8th-grade social studies	.74	.00
MAT language	.92	−.12	7th-grade science	.80	−.03
MAT SS language	.84	.07	8th-grade science	.73	.08
MAT arithmetic computation	.90	.21			
MAT arithetic problem	.84	.35			
MAT social studies	.75	−.05			
MAT SS social studies	.80	.36			
MAT science	.73	.19			
Canonical correlation	.90	.66		.90	.66
Factor redundancy	.54	.02		.57	.02
Total redundancy	.59			.61	

suggesting an interchangeability of their assessments. The authors seriously contended that the test scores' component could be reported to pupils and parents as the school's assessment of general academic development *in lieu of teachers' grades*. They thought that this single factor would be a sufficient report on their academic development of pupils. The second canonical components are essentially trivial. They do not capture the delightful principle encapsulated by the second principal component, namely that some students please tests more than they please second principal component, namely that some students please tests more than they please teachers, while other students please teachers more than they please tests. Note that the redundancy coefficients for the two batteries should have played an important role in modifying the authors' enthusiasm for a Procrustean reductionism in school records.

Newman (1972) followed 230 pupils in Cedar Rapids, Iowa from kindergarden through the sixth grade. The subjects had been predicted to be low achievers in first grade reading. In the first of two canonical correlation analyses, Newman modeled the dependency of first grade achievement on kindergarden reading readiness by means of the relationships between two canonical factors of 17 scales in an unusually complete assessment of readiness (including sex, age, two WISC, five Metropolitan, three Murphy-Durrell, two Thurstone, Bender, and Wepman) and two canonical factors of five first grade verbal scales (from the SAT battery). The first causal factor extracted only 16% of the readiness battery variance, and had a path coefficient of .68 for the first criterion factor, which extracted 54% of the achievement battery variance. These first factors were *g*-type variables. The second causal factor extracted only 9% of the readiness battery variance, and had a path coefficient of .44 for the second criterion factor, which extracted 19% of the achievement battery variance. As often happens with canonical factors, these second factors were not readily interpretable. The redundancy of achievement variance to readiness variance for this model was .29.

Newman's second canonical analysis employed two canonical factors each for the first-grade achievement battery and a five-scale sixth-grade achievement battery (ITBS tests). The first causal variable extracted 58% of the first-grade variance, and had a path coefficient of .72 for the first factor of the sixth-grade battery, which extracted 73% of the battery variance. Both these first factors were *g*-type variables. The second causal factor extracted only 14% of the first-grade variance, and had a path coefficient of .31 for the second sixth-grade factor, which extracted only 7% of the battery variance. Again, the second factors were difficult to interpret. The redundancy of sixth-grade achievement variance to first-grade achievement variance for this model was .39. These models had the great advantage of being based on longitudinal data. They demonstrated the long-term importance of a general intelligence construct in studies of school learning.

Canonical models usually have been diagrammed with path coefficients between matching causal and criterion canonical factors. Lohnes (1986) has suggested that in some cases the model might be simplified by dropping the criterion factors from the diagram, and extending the loadings on the causal factors to include the criterion variates. The loadings of criterion variates on the *k*th causal factor can be computed by multiplying their loadings on the *k*th criterion factor by the *k*th canonical correlation coefficient. Since discriminant analysis can be computed as a canonical regression of dummy codes for groups on the measurement variates, as seen in Section 7.8, it would be possible, in the same manner, to extend the loadings of the discriminant factors to include the group membership variates. The advantage of this idea is that it allows canonical correlation and discriminent analyses to be diagrammed in the conventions of path analysis in those cases where the analyst intends to argue a causal interpretation of the regressions involved.

As an example of this modification of canonical correlation analysis, Lohnes (1986, p. 187) presented two causal variables defined on nine independent measurement variates representing earlier school achievement, gender, and certain attitudes and adjustments to school. He termed the two causal components Aptitude and Scholasticism, and computed the loadings of four dependent measurements of achievement on them. These loadings of criterion variates on predictor components are exactly *structure* coefficients, i.e., correlations of the criterion variates with the predictor components. However, under the assumptions of a causal network theory for the data they are also path coefficients. The data for this example belonged to Keeves (1974), who had hypothesized a *performance cycle* in which past achievements modify present attitudes and school adjustments, which in turn modify future achievements, and so on. Keeves (1986) provides a full exposition of this theory, as well as demonstrations of several data-analytic methodologies for testing the theory on the data. Table 7.12, giving the canonical structural model of the performance cycle, reproduces Lohnes' solution. Aptitude emerged as the sole causal influence on 7th-year maths and science achievements, but its contribution to 7th-year liking for maths and liking for science was outstripped by the contribution of Scholasticism. Since 6th-year liking for maths and liking for school were involved in the operational definitions of both Aptitude and Scholasticism, some support for the performance cycle hypothesis could be seen in the model. The redundancies of the four-variates criterion battery on the two causal factors established an approximate 3:1 ratio of usefulness as explanatory principles between Aptitude and Scholasticism. The h^2 numbers in the table are *communalities* for the variates, i.e., proportions of observed variance on the variates explained by their loadings on the two factors. These are computed as the sums of squares of the loadings for the variates. For the four criterion variates, these are squared multiple correlation coefficients for their regressions on the two orthogonal predictor canonical components.

Many educational research studies are performed with intact classroom groups, rather than individual students, as the sampling unit. Usually the class mean (m) is employed as the observational unit in such studies. Lohnes (1972) undertook to demonstrate that valuable information characterizing each class may be recovered if measures of scatter (s), skew (g) and kurtosis (k) are included as observational units. Adopting R. B. Cattell's (1948) usage of the term *syntality* to denote the behavioral characteristics of a group of people which

TABLE 7.12 Canonical Structural Model of the Performance Cycle

Variate	Aptitude	Scholasticism	h^2
1. Gender (1 = male, 2 = female)	−.15	−.28	.10
2. Year 6 maths achievement	.94	−.12	.89
3. Year 6 science achievement	.82	−.06	.68
4. Year 6 liking for maths	.26	.54	.36
5. Year 6 liking for school	.27	.28	.15
6. Year 6 self-regard	.27	.05	.07
7. Academic motivation	.17	.88	.80
8. Attentiveness in maths	.43	−.02	.18
9. Attentiveness in science	.29	.08	.09
1. Year 7 maths achievement	.85	−.05	.72
2. Year 7 science achievement	.82	−.03	.67
3. Year 7 science attitude	.23	.56	.37
4. Year 7 liking for science	.33	.45	.31
Redundancy of criterion to factors	.387	.131	
Total criterion battery redundancy	.518		

differentiate it from other groups, as a counterpart term to *personality* as applied to individuals, Lohnes argued that the concept of syntality could be enriched if the research took into account the shape of a group's distribution on each measurement, as well as its location. He borrowed data for 219 second-grade classroom groups from the U.S. Office of Education Cooperative Reading Studies (Dykstra, 1968). These data were longitudinal and classroom was the unit of sampling, but fortunately test scores were available for the students in each classroom, making it possible to compute the four curve characteristics for each test in each class. Lohnes selected two readiness tests administered the first week of first grade [Pintner-Cunningham Primary General Abilities (P-C), and Durrell-Murphy Total Letters (D-ML)] and two achievement tests, the first given at the end of first grade [Stanford Paragraph Meaning (SPM1)] and the second given at the end of second grade [Stanford Paragraph Meaning (SPM2)]. In a canonical correlation design, the four descriptors of class syntality provided by each of the first three tests were grouped as the predictor battery, providing 12 predictor scores. The four syntality descriptors from the fourth test were taken as the criterion battery. Table 7.13 reports the results of the canonical correlation analysis for the canonical syntality study of 219 classes. Two useful canonical relationships were found. The most predictable criterion function correlated almost perfectly (.99) with the criterion *m*, but it also related significantly (−.61) to the criterion *g* measure. The first prediction function related strongly to the predictor *m* measures, but also related significantly to two of the three predictor *g* measures. The second criterion function correlated almost perfectly (.96) with the criterion *s*, but it also had some involvement (−.31) with the criterion *h* measure. The second prediction function related moderately to all three predictor *s* measures, one *g* measure and one *h* measure. Overall, it did seem that there

TABLE 7.13 Canonical Syntality Study of 219 Classes

Predictor	CF_1	CF_2	Criterion	CF_1	CF_2
P-C *m*	.79	.26	SPM2 *m*	.99	−.07
P-C *s*	−.26	.54	SPM2 *s*	−.10	.96
P-C *g*	−.21	.01	SPM2 *g*	−.61	.17
P-C *h*	.01	.01	SPM2 *h*	.08	−.31
D-ML *m*	.74	.10	Variance extracted	.34	.26
D-ML *s*	−.14	.62	Redundancy	.27	.12
D-ML *g*	−.67	.05			
D-ML *h*	.14	−.41			
SPM1 *m*	.95	−.19			
SPM1 *s*	.30	.64			
SPM1 *g*	−.56	.56			
SPM1 *h*	−.10	−.08			
First canonical $R = .89$			$R^2 = .79$		$\chi^2_{48} = 487$
2nd canonical $R = .67$			$R^2 = .45$		$\chi^2_{33} = 165$

was substantial information about the groups contained in the s measures beyond that supplied by the m measures, and at least a modest amount of additional information contained in the g and h measures.

EXERCISES

1. In a study attempting to determine the dimensions along which three curricular groups at the Blank Institute of Technology differ among one another, scores on three tests administered at time of admission were obtained for a random sample of 100 seniors in each of three groups: A. General Engineering; B. Physical Science; and C. Industrial Management.

The three tests used were: (1) SAT Mathematics Test, (2) SAT Verbal Test, and (3) Persuasiveness Scale on a personality inventory. The mean scores (suitably scaled down to reduce the arithmetic) on the three tests for each group were as follows:

	Test			
Group	**1**	**2**	**3**	n_g
A. General Engineering	6.14	5.30	1.62	100
B. Physical Science	6.69	5.84	1.21	100
C. Industrial Management	6.05	5.61	1.70	100

The within-groups and between-groups SSCP matrices were as follows:

$$\mathbf{W} = \begin{bmatrix} 167.69 & 65.24 & -12.18 \\ 65.24 & 269.64 & -61.39 \\ -12.18 & -61.39 & 141.62 \end{bmatrix}$$

and

$$\mathbf{B} = \begin{bmatrix} 23.96 & 13.95 & -18.08 \\ 13.95 & 14.71 & -10.36 \\ -18.08 & -10.36 & 13.64 \end{bmatrix}$$

Carry out a discriminant analysis for the data above, including significance tests to determine whether one or both of the discriminant functions are significant at the .05 level. Also plot the discriminant function means of the three groups (on a line or a plane, as the case may be), and try to interpret the dimension(s) of group differentiation.

(COMPUTATIONAL HINT: Carry five significant digits throughout your calculations. To facilitate this, it will be convenient to work with $\mathbf{W} \times 10^{-1}$ instead of $\mathbf{W}$ itself. Note the effect that this has on the eigenvalues.)

2. The means on three scales of a personality test for samples from three clinical groups were as follows:

Group	Scale 1	Scale 2	Scale 3	Sample Size
Anxiety neurosis	2.92	1.16	1.72	50
Psychopathic reaction	3.80	1.90	1.80	30
Obsessive-compulsive	4.70	1.60	2.10	20

The within-groups and between-groups SSCP matrices were as follows:

$$\mathbf{W} = \begin{bmatrix} 223.18 & 24.40 & 45.99 \\ 24.40 & 58.92 & 3.47 \\ 45.99 & 3.47 & 57.72 \end{bmatrix}$$

and

$$\mathbf{B} = \begin{bmatrix} 48.16 & 15.98 & 9.44 \\ 15.98 & 10.69 & 2.02 \\ 9.44 & 2.02 & 2.08 \end{bmatrix}$$

Carry out a discriminant analysis for the above data, including all relevant significance tests. (See the end of Exercise 1 for a computational hint.)

3. Do Exercise 2 by the canonical correlation approach. Note that you will first have to add $\mathbf{W}$ and $\mathbf{B}$ to get the total SSCP matrix $\mathbf{T}$, which becomes $\mathbf{S}_{pp}$ in this context. (See the numerical example in Section 7.8 for computation of $\mathbf{S}_{pc}$ and $\mathbf{S}_{cc}$.)

4. Prove that the second discriminant function (whose weights are the elements of the eigenvector $\mathbf{v}_2$ associated with the second largest eigenvalue λ_2 of $\mathbf{W}^{-1}\mathbf{B}$) has the largest discriminant-criterion value among those linear combinations of the X_i that are uncorrelated with the first discriminant function in the total sample.
[**Hint:** From Section 7.2 we know that each discriminant function has a relative (or conditional) maximum value for its discriminant criterion. We therefore need only show that Y_2 is uncorrelated with Y_1. Noting that this correlation is proportional to $\mathbf{v}_1'\mathbf{T}\mathbf{v}_2$ (where $\mathbf{T} = \mathbf{W} + \mathbf{B}$), we have to prove that $\mathbf{v}_1'\mathbf{T}\mathbf{v}_2 = 0$. This may be done by a slight modification of the proof of Theorem 3 in Section 5.5, starting as follows: From Eq. 7.9 we have

$$(\mathbf{B} - \lambda_i\mathbf{W})\mathbf{v}_i = \mathbf{0} \qquad \text{for each } i$$

Hence

$$\mathbf{B}\mathbf{v}_1 = \lambda_1\mathbf{W}\mathbf{v}_1 \qquad \text{and} \qquad \mathbf{B}\mathbf{v}_2 = \lambda_2\mathbf{W}\mathbf{v}_2$$

Premultiplying these equations by $\mathbf{v}_2'$ and $\mathbf{v}_1'$, respectively, taking the transposes of both sides of the first resulting equation (and remembering that both $\mathbf{B}$ and $\mathbf{W}$ are symmetric matrices), we arrive at a pair of equations whose left-hand sides are identical. The right-hand sides of these equations must, therefore, also be equal.]

5. A more nearly exact derivation of Eq. 7.37 for Wilks' Λ-criterion for testing the significance of canonical correlation coefficients begins as outlined below. Complete the derivation. If we regard the set of variables $Y_1, Y_2, \ldots, Y_q$ as the dependent variables, the expression for Λ given in Eq. 7.38 may equivalently be written as

$$\Lambda = |\mathbf{S}_{ey}|/|\mathbf{S}_{yy}|$$

where $\mathbf{S}_{ey}$ is the SSCP matrix of residual errors and $\mathbf{S}_{yy}$ is the usual SSCP matrix for the Y's. Since $\mathbf{S}_{ey}$ is analogous to the residual SS in the simple regression of Y on X, which is

$$\Sigma\,(y - \hat{y})^2 = \Sigma\, y^2 - (\Sigma\, xy)^2/\Sigma\, x^2$$

we may write

$$\mathbf{S}_{ey} = \mathbf{S}_{yy} - \mathbf{S}_{yx}\mathbf{S}_{xx}^{-1}\mathbf{S}_{xy}$$

Equation 7.38 then becomes

$$\begin{aligned}\Lambda &= |\mathbf{S}_{yy} - \mathbf{S}_{yx}\mathbf{S}_{xx}^{-1}\mathbf{S}_{xy}|/|\mathbf{S}_{yy}| \\ &= |\mathbf{S}_{yy}^{-1}||\mathbf{S}_{yy} - \mathbf{S}_{yx}\mathbf{S}_{xx}^{-1}\mathbf{S}_{xy}| \\ &= |\mathbf{I} - \mathbf{S}_{yy}^{-1}\mathbf{S}_{yx}\mathbf{S}_{xx}^{-1}\mathbf{S}_{xy}|\end{aligned}$$

Continue the derivation by utilizing the relation between the determinant of a square matrix and the eigenvalues of the latter.

6. Using the five cognitive variables of the HSB data set, do a discriminant analysis to determine the dimension(s) along which the centroids of the three HSP groups are best differentiated. Use the 5% significance level to test how many significant discriminant functions there are.

7. Using the three affective variables as well as the five cognitive variables, do a discriminant analysis between the boys and the girls. Which variables are most effective in differentiating the two sexes, and in which directions?

8. Do a canonical correlation analysis with the three affective variables as the first set, and the five cognitive variables as the second. Describe the s affective profiles that "go along" with the s cognitive proviles, respectively, where s is the number of significant canonical-variate pairs. (Use $\alpha = .05$.)

9. In the HSB data set, define four composite career-choice groups as follows:

Group I:	clerical (1), operative (8), sales (13)
Group II:	craftsman (2), farmer (3), technical (16)
Group III:	professional 1 (9)
Group IV:	professional 2 (10)

The numbers in parentheses indicate the code for each career as defined in the HSB data. The numbers of students in the four career groups should be $n_1 = 86$, $n_2 = 86$, $n_3 = 161$, and $n_4 = 94$, totaling $N = 427$. Do a discriminant analysis for the four groups defined above, using the three affective and five cognitive variables as predictors, and omitting the 173 students that do not belong to any of the four groups from your analysis. Use $\alpha = .05$ to determine the number of significant discriminant functions.

8

MULTIVARIATE ANALYSIS OF VARIANCE

The multivariate significance tests discussed up to this point, including discriminant analysis, may all be regarded as multivariate extensions of analysis of variance (ANOVA) as applied to one-factor experiments or one-way classification designs. Extensions to multivariate experiments involving more than one factor, that is, factorial experiments, may be made in a closely parallel manner. We describe in detail the multivariate analysis of variance (MANOVA) procedures for two-factor designs with equal cell frequencies. Extensions to other designs can be made in exactly the same way by anyone who is sufficiently well versed in the corresponding univariate ANOVA. Guidelines to effecting such extensions are given in Section 8.5.

In the discussions to follow, it is assumed that the reader has had some exposure to the basic concepts and terminology used in ANOVA for factorial designs in the univariate (one dependent variable) case, including the choice of correct error terms depending on whether a fixed-effects, random-effects, or mixed model is involved. Those who need to review these matters may refer to Glass and Hopkins (1984, chap. 18) or Hays (1981, chaps. 10–12).

8.1 TWO-FACTOR DESIGNS WITH MULTIPLE DEPENDENT VARIABLES: ADDITIVE COMPONENTS OF THE TOTAL SSCP MATRIX

Just as in the two-factor design with one dependent variable, we are here concerned with the analysis of data from experiments involving two treatment or classification variables whose levels are represented by the rows and columns of a two-way layout. For example, in a classroom learning experiment, each row may represent a different method for teaching shorthand, and each column, a different condition of massed or distributed practice. But instead of having one dependent or criterion variable by which to measure the outcome of the experiment, we now have two or more criterion measures, such as a speed score and an accuracy score in the shorthand-teaching experiment. Thus each experimental unit (or subject) will be associated with a set of p observations $X_1, X_2, \ldots, X_p$. Hence, for complete identification of a particular score, we need four subscripts, specifying, respectively, the variable number (α or $\beta = 1, 2, \ldots, p$), row number ($r = 1, 2, \ldots, R$), column number ($c = 1, 2, \ldots, C$), and subject number within each cell ($i = 1, 2, \ldots, n$). That is, $X_{\alpha rci}$ will denote the score on the αth dependent variable earned by the ith individual in the rth row, cth column cell [hereafter called the (r, c)-cell for short].

Preparatory to computing the several SSCP matrices that are the multivariate counterparts of the various SS's in univariate ANOVA, we find the several classes of total for each variable and assemble the members of each class in a vector. Thus

$$T_{\alpha rc.} = \sum_{i=1}^{n} X_{\alpha rci} \qquad \text{is the total in the } (r, c) \text{ cell} \quad \text{and}$$

$$\mathbf{t}'_{rc} = [T_{1rc}, T_{2rc}, \ldots, T_{prc}]$$

$$T_{\alpha r..} = \sum_{c=1}^{C} T_{\alpha rc.} \qquad \text{is the } r\text{th row total} \quad \text{and}$$

$$\mathbf{t}'_{r.} = [T_{1r.}, T_{2r.}, \ldots, T_{pr.}]$$

$$T_{\alpha .c.} = \sum_{r=1}^{R} T_{\alpha rc.} \qquad \text{is the } c\text{th column total} \quad \text{and}$$

$$\mathbf{t}'_{.c} = [T_{1.c}, T_{2.c}, \ldots, T_{p.c}]$$

$$T_{\alpha ...} = \sum_{r=1}^{R} T_{\alpha r..} = \sum_{c=1}^{C} T_{\alpha .c.} \qquad \text{is the grand total} \quad \text{and}$$

$$\mathbf{t}'_{..} = [T_{1..}, T_{2..}, \ldots, T_{p..}]$$

Corresponding to the total sum of squares in univariate ANOVA, we compute the total SSCP matrix $\mathbf{S}_t$, in accordance with Eq. 2.8:

$$\mathbf{S}_t = \mathbf{X}'\mathbf{X} - \mathbf{t}_{..}\mathbf{t}'_{..}/N \tag{8.1}$$

where $\mathbf{X}$ is the $N \times p$ data matrix and N is the total sample size, RCn, assuming that we have n subjects in each of the RC cells.

The within-cells SSCP matrix, $\mathbf{S}_w$, is computed by applying the same formula to each cell and then adding the RC individual-cell SSCP matrices. The row-, column-, and interaction-effect SSCP matrices are computed by variants of Eq. 2.8, and just as the total sum of squares in two-way ANOVA is the sum of the four additive components,

$$SS_t = SS_r + SS_c + SS_{rc} + SS_w$$

the total SSCP matrix in two-way MANOVA may be expressed as

$$\mathbf{S}_t = \mathbf{S}_r + \mathbf{S}_c + \mathbf{S}_{rc} + \mathbf{S}_w \tag{8.2}$$

The within-cells SSCP matrix, $\mathbf{S}_w$, as mentioned above, is computed by using Eq. 2.8 for each cell separately and then adding the resulting RC cell-by-cell SSCP matrices:

$$\mathbf{S}_w = \sum_{r=1}^{R} \sum_{c=1}^{C} (\mathbf{X}'_{rc}\mathbf{X}_{rc} - \mathbf{t}_{rc}\mathbf{t}'_{rc}/n) \tag{8.3}$$

The remaining three SSCP matrices (i.e., the effect SSCP matrices) on the right-hand side of Eq. 8.2 are computed by using variants of Eq. 2.8, in which the appropriate total vectors replace the data matrix $\mathbf{X}$. Thus

$$\mathbf{S}_r = \sum_{r=1}^{R} \mathbf{t}_{r.}\mathbf{t}'_{r.}/Cn - \mathbf{t}_{..}\mathbf{t}'_{..}/N \tag{8.4}$$

$$\mathbf{S}_c = \sum_{c=1}^{C} \mathbf{t}_{.c}\mathbf{t}'_{.c}/Rn - \mathbf{t}_{..}\mathbf{t}'_{..}/N \tag{8.5}$$

and

$$\mathbf{S}_{rc} = \sum_{r=1}^{R} \sum_{c=1}^{C} \mathbf{t}_{rc}\mathbf{t}'_{rc}/n - \sum_{r=1}^{R} \mathbf{t}_{r.}\mathbf{t}'_{r.}/Cn - \sum_{c=1}^{C} \mathbf{t}_{.c}\mathbf{t}'_{.c}/Rn + \mathbf{t}_{..}\mathbf{t}'_{..}/N \tag{8.6}$$

The interaction SSCP matrix, $\mathbf{S}_{rc}$, may alternatively be computed by first computing the among-cells SSCP matrix $\mathbf{S}_{\text{cells}}$ as

$$\mathbf{S}_{\text{cells}} = \sum_{r=1}^{R} \sum_{c=1}^{C} \mathbf{t}_{rc}\mathbf{t}'_{rc}/n - \mathbf{t}_{..}\mathbf{t}'_{..}/N$$

and subtracting both $\mathbf{S}_r$ and $\mathbf{S}_c$ from it; that is,

$$\mathbf{S}_{rc} = \mathbf{S}_{\text{cells}} - \mathbf{S}_r - \mathbf{S}_c \tag{8.6a}$$

Although Eq. 8.6a does not offer any computational saving over Eq. 8.6, it helps to remind us about the meaning of the interaction as the variation among cells over and above the variations attributable to the row and column effects. The reader should verify, using Eqs. 8.3 through 8.6, that the sum of the four component SSCP matrices is equal to the total SSCP matrix given in Eq. 8.1.

These SSCP matrices may be collected into a MANOVA summary table, together with their respective degrees of freedom, which are the same as those for the corresponding sums of squares in univariate ANOVA. Table 8.1 represents such a summary table.

TABLE 8.1 MANOVA Summary Table

Source	SSCP Matrix	Computed from Eq.	n.d.f.
Row effect	$\mathbf{S}_r$	8.4	$\nu_r = R - 1$
Column effect	$\mathbf{S}_c$	8.5	$\nu_c = C - 1$
Interaction	$\mathbf{S}_{rc}$	8.6	$\nu_{rc} = (R - 1)(C - 1)$
Within-cells	$\mathbf{S}_w$	8.3	$\nu_w = RC(n - 1)$
Total	$\mathbf{S}_t$	8.1	$\nu_t = nRC - 1$

8.2 SIGNIFICANCE TESTS IN MANOVA

The Λ criterion may again be used in carrying out the significance tests for the main effects and interaction effect in MANOVA, but a further generalization of the definition of this statistic must first be made. Recall that Λ was introduced in Chapter 4 as the ratio $|\mathbf{S}_w|/|\mathbf{S}_t|$, and was subsequently generalized to $|\mathbf{S}_e|/|\mathbf{S}_t|$ for use in connection with canonical correlation analysis in Chapter 7; that is, the earlier matrix $\mathbf{S}_w$ was there replaced by a more general matrix $\mathbf{S}_e$, of which the former was a special case. This time, it is the matrix whose determinant stands in the denominator of the Λ ratio that has to be generalized. That is, the total SSCP matrix $\mathbf{S}_t$ must now be replaced by a more general matrix, which, in effect, reduces to $\mathbf{S}_t$ in the situations previously considered. Let us therefore examine what $\mathbf{S}_t$ stood for in the problems treated so far.

In *K*-sample significance tests for group centroids (in which context Wilks' Λ was first introduced) as well as in discriminant analysis, we had the relation

$$\mathbf{S}_t = \mathbf{S}_w + \mathbf{S}_b \qquad \text{(or } \mathbf{T} = \mathbf{W} + \mathbf{B} \text{ in our earlier notation)}$$

In canonical analysis, we could have written

$$\mathbf{S}_t = \mathbf{S}_e + \mathbf{S}_{\text{can}}$$

where $\mathbf{S}_{\text{can}}$ stands for the SSCP due to canonical correlation. We did not do so, however, because our main concern at that point was to argue that the earlier within-groups SSCP matrix $\mathbf{S}_w$ could, conceptually at least, be replaced by the more general error SSCP matrix, $\mathbf{S}_e$. In each of these situations, therefore, the total SSCP matrix stood for the sum of the error SSCP matrix and the SSCP matrix attributable to the effect posited as an alternative to the *single* null hypothesis being tested—equality of the *K* population centroids, or zero canonical correlation in the population, as the case may be.

In higher-order designs, however, there are several null hypotheses to be tested. In the two-factor case, with which we are here concerned, the null hypotheses relate to the row effect, column effect, and interaction effect, represented by $\mathbf{S}_r$, $\mathbf{S}_c$, and $\mathbf{S}_{rc}$, respectively. It is clear from Eq. 8.2 that, in this case, the SSCP matrix for the effect being tested by any one of the null hypotheses plus the error SSCP matrix does not equal the total SSCP matrix, for $\mathbf{S}_t$ now has four additive components. Moreover, the error SSCP matrix itself will now depend on whether model I (fixed effects), model II (random effects), or model III (mixed model) applies to the particular experiment; either $\mathbf{S}_w$ or $\mathbf{S}_{rc}$ will be the appropriate error SSCP matrix. (In three-factor designs or more complicated two-factor designs such as when one factor is nested under the other, there will be a larger number of matrices from among which the appropriate error SSCP matrix must be chosen.)

The foregoing considerations suggest that $\mathbf{S}_t$, whose determinant formed the denominator of the Λ-ratio in its previous applications, must now be replaced by a matrix having the general form $\mathbf{S}_h + \mathbf{S}_e$, where $\mathbf{S}_h$ is the SSCP matrix for the effect being tested by a particular null hypothesis, and $\mathbf{S}_e$ is the error SSCP matrix appropriate for the particular effect under the applicable model. This is, in fact, the conclusion that emerges from the mathematical theory of likelihood-ratio criteria, which (as mentioned in Chapter 4) forms the basis for Wilks' Λ-ratio. We thus arrive at the completely general definition of the Λ-criterion, as follows:

$$\Lambda_h = \frac{|\mathbf{S}_e|}{|\mathbf{S}_h + \mathbf{S}_e|} \tag{8.7}$$

where the subscript h designates the particular null hypothesis being tested. Note that on substituting $\mathbf{S}_b$ for $\mathbf{S}_h$ and $\mathbf{S}_w$ for $\mathbf{S}_e$ in this expression, we get the Λ-ratio appropriate for K-sample (one-way MANOVA) significance tests, first introduced in Eq. 4.30. The appropriate substitutions for $\mathbf{S}_h$ and $\mathbf{S}_e$ in the case of two-factor designs with equal cell frequencies (or, more generally, cell frequencies proportional to the marginals) are summarized in Table 8.2. The substitution for $\mathbf{S}_h$ of course depends only on the hypothesis being tested, but that

TABLE 8.2 SSCP Matrices to be Substituted for $\mathbf{S}_h$ and $\mathbf{S}_e$ in Eq. 8.7 in Testing the Various Effects in Two-Way MANOVA with Equal (or Proportional) Cell Frequencies

		$\mathbf{S}_e$		
Effect	$\mathbf{S}_h$	I	II	III
Rows	$\mathbf{S}_r$	$\mathbf{S}_w$	$\mathbf{S}_{rc}$	$\mathbf{S}_{rc}$
Columns	$\mathbf{S}_c$	$\mathbf{S}_w$	$\mathbf{S}_{rc}$	$\mathbf{S}_w$
Interaction	$\mathbf{S}_{rc}$	$\mathbf{S}_w$	$\mathbf{S}_w$	$\mathbf{S}_w$

for $\mathbf{S}_e$ depends also on which model is applicable. In model III (mixed) it is assumed that the rows represent a fixed effect and the columns, a random effect.

As before, the actual computation of Λ can be made routine by utilizing the relationship between the determinant and eigenvalues of a matrix, provided that a computer program for solving eigenvalue problems is available. From Eq. 8.7 we get

$$1/\Lambda_h = |\mathbf{S}_e + \mathbf{S}_h|/|\mathbf{S}_e| = |\mathbf{I} + \mathbf{S}_e^{-1}\mathbf{S}_h|$$

Hence if the nonzero eigenvalues of the matrix $\mathbf{S}_e^{-1}\mathbf{S}_h$ are denoted by λ_1, λ_2, . . . , λ_t, we have

$$1/\Lambda_h = \prod_{i=1}^{t} (1 + \lambda_i)$$

or

$$\Lambda_h = 1/\prod_{i=1}^{t} (1 + \lambda_i) \tag{8.8}$$

which is formally identical with Eq. 7.12. The number t of nonzero eigenvalues is equal to ν_h [i.e., $\nu_r = R - 1$, $\nu_c = C - 1$ or $\nu_{rc} = (R - 1)(C - 1)$, as the case may be] or to p (the number of dependent variables), whichever is smaller. Using the eigenvalue approach to compute Λ is especially preferred if one wishes to follow up the significance tests with a discriminant analysis as described in Section 8.4.

After each Λ_h has been computed from Eq. 8.7 or 8.8, the significance test proceeds as before, using either Bartlett's V or Rao's R as the approximate test statistic. Minor details of the formulas for these statistics have to be modified as follows: The K (number of groups), which appeared in Eq. 4.32 for V and 4.33 for R, is now replaced by $\nu_h + 1$. Similarly, N is replaced by $\nu_e + \nu_h + 1$, where ν_e is equal to $\nu_w = RC(n - 1)$ or to $\nu_{rc} = (R - 1)(C - 1)$, as the case may be, according to the appropriate error SSCP matrix indicated in Table 8.2. (The reader should verify that these replacements are consistent with the replacements of $\mathbf{S}_b$ and $\mathbf{S}_t$ by $\mathbf{S}_h$ and $\mathbf{S}_e + \mathbf{S}_h$, respectively.) Thus the formulas for V and R applicable to two-factor designs assume the forms shown in Eqs. 8.9 and 8.10. (In fact, with the appropriate definitions of ν_e and ν_h, these formulas may be used in connection with a MANOVA for any type of design.)

For Bartlett's chi-square approximation, the test statistic is

$$\begin{aligned} V &= -(\nu_e + \nu_h - (p + \nu_h + 1)/2) \ln \Lambda_h \\ &= (\nu_e + \nu_h - (p + \nu_h + 1)/2) \ln \prod_{i=1}^{t} (1 + \lambda_i) \end{aligned} \tag{8.9}$$

which is distributed approximately as a chi-square with $p\nu_h$ degrees of freedom. If necessary, Schatzoff's correction factor may be applied to the selected percentile point of the relevant chi-square distribution to obtain the exact critical value for V, as described in Chapter 4.

The formula for Rao's R statistic now becomes as follows:

$$R = \frac{1 - \Lambda^{1/s}}{\Lambda^{1/s}} \frac{ms - p\nu_h/2 + 1}{p\nu_h} \tag{8.10}$$

with

$$m = \nu_e + \nu_h - (p + \nu_h + 1)/2$$

and

$$s = \sqrt{\frac{(p\nu_h)^2 - 4}{p^2 + \nu_h^2 - 5}}$$

R having an approximate F-distribution with $p\nu_h$ and $ms - p\nu_h/2 + 1$ degrees of freedom. As usual, when $p = 1$ (univariate ANOVA) or $p = 2$, or when $\nu_h = 1$ or 2 (corresponding to $K = 2$ or 3 in Table 4.2), R reduces to an exact F-variate with degrees of freedom as shown in that table, with the substitutions $K = \nu_h + 1$ and $N = \nu_e + \nu_h + 1$.

8.3 NUMERICAL EXAMPLE

An experiment was conducted for comparing two methods of teaching shorthand to female seniors in a vocational high school. Also of interest were the effects of distributed *versus* massed practice, represented by the following conditions:

C_1: 2 hours of instruction per day for 6 weeks

C_2: 3 hours of instruction per day for 4 weeks

C_3: 4 hours of instruction per day for 3 weeks

Ten subjects were assigned at random to each of the $2 \times 3 = 6$ treatment groups.

On completion of instruction, all 60 S's were given a standardized shorthand test with two subscores, X_1 = speed and X_2 = accuracy. Results are shown in Table 8.3.

Let us first compute the within-cells SSCP matrix, $\mathbf{S}_w$, because this is related most directly to the summary quantities listed at the foot of the columns of

TABLE 8.3 Speed (X_1) and Accuracy (X_2) Scores for 60 S's Taught Shorthand by Two Methods Under Three Conditions of Distributed Versus Massed Practice

	C_1		C_2		C_3	
	X_1	X_2	X_1	X_2	X_1	X_2
Method *A*	36	26	46	17	26	14
	34	22	34	21	31	14
	28	21	31	17	30	16
	34	23	31	18	34	16
	34	21	36	23	30	13
	29	19	26	19	27	13
	48	25	35	16	21	12
	28	20	33	19	31	15
	34	21	23	15	37	14
	38	20	30	14	29	14
ΣX_α	343	218	325	179	296	141
ΣX_α^2	12,077	4798	10,909	3271	8934	2.003
ΣX_1X_2	7553		5855		4204	
Method *B*	42	25	32	18	28	11
	47	24	39	19	28	10
	51	29	37	17	25	10
	35	25	31	17	22	12
	37	26	36	19	27	11
	44	28	32	19	25	12
	44	25	31	17	33	14
	49	24	41	21	31	13
	43	24	36	18	28	12
	36	26	40	20	23	11
ΣX_α	428	256	355	185	270	116
ΣX_α^2	18,586	6580	12,733	3439	7394	1360
ΣX_1X_2	10,970		6601		3153	

scores in each cell. Note that the expression in parentheses in Eq. 8.3 is the SSCP matrix for each cell treated as a single group. We may denote these cell-by-cell SSCP matrices by $\mathbf{S}_{11}$, $\mathbf{S}_{12}$, . . . , $\mathbf{S}_{23}$, and compute them separately, and then add them to obtain $\mathbf{S}_w$. The details for computing $\mathbf{S}_{11}$ in accordance with the expression in parentheses, with r and c both set equal to 1, are as follows:

$$\mathbf{S}_{11} = \mathbf{X}_{11}'\mathbf{X}_{11} - \mathbf{t}_{11}\mathbf{t}_{11}'/n$$

$$= \begin{bmatrix} 36 & 34 & \cdots & 38 \\ 26 & 22 & \cdots & 20 \end{bmatrix} \begin{bmatrix} 36 & 26 \\ 34 & 22 \\ \vdots & \vdots \\ 38 & 20 \end{bmatrix} - \begin{bmatrix} 343 \\ 218 \end{bmatrix} [343 \quad 218]/10$$

$$= \begin{bmatrix} 12{,}077 & 7{,}553 \\ 7{,}553 & 4{,}798 \end{bmatrix} - \begin{bmatrix} (343)^2 & (343)(218) \\ (218)(343) & (218)^2 \end{bmatrix} \div 10$$

$$= \begin{bmatrix} 312.1 & 75.6 \\ 75.6 & 45.6 \end{bmatrix}$$

Carrying out parallel computations for all six cells, we find the cell-by-cell SSCP matrices to be as follows:

$$\mathbf{S}_{11} = \begin{bmatrix} 312.1 & 75.6 \\ 75.6 & 45.6 \end{bmatrix} \quad \mathbf{S}_{12} = \begin{bmatrix} 346.5 & 37.5 \\ 37.5 & 66.9 \end{bmatrix} \quad \mathbf{S}_{13} = \begin{bmatrix} 172.4 & 30.4 \\ 30.4 & 14.9 \end{bmatrix}$$

$$\mathbf{S}_{21} = \begin{bmatrix} 267.6 & 13.2 \\ 13.2 & 26.4 \end{bmatrix} \quad \mathbf{S}_{22} = \begin{bmatrix} 130.5 & 33.5 \\ 33.5 & 16.5 \end{bmatrix} \quad \mathbf{S}_{23} = \begin{bmatrix} 104.0 & 21.0 \\ 21.0 & 14.4 \end{bmatrix}$$

Adding these six matrices, we get

$$\mathbf{S}_w = \sum_r \sum_c \mathbf{S}_{rc} = \begin{bmatrix} 1333.10 & 211.20 \\ 211.20 & 184.70 \end{bmatrix}$$

To compute the other SSCP matrices in accordance with Eqs. 8.4 through 8.6, it is convenient first to arrange the vectors of cell totals, marginal totals, and grand totals in the following pattern:

[343 218]	[325 179]	[296 141]	[964 538]
[428 256]	[355 185]	[270 116]	[1053 557]
[771 474]	[680 364]	[566 257]	[2017 1095]

For computing $\mathbf{S}_r$ and $\mathbf{S}_c$, we need only the marginal and grand-total vectors in the display above. Thus $\mathbf{S}_r$ is computed from Eq. 8.4 as

$$\mathbf{S}_r = \left\{ \begin{bmatrix} 964 \\ 538 \end{bmatrix} [964 \quad 538] + \begin{bmatrix} 1053 \\ 557 \end{bmatrix} [1053 \quad 537] \right\} \div (3)(10)$$

$$- \begin{bmatrix} 2017 \\ 1095 \end{bmatrix} [2017 \quad 1095] \div 60 = \begin{bmatrix} 132.02 & 28.18 \\ 28.18 & 6.02 \end{bmatrix}$$

Exactly parallel computations, following Eq. 8.5, lead to the matrix $\mathbf{S}_c$. The SSCP matrices for the two main effects are thus found to be

$$\mathbf{S}_r = \begin{bmatrix} 132.02 & 28.18 \\ 28.18 & 6.02 \end{bmatrix} \quad \text{and} \quad \mathbf{S}_c = \begin{bmatrix} 1055.03 & 1111.55 \\ 1111.55 & 1177.30 \end{bmatrix}$$

Computation of the interaction SSCP matrix $\mathbf{S}_{rc}$ involves the vectors of cell totals as well as those of the marginal and grand totals, as is evident from Eq. 8.6. Since we have already computed $\mathbf{S}_r$ and $\mathbf{S}_c$, it is more expedient to use Eq. 8.6a instead of Eq. 8.6. That is, we first compute $\mathbf{S}_{\text{cells}}$ as

$$\begin{aligned}\mathbf{S}_{\text{cells}} &= \left(\begin{bmatrix} 343 \\ 218 \end{bmatrix} [343 \quad 218] + \begin{bmatrix} 325 \\ 179 \end{bmatrix} [325 \quad 179] + \cdots \right. \\ &\quad \left. + \begin{bmatrix} 270 \\ 116 \end{bmatrix} [270 \quad 116] \right) \div 10 \\ &\quad - \begin{bmatrix} 2017 \\ 1095 \end{bmatrix} [2017 \quad 1095] \div 60 \\ &= \begin{bmatrix} 1495.09 & 1314.55 \\ 1314.55 & 1282.55 \end{bmatrix}\end{aligned}$$

We then subtract both $\mathbf{S}_r$ and $\mathbf{S}_c$ from this matrix to obtain

$$\begin{aligned}\mathbf{S}_{rc} &= \begin{bmatrix} 1495.09 & 1314.55 \\ 1314.55 & 1282.55 \end{bmatrix} - \begin{bmatrix} 132.02 & 28.18 \\ 28.18 & 6.02 \end{bmatrix} - \begin{bmatrix} 1055.03 & 1111.55 \\ 1111.55 & 1177.30 \end{bmatrix} \\ &= \begin{bmatrix} 308.04 & 174.82 \\ 174.82 & 99.23 \end{bmatrix}\end{aligned}$$

The four SSCP matrices, $\mathbf{S}_w$, $\mathbf{S}_r$, $\mathbf{S}_c$, and $\mathbf{S}_{rc}$, computed above are collected, together with the total SSCP matrix $\mathbf{S}_t$, in the MANOVA summary table below, following the pattern of Table 8.1. (Although $\mathbf{S}_t$ does not enter into the significance tests as such, it is advisable to compute this matrix independently from Eq. 8.1, as an arithmetic check; the four SSCP matrices computed above should sum to $\mathbf{S}_t$, as stated in Eq. 8.2. The calculation of $\mathbf{X}'\mathbf{X}$ amounts to adding, over cells, the three summary quantities shown in the bottom two lines of the six cells in Table 8.3. This is left as an exercise for the reader.)

The MANOVA summary table (Table 8.4) shows, in addition to the quantities indicated in Table 8.1, the determinants needed for calculating the three Λ_h values in accordance with Eq. 8.7, as well as the Λ-ratios themselves. Since the fixed-effects model (model I) is the most reasonable one to be regarded as applying in the experiment of this example,[1] $\mathbf{S}_w$ is taken as the error SSCP matrix for testing all three effects (see Table 8.2).

Exact significance tests are available for this example because p (the number of dependent variables) is equal to 2. Thus we need not use either of the approximate test statistics V and R, but may employ the relevant F-variate shown in Table 4.2, with the replacements for K and N noted earlier. That is, we use

$$\frac{1 - \Lambda_h^{1/2}}{\Lambda_h^{1/2}} \frac{\nu_e - 1}{\nu_h}$$

as an F-variate with $2\nu_h$ and $2(\nu_e - 1)$ degrees of freedom.

For testing the row effect (methods), we have

$$F_r = \frac{1 - \sqrt{.9077}}{\sqrt{.9077}} \frac{54 - 1}{1} = 2.63$$

This value falls a little short of the 95th percentile of the F-distribution with 2 d.f.'s in the numerator and 106 in the denominator, which is found to be about 3.10, by rough interpolation in Table E.4. Thus if we are using the "conventional" α-value of .05, we would conclude that there is not enough evidence to warrant rejecting the null hypothesis of "zero row effect" in the population. The implication is that methods A and B probably do not differ in their effectiveness when the shorthand acquisition is measured by the two subtests of our standardized test.

For the column effect (distributed versus massed practice), the test statistic takes the value

$$F_c = \frac{1 - \sqrt{.1341}}{\sqrt{.1341}} \frac{53}{2} = 45.85$$

which far exceeds even the 99.9th percentile of the F-distribution with 4 and 106 degrees of freedom. We may therefore conclude that there is strong evidence of nonchance differences due to the extent of distribution of practice. Examination of the column-total vectors indicates that among the three conditions studied, achievement as measured by both speed and accuracy scores steadily increases with increasing degree of distribution of practice. (The interpretation will often not be as clear-cut as in this example because the trends of increase may differ from one dependent variable to another. For this reason it will usually be instructive to follow the significance tests up with discriminant analyses, as described in the next section.)

[1]The row effect (methods) is obviously fixed; we could not possibly think of generalizing beyond the two methods used. The status of the column effect is somewhat ambiguous, but it seems more reasonable to regard the conditions stated at the outset of this section as constituting the entire population of interest, rather than as a random sample from the set of all possible schedules totaling 60 hours of instruction.

TABLE 8.4 MANOVA Summary Table

Source	SSCP Matrix	n.d.f.	$\lvert \mathbf{S}_h + \mathbf{S}_e \rvert$ or $\lvert \mathbf{S}_e \rvert$	Λ_h
Row effect (methods *A* and *B*)	$\begin{bmatrix} 132.02 & 28.18 \\ 28.18 & 6.02 \end{bmatrix}$	1	$\begin{vmatrix} 1465.11 & 239.38 \\ 239.38 & 190.72 \end{vmatrix} = 222{,}123$	.9077
Column effect (distribution of practice)	$\begin{bmatrix} 1055.03 & 1111.55 \\ 1111.55 & 1177.30 \end{bmatrix}$	2	$\begin{vmatrix} 2388.13 & 1322.75 \\ 1322.75 & 1362.00 \end{vmatrix} = 1502{,}966$	.1341
Interaction	$\begin{bmatrix} 308.04 & 174.82 \\ 174.82 & 99.23 \end{bmatrix}$	2	$\begin{vmatrix} 1641.14 & 386.02 \\ 386.02 & 283.93 \end{vmatrix} = 316{,}957$	.6361
Within-cells (error)	$\begin{bmatrix} 1333.10 & 211.20 \\ 211.20 & 184.70 \end{bmatrix}$	54	$\begin{vmatrix} 1333.10 & 211.20 \\ 211.20 & 184.70 \end{vmatrix} = 201{,}618$	—
Total SSCP	$\begin{bmatrix} 2828.18 & 1525.75 \\ 1525.75 & 1467.25 \end{bmatrix}$	59	— —	—

Attending to the interaction effect, the value of the test statistic is

$$F_{rc} = \frac{1 - \sqrt{.6361}}{\sqrt{.6361}} \frac{53}{2} = 6.73$$

which exceeds the 99.9th percentile (approximately 5.03) of the F-distribution with 4 and 106 degrees of freedom. We may conclude with high confidence that the relative effectiveness of the two methods depends on which one of the three conditions of distributed practice is actually used. Examination of the cell-total vectors suggests that method B is more effective than method A under the high- and medium-distributed practice conditions (C_1 and C_2), but that the reverse is true when the highly massed schedule (C_3: 4 hours' daily instruction for 3 weeks) is used. Again, the interpretation is clear-cut in this example because both criterion measures show the same trend; but we cannot expect this to be the case in general.

8.4 DISCRIMINANT ANALYSIS IN FACTORIAL DESIGNS

When a particular effect has been found significant in MANOVA, the question still remains of just how the several groups representing the levels of that effect differ in terms of the criterion variables used. It would be an exception rather than the rule to find the same trend or pattern of differences to be repeated for all the variables involved, as we did in the miniature example in Section 8.3. A modified version of discriminant analysis may help to answer this question by identifying the dimension or dimensions along which the relevant subgroups differ most conspicuously.

The modification of discriminant analysis that is required for this purpose is exactly parallel to the modification (or generalization) of the Λ-ratio which was made in the foregoing. That is, the coefficients of the discriminant function(s) associated with a particular effect are now obtained as the elements of the vector(s) satisfying

$$(\mathbf{S}_h - \lambda \mathbf{S}_e)\mathbf{v} = \mathbf{0} \qquad (8.11)$$

instead of the earlier Eq. 7.9:

$$(\mathbf{B} - \lambda \mathbf{W})\mathbf{v} = \mathbf{0}$$

(Observe that the parallel with the modification of Λ lies in the fact that the denominator of this ratio has now become $|\mathbf{S}_h + \mathbf{S}_e|$ instead of the previous $|\mathbf{S}_t| = |\mathbf{B} + \mathbf{W}|$ for significance testing in the usual multigroup discriminant analysis.)

Thus, assuming (as we shall ordinarily be justified in doing) that $\mathbf{S}_e$ is nonsingular, our problem amounts to finding the eigenvector(s) of the matrix $\mathbf{S}_e^{-1}\mathbf{S}_h$, for Eq. 8.11 may be replaced by its equivalent,

$$(\mathbf{S}_e^{-1}\mathbf{S}_h - \lambda\mathbf{I})\mathbf{v} = \mathbf{0} \tag{8.12}$$

This, incidentally, is the basis for our earlier assertion that it would be preferable to compute Λ_h via the eigenvalue(s) of $\mathbf{S}_e^{-1}\mathbf{S}_h$, as indicated in Eq. 8.8, rather than by the definitional formula, Eq. 8.7, especially when a follow-up discriminant analysis is to be carried out.

After the discriminant functions associated with each of the significant effects have been computed, the procedures for interpretation are exactly the same as for the case of ordinary multigroup discriminant analysis. In fact, since the number of "groups" representing the levels of each independent variable is generally fairly small, there is a good chance that only one or two discriminant functions will be significant for a given main effect. This means that the discriminant function means or centroids for the several levels can be plotted on graph paper, and the interpretation should be simpler than in situations where three or more significant functions are found—as is quite likely in multigroup problems with a large number of groups.

EXAMPLE 8.1. We illustrate the computation of discriminant functions in MANOVA in the context of the miniature example of Section 8.3—even though, in practice, there would hardly be any point in carrying out this follow-up analysis when all criterion measures show the same trend.

Since the row effect (methods) was not significant, we do not compute the discriminant function between methods A and B. Those for the column effect (extent of distributed practice) and interaction are the only ones of possible interest. Computations for these discriminant functions are shown below.

Since $\mathbf{S}_w$ is the error SSCP matrix for all effects in this problem, the matrices $\mathbf{S}_e^{-1}\mathbf{S}_h$ whose eigenvectors we need to find in order to get the desired discriminant functions are $\mathbf{S}_w^{-1}\mathbf{S}_c$ and $\mathbf{S}_w^{-1}\mathbf{S}_{rc}$. Referring to Table 8.4, where the several SSCP matrices have been collected, we first compute

$$\mathbf{S}_w^{-1} = \begin{bmatrix} 1333.10 & 211.20 \\ 211.20 & 184.70 \end{bmatrix}^{-1}$$

$$= \begin{bmatrix} .9161 & -1.0475 \\ -1.0475 & 6.6120 \end{bmatrix} \times 10^{-3}$$

Postmultiplying this by $\mathbf{S}_c$ as given in Table 8.4, we get

$$\mathbf{S}_w^{-1}\mathbf{S}_c = \begin{bmatrix} -.1978 & -.2149 \\ 6.2444 & 6.6200 \end{bmatrix}$$

as the matrix to be substituted in Eq. 8.12 for computing the discriminant functions associated with the column effect.

The characteristic equation $|\mathbf{S}_w^{-1}\mathbf{S}_c - \lambda\mathbf{I}| = 0$ is found to be

$$\lambda^2 - 6.4222\lambda + .0325 = 0$$

whose roots are

$$\lambda_1 = 6.4171 \qquad \text{and} \qquad \lambda_2 = .0051$$

At this point we may verify that computing Λ_c via the eigenvalues of $\mathbf{S}_w^{-1}\mathbf{S}_c$ in accordance with Eq. 8.8 yields the same result as that obtained from the definitional Eq. 8.7. We find

$$\Lambda_c = 1/(7.4171)(1.0051) = .1341$$

which agrees exactly (to four decimal places) with the value shown in Table 8.4, which was computed from the definitional formula.

The next step is to ascertain whether both or only the first of these eigenvalues will lead to significant discriminant functions. This is done, as described in Chapter 7, by examining the partial sums (in this case simply the second term) in the expression for Bartlett's approximate chi-square statistic, V. In the present example, it is evident by mere inspection that only the first eigenvalue is significant (for it accounts for over 99.9% of the trace of $\mathbf{S}_w^{-1}\mathbf{S}_c$). We carry out the test solely for illustrative purposes. Substituting the eigenvalues in Eq. 8.9 with $\nu_e = \nu_w = 54$, $\nu_h = \nu_c = 2$, and $p = 2$, we get

$$V = [54 + 2 - (2 + 2 + 1)/2](\ln 7.4171 + \ln 1.0051)$$

$$= (53.5)(2.0038 + 0.0051) = 107.21 + 0.27$$

From the description following Eq. 7.14, with K replaced by $\nu_h + 1$, it is seen that the two terms in the indicated sum should be compared to chi-squares with $p + \nu_h - 1$ and $p + \nu_h - 3$ degrees of freedom, respectively. For this example, these n.d.f's become 3 and 1, respectively. (Note that their sum is $4 = p\nu_h$, which is the n.d.f. for V itself, as stated after Eq. 8.9.) It is obvious, as we expected, that the second term, 0.27, falls far short of significance as a chi-square with one degree of freedom. Thus we compute only the first discriminant function, that associated with the larger eigenvalue 6.4171 of $\mathbf{S}_w^{-1}\mathbf{S}_c$.

We substitute $\lambda_1 = 6.4171$ in the expression $\mathbf{S}_w^{-1}\mathbf{S}_c - \lambda\mathbf{I}$, and find the adjoint of the resulting matrix. The outcome is

$$\mathbf{adj}(\mathbf{S}_w^{-1}\mathbf{S}_c - 6.4171\mathbf{I}) = \begin{bmatrix} .2029 & .2149 \\ -6.2444 & -6.6149 \end{bmatrix}$$

Normalizing either column of this matrix to unity by dividing the elements by the square root of the sum of the squares of the two elements, and further multiplying each element by -1 (so that the larger element will be positive), we find the normalized eigenvector associated with λ_1 to be

$$\mathbf{v}'_{c(1)} = [-.0325, .9995]$$

Thus, the single significant discriminant function for the column effect is

$$Y_c = -.0325X_1 + .9995X_2$$

For all practical purposes, the accuracy score alone suffices to differentiate among the three conditions of distribution of practice.

For computing the discriminant functions associated with the interaction effect, the matrix to be substituted for $\mathbf{S}_e^{-1}\mathbf{S}_h$ in Eq. 8.12 is

$$\mathbf{S}_w^{-1}\mathbf{S}_{rc} = \begin{bmatrix} .09907 & .05621 \\ .83324 & .47298 \end{bmatrix}$$

The characteristic equation is

$$\lambda^2 - .57205\lambda + .00002 = 0$$

Since the constant term is equal to zero within rounding error, we may take

$$\lambda_1 = .5721$$

to be the single nonvanishing eigenvalue.

The associated eigenvector, normalized to unity, is found to be

$$\mathbf{v}'_{rc} = [.1181, .9930]$$

Again, the discriminant function is practically equal to X_2 (the accuracy score) itself.

8.5 OTHER DESIGNS

In the foregoing we discussed the procedures of MANOVA in its simplest application: two-factor designs with equal cell frequencies. We now give a brief guide to making the requisite extensions of the methods of ANOVA to multivariate designs of other types. Simply stated, the basic principle is always this: Compute the SSCP matrix corresponding to each sum of squares in ANOVA; select the appropriate error SSCP matrix (following the same rule as in ANOVA) for testing each null hypothesis, and calculate Λ_h from Eq. 8.7 or 8.8.

For two-factor designs with cell frequencies unequal but proportional to the marginal frequencies, the modifications to the procedures described above are quite minor. Only slight changes in the computational formulas (8.3) to (8.6a) need to be introduced, and the subsequent steps remain exactly as they were in the equal-frequencies case. (In fact, the design with equal cell frequencies is simply a special case of the more general proportional-frequencies design, also known as the *orthogonal design*.)

Denoting the frequency, that is, the number of experimental units, in the (r, c)-cell by n_{rc}, we further define:

$$n_{r.} = \sum_{c=1}^{C} n_{rc} \qquad r\text{th row frequency}$$

$$n_{.c} = \sum_{r=1}^{R} n_{rc} \qquad c\text{th column frequency}$$

and

$$N = \sum_{r=1}^{R} n_{r.} = \sum_{c=1}^{C} n_{.c} \qquad \text{total sample size}$$

(Proportionality of cell frequencies to the marginals is said to hold when $n_{rc} = n_{r.}n_{.c}/N$ for all cells.) Then the modifications consist simply in replacing, in the computational formulas, all instances of *n occurring alone* by n_{rc}, all instances of nC by $n_{r.}$, those of nR by $n_{.c}$, and nRC by N. [It is understood, of course, that any subscripted n is included within the scope of summation with regard to the symbol(s) in the subscript.] Thus the formulas for $\mathbf{S}_w$ and $\mathbf{S}_{rc}$ become, by modifying Eqs. 8.3 and 8.6,

$$\mathbf{S}_w = \sum_{r=1}^{R}\sum_{c=1}^{C} [\mathbf{X}'_{rc}\mathbf{X}_{rc} - \mathbf{t}_{rc}\mathbf{t}'_{rc}/n_{rc}] \tag{8.13}$$

and

$$\mathbf{S}_{rc} = \sum_{r=1}^{R}\sum_{c=1}^{C} \mathbf{t}_{rc}\mathbf{t}'_{rc}/n_{rc} - \sum_{r=1}^{R} \mathbf{t}_{r.}\mathbf{t}'_{r.}/n_{r.} - \sum_{c=1}^{C} \mathbf{t}_{.c}\mathbf{t}'_{.c}/n_{.c} + \mathbf{t}_{..}\mathbf{t}'_{..}/N \tag{8.14}$$

respectively. The modified formulas for $\mathbf{S}_r$ and $\mathbf{S}_c$ should be obvious from that for $\mathbf{S}_{rc}$, since the latter contains all the "ingredients" for them.

When the cell frequencies are disproportionate, the problem becomes more involved; in fact, the significance tests are then only approximate. The two most commonly used methods are the method of least squares and that of unweighted means. We outline the latter method here. The least-squares approach leads to significance tests that are more nearly exact than does the unweighted-means method described below, but is computationally far more complex. (Of course with the widespread availability of computer software, this may no longer be a problem.) The least-squares approach (or the general linear model) is discussed in Chapter 9, but the handling of the disproportionate cell-frequency case is only indirectly alluded to, and a reference for its detailed treatment is given.

For carrying out the unweighted-means analysis in MANOVA, the first step is to compute the vectors of cell means for all the dependent variables. These cell-mean vectors are then used *as though* they were observation vectors based on one experimental unit per cell for computing all the SSCP matrices *except* $\mathbf{S}_w$. That is, Eqs. 8.4 to 8.6, with n set equal to 1 and N to RC, are used for computing $\mathbf{S}_r$, $\mathbf{S}_c$, and $\mathbf{S}_{rc}$. (Be sure to note that the various totals in these formulas are now obtained by the appropriate summing of *cell means,* not the original scores.)

The within-cells SSCP matrix $\mathbf{S}_w$ is computed in accordance with Eq. 8.3 using the original data $\mathbf{X}_r$. But this matrix is then "scaled down" by dividing it by the harmonic mean of the cell frequencies, thus making it commensurate with the other SSCP matrices that were computed by treating the cell means as though they were single observations. Division by the harmonic mean of the cell frequencies is, of course, the same as multiplication by the *reciprocal* of the harmonic mean, that is, the average of the reciprocal of the cell frequencies. The multiplicative factor is

$$1/\tilde{n} = \left[\sum_{r=1}^{R} \sum_{c=1}^{C} (1/n_{rc})\right]/RC \tag{8.15}$$

Thus the actual within-cells SSCP matrix $\mathbf{S}_w$ is replaced by a "scaled-down" matrix,

$$\tilde{\mathbf{S}}_w = (1/\tilde{n})\mathbf{S}_w \tag{8.16}$$

The analysis from this point on proceeds in exactly the same way as for the equal-frequencies design, except that the n.d.f. for $\mathbf{S}_w$ is $N - RC$ instead of the $RC(n - 1)$ shown in Table 8.1. It should be noted, however, that the analysis is now only approximate; hence the outcome must be interpreted with caution, especially if it is one of borderline significance.

Other designs commonly used in educational and psychological research include three-factor designs, randomized-blocks design, repeated-measures design, and nested-factors (or hierarchical) design. We cannot describe the multivariate extensions of all these analyses without making this chapter inordinately long. The reader should have little difficulty in arriving at the correct procedures if he or she is familiar with the corresponding univariate ANOVA methods and remembers the general principle stated at the outset of this section: To each sum of squares in ANOVA there corresponds an SSCP matrix in MANOVA. Thus, in a randomized blocks design, the highest-order interaction SSCP matrix replaces the within-cells SSCP matrix of the corresponding randomized groups design. [This is because a q-factor randomized-blocks design with m blocks is formally equivalent with a $(q + 1)$-factor randomized-groups design.] In a nested-factors design, different error SSCP matrices will have to be used for testing the effects involving nested and crossed factors. A general computer program capable of handling all the commonly used designs was developed by Bock (1963, 1965) and put into operational form by Clyde, Cramer, and Sherin (1966).

8.6 OTHER TEST CRITERIA

In Chapter 7 and in this one so far, we have used the Λ-ratio exclusively as the test criterion in carrying out significance tests. Although this is the oldest and most widely used criterion, it is by no means the only one available. The several alternative criteria that have been proposed are all functions of the eigenvalues of $\mathbf{S}_e^{-1}\mathbf{S}_h$, just as Λ is. Among them, the three best known ones are the following, where $\lambda_1, \lambda_2, \ldots, \lambda_p$ are the eigenvalues of $\mathbf{S}_e^{-1}\mathbf{S}_h$ in descending order of magnitude: (a) Hotelling's (1951) trace criterion

$$\tau = \sum_{i=1}^{p} (\mathbf{S}_e^{-1}\mathbf{S}_h)_{ii} = \sum_{i=1}^{p} \lambda_i \tag{8.17}$$

(b) Roy's (1957, chap. 6) largest root criterion

$$\theta = \lambda_1/(1 + \lambda_1) \tag{8.18}$$

which is the largest eigenvalue of $(\mathbf{S}_e + \mathbf{S}_h)^{-1}\mathbf{S}_h$, and (c) the Pillai–Bartlett trace criterion

$$V = \sum_{i=1}^{p} [(\mathbf{S}_e + \mathbf{S}_h)^{-1}\mathbf{S}_h]_{ii} = \sum_{i=1}^{p} [\lambda_i/(1 + \lambda_i)] \tag{8.19}$$

It is seen from the derivation of Eq. 8.12 that λ_i is also the value of the discriminant criterion for the ith discriminant function associated with the effect being tested. Thus large values of the λ_i indicate significant effects. It therefore follows that since τ, θ, and V are all increasing functions of λ_i, these test criteria are such that when their observed values exceed the upper $100(1 - \alpha)$ percentile points of their respective sampling distributions, the relevant null hypothesis is rejected at the $100\alpha\%$ level of significance. This is in contrast to the Λ-criterion that we have been using. In that case, observed values smaller than the 100α percentile point led to rejection of the null hypothesis, because

$$\Lambda = 1/(1 + \lambda_1)(1 + \lambda_2) \cdots (1 + \lambda_p)$$

is a *decreasing* function of the λ_i.

Pillai (1960) has derived the sampling distribution of τ (which he denotes as $U^{(s)}$ under the null hypothesis, and has tabled selected percentile points of the distribution, or rather the family of distributions for various combinations of values of the parameters m, n, and s, defined as follows:

$$m = (p - \nu_h - 1)/2 \qquad n = (\nu_e - p - 1)/2$$

and

$$s = \min(\nu_h, p)$$

Similarly, selected percentile points for the null distributions of θ are given by Heck (1960) and by Pillai (1960, 1965), also using m, n, and s as parameters. Finally, extensive tables of percentile points of the Pillai–Bartlett V have been constructed by Mijares (1964), again with m, n, and s as parameters. Pillai (1960) gives less extensive tables in the same volume in which his tabulation of the distributions of θ appears. In addition, he presents the following function of V:

$$\frac{V}{s - V} \frac{\nu_e - p + s}{b}$$

which approximately follows an F-distribution with sb and $s(\nu_e - p + s)$ degrees of freedom. Here

$$b = \max(\nu_h, p)$$

Note that when $\nu_h = 1$, $\mathbf{S}_e^{-1}\mathbf{S}_h$ has only one nonzero eigenvalue λ_1; therefore, all three of the test criteria introduced in this section, as well as the earlier criterion Λ, become simple functions of λ_1 and hence also of one another. Specifically,

$$\tau = \lambda_1 \qquad \theta = \lambda_1/(1 + \lambda_1) = V \qquad \text{and} \qquad \Lambda = 1/(1 + \lambda_1)$$

Solving the last relation for λ_1, we get

$$\lambda_1 = \frac{1 - \Lambda}{\Lambda}$$

But from Table 4.2, by replacing N by $\nu_e + \nu_h + 1 = \nu_e + 2$, it is seen that

$$\frac{1 - \Lambda}{\Lambda} \frac{\nu_e - p + 1}{p}$$

follows the F-distribution with p and $\nu_e - p + 1$ degrees of freedom. Consequently,

$$\tau \frac{\nu_e - p + 1}{p} \qquad \frac{\theta}{1 - \theta} \frac{\nu_e - p + 1}{p} \qquad \frac{V}{1 - V} \frac{\nu_e - p + 1}{p}$$

may all be referred to the same F-distribution, and the need for special tables disappears in the special case when $\nu_h = 1$.

When $\nu_h > 1$, however, not only are the tables referred to above necessary for using the largest-root and the two trace criteria, but the conclusions from the significance tests based on the several criteria may disagree. The question then arises: Which is the correct or ''best'' conclusion? Unfortunately, there is no unique answer to this question, because it all depends on the nature of the alternative hypothesis that happens to be true—that is, on the actual configuration of the population centroids.

Several studies have addressed this question. A Monte Carlo study by Schatzoff (1966b) used an index called the expected significance level to compare the relative sensitivities of six test statistics, including the four discussed here, over a wide variety of population structures. Roy's largest-root criterion θ was found to be the most sensitive when the population centroids differed mostly along a single dimension, but was otherwise the least sensitive. The other three criteria varied in their relative sensitivity under different conditions, except that when the ratio of sample size to number of variables (N/p) was relatively small, the Pillai–Bartlett trace criterion V did rather poorly regardless of the population structure. Thus, under most conditions, it was a toss-up between Wilks' Λ and Hotelling's trace criterion τ.

A more recent study by Olson (1976) arrived at a somewhat different conclusion. He recommended the Pillai–Bartlett V for general use because it was found to be the most robust under violation of the homogeneity of covariance matrices assumption, *especially for* relatively small sample sizes. When $\nu_e/\nu_h \geq 10p$, however, Olson found all but the largest-root criterion to be just about equally powerful. More specifically, he found, by synthesizing the results of several studies in the literature in addition to his own, that when the population centroids are largely confined to a single dimension (a condition referred to as a *concentrated* noncentrality structure), the criteria tend to be ranked θ, τ, Λ, V in descending order of power. Under conditions of *diffuse* noncentrality structure (i.e., when the population centroids differ almost equally in all dimensions) the ordering is reversed: V, Λ, τ, θ.

Based on simultaneous considerations of power and robustness, Olson considers the Pillai–Bartlett V to be generally superior to the other three. Thus there seems to be a considerable disagreement between Schatzoff's and Olson's recommendations. The disparity may be attributed partly, however, to the standards that one adopts in considering a sample ''small'' versus ''reasonably large'' relative to the number of variables used. For Olson, a sample is small unless the N/p ratio exceeds $10\nu_e$—and hence V is considered the best most of the time. For Schatzoff, on the other hand, N/p has to be quite small before the superiority of V manifests itself. Stevens (1979) claims that under concentrated noncentrality conditions, there is little difference among V, τ, and Λ, because the robustness advantage of V does not show up unless the heterogeneity of covariance matrices is exceptionally severe.

Before the reader despairs at the lack of consensus on the matter of choice of test statistic in MANOVA, it is well to remember that the condition of concentrated versus diffuse noncentrality structures refers to the population, as does the homogeneity/heterogeneity of covariance matrices. Since the corresponding conditions in the sample do not guarantee the same to hold in the population (even though the likelihood may be enhanced), the judicious approach may be to examine the conclusions based on all four criteria (which most computer packages give anyway), and if they differ, to incline toward that conclusion which is consonant with one's judgment of the relative seriousness of type I versus type II errors. Finally, there is the reassuring fact that when the sample sizes are very large, the tests based on Λ, τ, and V become asymptotically equivalent.

Studies illustrating the use of the various test criteria are described by Bock (1966), Bock and Haggard (1968), and Jones (1966).

Other Measures of Strength of Association

We saw in Chapter 4 that the likelihood-ratio criterion Λ is related to the multivariate correlation ratio—a measure of the strength of association—by the simple equation $\eta^2_{\text{mult}} = 1 - \Lambda$. It seems reasonable to expect, therefore, that each of the other significance-test criteria cited in this section is simply related to some other measure of strength of association. This is in fact the case.

In Hotelling's trace criterion, given by Eq. 8.17, each term is the square of what may be called the *effect size* as measured by each dependent variable. Considering the simplest case of the main effect of a factor (i.e., independent variable) with just two levels, the ith term of the first summational expression is proportional to the squared standardized difference between the means of the ith dependent variable in the two levels of that factor. Thus, Hotelling's trace criterion is simply the sum of the squared effect size as measured by each dependent variable separately. It is the sum of p separate univariate measures of association, one for each dependent variable in isolation.

Roy's largest root criterion is a little more complicated. Referring to the third unnumbered equation in Section 7.10 (p. 246) and solving it for μ_i^2, the square of the ith canonical correlation, we obtain

$$\mu_i^2 = \lambda_i/(1 + \lambda_i)$$

Hence, Roy's test criterion θ is the square of the first (i.e., the largest) canonical correlation between the set of dependent variables and the set of dummy variables that could be used to indicate a subject's membership in one of the levels of the factor under consideration. It may therefore be interpreted as the proportion of the generalized variance of the set of dependent variables that can be "attributed" to the factor via the first canonical variate.

Given the above interpretation of Roy's largest-root criterion, it follows that the Pillai–Bartlett trace criterion is the total proportion of the generalized variance of the set of dependent variables that is attributable to the factor via all of the canonical variates.

8.7 RESEARCH EXAMPLES

Cooley and Lohnes (1971, pp. 309–12) randomly sampled 39 subjects in each of eight interaction cells of a factorial design. The first independent variable was Gender (2 levels), and the second was College Plans (4 levels). The population was 1960 high school seniors, as represented in Project TALENT's national probability sample. Six dependent variables were provided by selected ability tests.

At the outset it is important to recognize that high positive correlations among the six tests indicate that the dependent measures are saturated with a latent source variable (Lohnes, 1966). This latent variable could be termed General Intellectual Development. Because of this saturation, it is likely that the single discriminant function will serve well to separate the two Gender levels, and also that a single discriminant function will serve to separate the four Plans levels. What is interesting is the comparison of these two discriminant functions. The research question is whether quite different factors of ability are required to provide the most-predictable-criterion for the two different independent factors of the design. From previous Project TALENT research (Cooley and Lohnes, 1968) it can be predicted that the discriminant function for the Gender variable will be bipolar with negative loadings for Information and Mathematics tests, on which males excelled as a group, and positive loadings on English Compositer and Reading Comprehension tests, on which females excelled as a group. For the College Plans variable it can be predicted that a general ability factor loading positively on all six dependent variables will be the most-predictable-criterion. If this is so, it will be interesting to see which test contributes most to the location of this General Intellectual Development factor.

Since this is *not* a randomized experiment, the assessment of Gender and Plans as the independent variables is quite arbitrary and justified only as a matter of convention, unless a strong *a priori* case can be made on external evidence for the assignment. Some educational psychologists might believe that a case could be made for the latent general ability variable as the true independent variable for College Plans, at least. *No causal interpretation* of statistically significant regressions will be possible on the internal logic of this research design, since both sets of variables were sampled simultaneously in an on-going educational ecology. On the other hand, causal interpretations might be rendered plausible by assembling appropriate external evidence and argument.

The MANOVA table (Table 8.5) indicates that there is no interaction between Gender and College Plans. This fortunate result obviates the need to peruse the

TABLE 8.5 Gender by College Plans MANOVA

Source	n.d.f.[a] (num)	n.d.f.[b] (den)	F	η^2	$\hat{\omega}^2_{mult,c}$
Gender	6	299	35.	.41	.384
College plans	18	846	6.5	.30	.259
Interaction	18	846	.8	.05	0

[a,b]The *n.d.f.*s indicated here are not those for the SSCP matrices as in Table 8.4, but those for the R statistic of Eq. 8.10 treated as an approximate F-variate with $p\nu_h$ degrees of freedom in the numerator and $ms - p\nu_h/2 + 1$ in the denominator. (See Eq. 8.10 for definitions of m and s.) Since $p = 6$ in this example, the *n.d.f.*s for the F for testing the Gender effect are (6)(1) = 6 in the numerator and 299 in the denominator.

interaction cell means, and permits us to focus on the Gender means and the Plans means separately.

The table for the Gender comparison (Table 8.6) makes it clear that the significant discrimination is concentrated on the two variables of Information I (males superior) and English Composite (females superior), although the signs of the discriminant function structure coefficients are in the hypothesized directions for Information II, Mathematics Composite, and Reading Comprehension. Note the unusual outcome for the Abstract Reasoning test, on which both Gender groups achieved the same mean. Talk about gender-fairness in a test!

The table for the College Plans comparison (Table 8.7) makes it clear that the discrimination is nicely spread across the six tests. The only useful discriminant function is clearly a General Intellectual Development factor, as shown by the positive structure coefficient for every test. It is interesting that the Mathematics composite is the test which contributes most to the orientation of the most-predictable-criterion. As a matter of fact, this finding can be used as a basis for arguing that the College Plans variable is correctly assigned as the independent variable, because the Mathematics Composite was heavily dependent on college-preparatory content which only the students in college preparatory curriculum tracks were likely to have the opportunity to study and learn. Notice that the two groups not contemplating college (Some Training and No Training) are not separated by the tests.

TABLE 8.6 Gender Comparisons

Dependent Variable	Male Mean	Female Mean	F^1_{304}	Structure Coefficient
Information part I	155	136	30.	−.47
Information part II	79	76	1.8	−.16
English composite	81	90	47.	.51
Reading comprehension	33	35	3.3	.10
Abstract reasoning	10	10	0.0	−.03
Mathematics composite	25	23	3.3	−.21

TABLE 8.7 College Plans Comparisons

Dependent Variable	Definite College	Maybe College	Some Training	No Post H.S. Training	F^3_{304}	Structure Coefficient
Information I	166	153	131	132	26.	.89
Information II	86	81	70	72	17.	.73
English composite	90	87	84	82	10.	.43
Reading compreh.	37	37	31	30	16.	.64
Abstract reasoning	10	10	9	9	5.5	.37
Mathematics computation	31	26	20	20	31.	.93

The MANOVA η^2 values, where $\eta^2 = 1 - \Lambda$ shown in Table 8.5, for each independent variable, indicates that the Gender discrimination is stronger than the Plans discrimination. Unfortunately these descriptive measures of the strength of association are upwardly biased estimates of population ρ^2 values and should be viewed cautiously since they have not been corrected for shrinkage. One way to correct η^2 for shrinkage is to treat it as an R^2 and apply Eq. 3.24 to it. Another way is to convert Λ to $\hat{\omega}^2_{\text{mult}}$ via Eq. 4.35, apply Eq. 4.36 to correct it for *its* upward bias and (if desired) to convert the corrected $\hat{\omega}^2_{\text{mult,c}}$ back to a corrected Λ and thence to a corrected η^2. The values of $\hat{\omega}^2_{\text{mult,c}}$ corresponding to the η^2 values given in Column 5 of Table 8.5 are shown in the last column there.

It may be concluded that the factorial MANOVA design of this study enabled the researchers to create an informative comparison of the regressions of six ability tests on the independent design factors of Gender and College Plans.

EXERCISES

1. An experiment was carried out to explore possible differences in grammatical usage among men and women with varying degrees of formal education. Thus the independent variables were education and sex. Three levels were used for education: (1) did not complete high school, (2) graduated from high school, and (3) attended college for two or more years.

   The dependent variables were measures of relative frequencies of use of (1) personal possessive pronouns, (2) nouns, and (3) quantifiers in tape-recorded stories told by S's in response to a set of 20 cartoons stripped of their captions. (More specifically, the scores were 20 arcsin $\sqrt{p}$, where p is the observed relative frequency for each grammatical category, and the multiplier 20 was used to get numbers of convenient orders of magnitude.)

   Shown below are the six cell-mean vectors for the three dependent variables, and (in italics) the cell frequencies. The within-cells SSCP matrix $\mathbf{S}_w$, based on the original scores (not shown), is also given (adapted from Jones, 1966). Carry out a complete MANOVA by the method of unweighted means for these data, including the construction of discriminant functions for those effects that are significant at the 1% level.

	Sex	
Education	Male	Female
1	[2.09, 4.53, 3.56] *8*	[1.38, 4.17, 3.38] *10*
2	[2.12, 5.35, 3.59] *11*	[1.47, 4.89, 3.12] *8*
3	[2.13, 5.94, 3.51] *9*	[1.74, 5.37, 3.27] *8*

$$\mathbf{S}_w = \begin{bmatrix} 6.9312 & 2.9520 & -.5040 \\ 2.9520 & 19.0512 & 1.5792 \\ -.5040 & 1.5792 & 7.3008 \end{bmatrix}$$

2. An experiment was conducted for comparing three different approaches to teaching arithmetic and language skills to fifth-grade pupils:

(1) "Traditional" [T]: lecture, discussion, and drill
(2) "Programmed" [P]: entire course taught by programmed instruction
(3) "Eclectic" [E]: a combination of the first two approaches, wherein the teacher does the major classroom exposition, and exercises and self-tests are handled through programmed material.

Since it is well known that boys and girls of this age differ in their relative performances in these two subjects, sex was used as a second factor.

Specifically, 60 boys and 60 girls were selected at random from the fifth grade of a large school, and were randomly assigned (20 pupils of each sex per group) to the three treatment groups.

At the end of five months of instruction, all 120 pupils were given a standardized achievement test in arithmetic (X_1) and a standardized achievement test in the language skills (X_2).

The total scores for each of the six subgroups on the two tests (in the order [X_1, X_2]) were as given in the cells of the following table, which also shows the marginal and the grand totals:

	T	*P*	*E*	
Boys	[514, 453]	[524, 473]	[536, 490]	[1574, 1416]
Girls	[458, 507]	[473, 518]	[494, 571]	[1425, 1596]
	[972, 960]	[997, 991]	[1030, 1061]	[2999, 3012]

The within-cells SSCP matrix for these data was found to be

$$\mathbf{S}_w = \begin{bmatrix} 1795.39 & 1230.56 \\ 1230.56 & 1753.50 \end{bmatrix}$$

(a) Carry out a MANOVA to test the significance of the two main effects and the interaction effect on the vector criterion variable, $[X_1, X_2]$. Use the 1% level of significance for all decisions.

(b) Construct the linear discriminant function(s) corresponding to the effect(s) found significant in part (a).

3. Draw a sample of 80 boys and 96 girls, with each sex distributed in the three high school programs (general, college prep, vocational) in the ratio 1:2:1, from the HSB data set of Appendix F. To assure approximate randomness of the sample, take the 300 cases with odd-numbered IDs, and from these select the first 176 cases that are distributed in the six cells in the stipulated ratios. Using the writing, math, and science achievement tests as the dependent variables, do a 2×3 two-way MANOVA with $\alpha = .01$. Note that the design is orthogonal because the cell frequencies are proportional in the two rows.

4. For each of the effects found significant in Exercise 3, do a discriminant analysis. Use $\alpha = .01$ to decide how many discriminant functions are significant for the HSP and sex $\times$ HSP effects. Interpret the results.

5. Double-classify all 600 students in the HSB data set with respect to SES and HSP. Do a two-way MANOVA using all five cognitive variables as dependent variables and taking $\alpha = .01$.

9

THE GENERAL LINEAR MODEL: A "NEW" TREND IN ANALYSIS OF VARIANCE[1]

More and more frequently these days one hears of the *general linear model* (or the general linear hypothesis) being used in connection with significance tests in analysis of variance (ANOVA) and analysis of covariance (ANCOVA) and their multivariate counterparts (MANOVA and MANCOVA). To most researchers in the behavioral and social sciences who received their training in statistics before the mid-1960s—and to some who received it more recently—this "new" trend must be quite bewildering. Gone, it seems, are the familiar F-ratios of the between- to within-groups mean squares in one-way ANOVA and their extensions in more complicated designs. Instead, one hears of (or sees on computer printouts) estimates and significance tests for various parameters in the linear model. It is difficult indeed for the uninitiated to make connections between the traditional F-tests and the "new" F-tests related to linear-model parameters or contrasts among them. It is the purpose of this chapter to help the reader make such connections.

In a sense, the current trend toward the linear-model approach to ANOVA and ANCOVA is not new at all (which is why we used the word "new" in quotes above). In the early writings of R. A. Fisher, the originator of ANOVA, it is evident that he initially approached the problem of multigroup significance testing via the multiple linear regression method—which, as we shall soon see, is essentially what the general linear model is. It was only (or at least mainly) because the calculations needed for the multiple regression approach were practically infeasible for all but the simplest designs in pre-computer days that Fisher invented the MS_b/MS_w approach.

[1]Grateful acknowledgment is extended to Drs. Samuel Krug and David Watterson for their helpful comments on an earlier version of this chapter.

We may thus say that the availability of high-speed electronic computers has spawned a flurry of revived interest in the multiple-regression approach, since tedious calculations were no longer an obstacle. Of course, many refinements were made on Fisher's original developments, and it was shown that all conceivable designs in ANOVA and ANCOVA could be handled by a single general model, differing from design to design only in minute technical detail. Hence the enormous esthetic appeal of the general linear model approach, despite the horrendous calculations associated with it.

In this chapter we consider in detail only the very simplest instances of the general linear model, and attempt to show the equivalence of testing results between this and the "traditional" MS_b/MS_w approach in as nontechnical and intuitive a manner as possible. Applications to more complicated designs will merely be indicated, with no attempt to carry out the solutions, just to give the reader a rough idea of how the generalizations are made. Those who are interested in delving further into the theoretical detail of the more complicated cases (which are, of course, the more useful instances) may refer to books by Mendenhall (1968), Searle (1971), Bock (1975), Finn (1974), and Lunneborg and Abbott (1983).

9.1 THE SIMPLEST CASE OF TWO GROUPS: A RUDIMENTARY APPROACH

The reader is no doubt aware that when the purpose is to test the significance of the difference between the means of two independent groups, a simple t-test will suffice. However, the square of a t-statistic with $N - 2$ degrees of freedom is an F-statistic with 1 degree of freedom in the numerator and $N - 2$ in the denominator (see, e.g., Glass & Hopkins, 1984, p. 270). Moreover, it can be shown that the square of the t-ratio for testing H_0: $\mu_1 = \mu_2$ is algebraically equivalent to

$$F = MS_b/MS_w$$

as computed for the two-group case (Edwards, 1968, p. 141). Thus if we show the equivalence of the significance test via a rudimentary linear model approach with the customary t-test, we will have shown the equivalence of the former with the F-test of ANOVA in its simplest possible instance.

The first step in our elementary linear-model formulation is to define a "dummy variable" X which indicates to which of the two groups a person belongs. This can be done by letting X take any two distinct values (e.g., 2 and 5, 3 and 10, etc.), one for each group. It is simplest, of course, to settle for $X = 1$ and 0. That is, X is so defined that every person in group 1 gets the "score" $X = 1$, and every member of group 2 gets $X = 0$.

Next, we denote the criterion variable by Y; this is the variable whose means $\bar{Y}_1$ and $\bar{Y}_2$ in the two groups we wish to compare. We are interested, that is, in whether $\bar{Y}_1$ and $\bar{Y}_2$ are significantly different from each other.

In our present "rudimentary" approach, we recast the question into the form of a simple linear regression problem for predicting Y from X. It should be intuitively clear that if there is a significant difference between $\bar{Y}_1$ and $\bar{Y}_2$, the two group means on Y, this will be reflected in a significant predictability of Y from the group membership variable. Or to put it the other way around, if knowledge of the X "score" of an individual (i.e., knowledge of whether the person is a member of group 1 or of group 2) adds nothing to the accuracy with which we can predict his or her Y score, there must not be a significant or stable difference between $\bar{Y}_1$ and $\bar{Y}_2$. Thus the two questions, (a) "Are $\bar{Y}_1$ and $\bar{Y}_2$ significantly different from each other?" and (b) "Is there a significant predictability of Y from X?" are seen to be closely interrelated. In fact, it can be shown (see Section 9.6) that the two questions are *identical,* provided that we interpret (b) to mean "Is the regression coefficient of Y on X significantly different from zero?" But first let us see the interrelation at a less technical level.

As given in Eq. 3.1, the linear regression equation of Y on X is

$$\hat{Y} = a + bX \tag{9.1}$$

where the constants a (the intercept) and b (the regression coefficient) are determined so as to minimize the sum of squared discrepancies

$$Q = \Sigma\,(Y - \hat{Y})^2$$

between actual and predicted Y scores in the total sample. In the present context, we are concerned only with the regression coefficient b, and it is shown in most statistics texts (e.g., Glass & Stanley 1970, pp. 570–571) that the value of b which minimizes the sum of squared errors Q is given by the formula

$$b = \frac{N\,\Sigma\,XY - (\Sigma\,X)(\Sigma\,Y)}{N\,\Sigma\,X^2 - (\Sigma\,X)^2} \tag{9.2}$$

Let us see how some of the sums ($\Sigma\,X$, $\Sigma\,X^2$, etc.) that occur in this formula simplify in the present problem. Recalling that X is a dummy variable taking the value 1 for each member of group 1 and the value 0 for group 2 members, it follows that in the sum $\Sigma\,X$, 1's will be added as many times as there are members of group 1, say n_1, and 0's will be added a number of times equaling the size of group 2, say n_2. That is,

$$\begin{aligned}\Sigma\,X &= \underbrace{(1 + 1 + \cdots + 1)}_{n_1 \text{ times}} + \underbrace{(0 + 0 + \cdots + 0)}_{n_2 \text{ times}} \\ &= n_1\end{aligned}$$

Similarly, since squaring 1's and 0's does not change their values, it follows that

$$\Sigma X^2 = \Sigma X = n_1$$

Next, how about ΣXY? This calls for our multiplying each person's Y score by his or her X score and summing these products. But since $X = 1$ for every group 1 member, the product XY for each person in group 1 will simply be his or her Y score itself. On the other hand, since $X = 0$ for everyone in group 2, the product XY will be 0 for each group 2 member, and such products will add nothing to the sum ΣXY for the total sample of $n_1 + n_2 = N$ people. Thus we see that

$$\Sigma XY = \sum_{\text{gr 1}} Y$$

the total Y score for group 1 alone. The remaining sum, ΣY, in the formula for b cannot be simplified in any way; it is the total Y score for the entire sample, including both group 1 and group 2 members.

Substituting the simplified expressions for ΣX, ΣX^2, and ΣXY (and leaving ΣY as is) in Eq. 7.2, we get

$$b = \frac{N \sum_{\text{gr 1}} Y - n_1(\Sigma Y)}{Nn_1 - n_1^2}$$

which, upon dividing numerator and denominator by n_1 and utilizing the fact that

$$\sum_{\text{gr 1}} Y/n_1 = \bar{Y}_1 \qquad \text{(the group 1 mean of } Y\text{)}$$

becomes

$$\begin{aligned} b &= \frac{N\bar{Y}_1 - \Sigma Y}{N - n_1} \\ &= \frac{N\bar{Y}_1 - \Sigma Y}{n_2} \end{aligned} \tag{9.3}$$

Now, since the grand total, ΣY, of Y is the sum of its totals for groups 1 and 2, and since the latter are each equal to the corresponding group mean multiplied by the group size (n_1 and n_2, respectively), we have

$$\Sigma Y = n_1\bar{Y}_1 + n_2\bar{Y}_2$$

Substituting this expression for ΣY in Eq. 9.3 and noting that $N - n_1 = n_2$, we obtain

$$\begin{aligned} b &= \frac{n_2\bar{Y}_1 - n_2\bar{Y}_2}{n_2} \\ &= \bar{Y}_1 - \bar{Y}_2 \end{aligned} \tag{9.4}$$

We thus see that the regression coefficient in Eq. 9.1 is precisely equal to the difference between the two group means on Y. Without going any further, we may therefore conclude that the two questions, "Are $\overline{Y}_1$ and $\overline{Y}_2$ significantly different?" and "Does X contribute significantly to the prediction of Y?" (i.e., is b significantly different from zero?) are indeed equivalent. To demonstrate that the t-tests for answering these questions produce identical results, however, requires (even for the simplest case) a little more algebraic manipulations than some readers may care to go through. Such a demonstration is therefore relegated to Section 9.6 and we shall here illustrate the equivalence by means of a numerical example.

EXAMPLE 9.1 Suppose that the subjects in two treatment groups in an experiment earned the following criterion scores (Y).

	Group 1 ($n_1 = 10$)	Group 2 ($n_2 = 12$)	Total sample ($N = 22$)
	16	4	
	18	10	
	6	9	
	12	13	
	11	11	
	12	9	
	23	13	
	19	9	
	7	5	
	11	8	
		7	
		10	
$\sum_{\text{gr } j} Y$	135	108	243 $= \Sigma Y$
$\overline{Y}_j$	13.5	9.0	11.05 $= \overline{Y}$
$\sum_{\text{gr } j} Y^2$	2085	1056	3141 $= \Sigma Y^2$

We want to test the null hypothesis that $\mu_1 = \mu_2$. The standard approach to doing this is, of course, to carry out a t-test for two independent means. The intermediate quantities needed for this purpose are shown at the foot of the data table. The next step is to compute the within-groups variance estimate, s_w^2:

$$s_w^2 = \frac{\left[\sum_{\text{gr 1}} Y^2 - \left(\sum_{\text{gr 1}} Y\right)^2 / n_1\right] + \left[\sum_{\text{gr 2}} Y^2 - \left(\sum_{\text{gr 2}} Y\right)^2 / n_2\right]}{n_1 + n_2 - 2}$$

$$= \frac{[2085 - (135)^2/10] + [1056 - (108)^2/12]}{20}$$

$$= 17.325$$

Then the required t-ratio is found as

$$t = \frac{\overline{Y}_1 - \overline{Y}_2}{\sqrt{s_w^2(1/n_1 + 1/n_2)}}$$

$$= \frac{13.5 - 9.0}{\sqrt{17.325(\frac{1}{10} + \frac{1}{12})}}$$

$$= 2.5250$$

with $n_1 + n_2 - 2 = 20$ degrees of freedom. (Normally, we would not calculate t to so many decimal places, but here we deliberately carry four places to show the extent of agreement with the result to be obtained later.) The difference, 4.5, between the two Y means is found to be significant at the 5% level, since $t_{20;.95} = 1.725$.

Now let us use our rudimentary linear model approach by computing the regression coefficient b in the regression equation for "predicting" Y from the group-membership variable X, and test it for significance. Although it was already shown in Eq. 9.4 that

$$b = \overline{Y}_1 - \overline{Y}_2$$

we shall compute b from its definitional formula to verify this result numerically. The various sums needed for using Eq. 9.2 are obtained as follows:

$$\Sigma X = \underbrace{(1 + 1 + \cdots + 1)}_{\text{10 times}} + \underbrace{(0 + 0 + \cdots + 0)}_{\text{12 times}} = 10$$

Similarly,

$$\Sigma X^2 = 10$$

and

$$\Sigma XY = (1)(16) + (1)(18) + \cdots + (1)(11) + (0)(4) + (0)(10) + \cdots + (0)(10) = 135$$

which is $\sum_{\text{gr 1}} Y$, as shown at the foot of the data table. Finally,

$$\Sigma Y = 135 + 108 = 243$$

Substituting these numerical values (and $N = 22$) in Eq. 9.2, we get

$$b = \frac{(22)(135) - (10)(243)}{(22)(10) - (10)^2}$$

$$= \frac{540}{120} = 4.5$$

which, of course, agrees with the value of $\overline{Y}_1 - \overline{Y}_2$.

We now wish to test whether b is significantly different from zero. Probably the reader is more familiar with the test of significance of the *correlation* coefficient than that of the *regression* coefficient. Since these two coefficients are related by the equation

$$r = b\sqrt{\Sigma x^2}/\sqrt{\Sigma y^2}$$

(where $x = X - \overline{X}$ and $y = Y - \overline{Y}$), the two significance tests are equivalent; that is, if r is significantly different from zero, then so is b, and vice versa. (See Section 9.7 for details.)

Substituting the values already found above in the formulas

$$\Sigma x^2 = \Sigma X^2 - (\Sigma X)^2/N$$

and

$$\Sigma y^2 = \Sigma Y^2 - (\Sigma Y)^2/N$$

we find

$$\Sigma x^2 = 10 - (10)^2/22 = 5.4545$$

and

$$\Sigma y^2 - 3141 - (243)^2/22 = 456.955$$

(The value of ΣY^2 comes from the lower-right hand corner of the data table.) Therefore, the correlation coefficient has the value

$$r = (4.5)\sqrt{5.4545}/\sqrt{456.955} = .4917$$

Now the formula for the t-statistic for testing the significance of a correlation coefficient is

$$t = \frac{r\sqrt{N - 2}}{\sqrt{1 - r^2}} \qquad \text{with d.f.} = N - 2\ (= 20 \text{ in this case})$$

Substituting the value of r just obtained in this formula, we obtain

$$t = \frac{(.4917)\sqrt{20}}{\sqrt{1 - (.4917)^2}} = 2.5253$$

This value is clearly equal, within rounding error, to the value 2.5250 computed earlier from the formula for the t-statistic for testing the significance of the difference between two independent means. We have thus verified numerically, not only that

$$b = \overline{Y}_1 - \overline{Y}_2$$

but also that the t-tests for the significance of b and of $\overline{Y}_1 - \overline{Y}_2$ by the respective formulas (which are formally quite different) yield identical results.

The demonstration just completed should convince even the most skeptical reader that, at least in the two-group case, a test of the significance of the difference between means can be equivalently carried out by a linear regression model approach.

9.2 THE TWO-GROUP CASE REVISITED: STANDARD APPROACH

One problem with the rudimentary approach taken in Section 9.1 is that it does not readily generalize to multigroup cases, much less to higher-order designs. In contrast, the standard approach, although more complicated at first glance, has the advantage of being formally generalizable to any ANOVA design.

The starting point for the standard approach is the linear structural model for the design. The simplest form in which this can be written for the present example is

$$Y_{ij} = u_j + \epsilon_{ij} \qquad (i = 1, 2, \ldots, n_j; \quad j = 1, 2)$$

which simply states that an individual's Y-score is the sum of the mean (μ_j) of the population to which he belongs and an error term peculiar to the individual—certainly an intuitively appealing statement. Unfortunately, however, this simple form of the structural equation does not explicitly specify the type of design involved (whether it is a one-way or a higher-order design, etc.). To make this explicit, it is necessary to introduce the *effect parameters* α_j, defined as the deviation of the jth population mean from the overall mean; that is,

$$\alpha_j = \mu_j - \mu$$

Upon expressing μ_j, from this definition, as $\mu + \alpha_j$, the structural equation can be rewritten as

$$Y_{ij} = \mu + \alpha_j + \epsilon_{ij} \qquad (i = 1, 2, \ldots, n_j; \quad j = 1, 2) \qquad (9.5)$$

This is the form that we shall use in the sequel.

Let us now denote the sum of the first two terms in the right-hand side of Eq. 9.5 by $\hat{Y}_{ij}$, omitting the error term ϵ_{ij}:

$$\hat{Y}_{ij} = \mu + \alpha_j \qquad (i = 1, 2, \ldots, n_j;\; j = 1, 2) \tag{9.6}$$

This equation may be rewritten as a single equation for each person in the total sample, numbered consecutively from 1 through N ($= n_1 + n_2$), instead of using a two-phase enumeration—1 through n_1 for group 1 members and 1 through n_2 for group 2 members—by the following device: We introduce *two* dummy variates X_1, X_2, indicating group membership; that is,

$$X_1 = \begin{cases} 1 & \text{for group 1 members} \\ 0 & \text{for group 2 members} \end{cases}$$

$$X_2 = \begin{cases} 0 & \text{for group 1 members} \\ 1 & \text{for group 2 members} \end{cases}$$

Equation 9.6 then becomes

$$\hat{Y}_i = \mu + \alpha_1 X_{1i} + \alpha_2 X_{2i} \qquad (i = 1, 2, \ldots, N) \tag{9.7}$$

as the reader may easily verify. (Multiplying each of the effect parameters α_1 and α_2 by the appropriate one of the dummy variates X_1 and X_2, respectively, results in automatically selecting the correct one of the two α_j's, depending on which group a particular individual belongs to.)

The reader will recognize that Eq. 9.7 has the form of a *multiple regression equation* of Y on two "predictors," X_1 and X_2. (This is in contrast to our rudimentary approach in Section 9.1, which involved only a simple regression equation of Y on a single "predictor," X.) The reason α_1 and α_2 are used instead of the customary b_1 and b_2 (or β_1 and β_2) for the regression weights is that we need to reserve b_j for weights associated with covariates and β_j for the effects of the second factor in a two-way (or higher) design. Estimation of the parameters μ, α_1, α_2 is again done by invoking the principle of least squares—that is, minimizing the sum of squared discrepancies

$$Q = \Sigma\,(Y_i - \hat{Y}_i)^2$$

between actual and predicted Y scores. A comparison of Eqs. 9.5 and 9.6 shows that an alternative expression for Q is

$$Q = \sum_{j=1}^{2} \sum_{i=1}^{n_j} \epsilon_{ij}^2$$

The equations from which the estimates m, a_1, and a_2 of μ, α_1, and α_2 are obtained are, of course, the normal equations that were derived in Chapter 3 and given in Eq. 3.10. In the present notation, they become

$$
\begin{aligned}
Nm \quad &+ (\Sigma\, X_{1i})a_1 &&+ (\Sigma\, X_{2i})a_2 &&= \Sigma\, Y_i \\
(\Sigma\, X_{1i})m &+ (\Sigma\, X_{1i}^2)a_1 &&+ (\Sigma\, X_{1i}X_{2i})a_2 &&= \Sigma\, X_{1i}Y_i \\
(\Sigma\, X_{2i})m &+ (\Sigma\, X_{2i}X_{1i})a_1 &&+ (\Sigma\, X_{2i}^2)a_2 &&= \Sigma\, X_{2i}Y_i
\end{aligned}
$$

From the definitions of X_1 and X_2 given above, all the sums except $\Sigma\, Y_i$ in this set of equations simplify as follows:

$$
\begin{aligned}
&\Sigma\, X_{1i} = n_1 \qquad \Sigma\, X_{2i} = n_2 \\
&\Sigma\, X_{1i}^2 = n_1 \qquad \Sigma\, X_{2i}^2 = n_2 \\
&\Sigma\, X_{1i}Y_i = \sum_{\text{gr 1}} Y_i \qquad \text{(i.e., the } Y \text{ total for group 1)} \\
&\Sigma\, X_{1i}Y_i = \sum_{\text{gr 2}} Y_i \qquad \text{(i.e., the } Y \text{ total for group 2)}
\end{aligned}
$$

(These come about in exactly the same way as $\Sigma\, X$, $\Sigma\, X^2$, and $\Sigma\, XY$ were simplified in our previous approach using only one dummy variate X.) Finally, it can be seen that

$$\Sigma\, X_{1i}X_{2i} = 0$$

since for *every* individual in the total sample, either X_1 or X_2 must be zero, and hence the product X_1X_2 vanishes for everyone.

Substituting these special values for the various sums in our normal equations, and using the abbreviations T_1, T_2 and T for the Y totals in group 1, group 2, and the total sample, respectively, the equations simplify to

$$
\begin{aligned}
Nm + n_1a_1 + n_2a_2 &= T \\
n_1m + n_1a_1 \qquad\quad &= T_1 \\
n_2m \qquad\quad + n_2a_2 &= T_2
\end{aligned}
\tag{9.8}
$$

These, then, are the normal equations for the special multiple regression problem into which the two-group analysis-of-variance problem (or the t-test problem) was recast by the standard approach of using two instead of one dummy variate. (In general, as many dummy variates would be used as there are groups.)

At first glance, it appears that there are three equations in three unknowns, and hence that these equations may be solved for the unknowns m, a_1, a_2 by any of several methods for solving a set of simultaneous linear equations. Upon closer scrutiny, however, we find an anomaly in this set of equations. Namely, when we add the respective members of the last two equations, we get

$$(n_1 + n_2)m + n_1a_1 + n_2a_2 = T_1 + T_2$$

or, since

$$n_1 + n_2 = N \quad \text{and} \quad T_1 + T_2 = T$$

$$Nm + n_1a_1 + n_2a_2 = T$$

which is none other than the first equation of the set! In other words, what appeared to be a set of three equations actually contains only two *independent* equations; the first is merely the sum of the other two equations and hence adds no new constraints on the unknowns. Effectively, we have only two equations in three unknowns. Thus no unique solution exists. So what do we do?

Statisticians have developed several alternative procedures for coping with this anomaly (technically known as "deficient rank") in the set of normal equations for the special multiple regression problem arising from the general linear-model approach to the analysis of variance. Perhaps the simplest procedure, and the one we shall adopt here, is to introduce a new constraint on the parameters. In fact, at least in the case of *fixed-effects* models, a natural constraint implicitly exists between α_1 and α_2. Recall that

$$\alpha_1 = \mu_1 - \mu$$

and

$$\alpha_2 = \mu_2 - \mu$$

Hence

$$\alpha_1 + \alpha_2 + \mu_1 + \mu_2 - 2\mu$$

But if the levels of the factor (the treatment or classification variable) are "fixed"—that is, when the levels represented in the experiment exhaust the entire set of levels that we are interested in—it follows that[2]

$$\mu = \frac{\mu_1 + \mu_2}{2}$$

and hence that

$$\alpha_1 + \alpha_2 = \mu_1 + \mu_2 - 2\left(\frac{\mu_1 + \mu_2}{2}\right) = 0$$

By discarding the first equation from the set (7.8) and adding the constraint $a_1 + a_2 = 0$ (corresponding to the constraint $\alpha_1 + \alpha_2 = 0$ in the population) to the system instead, we have a new set of equations,

[2]If the factor is a classification variable (e.g., right-handed versus left-handed people) and the proportions of individuals in the two categories are unequal, say π_1 and π_2 (where $\pi_1 + \pi_2 = 1$), the overall population mean would be a weighted mean of the two subpopulation means:

$$\mu = \pi_1\mu_1 + \pi_2\mu_2$$

Hence the natural constraint on the α_j's would be

$$\pi_1\alpha_1 + \pi_2\alpha_2 = 0$$

instead of

$$\alpha_1 + \alpha_2 = 0$$

$$\begin{aligned} n_1m + n_1a_1 \qquad\qquad &= T_1 \\ n_2m \qquad\quad + n_2a_2 &= T_2 \\ a_1 + \quad a_2 &= 0 \end{aligned} \tag{9.9}$$

This set does not suffer from the deficient rank problem, and we have three independent equations in three unknowns, and hence can get a unique solution for m, a_1, and a_2, respectively. They turn out to be

$$m = (\bar{Y}_1 + \bar{Y}_2)/2$$

$$a_1 = (\bar{Y}_1 - \bar{Y}_2)/2$$

$$a_2 = (\bar{Y}_2 - \bar{Y}_1)/2$$

Thus we see again that tests of the significance of a_1 and a_2 amount to a test of the significance of the difference between $\bar{Y}_1$ and $\bar{Y}_2$, a conclusion we already arrived at using the rudimentary approach in Section 9.1.

The numerical values of the estimates for Example 9.1 are

$$m = (13.5 + 9.0)/2 = 11.25$$

$$a_1 = (13.5 - 9.0)/2 = 2.25$$

$$a_2 = (9.0 - 13.5)/2 = -2.25$$

(Note that m is not equal to the grand mean of the total sample, which is 11.05, but is the *unweighted mean* of the two group means. When the sizes of the two groups are equal, the unweighted mean will of course coincide with the grand mean.) The reader who has difficulty solving Eqs. 9.9 in literal form may make numerical substitutions for the coefficients and constant terms to obtain

$$\begin{aligned} 10m + 10a_1 \qquad\qquad &= 135 \\ 12m \qquad\quad + 12a_2 &= 108 \\ a_1 + \quad a_2 &= 0 \end{aligned}$$

and solve this specific set of equations, or at least verify that the values given above for m, a_1, and a_2 do satisfy them.

9.3 MATRIX FORMULATION: PARAMETER ESTIMATION

It was mentioned at the outset of Section 9.2 that the standard approach, although more complicated than our rudimentary method, is more readily generalizable to multigroup and factorial designs. To show this it is necessary to express Eq.

9.5 in matrix notation. This is done by defining the N-dimensional vector of Y-scores, called the *observation vector,*

$$\mathbf{Y} = \begin{bmatrix} Y_{11} \\ Y_{21} \\ \vdots \\ Y_{n_1 1} \\ Y_{12} \\ \vdots \\ Y_{n_2 2} \end{bmatrix}$$

and equating it to the product of an $N \times 3$ *design matrix* (so called because it specifies the ANOVA design involved)

$$\mathbf{X} = \begin{bmatrix} 1 & 1 & 0 \\ 1 & 1 & 0 \\ \vdots & \vdots & \vdots \\ 1 & 1 & 0 \\ 1 & 0 & 1 \\ \vdots & \vdots & \vdots \\ 1 & 0 & 1 \end{bmatrix}$$

times the three-dimensional *parameter vector*

$$\boldsymbol{\Theta} = \begin{bmatrix} \mu \\ \alpha_1 \\ \alpha_2 \end{bmatrix}$$

plus the N-dimensional *error vector*

$$\boldsymbol{\varepsilon} = \begin{bmatrix} \epsilon_{11} \\ \epsilon_{21} \\ \vdots \\ \epsilon_{n_1 1} \\ \epsilon_{12} \\ \vdots \\ \epsilon_{n_2 2} \end{bmatrix}$$

Thus, in extended matrix notation, the N equations (one for each subject in the total sample) represented by Eq. 9.5 assume the form

$$\begin{bmatrix} Y_{11} \\ Y_{21} \\ \vdots \\ Y_{n_1 1} \\ Y_{12} \\ \vdots \\ Y_{n_2 2} \end{bmatrix} = \begin{bmatrix} 1 & 1 & 0 \\ 1 & 1 & 0 \\ \vdots & \vdots & \vdots \\ 1 & 1 & 0 \\ 1 & 0 & 1 \\ \vdots & \vdots & \vdots \\ 1 & 0 & 1 \end{bmatrix} \begin{bmatrix} \mu \\ \alpha_1 \\ \alpha_2 \end{bmatrix} + \begin{bmatrix} \epsilon_{11} \\ \epsilon_{21} \\ \vdots \\ \epsilon_{n_1 1} \\ \epsilon_{12} \\ \vdots \\ \epsilon_{n_2 2} \end{bmatrix} \tag{9.10}$$

$$\mathbf{Y} \qquad\qquad \mathbf{X} \qquad\qquad \boldsymbol{\Theta} \qquad\qquad \boldsymbol{\varepsilon}$$

The design matrix $\mathbf{X}$ always contains N 1's in the first column, and for this example, n_1 1's followed by n_2 0's in the second column, and n_1 0's followed by n_2 1's in the third. Thus, by carrying out the indicated matrix multiplication $\mathbf{X}\boldsymbol{\Theta}$ and adding the vector $\boldsymbol{\varepsilon}$ to the product (which is an N-dimensional column vector), it is readily seen that the first n_1 elements of the vector resulting from the operations indicated on the right-hand side of this equation are equal to

$$\mu + \alpha_1 + \epsilon_{i1} \qquad (i = 1, 2, \ldots, n_1)$$

while the last n_2 elements are

$$\mu + \alpha_2 + \epsilon_{i2} \qquad (i = 1, 2, \ldots, n_2)$$

Hence the single matrix equation (9.10) stands for the N equations

$$Y_{i1} = \mu + \alpha_1 + \epsilon_{i1} \qquad (i = 1, 2, \ldots, n_1)$$

$$Y_{i2} = \mu + \alpha_2 + \epsilon_{i2} \qquad (i = 1, 2, \ldots, n_2)$$

represented by Eq. 9.5.

Written in symbolic matrix form, Eq. 9.10 assumes the compact form

$$\mathbf{Y} = \mathbf{X}\boldsymbol{\Theta} + \boldsymbol{\varepsilon} \tag{9.11}$$

It is this form that is customarily known as the general linear model and which is readily generalizable to analyses of variance and covariance of any design. Of course, the design matrix $\mathbf{X}$ and the parameter vector $\boldsymbol{\Theta}$ will be much more complex for complicated designs. (In particular, for analysis of covariance, $\mathbf{X}$ will not consist entirely of 1's and 0's, but will have one or more columns listing the N subjects' actual scores on the covariates.) But the point is that all manners of design are *formally* represented by the same equation, (9.11). Thus once the solutions (parameter estimation and significance testing) have been developed for this general equation, they are applicable to any problem in the rubric of analyses of variance and covariance. Herein lies the appeal of the general linear model approach.

The solution of the parameter-estimation phase invokes the principle of least squares, which, as stated on page 302, calls for minimizing the quantity

$$Q = \sum_{j=1}^{2} \sum_{i=1}^{n_j} \epsilon_{ij}^2$$

In our present matrix notation, this becomes

$$\begin{aligned} Q &= \boldsymbol{\varepsilon}'\boldsymbol{\varepsilon} \\ &= (\mathbf{Y} - \mathbf{X\Theta})'(\mathbf{Y} - \mathbf{X\Theta}) \end{aligned}$$

since it follows from Eq. 9.11 that

$$\boldsymbol{\varepsilon} = \mathbf{Y} - \mathbf{X\Theta}$$

The symbolic vector derivative of Q with respect to $\mathbf{\Theta}$ (see Appendix C) is then computed, and the result set equal to a null vector $\mathbf{O}$, to get the normal equations in matrix form. The reader may verify, by carrying out the indicated matrix multiplications below, that the normal equations (9.8) assume the form

$$(\mathbf{X'X})\hat{\mathbf{\Theta}} = \mathbf{X'Y} \qquad (9.12)$$

where $\hat{\mathbf{\Theta}}$ is the estimated parameter vector

$$\begin{bmatrix} m \\ a_1 \\ a_2 \end{bmatrix}$$

Now, if $\mathbf{X'X}$ were a nonsingular matrix (as it is in the case of ordinary multiple regression problems) it would have an inverse $(\mathbf{X'X})^{-1}$, so Eq. 9.12 could be immediately solved by premultiplying both sides by $(\mathbf{X'X})^{-1}$ to get

$$\hat{\mathbf{\Theta}} = (\mathbf{X'X})^{-1}(\mathbf{X'Y})$$

But in fact $\mathbf{X'X}$ *is* singular:

$$\mathbf{X'X} = \begin{bmatrix} 1 & 1 & \cdots & 1 & 1 & \cdots & 1 \\ 1 & 1 & \cdots & 1 & 0 & \cdots & 0 \\ 0 & 0 & \cdots & 0 & 1 & \cdots & 1 \end{bmatrix} \begin{bmatrix} 1 & 1 & 0 \\ 1 & 1 & 0 \\ \vdots & \vdots & \vdots \\ 1 & 1 & 0 \\ 1 & 0 & 1 \\ \vdots & \vdots & \vdots \\ 1 & 0 & 1 \end{bmatrix} = \begin{bmatrix} N & n_1 & n_2 \\ n_1 & n_1 & 0 \\ n_2 & 0 & n_2 \end{bmatrix}$$

whose first row is the sum of the other two rows. (This is, of course, the same thing as the first of Eqs. 9.8 being the sum of the other two equations in the set—an anomaly which we noted earlier.) Therefore, Eq. 9.12 cannot be solved in this simple manner; in fact, no unique solution exists.

In Section 9.2 the problem of having only two independent equations in a set of three was resolved by the expedient of adding a constraint on the parameters (i.e., an extra equation was added to the set). Another, and in many ways a more desirable way to solve the problem is to carry out a *reparameterization*. That is, instead of trying to estimate the original three parameters μ, α_1, α_2 themselves, we choose to estimate some two linear combinations of the parameters. In general, the maximum number of linear combinations that can be estimated is equal to the rank of the matrix $\mathbf{X}'\mathbf{X}$—that is, the number of linearly independent rows (or columns) it has, which is the same as the number of independent equations in the set to which Eqs. 9.8 generalizes. (In general, this is equal to the number of groups, or cells, in the design.)

There are, of course, an infinite number of ways in which we could choose the linear combinations of the parameters to estimate, but we would be guided, for one thing, by the meaningfulness of the chosen linear combinations. For example, we might be interested in estimating an "overall level" parameter, $\lambda = \mu + (\alpha_1 + \alpha_2)/2$, and the difference between the two effect parameters, $\delta = \alpha_1 - \alpha_2$. These are clearly linear combinations of the original parameters; that is,

$$\lambda = 1 \cdot \mu + \tfrac{1}{2} \cdot \alpha_1 + \tfrac{1}{2} \cdot \alpha_2$$

$$\delta = 0 \cdot \mu + 1 \cdot \alpha_1 + (-1) \cdot \alpha_2$$

In matrix notation, these equations become

$$\begin{bmatrix} \lambda \\ \delta \end{bmatrix} = \begin{bmatrix} 1 & \frac{1}{2} & \frac{1}{2} \\ 0 & 1 & -1 \end{bmatrix} \begin{bmatrix} \mu \\ \alpha_1 \\ \alpha_2 \end{bmatrix}$$

The matrix containing the coefficients (or combining weights) that define the new parameters as linear combinations of the original ones is called the *contrast matrix*, because usually all but the first of the combinations are of a special type known as *contrasts*.[3] If we denote the contrast matrix by $\mathbf{C}$ and the vector of new parameters by $\boldsymbol{\Theta}^*$, the transformation equation may be symbolized as

$$\boldsymbol{\Theta}^* = \mathbf{C}\boldsymbol{\Theta} \qquad (9.13)$$

[3] A contrast is a special type of linear combination with the property that the weights sum to zero. Thus $2a_1 - a_2 - a_3$ and $3a_1 + 3a_2 - 2a_3 - 2a_4 - 2a_5$ are contrasts. A difference $a_1 - a_2$ is the simplest example of a contrast. Contrasts in general may, therefore, be regarded as generalizations of a difference.

Once we have appropriately defined the new set of parameters, we may rewrite the structual equation (i.e., the linear model), Eq. 9.11, in terms of these new parameters as

$$\mathbf{Y} = \mathbf{K}\boldsymbol{\Theta}^* + \boldsymbol{\varepsilon} \tag{9.14}$$

with a new matrix **K**, called the *basis* (of the design) in place of the design matrix **X**. How is **K** determined? It turns out that[4]

$$\mathbf{K} = (\mathbf{XC}')(\mathbf{CC}')^{-1} \tag{9.15}$$

Next, the normal equations, from which the least-squares estimates of the new parameters are to be found, may be written in exact analogy with Eq. 9.12. We need only replace the design matrix **X** by the basis matrix **K**, and the original parameter vector $\boldsymbol{\Theta}$ by the new parameter vector $\boldsymbol{\Theta}^*$ and its estimate $\hat{\boldsymbol{\Theta}}^*$. We then get

$$(\mathbf{K}'\mathbf{K})\hat{\boldsymbol{\Theta}}^* = \mathbf{K}'\mathbf{Y} \tag{9.16}$$

The difference between this equation and Eq. 9.12, which it replaces in the reparameterized system, is that whereas $\mathbf{X}'\mathbf{X}$ was singular, $\mathbf{K}'\mathbf{K}$ is nonsingular by virtue of the way in which the constrast matrix **C** is defined (i.e., so that it will have rank equal to its number of rows. (One further condition that **C** must satisfy will be pointed out later.) Thus we are assured that a unique solution to Eq. 9.16 exists, in the form

$$\hat{\boldsymbol{\Theta}}^* = (\mathbf{K}'\mathbf{K})^{-1}(\mathbf{K}'\mathbf{Y}) \tag{9.17}$$

This, of course, was the purpose of reparameterizing.

At this point the reader may get an uneasy feeling that we have strayed away from the original problem. We started out by intending to estimate the three parameters, μ, α_1, and α_2 (more particularly, the last two) because testing the significance of α_1 and α_2 was found to be equivalent to testing the significance of $\bar{Y}_1 - \bar{Y}_2$ (refer back to page 305, solution of Eqs. 9.9)—and testing the significance of differences among means is, of course, the basic purpose of ANOVA. We invoked the least-squares principle to obtain the normal equations

[4]For those with more than a passing familiarity with matrix algebra, this relation results from the fact that

$$\begin{aligned}\mathbf{K}\boldsymbol{\Theta}^* &= \mathbf{K}(\mathbf{C}\boldsymbol{\Theta})\\ &= (\mathbf{KC})\boldsymbol{\Theta}\end{aligned}$$

whence, comparing with Eq. 9.11, it must be the case that

$$\mathbf{KC} = \mathbf{X}$$

Then

$$\mathbf{K}(\mathbf{CC}') = \mathbf{XC}'$$

and since $\mathbf{CC}'$ is nonsingular,

$$\mathbf{K} = (\mathbf{XC}')(\mathbf{CC}')^{-1}$$

(the earlier Eqs. 9.8, converted into Eq. 9.12 in matrix notation), purportedly to solve for estimates of μ, α_1, and α_2. But we found these equations to be *insoluble* because of the deficient-rank problem—the matrix counterpart of the problem of having only two independent equations in the set of three equations, (9.8). So we resorted to the "gimmick" of reparameterizing—essentially, electing to estimate a new set of parameters (defined as linear combinations of the original set) that *can* be estimated. Sure enough, this device leads to a new set of normal equations, (9.17), for the new parameters, that does have a unique solution—provided that certain rules are observed in defining the new parameters. But what good is it to have this solution—estimates of the new parameters—if it does not bear on our original problem? The answer is that the solution does indeed bear on the original purpose of significance testing.

Consider the second of the new parameters, $\delta = \alpha_1 - \alpha_2$, chosen above. Recalling that $\alpha_1 = \mu_1 - \mu$ and $\alpha_2 = \mu_2 - \mu$, it follows that

$$\delta = (\mu_1 - \mu) - (\mu_2 - \mu) = \mu_1 - \mu_2$$

Thus, estimating δ does in fact relate very directly to our purpose of testing the significance of the difference between $\overline{Y}_1$ and $\overline{Y}_2$. In multigroup designs with J groups, the new parameter $\delta = \alpha_1 - \alpha_2$ generalizes to a set of new parameters

$$\begin{aligned} \delta_1 &= \alpha_1 - \alpha_J \\ \delta_2 &= \alpha_2 - \alpha_J \\ &\vdots \\ \delta_{j-1} &= \alpha_{J-1} - \alpha_J \end{aligned}$$

each of which is equal to the difference between one of the first $J - 1$ population means and the last (Jth) population mean.

The foregoing discussion should not only allay the reader's apprehension that the process of reparameterization may lead us astray from our original problem. It should also convince the reader that the requirement (see p. 310, discussion following Eq. 9.16) of choosing a contrast matrix $\mathbf{C}$ "so that it will have rank equal to its number of rows" is not difficult to meet in practice. One may always satisfy this condition by taking the $J \times (J + 1)$ matrix

$$\mathbf{C} = \begin{bmatrix} 1 & 1/J & 1/J & \cdots & 1/J & 1/J \\ 0 & 1 & 0 & \cdots & 0 & -1 \\ 0 & 0 & 1 & \cdots & 0 & -1 \\ \vdots & & & & & \\ 0 & 0 & 0 & \cdots & 1 & -1 \end{bmatrix}$$

as the contrast matrix. It is only when the researcher is specifically interested in more elaborate contrasts than the simple differences of each of the first $J - 1$ means and the Jth that he or she needs to make sure that the resulting contrast matrix has full row rank (i.e., that all of its rows are linearly independent, or that *none* of its rows is a linear combination of the other rows). And even then, as long as the researcher is sure that he or she is not asking redundant questions (i.e., a set of questions such that the answers to some of them would logically imply the answer to the others), the mathematical requirements for **C** will automatically be satisfied—subject only to one further requirement (also easily met) that will be discussed later.

Finally, it should be reiterated that the purpose of recasting ANOVA into a regression approach is to achieve generality. In the simple, two-group example which we have been carrying for illustrative purposes, nothing is gained by using this more complicated (for this example) approach. The point is that the same approach—solving Eq. 9.16 in the form of Eq. 9.17—suffices for *all* ANOVA (and ANCOVA) designs, thus eliminating the need for special computational formulas for different designs. This is the advantage of the general linear model.

EXAMPLE 9.2 Let us now apply the foregoing developments to the numerical example given in Section 9.2. The design matrix, shown at the beginning of this section, is

$$\mathbf{X} = \begin{bmatrix} 1 & 1 & 0 \\ 1 & 1 & 0 \\ \vdots & \vdots & \vdots \\ 1 & 1 & 0 \\ 1 & 0 & 1 \\ \vdots & \vdots & \vdots \\ 1 & 0 & 1 \end{bmatrix} \begin{matrix} \left.\vphantom{\begin{matrix}1\\1\\\vdots\\1\end{matrix}}\right\} n_1 \ (= 10) \text{ rows} \\ \left.\vphantom{\begin{matrix}1\\\vdots\\1\end{matrix}}\right\} n_2 \ (= 12) \text{ rows} \end{matrix}$$

For reparameterizing to $\lambda = \mu + (\alpha_1 + \alpha_2)/2$ and $\delta = \alpha_1 - \alpha_2$, the contrast matrix is

$$\mathbf{C} = \begin{bmatrix} 1 & \frac{1}{2} & \frac{1}{2} \\ 0 & 1 & -1 \end{bmatrix}$$

In order to determine the basis matrix **K** in accordance with Eq. 9.15, we need to get the matrix products $\mathbf{XC}'$ and $\mathbf{CC}'$ (and then the inverse of the latter). The results are

$$\mathbf{XC}' = \begin{bmatrix} 1 & 1 & 0 \\ 1 & 1 & 0 \\ \vdots & \vdots & \vdots \\ 1 & 1 & 0 \\ 1 & 0 & 1 \\ \vdots & \vdots & \vdots \\ 1 & 0 & 1 \end{bmatrix} \begin{bmatrix} 1 & 0 \\ \frac{1}{2} & 1 \\ \frac{1}{2} & -1 \end{bmatrix} = \begin{bmatrix} \frac{3}{2} & 1 \\ \frac{3}{2} & 1 \\ \vdots & \vdots \\ \frac{3}{2} & 1 \\ \frac{3}{2} & -1 \\ \vdots & \vdots \\ \frac{3}{2} & -1 \end{bmatrix}$$

and

$$\mathbf{CC}' = \begin{bmatrix} 1 & \frac{1}{2} & \frac{1}{2} \\ 0 & 1 & -1 \end{bmatrix} \begin{bmatrix} 1 & 0 \\ \frac{1}{2} & 1 \\ \frac{1}{2} & -1 \end{bmatrix} = \begin{bmatrix} \frac{3}{2} & 0 \\ 0 & 2 \end{bmatrix}$$

whence

$$(\mathbf{CC}')^{-1} = \begin{bmatrix} \frac{2}{3} & 0 \\ 0 & \frac{1}{2} \end{bmatrix}$$

(since the inverse of a diagonal matrix is simply the diagonal matrix whose diagonal elements are the reciprocals of those of the original matrix).

Thus, from Eq. 9.17, we obtain the basis matrix

$$\mathbf{K} = (\mathbf{XC}')(\mathbf{CC}')^{-1} = \begin{bmatrix} \frac{3}{2} & 1 \\ \vdots & \vdots \\ \frac{3}{2} & 1 \\ \frac{3}{2} & -1 \\ \vdots & \vdots \\ \frac{3}{2} & -1 \end{bmatrix} \begin{bmatrix} \frac{2}{3} & 0 \\ 0 & \frac{1}{2} \end{bmatrix} = \begin{bmatrix} 1 & \frac{1}{2} \\ 1 & \frac{1}{2} \\ \vdots & \vdots \\ 1 & \frac{1}{2} \\ 1 & -\frac{1}{2} \\ \vdots & \vdots \\ 1 & -\frac{1}{2} \end{bmatrix}$$

The reparameterized structural equations are, from Eq. 9.14,

$$Y_{i1} = \lambda + \delta/2 + \epsilon_{i1} \qquad (i = 1, 2, \ldots, n_1)$$

$$Y_{i2} = \lambda - \delta/2 + \epsilon_{i2} \qquad (i = 1, 2, \ldots, n_2)$$

The reader should verify that when the new parameters are expressed in terms of the original ones, these revert back to Eq. 9.5, that is,

$$Y_{i1} = \mu + \alpha_1 + \epsilon_{i1}$$

$$Y_{i2} = \mu + \alpha_2 + \epsilon_{i2}$$

Thus, nothing has in fact been changed by the process of reparameterization (which is as it should be); the process is merely a mathematical device to ensure that the normal equations will be uniquely solvable.

We next need to compute **K′K** and **K′Y** in order to write and solve the normal equations:

$$\mathbf{K'K} = \begin{bmatrix} 1 & 1 & \cdots & 1 & 1 & \cdots & 1 \\ \frac{1}{2} & \frac{1}{2} & \cdots & \frac{1}{2} & -\frac{1}{2} & \cdots & -\frac{1}{2} \end{bmatrix} \begin{bmatrix} 1 & \frac{1}{2} \\ 1 & \frac{1}{2} \\ \vdots & \vdots \\ 1 & \frac{1}{2} \\ 1 & -\frac{1}{2} \\ \vdots & \vdots \\ 1 & -\frac{1}{2} \end{bmatrix} = \begin{bmatrix} N & \dfrac{n_1 - n_2}{2} \\ \dfrac{n_1 - n_2}{2} & \dfrac{N}{4} \end{bmatrix}$$

or, for our numerical example with $n_1 = 10$, $n_2 = 12$, and $N = 22$,

$$\mathbf{K'K} = \begin{bmatrix} 22 & -1 \\ -1 & \frac{11}{2} \end{bmatrix}$$

$$\mathbf{K'Y} = \begin{bmatrix} 1 & 1 & \cdots & 1 & 1 & \cdots & 1 \\ \frac{1}{2} & \frac{1}{2} & \cdots & \frac{1}{2} & -\frac{1}{2} & \cdots & -\frac{1}{2} \end{bmatrix} \begin{bmatrix} Y_{11} \\ Y_{21} \\ \vdots \\ Y_{n_1 1} \\ Y_{12} \\ \vdots \\ Y_{n_2 2} \end{bmatrix} = \begin{bmatrix} T \\ \dfrac{T_1 - T_2}{2} \end{bmatrix}$$

where T_1 (= 135) and T_2 (= 108) are the two group totals of Y, and T (= 243) is the grand total. Thus the normal equations, in matrix form, are

$$\begin{bmatrix} N & \dfrac{n_1 - n_2}{2} \\ \dfrac{n_1 - n_2}{2} & \dfrac{N}{4} \end{bmatrix} \begin{bmatrix} \lambda \\ \delta \end{bmatrix} = \begin{bmatrix} T \\ \dfrac{T_1 - T_2}{2} \end{bmatrix}$$

or, numerically,

$$\begin{bmatrix} 22 & -1 \\ -1 & \frac{11}{2} \end{bmatrix} \begin{bmatrix} \lambda \\ \delta \end{bmatrix} = \begin{bmatrix} 243 \\ \frac{27}{2} \end{bmatrix}$$

Written out in ordinary algebraic notation, this becomes

$$22\lambda - \delta = 243$$

$$-\lambda + 5.5\delta = 13.5$$

which may be solved by any method for solving a set of simultaneous linear equations to get $\lambda = 11.25$ and $\delta = 4.5$. We shall, however, illustrate the matrix solution given in Eq. 9.17. First, we find

$$(\mathbf{K}'\mathbf{K})^{-1} = \frac{1}{120} \begin{bmatrix} \frac{11}{2} & 1 \\ 1 & 22 \end{bmatrix}$$

whence

$$\hat{\mathbf{\Theta}}^* = \begin{bmatrix} \hat{\lambda} \\ \hat{\delta} \end{bmatrix} = \frac{1}{120} \begin{bmatrix} \frac{11}{2} & 1 \\ 1 & 22 \end{bmatrix} \begin{bmatrix} 243 \\ \frac{27}{2} \end{bmatrix} = \frac{1}{120} \begin{bmatrix} 1350 \\ 540 \end{bmatrix} = \begin{bmatrix} 11.25 \\ 4.5 \end{bmatrix}$$

Recalling that $\delta = \alpha_1 - \alpha_2$, and hence $\hat{\delta} = a_1 - a_2$, we see that the value $\hat{\delta} = 4.5$ is consistent with the results $a_1 = 2.25$ and $a_2 = -2.25$ obtained earlier by solving Eqs. 9.9. The reader may be puzzled as to why the value 11.25 for $\hat{\lambda} = m + (a_1 + a_2)/2$ coincides with the value earlier obtained for m itself. But when it is recalled that the previous solutions for m, a_1 and a_2 were contingent on the side condition [the third equation in set (9.9)] $a_1 + a_2 = 0$, the apparent paradox disappears; under this condition, $m + (a_1 + a_2)/2$ reduces to m itself.

It was earlier stated (p. 309) that the choice of the contrast matrix **C** was largely dictated by the interests of the researcher. "What sorts of linear combinations of the parameters would it be meaningful to estimate?" is the main question that one should ask oneself in constructing an appropriate contrast matrix. However, we are not entirely free to choose just any contrast matrix. It was already mentioned that **C** must have rank equal to its number of rows in order that **CC**′ be nonsingular. This means that none of the rows of **C** must be a linear combination of the other rows. We elaborate on this requirement a little further now.

For example, in a one-way ANOVA with three groups the design matrix **X** has four columns (corresponding to the original parameters, μ, α_1, α_2, and α_3, of the structural equation), but only three of these columns are linearly independent: the first column is the sum of the other three. Hence **X′X**, although of order 4×4, has rank 3. That is, the set of four normal equations for estimating the parameters contains only three independent equations. Thus any contrast matrix **C** must be of order at most 3 × 4, *with all rows linearly independent*. The matrix

$$\begin{bmatrix} 0 & 1 & -1 & 0 \\ 0 & 1 & 0 & -1 \\ 0 & 0 & 1 & -1 \end{bmatrix}$$

for example, is therefore *unacceptable* as a contrast matrix, since the third row is a linear combination of the first two:

$$[0, 0, 1, -1] = -[0, 1, -1, 0] + [0, 1, 0, -1]$$

In other words, this matrix has rank 2, which is less than its number of rows.

The foregoing restriction on **C**—that it must have rank equal to its number of rows (or "full row rank")—is an "innocuous" one. That is, no one would be really interested in estimating a set of new parameters (linear combinations of the original ones) that are *not* linearly independent among themselves, since at least one of them would be totally redundant. For example, if the matrix above were used as a contrast matrix despite its failing to meet the requirement of full row rank, we would be estimating new parameters δ_1, δ_2, and δ_3, defined, in accordance with Eq. 9.13, as

$$\begin{bmatrix} \delta_1 \\ \delta_2 \\ \delta_3 \end{bmatrix} = \begin{bmatrix} 0 & 1 & -1 & 0 \\ 0 & 1 & 0 & -1 \\ 0 & 0 & 1 & -1 \end{bmatrix} \begin{bmatrix} \mu \\ \alpha_1 \\ \alpha_2 \\ \alpha_3 \end{bmatrix} = \begin{bmatrix} \alpha_1 - \alpha_2 \\ \alpha_1 - \alpha_3 \\ \alpha_2 - \alpha_3 \end{bmatrix}$$

But clearly, once we have estimates of $\delta_1 = \alpha_1 - \alpha_2$ and $\delta_2 = \alpha_2 - \alpha_3$, we can automatically get an estimate of $\delta_3 = \alpha_2 - \alpha_3$ from $\delta_2 - \delta_1 = (\alpha_1 - \alpha_3) - (\alpha_1 - \alpha_2)$. Thus it is absolutely pointless to seek a separate estimate of δ_3 in addition to estimates of δ_1 and δ_2. It is in this sense, then, that the restriction that a contrast matrix **C** be of full row rank is innocuous (indeed, vacuous): This requirement is already tacitly implied in the dictum that the new parameters should (all) be meaningful and of interest to the researcher. The researcher need only make sure that he or she is not seeking to estimate any new parameter that would be redundant, given estimates of the others.

There is, however, another restriction on **C** (as forewarned on p. 312) that is somewhat more subtle than the one above. This is the requirement that the rows of **C** must be linear combinations of the rows of the design matrix **X**. Loosely speaking, this means that the linear combinations of the original parameters that define the new parameters must be "naturally suggested" by the design of the experiment. The experimental data cannot answer questions that are not at least implicitly built into the experimental design! Put in this way, this requirement, too, seems innocuous and commonsensical. Yet, why did the foregoing development not seem explicitly to call for this requirement? [Note, by contrast, that the full-row-rank requirement on **C** *was* explicitly invoked. Otherwise, $\mathbf{CC}'$ would be singular, so that $(\mathbf{CC}')^{-1}$ would not exist, and Eq. 9.15 could not be used to define the basis matrix **K** to replace the design matrix **X** in the reparameterized system.] Before answering this question, let us show that the **C** matrix chosen in our numerical example does indeed satisfy this extra condition.

Recall that the design matrix for our example was

$$\mathbf{X} = \begin{bmatrix} 1 & 1 & 0 \\ 1 & 1 & 0 \\ \vdots & \vdots & \vdots \\ 1 & 1 & 0 \\ 1 & 0 & 1 \\ \vdots & \vdots & \vdots \\ 1 & 0 & 1 \end{bmatrix}$$

and that the contrast matrix **C** we chose was

$$\mathbf{C} = \begin{bmatrix} 1 & \frac{1}{2} & \frac{1}{2} \\ 0 & 1 & -1 \end{bmatrix}$$

The first row of **C** is the average of the first and $(n_1 + 1)$th row of **X** (or, for that matter, the average of any one of the first n_1 rows and any one of the last n_2 rows of **X**):

$$\left[1, \tfrac{1}{2}, \tfrac{1}{2}\right] = \frac{[1, 1, 0] + [1, 0, 1]}{2}$$

$$= \tfrac{1}{2}[1, 1, 0] + \tfrac{1}{2}[1, 0, 1]$$

The second row of **C** is the difference between the first and $(n_1 + 1)$th rows of **X**:

$$[0, 1, -1] = [1, 1, 0] - [1, 0, 1]$$

So the rows of the **C** we chose were in fact linear combinations of the rows of **X**.

But, once again, why was this condition necessary when we seemed to get by without it in our foregoing development? The answer lies in a tacit assumption involved in going from the original structural equation (9.11):

$$\mathbf{Y} = \mathbf{X}\boldsymbol{\Theta} + \boldsymbol{\varepsilon}$$

to the reparameterized structural equation (9.14):

$$\mathbf{Y} = \mathbf{K}\boldsymbol{\Theta}^* + \boldsymbol{\varepsilon}$$

As stated in footnote 4, (page 310), this requires that **K**, **C**, and **X** stand in the relation

$$\mathbf{KC} = \mathbf{X}$$

Once this relation is *assumed,* the expression (9.15) for **K** follows logically. But this relation *does* impose a condition on **C**; it says that the rows of **X** must be expressible as linear combinations of the rows of **C**. However, since **X** exists before **C** does, it makes more sense to state the requirement the other way around: that *the rows of* **C** *be linear combinations of the rows of* **X**. This is known as the *estimability condition* of the new parameters $\boldsymbol{\Theta}^*$. Unless this condition is satisfied, the basis matrix **K** formally defined by Eq. 9.15 will not enable us to retrieve the design matrix **X** from the relation $\mathbf{KC} = \mathbf{X}$. This means that the new parameters $\boldsymbol{\Theta}^* = \mathbf{C}\boldsymbol{\Theta}$ cannot be estimated from the data of our experiment. This is the subtle way in which mathematics (matrix algebra in this case) tells us that we cannot expect our data to answer questions that our experiment was not designed to answer.

To clarify the point, we give an example of a potential contrast matrix which satisfies the full-row-rank condition but does not fulfill the estimability condition. Returning to the context of our two-group example, consider a proposed contrast matrix

$$\mathbf{C} = \begin{bmatrix} 1 & 0 & 0 \\ 0 & 1 & -1 \end{bmatrix}$$

which corresponds to an intention to estimate μ and $\alpha_1 - \alpha_2$. The rows of **C** are linearly independent. (Two rows are linearly dependent only if they are identical or proportional.) Hence **C** has full row rank, 2. However, the first row of **C** cannot be expressed as a linear combination of the rows of **X**; that is, the estimability condition is not satisfied. Despite this fact, we can (incorrectly) go through the motions of determining the basis matrix **K**, since $\mathbf{CC}'$ is nonsingular. We find

$$\mathbf{XC}' = \begin{bmatrix} 1 & 1 & 0 \\ 1 & 1 & 0 \\ \vdots & \vdots & \vdots \\ 1 & 1 & 0 \\ 1 & 0 & 1 \\ \vdots & \vdots & \vdots \\ 1 & 0 & 1 \end{bmatrix} \begin{bmatrix} 1 & 0 \\ 0 & 1 \\ 0 & -1 \end{bmatrix} = \begin{bmatrix} 1 & 1 \\ 1 & 1 \\ \vdots & \vdots \\ 1 & 1 \\ 1 & -1 \\ \vdots & \vdots \\ 1 & -1 \end{bmatrix}$$

and

$$\mathbf{CC}' = \begin{bmatrix} 1 & 0 & 0 \\ 0 & 1 & -1 \end{bmatrix} \begin{bmatrix} 1 & 0 \\ 0 & 1 \\ 0 & -1 \end{bmatrix} = \begin{bmatrix} 1 & 0 \\ 0 & 2 \end{bmatrix}$$

Hence, in accordance with Eq. 9.15, we get

$$\mathbf{K} = \begin{bmatrix} 1 & 1 \\ 1 & 1 \\ \vdots & \vdots \\ 1 & 1 \\ 1 & -1 \\ \vdots & \vdots \\ 1 & -1 \end{bmatrix} \begin{bmatrix} 1 & 0 \\ 0 & \frac{1}{2} \end{bmatrix} = \begin{bmatrix} 1 & \frac{1}{2} \\ 1 & \frac{1}{2} \\ \vdots & \vdots \\ 1 & \frac{1}{2} \\ 1 & -\frac{1}{2} \\ \vdots & \vdots \\ 1 & -\frac{1}{2} \end{bmatrix}$$

That this a spurious result becomes clear when we try to retrieve **X** from the product **KC**:

$$\mathbf{KC} = \begin{bmatrix} 1 & \frac{1}{2} \\ 1 & \frac{1}{2} \\ \vdots & \vdots \\ 1 & \frac{1}{2} \\ 1 & -\frac{1}{2} \\ \vdots & \vdots \\ 1 & -\frac{1}{2} \end{bmatrix} \begin{bmatrix} 1 & 0 & 0 \\ 0 & 1 & -1 \end{bmatrix} = \begin{bmatrix} 1 & \frac{1}{2} & -\frac{1}{2} \\ 1 & \frac{1}{2} & -\frac{1}{2} \\ \vdots & \vdots & \vdots \\ 1 & \frac{1}{2} & -\frac{1}{2} \\ 1 & -\frac{1}{2} & \frac{1}{2} \\ \vdots & \vdots & \vdots \\ 1 & -\frac{1}{2} & \frac{1}{2} \end{bmatrix}$$

which is *not* equal to the design matrix **X**. Therefore, the proposed **C** does not qualify as a contrast matrix: The reparameterization it represents is not acceptable, in that the "new" parameters (note that the first is simply one of the original parameters, μ) cannot be estimated by data from an experiment with design matrix **X**.

Once again, however, the researcher need not be unduly concerned that it may be difficult to construct a contrast matrix that satisfies the estimability condition—just as it was unnecessary to worry about the full-row-rank condition as long as he or she is asking nonredundant questions. The fact is that the estimability condition is guaranteed to be satisfied if the researcher adheres to the following rules of thumb:

1. As the first row of **C**, take the average of all or a subset of the rows of **X**. (This gives some sort of "overall level" parameter.)
2. As the remaining $J - 1$ rows ($J - 1$ being equal to the between-cells degrees of freedom in the design), use contrasts of the effect parameters that are linearly independent. (Incidentally, they need *not* be orthogonal, just linearly independent.)

SUMMARY

Since this section—dealing with the parameter-estimation aspect of ANOVA in the matrix formulation of the general linear model—was a lengthy one, we recapitulate its main points in itemized form for ready reference.

1. The starting point of the matrix formulation is the structural model

$$\mathbf{Y} = \mathbf{X\Theta} + \boldsymbol{\varepsilon} \tag{9.11}$$

where **Y** is the observation vector, listing the criterion scores for all observational units (usually subjects) in the entire sample; **X** is the design matrix that specifies the experimental design involved; $\mathbf{\Theta}$ is the parameter vector listing the grand mean (μ) of Y in the total population and the various effect parameters, as well as regression coefficients if the design calls for an analysis of covariance; and $\boldsymbol{\varepsilon}$ is the error vector representing lack of fit of the model.

2. The least-squares principle is then invoked in the hope of estimating the parameters listed in $\mathbf{\Theta}$ so as to minimize the sum of squared errors

$$Q = \boldsymbol{\varepsilon}'\boldsymbol{\varepsilon} = (\mathbf{Y} - \mathbf{X\Theta})'(\mathbf{Y} - \mathbf{X\Theta})$$

Unfortunately, this leads to a set of normal equations

$$(\mathbf{X}'\mathbf{X})\hat{\mathbf{\Theta}} = \mathbf{X}'\mathbf{Y} \tag{9.12}$$

which is of deficient rank. That is, $\mathbf{X}'\mathbf{X}$ is singular, so that Eq. 9.12 is not amenable to the usual multiple-regression solution

$$\hat{\mathbf{\Theta}} = (\mathbf{X}'\mathbf{X})^{-1}(\mathbf{X}'\mathbf{Y})$$

to yield the estimated parameter values $\hat{\mathbf{\Theta}}$.

3. One way of resolving this apparent impasse is to reparameterize.[5] That is, instead of trying to estimate the original parameters listed in $\mathbf{\Theta}$, we define a new set of parameters $\mathbf{\Theta}^*$ that are linear combinations of the original parameters by means of a transformation equation

$$\mathbf{\Theta}^* = \mathbf{C\Theta} \tag{9.13}$$

[5]Another way is to add certain constraints on the parameters, as was done in Section 9.2. Yet another way is to utilize a *generalized inverse* of $\mathbf{X}'\mathbf{X}$, which, as shown in Section 5.7, does not require $\mathbf{X}'\mathbf{X}$ to be nonsingular. The interested reader may refer to Searle (1971) for an exposition of this approach.

The transformation matrix **C**, called the contrast matrix, must satisfy two conditions:

(a) It must be of full row rank. That is, none of its rows may be expressible as a linear combination of the other rows. (If **C** violates this condition, the researcher is trying to estimate redundant parameters, which is pointless.) One consequence of this condition is that the number of rows of **C** must never exceed the rank of $\mathbf{X}'\mathbf{X}$, or the number of groups in the design.

(b) Its rows must be expressible as linear combinations of the rows of the design matrix **X**. This is called the *estimability condition;* unless it is satisfied, the proposed new parameters simply cannot be estimated from the data of the experiment with design matrix **X**. However, these conditions are not as restrictive as they may seem. It was pointed out that, as long as the new parameters are nonredundant, condition (a) will be satisfied. Rules of thumb were given to ensure that condition (b) will also be met.

4. Upon choosing an appropriate contrast matrix **C** in line with the researcher's interests, but subject to conditions (a) and (b), the structural equation in the reparameterized system becomes

$$\mathbf{Y} = \mathbf{K}\boldsymbol{\Theta}^* + \boldsymbol{\varepsilon} \tag{9.14}$$

where a new matrix **K**, called the basis of the design, replaces the design matrix **X** in Eq. 9.11. The basis matrix is computed from the relation

$$\mathbf{K} = (\mathbf{X}\mathbf{C}')(\mathbf{C}\mathbf{C}')^{-1} \tag{9.15}$$

which is computable since condition (a) for the contrast matrix assures $\mathbf{C}\mathbf{C}'$ to be nonsingular.

5. Again invoking the least-squares principle, the equations for estimating the new parameters $\boldsymbol{\Theta}^*$ are found to be

$$(\mathbf{K}'\mathbf{K})\hat{\boldsymbol{\Theta}}^* = \mathbf{K}'\mathbf{Y} \tag{9.16}$$

which is formally identical with Eq. 9.12, upon replacing **X** by **K** and $\hat{\boldsymbol{\Theta}}$ by $\hat{\boldsymbol{\Theta}}^*$. This system of equations does not suffer froam the deficient-rank problem (i.e., $\mathbf{K}'\mathbf{K}$ is nonsingular) and may immediately be solved for $\hat{\boldsymbol{\Theta}}^*$ by the formula

$$\hat{\boldsymbol{\Theta}}^* = (\mathbf{K}'\mathbf{K})^{-1}(\mathbf{K}'\mathbf{Y}) \tag{9.17}$$

This completes the parameter-estimation aspect of ANOVA. It should be emphasized once again that the outline above holds not only for the simple, two-group example we have explicitly dealt with in this and previous sections for illustrative purposes, but for any design whatsoever, including analysis of covariance. In the next section we address ourselves to the significance-testing aspect.

9.4 MATRIX FORMULATION: SIGNIFICANCE TESTING

[**Note:** A thorough understanding of this section requires somewhat greater mathematical and statistical sophistication than was needed for the previous sections. In particular, it is essential to have a working knowledge of the algebra of expected values,[6] which the reader may gain by referring, for example, to Hays (1981, pp. 625–632). However, the reader who does not have the time or inclination to acquire such background knowledge may skim the earlier parts of the section and focus on the numerical example and subsequent discussions, which together should convey a general idea of what is involved.]

The parameter-estimation aspect of ANOVA discussed in Section 9.3 may have been a novel concept for many readers—one that is not customarily associated with analysis of variance, whose main purpose has been the testing of hypotheses concerning population means. It should occur to the reader, however, that once the point estimates derived above are supplemented by estimates of their standard errors, we have the ways and means not only for carrying out the desired significance tests but also for interval estimation (i.e., the construction of confidence intervals for the parameters). (Although we shall not concern ourselves directly with interval estimation here, the reader is probably aware that the information gained from confidence intervals contains the conclusions obtainable from significance tests, and then some more.) This is the main reason why the general linear model approach to ANOVA is carried out in two phases: first estimating the (reparameterized) parameters, and then testing their significance by deriving their standard errors (or, more directly, the variance–covariance matrix of their joint sampling distribution). The latter, then, is the objective of this section.

We first note, from Eq. 9.17, that the (point) estimates of the new parameters are linear combinations of the observations:

$$\hat{\mathbf{\Theta}}^* = (\mathbf{K}'\mathbf{K})^{-1}(\mathbf{K}'\mathbf{Y}) \qquad (9.17)$$

To see this more clearly, we may rewrite the right-hand side of the parameter-estimation equation as

$$[(\mathbf{K}'\mathbf{K})^{-1}\mathbf{K}']\mathbf{Y}$$

and denote the coefficient matrix $(\mathbf{K}'\mathbf{K})^{-1}\mathbf{K}'$ by $\mathbf{A}$ for short. This is a $J \times N$ matrix, since $\mathbf{K}$ is $N \times J$, where N is the total sample size and J is the number of parameters in the reparameterized system, generally equal to the number of groups in the design. (Recall that $J = 2$ in our example.) Thus the first element of $\hat{\mathbf{\Theta}}^*$ is

$$\hat{\Theta}_1^* = a_{11}Y_1 + a_{12}Y_2 + \cdots + a_{1N}Y_N$$

[6]An expected value of a random variable is the mean of its sampling distribution, which may be characterized as the "long-run average" on repeated sampling.

The jth element is computed by using the jth row elements, $a_{j1}, a_{j2}, \ldots, a_{jN}$, as the combining weights in the linear combination of the Y_is.

Hence, in order to determine the variance–covariance matrix $\mathbf{V}(\hat{\mathbf{\Theta}}^*)$ of the parameters, we must first express the expected values and variances of the observations Y_i *in terms of* quantities that appear in the reparameterized structural equation,

$$\mathbf{Y} = \mathbf{K\Theta}^* + \boldsymbol{\varepsilon} \tag{9.14}$$

The expected value of $\mathbf{Y}$ (i.e., the vector of expected values of $Y_1, Y_2, \ldots, Y_N$) follows immediately from this equation; that is,

$$\begin{aligned}\mathbf{E(Y)} &= \mathbf{E}(\mathbf{K\Theta}^* + \boldsymbol{\varepsilon}) \\ &= \mathbf{K\Theta}^*\end{aligned} \tag{9.18}$$

the second step following from the facts that $\mathbf{E}(\mathbf{K\Theta}^*) = \mathbf{K\Theta}^*$ (since $\mathbf{K}$ and $\mathbf{\Theta}^*$ involve only fixed quantities) and $\mathbf{E}(\boldsymbol{\varepsilon}) = \mathbf{O}$ by definition of random errors.

Then the variances of Y_i's (or, more precisely, the variance–covariance matrix of the joint distribution of the Y_i's) is obtained as

$$\begin{aligned}\mathbf{V(Y)} &= \mathbf{E[Y - E(Y)][Y - E(Y)]}' \\ &= \mathbf{E}(\mathbf{Y} - \mathbf{K\Theta}^*)(\mathbf{Y} - \mathbf{K\Theta}^*)' \\ &= \mathbf{E}(\boldsymbol{\varepsilon\varepsilon}') \\ &= \sigma^2\mathbf{I}_N\end{aligned} \tag{9.19}$$

Here we have utilized the basic assumption of analysis of variance, that the errors are *independently* distributed with zero means and a *common* variance σ^2. The final result, $\sigma^2\mathbf{I}_N$ (where $\mathbf{I}_N$ is the identity matrix of order N), is an $N \times N$ diagonal matrix with all diagonal elements equal to σ^2. This simply reflects the basic assumption just stated. Since ϵ_i is the only random variable in the structural equation for Y_i, the variance of each Y_i is equal to that of the corresponding ϵ_i; hence all Y_i's have the same variance, σ^2. Also, the covariance of Y_i and $Y_{i'}$ for any pair of individuals $i \neq i'$ is the covariance of ϵ_i and $\epsilon_{i'}$, which is zero. Thus the variance–covariance matrix of the Y_i's is a diagonal matrix with common diagonal element σ^2:

$$\begin{bmatrix} \sigma^2 & 0 & 0 & \cdots & 0 \\ 0 & \sigma^2 & 0 & \cdots & 0 \\ 0 & 0 & \sigma^2 & \cdots & 0 \\ \vdots & & & \ddots & \\ 0 & 0 & 0 & \cdots & \sigma^2 \end{bmatrix}$$

Although our main objective is to determine the variance–covariance matris of $\hat{\mathbf{\Theta}}^*$, let us first get the expected-value vector $\mathbf{E}(\hat{\mathbf{\Theta}}^*)$ of $\hat{\mathbf{\Theta}}^*$. For unless we are assured that $\hat{\mathbf{\Theta}}^*$ is an unbiased estimator of $\mathbf{\Theta}^*$, knowing its standard error will be of no avail for significance testing. Resorting once again to Eq. 9.17, we find

$$\begin{aligned}\mathbf{E}(\hat{\mathbf{\Theta}}^*) &= \mathbf{E}[(\mathbf{K}'\mathbf{K})^{-1}\mathbf{K}'\mathbf{Y}] \\ &= (\mathbf{K}'\mathbf{K})^{-1}\mathbf{K}'\mathbf{E}(\mathbf{Y})\end{aligned}$$

the second step following because $(\mathbf{K}'\mathbf{K})^{-1}\mathbf{K}'$ involves only fixed constants. Substituting expression (9.18) for $\mathbf{E}(\mathbf{Y})$ in this equation, we get

$$\begin{aligned}\mathbf{E}(\hat{\mathbf{\Theta}}^*) &= (\mathbf{K}'\mathbf{K})^{-1}\mathbf{K}'(\mathbf{K}\mathbf{\Theta}^*) \\ &= (\mathbf{K}'\mathbf{K})^{-1}(\mathbf{K}'\mathbf{K})\mathbf{\Theta}^* \\ &= \mathbf{\Theta}^*\end{aligned} \tag{9.20}$$

Thus we have verified that $\hat{\mathbf{\Theta}}^*$ is indeed an unbiased estimator of $\mathbf{\Theta}^*$.

We are now ready to compute the variance–covariance matrix of $\hat{\mathbf{\Theta}}^*$. Since, as indicated earlier, each element of $\hat{\mathbf{\Theta}}^*$ is a linear combination of the observations Y_i, the variance of the jth element, say, is obtained from a formula analogous to one that was derived as Eq. 7.2 for $SS_w(Y)$ in connection with discriminant analysis; that is,

$$\text{var}\,(\hat{\Theta}_j^*) = \mathbf{a}_j'\mathbf{V}(\mathbf{Y})\mathbf{a}_j$$

where $\mathbf{a}_j'$ is the jth row of the coefficient matrix $(\mathbf{K}'\mathbf{K})^{-1}\mathbf{K}'$ in Eq. 9.17, temporarily denoted $\mathbf{A}$ for short, and $\mathbf{a}_j$ is $\mathbf{a}_j'$ written as a column vector (which is *not* the jth column of $\mathbf{A}$, it should be noted). Similarly, it may be shown that the covariance between the jth and kth elements of $\hat{\mathbf{\Theta}}^*$ is

$$\text{cov}\,(\hat{\Theta}_j^*, \hat{\Theta}_k^*) = \mathbf{a}_j'\mathbf{V}(\mathbf{Y})\mathbf{a}_k$$

where $\mathbf{a}_j'$ is as defined above, and $\mathbf{a}_k$ is the kth *row* of $\mathbf{A}$ written as a column vector.

Putting these two results together, we obtain the variance–covariance matrix of the elements of $\hat{\mathbf{\Theta}}^*$ as follows:

$$\mathbf{V}(\hat{\mathbf{\Theta}}^*) = \mathbf{A}\mathbf{V}(\mathbf{Y})\mathbf{A}'$$

Now replacing $\mathbf{A}$ by what it stands for, and also substituting the last expression in Eq. 7.19 for $\mathbf{V}(\mathbf{Y})$, we finally get

$$\begin{aligned}\mathbf{V}(\hat{\mathbf{\Theta}}^*) &= [(\mathbf{K}'\mathbf{K})^{-1}\mathbf{K}']\sigma^2\mathbf{I}_N[(\mathbf{K}'\mathbf{K})^{-1}\mathbf{K}']' \\ &= \sigma^2(\mathbf{K}'\mathbf{K})^{-1}\mathbf{K}'\mathbf{K}(\mathbf{K}'\mathbf{K})'^{-1} \\ &= \sigma^2(\mathbf{K}'\mathbf{K})^{-1}\end{aligned} \tag{9.21}$$

where we have utilized the fact that $\mathbf{K'K}$ is symmetric, and hence that

$$(\mathbf{K'K})'^{-1} = (\mathbf{K'K})^{-1}$$

Although Eq. 9.21 gives the exact variance–covariance matrix of $\hat{\boldsymbol{\Theta}}^*$, it cannot be used in practice because the population variance σ^2 is unknown. We must therefore replace σ^2 by its unbiased estimate $\hat{\sigma}^2$ in order to have a usable formula. It is well known that an unbiased estimate of σ^2 is provided by the pooled within-group mean square, MS_w, in the analysis of variance. We shall here deliberately compute this by a more complicated formula than the usual

$$MS_w = \frac{1}{N - J} \sum_{j=1}^{J} \sum_{i=1}^{n_j} (Y_{ij} - \overline{Y}_j)^2$$

in order to have an expression that generalizes to more complex designs.

The approach is as follows: We first calculate the *raw* total sum of squares $\mathbf{S}_T$, which is simply the sum of the squared criterion scores in the total sample. In matrix notation, this is

$$\mathbf{S}_T = \mathbf{Y'Y} \tag{9.22}$$

Then we find the *raw* model (or between groups) sum of squares $\mathbf{S}_B$ from

$$\begin{aligned} \mathbf{S}_B &= (\mathbf{K}\hat{\boldsymbol{\Theta}}^*)'(\mathbf{K}\hat{\boldsymbol{\Theta}}^*) \\ &= \hat{\boldsymbol{\Theta}}^{*\prime}(\mathbf{K'K})\hat{\boldsymbol{\Theta}}^* \\ &= \hat{\boldsymbol{\Theta}}^{*\prime}(\mathbf{K'Y}) \end{aligned} \tag{9.23}$$

the last expression following from the fact that

$$\hat{\boldsymbol{\Theta}}^* = (\mathbf{K'K})^{-1}\mathbf{K'Y}$$

from Eq. 9.17.

It should be noted that the quantities $\mathbf{S}_T$ and $\mathbf{S}_B$ are not the customary SS_t and SS_b in ANOVA—which are the deviation-score total and between-groups sums of squares, respectively. $\mathbf{S}_T$ and $\mathbf{S}_B$, on the other hand, are *raw-score* sums of squares.[7] Nevertheless, when the difference $\mathbf{S}_T - \mathbf{S}_B$ is formed, it will represent a deviation-score sum of squares, due to the relations

$$SS_t = \mathbf{S}_T - (\Sigma\Sigma\, Y_{ij})^2/N$$

and

$$SS_b = \mathbf{S}_B - (\Sigma\Sigma\, Y_{ij})^2/N$$

[7]We also note, in passing, that although these are scalar quantities, we deliberately use matrix symbols so that the same formulas will continue to hold in the MANOVA case, when $\mathbf{S}_T$ and $\mathbf{S}_B$ will be $p \times p$ matrices (where p is the number of criterion variables). It might further be pointed out that we have used $\mathbf{Y}$, $\boldsymbol{\Theta}$, $\boldsymbol{\Theta}^*$, and so on, instead of $\mathbf{y}$, $\boldsymbol{\theta}$, $\boldsymbol{\theta}^*$—even though they stand for vectors rather than matrices in the univariate case—for the same reason: In the multivariate case, they will stand for matrices.

That is, the term $-(\Sigma\Sigma\ Y^2)/N$ common to SS_t and SS_b will cancel out upon subtraction. Hence the within-groups sum of squares SS_w (or more generally, the error sum of squares SS_e) may be obtained equivalently either from the customary

$$SS_w = SS_t - SS_b$$

or from the difference of the two raw-score sums of squares, $\mathbf{S}_T$ and $\mathbf{S}_B$. We thus have

$$SS_e = \mathbf{S}_E = \mathbf{S}_T - \mathbf{S}_B$$

Upon substituting expressions (9.22) and (9.23) for $\mathbf{S}_T$ and $\mathbf{S}_B$, respectively, in this equation, we obtain

$$\mathbf{S}_E = \mathbf{Y}'\mathbf{Y} - \hat{\boldsymbol{\Theta}}^{*\prime}(\mathbf{K}'\mathbf{Y}) \tag{9.24}$$

The final step in computing $\hat{\sigma}^2$ (which is just MS_w in simple designs) by this more complicated—but vastly more general—method is to divide $\mathbf{S}_E$ by its degrees of freedom $N - J$. [More generally, if $\hat{\boldsymbol{\Theta}}^*$ has fewer than the maximal number of estimable parameters J, say I $(< J)$, $\mathbf{S}_E$ will have $N - I$ degrees of freedom.] We then get

$$\hat{\sigma}^2 = \frac{1}{N - J}[\mathbf{Y}'\mathbf{Y} - \hat{\boldsymbol{\Theta}}^{*\prime}(\mathbf{K}\mathbf{Y})] \tag{9.25}$$

as the estimate to replace σ^2 in expression (9.21) for $\mathbf{V}(\hat{\boldsymbol{\Theta}}^*)$. Denoting the *estimated* variance–covariance matrix of $\hat{\boldsymbol{\Theta}}^*$ by $\hat{\mathbf{V}}(\hat{\boldsymbol{\Theta}}^*)$, we finally get a usable formula,

$$\hat{\mathbf{V}}(\hat{\boldsymbol{\Theta}}^*) = \frac{1}{N - J}[\mathbf{Y}'\mathbf{Y} - \hat{\boldsymbol{\Theta}}^{*\prime}(\mathbf{K}'\mathbf{Y})](\mathbf{K}'\mathbf{K})^{-1} \tag{9.26}$$

The diagonal elements of this matrix are the estimated sampling variances of the successive elements of $\hat{\boldsymbol{\Theta}}^*$; hence, their square roots provide estimates of the standard errors of the new parameters, and significance testing (or construction of confidence intervals) becomes feasible. However, there is one further complication. The off-diagonal elements of $\hat{\mathbf{V}}(\hat{\boldsymbol{\Theta}}^*)$ will, in general, be nonzero, which means that the new (reparameterized) parameters—although linearly independent by definition (see p. 316)—are not always *statistically* independent. Loosely speaking, even though the questions asked by the significance tests of the various new parameters are not *deterministically* redundant, they are (in general) *probabilistically* redundant. For example, even though $\theta_1^* = 0$ and $\theta_2^* = 0$ will never *strictly* imply that $\theta_3^* = 0$ (say), they may increase the likelihood that θ_3^* is zero.

In the two-group case, the complication mentioned above is inessential: The only null hypothesis we would be really interested in testing is that Θ_2^* is zero; that is,

$$\delta = \alpha_1 - \alpha_2 = 0$$

The other null hypothesis, $\theta_1^* = 0$, or

$$\lambda = \mu + (\alpha_1 + \alpha_2)/2 = 0$$

is almost certain to be rejected, and is thus of no intrinsic interest (although interval estimation of λ may well be of interest). In multigroup cases, however, the complication does pose a genuine problem (and even more so in higher-order designs). In these cases we are interested in testing several null hypotheses implied by the reparameterized parameter vector (or matrix, when there are several criterion variables) $\mathbf{\Theta}^*$. If these parameters are not statistically independent (i.e., if they are correlated), as indicated by $\hat{\mathbf{V}}(\hat{\mathbf{\Theta}}^*)$ not being a diagonal matrix, then the several significance tests are interdependent, which "muddies up" the adopted significance level. This problem and its solution will be discussed further after we illustrate the foregoing developments in a context free of the problem—the simple, two-group case we have been using throughout. We mentioned the problem at this stage merely to forewarn the reader that the significance tests will not always be quite as straightforward as in the numerical example we now present.

EXAMPLE 9.3. For testing the null hypothesis

$$H_0: \alpha_1 - \alpha_2 = 0$$

we need to determine the estimated sampling variance–covariance matrix $\hat{\mathbf{V}}(\hat{\mathbf{\Theta}}^*)$ in accordance with Eq. 9.26. In particular, we need the second diagonal element of this matrix, whose square root is the estimated standard error, $s_{\alpha_1-\alpha_2}$, of the unbiased estimate $a_1 - a_2$ of $\alpha_1 - \alpha_2$. Once this is found, we may carry out the significance test by computing either

$$t = \frac{a_1 - a_2}{s_{\alpha_1 - \alpha_2}}$$

and referring to the t-distribution with $N - J$ degrees of freedom, or

$$F = \frac{(a_1 - a_2)^2}{s^2_{\alpha_1 - \alpha_2}}$$

and referring to the F-distribution with 1 and $N - J$ degrees of freedom.

In order to compute $\hat{\mathbf{V}}(\hat{\mathbf{\Theta}}^*)$ from Eq. 9.26, we first need to get $\mathbf{S}_T$ as defined by Eq. 9.22.

$$\mathbf{S}_T = \mathbf{Y}'\mathbf{Y} \tag{9.22}$$

This is simply the sum of the squared criterion scores for the total sample, and has already been displayed (as ΣY^2) in the data table on page 298:

$$\mathbf{Y}'\mathbf{Y} = 3141$$

Next, we need to compute $\mathbf{S}_B$ in accordance with Eq. 9.23:

$$\mathbf{S}_B = \hat{\mathbf{\Theta}}^{*\prime}(\mathbf{K}'\mathbf{Y}) \tag{9.23}$$

We have already found, for our example (see p. 315) that

$$\hat{\mathbf{\Theta}}^* = \begin{bmatrix} 11.25 \\ 4.50 \end{bmatrix}$$

Also,

$$\mathbf{K}'\mathbf{Y} = \begin{bmatrix} T \\ \dfrac{T_1 - T_2}{2} \end{bmatrix}$$

has been computed on the same page as

$$\mathbf{K}'\mathbf{Y} = \begin{bmatrix} 243 \\ \frac{27}{2} \end{bmatrix}$$

We therefore get

$$\hat{\mathbf{\Theta}}^{*\prime}(\mathbf{K}'\mathbf{Y}) = [11.25, 4.50] \begin{bmatrix} 243 \\ \frac{27}{2} \end{bmatrix}$$

$$= 2794.5$$

Finally, $(\mathbf{K}'\mathbf{K})^{-1}$ has been found to be

$$(\mathbf{K}'\mathbf{K})^{-1} = \frac{1}{120} \begin{bmatrix} \frac{11}{2} & 1 \\ 1 & 22 \end{bmatrix}$$

Putting these three numerical results together and recalling that $N - J = 20$ for our example, we compute $\hat{\mathbf{V}}(\hat{\mathbf{\Theta}}^*)$ in accordance with Eq. 9.26, as follows:

$$\hat{\mathbf{V}}(\hat{\boldsymbol{\Theta}}^*) = \frac{1}{N-J}[\mathbf{Y}'\mathbf{Y} - \hat{\boldsymbol{\Theta}}^{*\prime}(\mathbf{K}'\mathbf{Y})](\mathbf{K}'\mathbf{K})^{-1}$$

$$= \frac{1}{20}[3141 - 2794.5]\frac{1}{120}\begin{bmatrix} \frac{11}{2} & 1 \\ 1 & 22 \end{bmatrix}$$

$$= (17.325)\frac{1}{120}\begin{bmatrix} 5.5 & 1 \\ 1.0 & 22 \end{bmatrix}$$

$$= (17.325)\begin{bmatrix} .045833 & .008333 \\ .008333 & .183333 \end{bmatrix}$$

$$= \begin{bmatrix} .7941 & .1444 \\ .1444 & 3.1762 \end{bmatrix}$$

Before proceeding to the significance test itself, we pause to note that the numerical value 17.325 for

$$\frac{1}{N-J}[\mathbf{Y}'\mathbf{Y} - \hat{\boldsymbol{\Theta}}^{*\prime}(\mathbf{K}'\mathbf{Y})]$$

agrees with that for the within-groups variance estimate, s_w^2, calculated on page 299 in connection with the usual t-test for the two-group case. This is the quantity which, for more complex designs, generalizes to $\hat{\sigma}^2$ as given in Eq. 9.25.

We now come to the crux of our problem: testing the null hypothesis, $\alpha_1 - \alpha_2 = 0$. As indicated above, the second diagonal element of $\hat{\mathbf{V}}(\hat{\boldsymbol{\Theta}}^*)$, 3.1762, is the square of the standard error, $s_{\alpha_1-\alpha_2}$, of $a_1 - a_2$. We already know (see the second element of $\hat{\boldsymbol{\Theta}}^*$ displayed above) that $a_1 - a_2 = 4.50$. Thus we may compute

$$t = \frac{4.5}{\sqrt{3.1762}} = 2.5250$$

which agrees with the t-value obtained earlier (p. 299) by the straightforward t-test approach. This value, as already noted, is significant at the 5% level of significance as a t with $N - J = 20$ degrees of freedom. We therefore reject

$$H_0\text{: } \alpha_1 - \alpha_2 = 0$$

which is the same as the null hypothesis that $\mu_1 = \mu_2$ (see p. 311).

The reader may have been overwhelmed by the complexity of the significance testing procedure illustrated above, compared to the ordinary t-test. Of course, no one would carry out a simple two-sample significance test in this manner. The reader is reminded, once again, that our purpose is to explain the rationale and mechanics of a general method applicable to any ANOVA or MANOVA design. Still, he or she may wonder why a 2 × 2 matrix $\hat{\mathbf{V}}(\hat{\mathbf{\Theta}}^*)$ had to be computed when the only element used was the second diagonal element, 3.1762.

In point of fact, the procedure could have been simplified by eliminating the first new parameter $\lambda = \mu + (\alpha_1 + \alpha_2)/2$ (i.e., the first element of $\mathbf{\Theta}^*$) from the outset. Then the "vector" of new parameters $\mathbf{\Theta}^*$ would have reduced to a scalar $\alpha_1 - \alpha_2$; the transformation matrix **C** in Eq. 9.13 would have been a row vector [0, 1, −1]; the basis **K** defined by Eq. 9.15 would be an N-dimensional column vector instead of an $N \times 2$ matrix; and so on down the line until $\hat{\mathbf{V}}(\hat{\mathbf{\Theta}}^*)$ emerges as the scalar 3.1762. But we would not have had an example with which to illustrate genuine matrix procedures. (Nevertheless, the reader may find it instructive to work through this reduced routine; he or she will then be computing all the quantities that are involved in the ordinary t-test—and nothing else—in a slightly different way.) In larger designs, the J groups (or cells), even the deletion of the general level parameter

$$\lambda = \mu + (\alpha_1 + \alpha_2 + \cdots + \alpha_J)/J$$

will leave $\mathbf{\Theta}^*$ as a $(J - 1)$-dimensional column vector, and $\hat{\mathbf{V}}(\hat{\mathbf{\Theta}}^*)$ will emerge as a $(J - 1) \times (J - 1)$ matrix, all of whose diagonal elements would be used in significance tests.

In order not to leave the reader with the impression that the overall level parameter λ [and consequently also the first diagonal element of $\hat{\mathbf{V}}(\hat{\mathbf{\Theta}}^*)$] is a completely useless appendage, we point out that an interval estimation of λ may often be of considerable interest. That is, even though it is usually a foregone conclusion that $\lambda \neq 0$ (unless, for example, the observations are deviations from some prior standard)—and hence there is no point in testing H_0: $\lambda = 0$—it may still be of value to construct a confidence interval for λ. This would tell us (for instance) the upper and lower limits of the effect the treatments, on the average, may be expected to have compared to a control population with no treatment at all. In our numerical example, the estimate of λ (the first element of $\hat{\mathbf{\Theta}}^*$) was 11.25, and the first diagonal element of $\hat{\mathbf{V}}(\hat{\mathbf{\Theta}}^*)$ was .7941. The 95% confidence interval for λ is, therefore,

$$[11.25 - 2.086\sqrt{.7941},\ 11.25 + 2.086\sqrt{.7941}]$$

(where 2.086 is the 97.5% point of the t-distribution with 20 d.f.) or

$$[9.39,\ 13.11]$$

That is, we may be 95% sure that the grand mean for both treatments combined (assuming that $\alpha_1 + \alpha_2 = 0$) lies between 9.39 and 13.11.

Having accounted for the *diagonal* elements of $\hat{\mathbf{V}}(\hat{\mathbf{\Theta}}^*)$, the question remains: Of what use are the off-diagonal elements? In truth, they are only of ancillary interest. They indicate the extent to which the parameters listed in $\mathbf{\Theta}^*$ are statistically interdependent (or correlated), and hence the degree to which we can trust the results of significance tests carried out in the manner illustrated above. In the two-group case, as mentioned earlier, there is no problem: There is only one significance test of intrinsic interest. But in multigroup and higher-order designs, the problem posed by the correlatedness of the new parameters can become serious. If we blatantly carry out the several significance tests using the square roots of the diagonal elements as standard errors of their respective parameters, we may expose ourselves to compounded type I errors (rejecting null hypotheses that are true).

In the present example—even if we had been genuinely interested in testing H_0: $\lambda = 0$—the problem would have been minimal. The off-diagonal elements (.1444) were very small. Their smallness can be gauged better by converting them (which are covariances) to correlations: The correlation between $\hat{\lambda}$ $[= m + (a_1 + a_2)/2]$ and $\hat{\delta}$ $(= a_1 - a_2)$ is

$$r_{12} = \frac{.1444}{\sqrt{(.7941)(3.1762)}} = .091$$

which is surely negligible. (It will be shown below that even this small correlation is due only to the unequal group sizes, $n_1 = 10$, $n_2 = 12$, in our example.)

In multigroup (or higher-order) designs, the off-diagonal elements—coverted to correlations (by dividing by the square root of the product of the two corresponding diagonal elements)—serve as a warning of the extent to which the significance tests, if carried out as illustrated above, would be invalid. The higher the correlations, the more suspect are the results of the significance tests. Only when all the correlations are zero [i.e., when $\hat{\mathbf{V}}(\hat{\mathbf{\Theta}}^*)$ is a diagonal matrix] are the significance tests truly valid without further modification.

Two conditions must be satisfied in order for the new parameters to be pairwise uncorrelated. First, the contrast matrix $\mathbf{C}$ must be *row-wise orthogonal;* second, the group sizes, n_j, must all be equal. Row-wise orthogonality means that, when the corresponding elements of any pair of rows are multiplied and the products added, the sum is zero. The $\mathbf{C}$ matrix of our numerical example does satisfy this condition. Recall that

$$\mathbf{C} = \begin{bmatrix} 1 & \frac{1}{2} & \frac{1}{2} \\ 0 & 1 & -1 \end{bmatrix}$$

the sum of the products of corresponding elements in the two rows is

$$(1)(0) + (\tfrac{1}{2})(1) + (\tfrac{1}{2})(-1) = 0$$

So it was only the fact that $n_1 \neq n_2$ that kept $\hat{\mathbf{V}}(\hat{\mathbf{\Theta}}^*)$ from being a diagonal matrix in this case. Inspection of Eq. 9.26 shows that $\hat{\mathbf{V}}(\hat{\mathbf{\Theta}}^*)$ will be a diagonal matrix whenever $\mathbf{K}'\mathbf{K}$ is, since the inverse of a diagonal matrix is also diagonal.

But the expression for $\mathbf{K'K}$, displayed on page 314, is

$$\mathbf{K'K} = \begin{bmatrix} N & \dfrac{n_1 - n_2}{2} \\ \dfrac{n_1 - n_2}{2} & \dfrac{N}{4} \end{bmatrix}$$

which shows that $\mathbf{K'K}$ would have been diagonal if n_1 and n_2 had been equal.

For multigroup designs, a convenient set of contrasts to use for ensuring row-wise orthogonality of $\mathbf{C}$ is the set of *Helmert contrasts*. With J groups, this is a set of $J - 1$ contrasts with the following coefficients, respectively:

$$\begin{matrix} J-1 & -1 & -1 & -1 & \cdots & -1 & -1 \\ 0 & J-2 & -1 & -1 & \cdots & -1 & -1 \\ 0 & 0 & J-3 & -1 & \cdots & -1 & -1 \\ \vdots & & & & & & \\ 0 & 0 & 0 & 0 & \cdots & 1 & -1 \end{matrix}$$

For example, when $J = 4$, the set of three Helmert contrasts has the coefficients

$$\begin{matrix} 3 & -1 & -1 & -1 \\ 0 & 2 & -1 & -1 \\ 0 & 0 & 1 & -1 \end{matrix}$$

Preceding this set of rows with one that defines the overall level, $\mu + (\alpha_1 + \alpha_2 + \alpha_3 + \alpha_4)/4$ the full contrast matrix in this case would be

$$\mathbf{C} = \begin{bmatrix} 1 & \frac{1}{4} & \frac{1}{4} & \frac{1}{4} & \frac{1}{4} \\ 0 & 3 & -1 & -1 & -1 \\ 0 & 0 & 2 & -1 & -1 \\ 0 & 0 & 0 & 1 & -1 \end{bmatrix}$$

The reader may verify that for each of the six pairs of rows that may be chosen from this matrix, the sum of the corresponding-elements products is zero. If this contrast matrix is used, *and* if $n_1 = n_2 = n_3 = n_4$, the diagonality of $\hat{\mathbf{V}}(\hat{\boldsymbol{\Theta}}^*)$ (i.e., the uncorrelatedness of the new parameters in $\boldsymbol{\Theta}^*$) will be guaranteed.

What happens, however, if the researcher is *not* interested in the questions implied by the Helmert contrasts (which essentially ask, "Is the first group mean $\overline{Y}_1$ significantly different from the average of the remaining group means? Is $\overline{Y}_2$ significantly different from the average of $\overline{Y}_3, \overline{Y}_4, \ldots, \overline{Y}_J$?" and so on)? Of course, there are other sets of orthogonal contrasts that may be used, but often the research questions cannot be expressed in terms of any orthogonal set.

(Recall that row-wise orthogonality was *not* one of the requirements in constructing the contrast matrix **C**.) The answer is that a further transformation becomes necessary, which, essentially, creates "re-reparameterized" parameters that *are* mutually uncorrelated. After this re-transformation is done, the significance tests may be carried out in the manner illustrated in our numerical example. The details of this orthogonalization transformation are too technical to go into here, and the interested reader is referred to Finn (1974, pp. 297–304). Fortunately, most commercially available computer programs for MANOVA using the general linear model—such as the SAS GLM Procedure and Finn and Bock's (1985) MULTIVARIANCE—make this transformation automatically. The user need not worry about the nonorthogonality of the contrasts that he or she selects for the first transformation matrix **C** for reparameterization. Our main purpose in discussing the nonorthogonality problem was to point out that the significance tests cannot generally be carried out exactly in the manner illustrated in our numerical example. The example should have served to focus on the basic rationale without the encumbrance of the additional problem generated by the nonorthogonality of the contrasts often selected by the researcher.

9.5 HIGHER-ORDER, MULTIVARIATE, AND COVARIANCE DESIGNS

As mentioned at the outset of this chapter, we do not intend to go into any detail with respect to the more complicated designs in ANOVA, MANOVA, and ANCOVA. As a matter of fact, as far as the parameter-estimation aspect is concerned, little needs to be added to what was discussed in Section 9.4 in conjunction with the two-group case. The only difference is that the design matrix becomes more complicated, the vectors become longer (or, in the case of MANOVA, become matrices—one column for each dependendent variable), and the computations become more complicated, almost certainly requiring recourse to a computer. This, as repeatedly pointed out already, is of course the beauty of the general linear model approach. The same set of formulas hold for the most complex as well as the simplest case. Thus, in this section we shall merely display the design matrices for three complex designs to give the reader a flavor of the extensions, and in one case shall also indicate a possible contrast matrix.

Although repeated-measures designs are very important because they are widely used in longitudinal studies, we do not discuss them in this book. This is because, except for the "pure" repeated measures design (e.g., when only one group is studied longitudinally) there are some unsolved problems concerning the conditions under which the analysis is valid, and it would take us too far afield to address these. In fact, for multigroup repeated-measures designs (variously termed split-plot, mixed, or between- and within-subjects designs) the necessary *and* sufficient condition for validity of the analysis was only recently given (Huynh and Mandeville, 1979) even for univariate ANOVA, and the condition does not seem readily generalizable to the multivariate case. Of course, for pure repeated-measures designs (i.e., those with one or more within-subject factors only) there is no problem. If there are q within-subject factors, we need only consider the subjects as constituting an additional factor and treat the design as a $(q + 1)$-factor design with one observation per cell.

In the significance-testing phase, the extension of the techniques to higher-order designs requires somewhat more essential modifications (one of which—a second transformation to orthogonalize the contrasts—was mentioned above); for MANOVA the modifications needed are even more drastic. Nevertheless, all these modifications are in the realm of technical refinements and the general rationale of the approach remains unchanged.

Higher-Order Designs

We illustrate the design matrix and a possible contrast matrix in the context of the simplest nontrivial higher-order design, a 2×3 design, and indicate the general principles for extension to other higher-order designs.

The structural equation for a 2×3 design (with two levels in factor A and three in factor B), including the interaction effect, is

$$Y_{ijk} = \mu + \alpha_j + \beta_k + (\alpha\beta)_{jk} + \epsilon_{ijk} \tag{9.27}$$
$$(i = 1, 2, \ldots, n_{jk}; \quad j = 1, 2; \quad k = 1, 2, 3)$$

Writing a separate equation for members of each of the six (A_j, B_k)-cells, this equation becomes

$$\begin{aligned}
Y_{i11} &= \mu + \alpha_1 + \beta_1 + (\alpha\beta)_{11} + \epsilon_{i11} \\
Y_{i12} &= \mu + \alpha_1 + \beta_2 + (\alpha\beta)_{12} + \epsilon_{i12} \\
Y_{i13} &= \mu + \alpha_1 + \beta_3 + (\alpha\beta)_{13} + \epsilon_{i13} \\
Y_{i21} &= \mu + \alpha_2 + \beta_1 + (\alpha\beta)_{21} + \epsilon_{i21} \\
Y_{i22} &= \mu + \alpha_2 + \beta_2 + (\alpha\beta)_{22} + \epsilon_{i22} \\
Y_{i23} &= \mu + \alpha_2 + \beta_3 + (\alpha\beta)_{23} + \epsilon_{i23}
\end{aligned}$$

Thus the design matrix in this case will have

$$N\left(= \sum_j \sum_k n_{jk}\right)$$

rows and 12 columns. There will be $6(= 2 \times 3)$ sets of repeated rows, numbering $n_{11}, n_{12}, \ldots, n_{23}$, rows, respectively (i.e., one set of repeated rows for each cell in the design). To save space, and also to facilitate grasping the pattern, each set of repeated rows will be indicated by a single row of column vectors of appropriate dimensionality, $n_{11}, n_{12}, \ldots, n_{23}$. The design matrix may then be written as

$$\mathbf{X} = \begin{bmatrix} \mathbf{1} & \mathbf{1} & \mathbf{0} & \mathbf{1} & \mathbf{0} & \mathbf{0} & \mathbf{1} & \mathbf{0} & \mathbf{0} & \mathbf{0} & \mathbf{0} & \mathbf{0} \\ \mathbf{1} & \mathbf{1} & \mathbf{0} & \mathbf{0} & \mathbf{1} & \mathbf{0} & \mathbf{0} & \mathbf{1} & \mathbf{0} & \mathbf{0} & \mathbf{0} & \mathbf{0} \\ \mathbf{1} & \mathbf{1} & \mathbf{0} & \mathbf{0} & \mathbf{0} & \mathbf{1} & \mathbf{0} & \mathbf{0} & \mathbf{1} & \mathbf{0} & \mathbf{0} & \mathbf{0} \\ \mathbf{1} & \mathbf{0} & \mathbf{1} & \mathbf{1} & \mathbf{0} & \mathbf{0} & \mathbf{0} & \mathbf{0} & \mathbf{0} & \mathbf{1} & \mathbf{0} & \mathbf{0} \\ \mathbf{1} & \mathbf{0} & \mathbf{1} & \mathbf{0} & \mathbf{1} & \mathbf{0} & \mathbf{0} & \mathbf{0} & \mathbf{0} & \mathbf{0} & \mathbf{1} & \mathbf{0} \\ \mathbf{1} & \mathbf{0} & \mathbf{1} & \mathbf{0} & \mathbf{0} & \mathbf{1} & \mathbf{0} & \mathbf{0} & \mathbf{0} & \mathbf{0} & \mathbf{0} & \mathbf{1} \end{bmatrix} \begin{matrix} (A_1B_1) \\ (A_1B_2) \\ (A_1B_3) \\ (A_2B_1) \\ (A_2B_2) \\ (A_2B_3) \end{matrix}$$

where it is understood that $\mathbf{X}$ is an $N \times 12$ matrix, not 6×12. For example, if there are 5 subjects in ech cell, $\mathbf{X}$ is a 30×12 matrix, with 5 repeated rows of each type that is explicitly displayed. [The (A_j, B_k) written to the right of $\mathbf{X}$ identifies the cell to which each row of $\mathbf{X}$ corresponds.] The parameter vector $\boldsymbol{\Theta}$ will be a 12-dimensional column vector which is the transpose of

$$\boldsymbol{\Theta}' = [\mu, \alpha_1, \alpha_2, \beta_1, \beta_2, \beta_3, (\alpha\beta)_{11}, (\alpha\beta)_{12}, \ldots, (\alpha\beta)_{23}]$$

With these definitions of the design matrix $\mathbf{X}$ and parameter vector $\boldsymbol{\Theta}$, respectively, and obvious definitions for the observation vector $\mathbf{Y}$ and error vector $\boldsymbol{\varepsilon}$, the reader may verify that the N instances of the structural equation (the six distinct types of which were displayed after Eq. 9.27, assume exactly the form of Eq. 9.11 written earlier for the two-group example, that is,

$$\mathbf{Y} = \mathbf{X}\boldsymbol{\Theta} + \boldsymbol{\varepsilon} \tag{9.11}$$

The effect of forming the product $\mathbf{X}\boldsymbol{\Theta}$ is, as before, to pick out for members in each cell the appropriate ones of the 12 parameters to appear in the structural equation for their Y-scores. For example, the score of any member of the (A_1, B_2)-cell, that is, any element of $\mathbf{Y}$ from the $(n_{11} + 1)$th through the $(n_{11} + n_{12})$th, will be expressed as the sum of those four parameters listed in $\boldsymbol{\Theta}$ that correspond, in ordinal position, to the four $\mathbf{1}$'s in the second row of $\mathbf{X}$, plus the appropriate element of $\boldsymbol{\varepsilon}$. Thus the first, second, fifth, and eighth elements in $\boldsymbol{\Theta}$ [i.e., μ, α_1, β_2, and $(\alpha\beta)_{12}$] will appear in the structural equation for Y_{i12}.

The next step is to construct the contrast matrix $\mathbf{C}$ for defining a new parameter matrix $\boldsymbol{\Theta}^*$ as in Eq. 9.13: $\boldsymbol{\Theta}^* = \mathbf{C}\boldsymbol{\Theta}$. How many new parameters (and hence, how many rows in **C**) can we have at the most? The simple answer, as before, is "as many as there are cells in the design," which is six in this case.[8]

One way to generate a suitable contrast matrix is as follows:

1. In the first row of a 6 × 12 matrix, write the coefficients for a general level parameter:

$$[1, \tfrac{1}{2}, \tfrac{1}{2}, \tfrac{1}{3}, \tfrac{1}{3}, \tfrac{1}{3}, \tfrac{1}{6}, \tfrac{1}{6}, \tfrac{1}{6}, \tfrac{1}{6}, \tfrac{1}{6}, \tfrac{1}{6}]$$

2. Write the submatrix showing the genuine contrast(s) for each factor separately; that is, the total contrast matrix for each factor, *less* the first row and first column. For clarifying the procedure, we list below the total contrast matrices, with the first row and column stricken out. The symbols $\mathbf{C}_a$ and $\mathbf{C}_b$ stand for the remaining submatrices, which are all we need. [They would be of order $(J - 1) \times J$ and $(K - 1) \times K$ if factor A had J levels and B, K levels.]

$$\mathbf{C}_a = \begin{bmatrix} 1 & \frac{1}{2} & \frac{1}{2} \\ 0 & 1 & -1 \end{bmatrix} \qquad \mathbf{C}_b = \begin{bmatrix} 1 & \frac{1}{3} & \frac{1}{3} & \frac{1}{3} \\ 0 & 1 & 0 & -1 \\ 0 & 0 & 1 & -1 \end{bmatrix}$$

3. Transcribe $\mathbf{C}_a$ and $\mathbf{C}_b$, respectively, into the second row and the third and fourth rows of the 6 × 12 matrix—the former to fall in the second and third columns (under the $\frac{1}{2}$, $\frac{1}{2}$ of the first row), and the latter in the fourth through

[8]The curious reader may wonder how this number is related to the number of independent equations among the set of 12 normal equations that could be written in this case after the fashion of Eqs. 9.8. The answer is as follows. Denoting the mth column of $\mathbf{X}$ by $\mathbf{X}_m$, we note that the following six dependencies or relations hold among the 12 columns:

$$\begin{aligned} \mathbf{X}_2 + \mathbf{X}_3 &= \mathbf{X}_1 \\ \mathbf{X}_4 + \mathbf{X}_5 + \mathbf{X}_6 &= \mathbf{X}_1 \\ \mathbf{X}_7 + \mathbf{X}_8 + \mathbf{X}_9 + \mathbf{X}_{10} + \mathbf{X}_{11} + \mathbf{X}_{12} &= \mathbf{X}_1 \\ \mathbf{X}_7 + \mathbf{X}_8 + \mathbf{X}_9 &= \mathbf{X}_2 \\ \mathbf{X}_7 + \mathbf{X}_{10} &= \mathbf{X}_4 \\ \mathbf{X}_8 + \mathbf{X}_{11} &= \mathbf{X}_5 \end{aligned}$$

(There are other relations among the columns, such as $\mathbf{X}_{10} + \mathbf{X}_{11} + \mathbf{X}_{12} = \mathbf{X}_3$, for example. But this one follows automatically from the first, third, and fourth relations above, and hence is redundant. Careful examination will reveal that no further *independent* relations hold among the 12 columns beyond the six listed above.) With six independent relations holding among 12 columns, we say that there are only 12 − 6 = 6 *independent* columns in $\mathbf{X}$, or that the rank of $\mathbf{X}$ is 6. Hence the rank of $\mathbf{X}'\mathbf{X}$ (the coefficient matrix for the set of normal equations in matrix form, as in Eq. 9.12 is also 6. Therefore, there are only six independent normal equations in the set to which Eq. 9.8 generalizes in this case. Thus an admissable contrast matrix can have at most six rows.

sixth columns (under $\frac{1}{3}$, $\frac{1}{3}$, $\frac{1}{3}$ of the first row). The partially completed 6×12 matrix now looks like this:

$$\left[\begin{array}{c|cc|ccc|cccccc}
1 & \frac{1}{2} & \frac{1}{2} & \frac{1}{3} & \frac{1}{3} & \frac{1}{3} & \frac{1}{6} & \frac{1}{6} & \frac{1}{6} & \frac{1}{6} & \frac{1}{6} & \frac{1}{6} \\
 & 1 & -1 & & & & & & & & & \\
 & & & 1 & 0 & -1 & & & & & & \\
 & & & 0 & 1 & -1 & & & & & & \\
\hline
 & & & & & & & & & & & \\
 & & & & & & & & & & &
\end{array}\right]$$

4. The 1 and -1 in the second row are regarded as having been obtained by subtracting the average of the fourth through sixth rows of matrix **X** (p. 335) from the average of the first through third rows. Thus the next three entries of the second row of **C** are zeros. However, the last six entries will be

$$\frac{1}{3} \quad \frac{1}{3} \quad \frac{1}{3} \quad -\frac{1}{3} \quad -\frac{1}{3} \quad -\frac{1}{3}$$

Similarly, the 1 and -1 in the third row of **C** come from subtracting the average of the third and sixth rows of **X** from the average of the first and fourth rows. Therefore, the last six entries of the third row of **C** will be

$$\frac{1}{2} \quad 0 \quad -\frac{1}{2} \quad \frac{1}{2} \quad 0 \quad -\frac{1}{2}$$

Similarly, the last six entries of the fourth row of **C** are

$$0 \quad \frac{1}{2} \quad -\frac{1}{2} \quad 0 \quad \frac{1}{2} \quad -\frac{1}{2}$$

The partially completed 6×12 matrix now becomes

$$\left[\begin{array}{c|cc|ccc|cccccc}
1 & \frac{1}{2} & \frac{1}{2} & \frac{1}{3} & \frac{1}{3} & \frac{1}{3} & \frac{1}{6} & \frac{1}{6} & \frac{1}{6} & \frac{1}{6} & \frac{1}{6} & \frac{1}{6} \\
 & 1 & -1 & & & & \frac{1}{3} & \frac{1}{3} & \frac{1}{3} & -\frac{1}{3} & -\frac{1}{3} & -\frac{1}{3} \\
 & & & 1 & 0 & -1 & \frac{1}{2} & 0 & -\frac{1}{2} & \frac{1}{2} & 0 & -\frac{1}{2} \\
 & & & 0 & 1 & -1 & 0 & \frac{1}{2} & -\frac{1}{2} & 0 & \frac{1}{2} & -\frac{1}{2} \\
\hline
 & & & & & & & & & & & \\
 & & & & & & & & & & &
\end{array}\right]$$

where the remaining blank positions in the second through fourth rows are understood to be zeros (deliberately left out to highlight the pattern so far).

5. The first six elements of the last two rows are also zeros. It remains only to fill the last six positions of these rows, which is done as follows[9]:

(a) Multiply the first row of $\mathbf{C}_b$ by the first element of $\mathbf{C}_a$ (i.e., 1 in this case). To this result (which is just the first row of $\mathbf{C}_b$ itself in this case) augment the product of the first row of $\mathbf{C}_b$ by the second element (-1 in this case) of $\mathbf{C}_a$. The combined result is a six-dimensional row vector.

$$[1, 0, -1, -1, 0, 1]$$

(b) Repeat the same procedure as above with the second row of $\mathbf{C}_b$, resulting in

$$[0, 1, -1, 0, -1, 1]$$

Putting together these two six-element row vectors, we have the lower right-hand 2 × 6 submatrix of the 6 × 12 contrast matrix **C**, and the final result is

$$\mathbf{C} = \left[\begin{array}{c|cc|ccc|cccccc}
1 & \frac{1}{2} & \frac{1}{2} & \frac{1}{3} & \frac{1}{3} & \frac{1}{3} & \frac{1}{6} & \frac{1}{6} & \frac{1}{6} & \frac{1}{6} & \frac{1}{6} & \frac{1}{6} \\
0 & 1 & -1 & 0 & 0 & 0 & \frac{1}{3} & \frac{1}{3} & \frac{1}{3} & -\frac{1}{3} & -\frac{1}{3} & -\frac{1}{3} \\
0 & 0 & 0 & 1 & 0 & -1 & \frac{1}{2} & 0 & -\frac{1}{2} & \frac{1}{2} & 0 & -\frac{1}{2} \\
0 & 0 & 0 & 0 & 1 & -1 & 0 & \frac{1}{2} & -\frac{1}{2} & 0 & \frac{1}{2} & -\frac{1}{2} \\
\hline
0 & 0 & 0 & 0 & 0 & 0 & 1 & 0 & -1 & -1 & 0 & 1 \\
0 & 0 & 0 & 0 & 0 & 0 & 0 & 1 & -1 & 0 & -1 & 1
\end{array}\right]$$

[9]The somewhat belabored instructions for this step are to provide for an easier transition to more complicated cases. For exammple, in a 3 × 3 design with factor A quantitative, so that the set of Helmert contrasts could be of interest, we might have

$$\mathbf{C}_a = \begin{bmatrix} 2 & -1 & -1 \\ 0 & 1 & -1 \end{bmatrix}$$

Assuming $\mathbf{C}_b$ to remain as above, the lower right-hand segment to be completed in this step would now be a 4 × 9 submatrix, as follows:

$$\begin{bmatrix} (2)\begin{bmatrix} 1 & 0 & -1 \\ 0 & 1 & -1 \end{bmatrix} & (-1)\begin{bmatrix} 1 & 0 & -1 \\ 0 & 1 & -1 \end{bmatrix} & (-1)\begin{bmatrix} 1 & 0 & -1 \\ 0 & 1 & -1 \end{bmatrix} \\ (0)\begin{bmatrix} 1 & 0 & -1 \\ 0 & 1 & -1 \end{bmatrix} & (1)\begin{bmatrix} 1 & 0 & -1 \\ 0 & 1 & -1 \end{bmatrix} & (-1)\begin{bmatrix} 1 & 0 & -1 \\ 0 & 1 & -1 \end{bmatrix} \end{bmatrix}$$

$$= \left[\begin{array}{ccc|ccc|ccc}
2 & 0 & -2 & -1 & 0 & 1 & -1 & 0 & 1 \\
0 & 2 & -2 & 0 & -1 & 1 & 0 & -1 & 1 \\
\hline
0 & 0 & 0 & 1 & 0 & -1 & -1 & 0 & 1 \\
0 & 0 & 0 & 0 & 1 & -1 & 0 & -1 & 1
\end{array}\right]$$

The reader should recall from Chapter 2 that the combination of two matrices in the manner illustrated above—multiplying the *second* matrix (of order $p \times q$, say) by *each* element of the first matrix (of order $m \times n$) in turn, and displaying the results as an $mp \times nq$ matrix—is called forming the *Kronecker product* of the two matrices, symbolized by $\mathbf{A} \otimes \mathbf{B}$. Thus the last matrix above is denoted by $\mathbf{C}_a \otimes \mathbf{C}_b$.

Having defined an acceptable contrast matrix **C**, we can now determine the new parameter matrix $\mathbf{\Theta}^*$ in accordance with Eq. 9.13,

$$\mathbf{\Theta}^* = \mathbf{C\Theta}$$

and the rest follows exactly as outlined in the summary of Section 9.4 so far as the parameter-estimation phase is concerned. That is, we compute the basis matrix following Eq. 9.15:

$$\mathbf{K} = (\mathbf{XC}')(\mathbf{CC}')^{-1}$$

Estimates of the new parameters are then computed from the reparameterized normal equation, (9.16):

$$(\mathbf{K}'\mathbf{K})\hat{\mathbf{\Theta}}^* = \mathbf{K}'\mathbf{Y}$$

whose solution is

$$\hat{\mathbf{\Theta}}^* = (\mathbf{K}'\mathbf{K})^{-1}(\mathbf{K}'\mathbf{Y}) \tag{9.17}$$

This completes the parameter-estimation phase.

The significance-testing aspect for all except the completely orthogonal design (orthogonal contrasts and equal cell sizes) requires a further reparameterization, as already mentioned. Apart from this complication, which is too technical for us to go into here, everything follows the pattern discussed in Section 9.4. Hence we shall not discuss this matter further here.

Multivariate Designs

When more than one criterion variable is to be considered simultaneously, we have a multivariate design. For example, if we measure performance by speed and accuracy scores, we are dealing with a bivariate problem. Any univariate ANOVA design can be generalized to a multivariate (MANOVA) design with minor modifications through the parameter-estimation phase in the general linear model approach. The significance-testing aspect requires further modifications beyond the ''re-reparameterization'' referred to earlier, part of which relates to different test statistics that have to be used, as discussed in Section 8.6.

Suppose that we have a set of p criterion variables we wish to use simultaneously in an experiment using a one-way MANOVA design. To save space we let $p = 2$ in the following exposition, but the discussions hold for any p. Also, we shall explicitly deal with the two-group case again. The observation vector **Y** and error vector $\boldsymbol{\varepsilon}$ of Section 9.4 now become $N \times 2$ matrices, and the parameter vector $\mathbf{\Theta}$ becomes a 3×2 matrix [in general, a $(J + 1) \times p$

matrix]. The design matrix **X** will remain unchanged. The structural equation in extended matrix form, corresponding to Eq. 9.10, is

$$\begin{bmatrix} Y_{11}^{(1)} & Y_{11}^{(2)} \\ \vdots & \\ Y_{n_1 1}^{(1)} & Y_{n_1 1}^{(2)} \\ Y_{12}^{(1)} & Y_{12}^{(2)} \\ \vdots & \\ Y_{n_2 2}^{(1)} & Y_{n_2 2}^{(2)} \end{bmatrix} = \begin{bmatrix} 1 & 1 & 0 \\ \vdots & \vdots & \vdots \\ 1 & 1 & 0 \\ 1 & 0 & 1 \\ \vdots & \vdots & \vdots \\ 1 & 0 & 1 \end{bmatrix} \begin{bmatrix} \mu^{(1)} & \mu^{(2)} \\ \alpha_1^{(1)} & \alpha_1^{(2)} \\ \alpha_2^{(1)} & \alpha_2^{(2)} \end{bmatrix} + \begin{bmatrix} \epsilon_{11}^{(1)} & \epsilon_{11}^{(2)} \\ \vdots & \vdots \\ \epsilon_{n_1 1}^{(1)} & \epsilon_{n_1 1}^{(2)} \\ \epsilon_{12}^{(1)} & \epsilon_{12}^{(2)} \\ \vdots & \\ \epsilon_{n_2 2}^{(1)} & \epsilon_{n_2 2}^{(2)} \end{bmatrix}$$

where the superscripts in parentheses indicate the criterion variable number.

In symbolic matrix form, however, this assumes exactly the same form as Eq. (9.11), that is,

$$\mathbf{Y} = \mathbf{X\Theta} + \boldsymbol{\varepsilon}$$

the only difference being that $\mathbf{Y}$, $\mathbf{\Theta}$, and $\boldsymbol{\varepsilon}$ are now matrices instead of vectors. Consequently, all the developments in Section 9.4 continue to hold without any further modifications except those that flow naturally from the above changes. The contrast matrix **C** remains unchanged, and hence also the basis matrix **K** of Eq. 9.15, which involves only the invariant **X** and **C** matrices in its computation:

$$\mathbf{K} = (\mathbf{XC}')(\mathbf{CC}')^{-1}$$

Of course, the estimated new parameter matrix $\hat{\mathbf{\Theta}}^*$ computed in accordance with Eq. 9.17,

$$\hat{\mathbf{\Theta}}^* = (\mathbf{K}'\mathbf{K})^{-1}(\mathbf{K}'\mathbf{Y})$$

will be a 2×2 (or, in general, a $J \times p$) matrix instead of a two-dimensional column vector.

In the significance-testing phase, the developments in Section 9.5 hold, with the modifications described below, up to but exclusive of the actual tests—provided that the design is completely orthogonal (i.e., if the contrasts are orthogonal and the groups are of equal size). If the design is not completely orthogonal, a further re-parameterization to achieve orthogonality is needed (just as in the univariate case), after which the procedures of Section 9.5 may be applied, again exclusive of the actual tests.

One essential modification in the multivariate situation is that instead of a population variance σ^2 of the single criterion variable Y in the univariate case, we have a population variance–covariance matrix $\mathbf{\Sigma}$ of order equal to the number of criterion variables, p. Consequently, the variance–covariance matrix $\mathbf{V}(\hat{\mathbf{\Theta}}^*)$

of the new parameters is given by a generalized form of Eq. 9.21, instead of that equation itself. Denoting the matrix $(\mathbf{K}'\mathbf{K})^{-1}$ which appears in Eq. 9.21 by $\mathbf{H}$ for short, the generalized form of Eq. 9.21 for the two-group, two-variable case ($J = p = 2$) is

$$\mathbf{V}(\hat{\boldsymbol{\Theta}}^*) = \begin{bmatrix} h_{11}\boldsymbol{\Sigma} & h_{12}\boldsymbol{\Sigma} \\ h_{21}\boldsymbol{\Sigma} & h_{22}\boldsymbol{\Sigma} \end{bmatrix} \tag{9.28}$$

where the h_{st} are the elements of $\mathbf{H} = (\mathbf{K}'\mathbf{K})^{-1}$. Note that when $p = 1$ (i.e., in the univariate case), the matrix $\boldsymbol{\Sigma}$ becomes a scalar σ^2, which may be factored out of the matrix to yield

$$\mathbf{V}(\hat{\boldsymbol{\Theta}}^*) = \mathbf{H}\sigma^2$$

which is precisely Eq. 9.21. The formal similarity of Eq. 9.28 to Eq. 9.21 is enhanced if we adopt the Kronecker product notation (see footnote on p. 338) for we may then write

$$\mathbf{V}(\hat{\boldsymbol{\Theta}}^*) = (\mathbf{K}'\mathbf{K})^{-1} \otimes \boldsymbol{\Sigma} \tag{9.29}$$

For estimating $\boldsymbol{\Sigma}$ from the sample in order to have a usable formula to replace Eq. 9.29, we go through the steps of computing $\mathbf{S}_T$ and $\mathbf{S}_B$ (which are now $p \times p$ matrices instead of scalars) from Eqs. 9.22 and 9.23, respectively, which need no modification. We then obtain the estimate of $\boldsymbol{\Sigma}$ from Eq. 9.25, modified only to the extent that the left-hand side is changed from $\hat{\sigma}^2$ to $\hat{\boldsymbol{\Sigma}}$, a $p \times p$ matrix. The final result, $\hat{\mathbf{V}}(\hat{\boldsymbol{\Theta}}^*)$, will be given by a generalized form of Eq. 9.26, obtained by substituting $\hat{\boldsymbol{\Sigma}}$ for $\boldsymbol{\Sigma}$ in Eq. 9.29; that is,

$$\hat{\mathbf{V}}(\hat{\boldsymbol{\Theta}}^*) = \frac{1}{N - J}(\mathbf{K}'\mathbf{K})^{-1} \otimes [\mathbf{Y}'\mathbf{Y} - \hat{\boldsymbol{\Theta}}^{*\prime}(\mathbf{K}'\mathbf{Y})] \tag{9.30}$$

Up to this point, the modifications—although fairly extensive—are only formal in nature (especially when the Kronecker product notation is used). It is what we do with the diagonal elements, or more precisely the diagonal submatrices, of $\hat{\mathbf{V}}(\hat{\boldsymbol{\Theta}}^*)$ that sets the multivariate extension considerably apart from its univariate counterpart. The actual test procedures are those discussed in Section 8.6.

In concluding this subsection, we point out that for higher-order multivariate designs, both the modifications discussed in the preceding subsection and those just described become necessary. As was pointed out in Section 8.4, it is also often desirable to follow up a MANOVA by a discriminant analysis for each significant effect observed, which is routinely done in Finn and Bock's (1985) MULTIVARIANCE program alluded to earlier.

Covariance Analysis

In an experiment using the ANOVA design, it may sometimes happen that truly random assignment of subjects to treatment groups is infeasible. For instance, existing intact classes may have to be taught by two methods for teaching some subject matter in an experiment seeking to compare their relative merits.[10] We may then find that the two classes differ significantly in some relevant attribute, say IQ, and hence it may become questionable whether any differences in criterion performance observed are due to the teaching methods or to IQ differences (or both). This may be true even if random assignment had been made to the treatment groups. The two (or more) groups may happen to differ considerably in mean IQ, or some other relevant variable, called the covariate. In such situations covariance analysis is a useful tool. It allows us to test whether method differences remain after the effect of different covariate means in the two classes has been "partialed out."

The structural model for a one-way covariance analysis is

$$Y_{ij} = \mu + \alpha_j + b_j(X_{ij} - \bar{X}) + \epsilon_{ij} \quad (i = 1, 2, \ldots, n_j;\ j = 1, 2, \ldots, J) \qquad (9.31)$$

where b_j is the regression slope of Y on the covariate X in the jth group, and $\bar{X}$ is the grand mean of X. For the two-group case, this may be written in extended matrix notation as

$$\begin{bmatrix} Y_{11} \\ \vdots \\ Y_{n_1 1} \\ Y_{12} \\ \vdots \\ Y_{n_2 2} \end{bmatrix} = \begin{bmatrix} 1 & 1 & 0 & x_{11} & 0 \\ \vdots & \vdots & \vdots & \vdots & \vdots \\ 1 & 1 & 0 & x_{n_1 1} & 0 \\ 1 & 0 & 1 & 0 & x_{12} \\ \vdots & \vdots & \vdots & \vdots & \vdots \\ 1 & 0 & 1 & 0 & x_{n_2 2} \end{bmatrix} \begin{bmatrix} \mu \\ \alpha_1 \\ \alpha_2 \\ b_1 \\ b_2 \end{bmatrix} + \begin{bmatrix} \epsilon_{11} \\ \vdots \\ \epsilon_{n_1 1} \\ \epsilon_{12} \\ \vdots \\ \epsilon_{n_2 2} \end{bmatrix}$$

where $x_{ij} = X_{ij} - \bar{X}$, the deviation of the covariate score of the ith member of group j from the grand mean of the covariate.

Again, this equation is seen to reduce, in symbolic matrix form, to Eq. 9.11:

$$\mathbf{Y} = \mathbf{X}\boldsymbol{\Theta} + \boldsymbol{\varepsilon}$$

The crucial difference, of course, is that the "design matrix" **X** now includes actual covariate scores (in deviation form) besides group-membership dummy variates taking values 1 or 0.

The nature of covariance analysis requires a slightly different testing (and hence also parameter-estimation) procedure from straight ANOVA. We must first test whether $b_1 = b_2$. The contrast matrix for this purpose is (in the two-group case) the row vector

[10]Strictly speaking, in such cases several classes should be chosen, and half of them randomly assigned to one teaching method, the other half to the other method. The classes would then be used as the unit of analysis. But often it is impossible to get a sufficient number of classes for this approach. One is then forced to rely on the quasi-random nature of assignment of pupils to classes, and use the two-class approach.

$$\mathbf{C} = [0, 0, 0, 1, -1]$$

Depending on the outcome of this test, the parameter vector is left as is (when it is concluded that $b_1 \neq b_2$) or modified to the transpose of

$$\mathbf{\Theta}' = [\mu, \alpha_1, \alpha_2, b, b]$$

The final step requires testing whether $\alpha_1 = \alpha_2$ when the effect of b (or b_1 and b_2) is partialed out. Essentially, this consists in forming "residualized" matrices corresponding to $\mathbf{S}_T$ and $\mathbf{S}_B$, after the parameter vector exclusive of the regression coefficients has been transformed in the manner described in Section 9.4. In principle, this means that wherever $\mathbf{Y}$ appears (explicitly, or implicitly through $\hat{\mathbf{\Theta}}^*$) in the formulas for $\mathbf{S}_T$ and $\mathbf{S}_B$, it is replaced by the vector of residuals $\mathbf{Y} - \hat{\mathbf{Y}}$ of Y from its regression on the covariate X. In practice, however, the step of residualizing the criterion scores themselves is side-stepped, and adjustments are made directly on the sums of squares.

9.6 THEORETICAL SUPPLEMENT: EQUIVALENCE OF t-TESTS FOR TESTING $\mu_1 = \mu_2$ AND FOR TESTING $\beta = 0$

We have already seen that the sample regression weight in question is given by

$$b = \bar{Y}_1 - \bar{Y}_2 \tag{9.4}$$

where $\bar{Y}_1$ is the criterion mean of the group for which $X = 1$, and $\bar{Y}_2$ is that of the group with $X = 0$.

The t-statistic for testing the significance of b (i.e., for testing whether b is significantly different from 0) is easily derivable from the more familiar t-statistic for the significance of the correlation coefficient r, namely,

$$t = \frac{r\sqrt{N-2}}{\sqrt{1-r^2}}$$

We need only express r in terms of b by the relation

$$r = bs_x/s_y$$

which, substituted in the t formula, yields

$$\begin{aligned} t &= \frac{(bs_x/s_y)\sqrt{N-2}}{\sqrt{1 - b^2s_x^2/s_y^2}} \\ &= \frac{bs_x\sqrt{N-2}}{\sqrt{s_y^2 - b^2s_x^2}} \end{aligned} \tag{9.32}$$

Now the variance s_x^2 (in the total sample) for the dummy variate X is

$$s_x^2 = \frac{\sum X^2 - (\sum X)^2/N}{N - 1}$$

$$= \frac{n_1 - n_1^2/N}{N - 1}$$

$$= \frac{n_1 n_2}{N(N - 1)}$$

Substituting this and expression (9.4) for b in (9.32) gives

$$t = \frac{(\bar{Y}_1 - \bar{Y}_2)\sqrt{\dfrac{n_1 n_2}{N(N - 1)}}\sqrt{N - 2}}{\sqrt{s_y^2 - (\bar{Y}_1 - \bar{Y}_2)^2 \dfrac{n_1 n_2}{N(N - 1)}}}$$

or (9.33)

$$t = \frac{(\bar{Y}_1 - \bar{Y}_2)\sqrt{\dfrac{n_1 n_2 (N - 2)}{N}}}{\sqrt{(N - 1)s_y^2 - \dfrac{n_1 n_2}{N}(\bar{Y}_1 - \bar{Y}_2)^2}}$$

The first term under the radical in the denominator of the right-hand expression is readily recognized to be the total sum of squares (SS_t) for Y. The second term turns out to be the between-groups sum of squares for the two-group case. For, from the usual formula for SS_b, we know that

$$SS_b = \sum_{j=1}^{2} n_j(\bar{Y}_j - \bar{Y})^2$$

in which we may substitute the expression

$$\bar{Y} = \frac{n_1\bar{Y}_1 + n_2\bar{Y}_2}{N}$$

for the grand mean $\bar{Y}$, to obtain, successively,

$$SS_b = n_1\left(\bar{Y}_1 - \frac{n_1\bar{Y}_1 + n_2\bar{Y}_2}{N}\right)^2 + n_2\left(\bar{Y}_2 - \frac{n_1\bar{Y}_1 + n_2\bar{Y}_2}{N}\right)^2$$

$$= \frac{n_1}{N^2}(N\bar{Y}_1 - n_1\bar{Y}_1 - n_2\bar{Y}_2)^2 + \frac{n_2}{N^2}(N\bar{Y}_2 - n_1\bar{Y}_1 - n_2\bar{Y}_2)^2$$

$$= \frac{n_1 n_2^2}{N^2} (\overline{Y}_1 - \overline{Y}_2)^2 + \frac{n_2 n_1^2}{N^2} (\overline{Y}_2 - \overline{Y}_1)^2$$

$$= \frac{n_1 n_2}{N} (\overline{Y}_1 - \overline{Y}_2)^2$$

Consequently the expression under the radical in Eq. 9.33 becomes

$$SS_t - SS_b = SS_w$$

the within-groups sum of squares of $\overline{Y}$. Hence the entire expression for t becomes

$$t = \frac{(\overline{Y}_1 - \overline{Y}_2)\sqrt{\dfrac{n_1 n_2 (N-2)}{N}}}{\sqrt{SS_w}}$$

or

$$t = \frac{\overline{Y}_1 - \overline{Y}_2}{\sqrt{\dfrac{SS_w}{N-2} \dfrac{N}{n_1 n_2}}}$$

$$= \frac{\overline{Y}_1 - \overline{Y}_2}{\sqrt{MS_w(1/n_1 + 1/n_2)}}$$

which is precisely the familiar expression for the t-statistic for testing the significance of the difference between two independent means.

9.7 RESEARCH EXAMPLES

Textbooks by Finn (1974) and Bock (1975) provided early and exemplary expositions of the general linear model approach to the analysis of unbalanced MANOVA designs, with and without covariates. Also, those authors were responsible for the first general linear model program package which achieved widespread distribution. Their MULTIVARIANCE program package continues to be one of the best tools for performing such analyses (Finn and Bock, 1985). Finn (1972b) represented the first research report in the education literature which demonstrated a five-way fixed-effects exact nonorthogonal multivariate analysis of variance. The randomized experiment was designed to probe the influence of expectations teachers held for students on teacher ratings of essays written by students. The *expectations* were stereotyped responses of teachers to information about sex, race, and ability level of the student writers. Four pairs of essays were used as stimuli presentations, and these were randomly paired with fictional sex, race, and ability level identifiers. Subjects whose essay ratings were usable were 300 volunteer elementary teachers, some from a large urban public school system and the rest from suburban systems around the urban center. Thus there were five main effects: (1) locale, (2) essay pair, (3) writer sex, (4) writer race, (5) writer ability level. The experiment was multivariate on the criterion side because there were two correlated rating scores, one for each essay, assigned by the teachers. Natural attrition of subjects for a great variety of reasons was the reason that a planned orthogonal MANOVA emerged as a nonorthogonal MANOVA. Since it is practically impossible to hold all subjects in a social science experiment, it is to be expected that almost all such experiments will end up requiring the GLM computational scheme to accomodate unequal cell sample sizes.

The Selected MANOVA Tests Table (Table 9.1) summarizes the multivariate significance test outcomes. Contrary to the prevalent theory of teacher-expectancy effects on student achievement, the main effect for ability level was not significant. However, the locale × ability interaction was significant, and as revealed by the Estimated Ability Differences for Locales Table (Table 9.2), the reality was that urban teachers provided quite disparate average ratings for students identified as high in ability and low in ability, whereas suburban teachers provided quite similar average ratings for students identified as the two ability levels. The implication was that urban teachers reacted to essays in reaction to their stereotypes for ability levels, while suburban teachers were able to avoid

TABLE 9.1 Selected MANOVA Tests

Source of Variation	n.d.f.	Decision (α = .05)
Locale (L)	1	Not significant
Race (R), eliminating L	1	Not significant
Sex (S), eliminating L, R	1	Not significant
Ability (A), eliminating L, R, S, E	1	Not significant
Essay pair (E), eliminating L, R, S, A	3	Significant
Locale × race, eliminating all above	1	Not significant
Locale × sex, eliminating all above	1	Significant
Locale × ability, eliminating all above	1	Significant
Locale × essay, eliminating all above	3	Not significant
Race × sex, eliminating all above	1	Not significant
Race × ability, eliminating all above	1	Not significant
Race × essay, eliminating all above	3	Not significant
Sex × ability, eliminating all above	1	Not significant
Sex × essay, eliminating all above	3	Not significant
Ability × essay, eliminating all above	3	Not significant
Locale × race × sex, eliminating all above	1	Significant
All other three-way interactions	1 or 3	Not significant
Residual	15	Not significant
Within-groups	237	

TABLE 9.2 Estimated Ability Differences for Locales

Source of High–Low Ability Differences	Essay 1	Essay 2
Urban teachers (N = 113)	8.41 (2.48)	5.84 (2.70)
Suburban teachers (N = 187)	.36 (1.92)	.83 (2.09)

Standard errors for mean differences in parentheses.

stereotypical reactions and to react to the actual qualities of the essays. Thus Finn's qualified answer to the question of whether teacher expectancies impact on student achievement was that for urban teachers they do and for suburban teachers they don't. This very important finding severely restricts the generality of the theory of teacher expectancy. It suggests that the theory applies to urban teachers and does not apply to suburban teachers, raising the question of what characteristics of these classes of teachers and/or the environments in which they teach can account for the difference in their behaviors. Finn provided considerable information about the two groups of teachers and schools to fuel speculation on this question.

Finn also found a significant locale × sex interaction and a significant locale × race × sex interaction. The Estimated Sex Differences by Race within Locales Table (Table 9.3) reveals that again the stereotype influences were much stronger in the average ratings provided by urban teachers than they were in the average ratings provided by suburban teachers. Urban teachers tended to act out an expectation that black females write better than do black males in rating essay 1, and to act out an expectation that white males are better than white females in rating both essays.

Finn (1980) reported a cross-national study (England, Sweden, U.S.) of sex differences in school achievement test scores. This was an ecological survey rather than a randomized experiment, but the facility of the GLM analysis, via MULTIVARIANCE to cope with unequal sample sizes in a non-orthogonal MANOVA setup was crucial to the work. Random samples of 14-year-olds were obtained from the International Association for the Evaluation of Educational Achievement (IEA) data files. Twelve criterion measurements were collected for very large samples of subjects (2,777 pupils in 47 one-sex and 88

TABLE 9.3 Estimated Sex Differences by Race within Locales

Source of Male–Female Differences	Essay 1	Essay 2
Urban teachers rating blacks	−5.52 (3.48)	1.81 (3.78)
Urban teachers rating whites	6.24 (3.58)	13.32 (3.90)
Suburban teachers on blacks	1.70 (2.71)	1.68 (2.95)
Suburban teachers on whites	−1.38 (2.73)	−4.03 (2.97)

Standard errors within parentheses.

coed schools in England; 2,324 pupils in 95 coed schools in Sweden, which had no one-sex schools; 5,193 pupils in 8 one-sex and 118 coed schools in the U.S.). MANOVA significance tests were employed to demonstrate that the main effect for Sex was significant, and that there was no significant interaction of Sex with Grade in School. The subjects were found in two grade levels in each country, and it had been expected that sex differences would be larger in the higher grade than in the lower, warranting the proposition that schooling was operating to increase sex differences as grade increased. The failure of this hypothesis was judged by the author to be noteworthy.

Finn provided a great deal of information about the schools in the three countries which can support speculation about how sex differences are related to school characteristics. However, just the task of displaying sex differences across three countries and two types of schools on 12 criteria is challenging. Finn solved the problem with an ingenious table of standardized male-minus-female mean differences (using the standard deviation of scores within a grade and a country as the metric for each difference), an edited version of which is given as the Standardized M-F Differences by Grade and School Type Table (Table 9.4). Notice the huge sex differences in favor of boys for the science subjects in the U.S. 9th grade one-sex schools (.60 for biology, .73 for chemistry, 1.26 for physics, and 1.30 for science practical).

TABLE 9.4 Standardized Male-Female Differences by Grade and School Type

Country:	England				Sweden		U.S.		
School Type:	One-sex		Coed		Coed		One-sex	Coed	
Grade:	8	9	8	9	7	8	9	8	9
Scale									
Reading comprehension	−36	−41	08	08	−04	−09	22	09	−02
Word knowledge	−03	−21	−09	20	−12	−02	37	−04	02
Hours reading for pleasure	−37	−33	−20	−22	−63	−50	−15	−21	−27
Reading attitude	−44	−39	−46	−36	−31	−31	−28	−37	−44
Biology	−08	−42	04	09	23	23	60	21	20
Chemistry	−20	−21	15	38	28	21	73	31	30
Physics	63	36	50	75	61	80	126	65	62
Science practical	61	24	38	69	49	58	130	26	46
Like/dislike science	46	13	36	54	38	31	59	41	44
Science activities	68	53	51	60	81	78	73	61	70
Science in the world	−18	−24	−10	14	14	04	−09	−12	01
Expectations	27	07	21	37	25	26	15	27	31

Decimal points omitted.

EXERCISES

1. Do the MANOVA of Exercise 3 at the end of Chapter 8 by the general linear model approach, using a GLM program in whatever computer package is available to you. Note that both the SAS and SPSSx packages give the user a choice among several types of error SSCP matrices. Try out all the different types available and see which one gives the same results as does the classical approach used in Chapter 8.

2. Do the MANOVA of Exercise 5 at the end of Chapter 8 by the general linear model approach. Note that since this problem involves a nonorthogonal design, the unweighted means analysis was done in Chapter 8 and this is only an approximate method. Hence even using the error SSCP matrix found in Exercise 1 to yield results identical to those given by the classical approach will not, in this problem, yield the same results. The results obtained by the GLM approach are, of course, more accurate.

10

APPLICATIONS TO CLASSIFICATION PROBLEMS

In one sense, the problem of classification is as old as science itself. Whether it be minerals, plants, or anthropological specimens, their classification into species, classes, and other taxonomic groups has always been a method for introducing order into an unstructured field. However, the systematic application of statistical (especially multivariate) techniques to the problem is a relatively recent development.

Probably the earliest such application was that by Tildesley (1921), who used Karl Pearson's "coefficient of racial likeness" for classifying prehistoric skeletal remains into racial groups on the basis of several anthropometric measurements. Discriminant analysis has been used for dealing with classification problems in a variety of fields, including politics (classifying U.S. Senators into "conservative," "progressive," and other groups on the basis of their voting records); anthropology (classifying Egyptian skulls into one of four series); and educational guidance ("classifying" college students into one of several curricular groups on the basis of a set of test scores). (For references, see Tatsuoka and Tiedeman, 1954.)

All of the examples cited above presume the existence of well-defined groups and deal with deciding how to classify as yet "unlabeled" individuals into one or another of these groups. This is the sense in which we shall use the term *classification problem*. The prior task of discovering or establishing the system of groups, which may, for distinction, be called *taxonomic problems*, will not be treated here. Statistical approaches to the taxonomic problem include Karl Pearson's (1894) "dissection of frequency curves" into normal components, Stephenson's (1953) inverted or *Q*-model factor analysis, McQuitty's (1955) pattern analysis, and Cattell and Coulter's (1966) taxonome method. More generally, the whole field of cluster analysis is devoted to taxonomic problems.

10.1 CLASSIFICATION AND THE CONCEPT OF RESEMBLANCE

The classification problems, as delineated above, amounts to seeking an answer to the question: Which of these several groups does this individual "resemble" the most, in terms of a specified set of measurable characteristics? That is, we have at hand a sample from each of K well-defined populations (such as clincal diagnostic categories, occupational groups, or curricular groups), with measures for each individual on p variables that are known or deemed to be important in differentiating among the several populations or groups. Subsequently, we encounter a person whose group membership is unknown, but for whom we have measures on these same p variables, and we wish to classify him as a member of one or another of these K groups—the one with which he shows greatest "resemblance" in terms of these p measures.

The crux of the matter obviously lies in how we define "resemblance" in this context. Various measures of *profile* (or *pattern*) *similarity* and of distance (i.e., *dissimilarity*) have been proposed in the literature (Mahalanobis, 1936; Cattell, 1949a; Du Mas, 1949; Cronbach & Gleser, 1952.) We here choose the familiar χ^2-statistic, introduced in Chapter 4 (Eq. 4.7), to serve as a measure of dissimilarity. This is a reasonable choice, since the larger the χ^2-value of an individual with reference to a given group, the farther away (in the generalized-distance sense) is the point $[X_{1i}, X_{2i}, \ldots, X_{pi}]$ representing his or her set of scores from the centroid $[\bar{X}_{1k}, \bar{X}_{2k}, \ldots, \bar{X}_{pk}]$ of that group. Thus he or she may be said to be the more deviant from the "average member" of that group, the larger the χ^2-value. Conversely, an individual with a small χ^2-value with reference to a group is "closer" to the average member of that group, and may hence be said to resemble that group. Furthermore, if the reference group is adequately describable by means of a multivariate normal distribution of the p variables, then knowledge of a person's χ^2-value allows us to estimate the percentage of individuals in the group who are "closer to" or "farther from" the group centroid than is that person. This is because, as shown in Chapter 4, the χ^2-value determines the particular percentage ellipsoid on which a given point lies.

Thus a simple classification scheme, which may be called the *minimum chi-square rule,* would be as follows: Compute the χ^2-value of the unclassified individual with respect to each of the K groups, and assign him or her to that group with respect to which his or her χ^2-value is the smallest. This rule has the property of minimizing the probability of misclassifications when the K populations have multivariate normal distributions with equal covariance matrices. If this common matrix $\boldsymbol{\Sigma}$ is known, it would, of course, be used in computing each of the K χ^2-values

$$\chi^2_{ik} = \mathbf{x}'_{ik}\boldsymbol{\Sigma}^{-1}\mathbf{x}_{ik}$$

for individual i, where $\mathbf{x}_{ik}$ is the vector of his p scores in deviation form—deviations from the kth population centroid $\boldsymbol{\mu}_k' = [\mu_{1k}, \mu_{2k}, \ldots, \mu_{pk}]$ if this is known; deviations from the kth sample centroid $\overline{X}_k' = [\overline{X}_{1k}, \overline{X}_{2k}, \ldots, \overline{X}_{pk}]$ if $\boldsymbol{\mu}_k$ is unknown.

Since, in practice, both $\boldsymbol{\Sigma}$ and the $\boldsymbol{\mu}_k$ are usually unknown, sample estimates have to be used for both the covariance matrix and the centroids. In this case, $\boldsymbol{\Sigma}$ is replaced by its within-groups estimate $\mathbf{S}_w/(N - K) = \mathbf{C}_w$, where $\mathbf{S}_w$ is the within-groups SSCP matrix, and $N = n_1 + n_2 + \cdots + n_k$. Thus the formula for χ^2_{ik} that is of greatest practical use becomes

$$\chi^2_{ik} = [X_{1i} - \overline{X}_{1k}, X_{2i} - \overline{X}_{2k}, \ldots, X_{pi} - \overline{X}_{pk}]\, \mathbf{C}_w^{-1} \times [X_{1i} - \overline{X}_{1k}, \ldots, X_{pi} - \overline{X}_{pk}]' = \mathbf{x}_{ik}' \mathbf{C}_w^{-1} \mathbf{x}_{ik} \tag{10.1}$$

EXAMPLE 10.1. In the numerical examples throughout this chapter, we utilize the data first introduced in Chapter 4 (p. 95) comprising the results of administering a three-scale activity preference questionnaire (APQ) to samples from three job categories of an airline company. We suppose that the company's personnel officer is faced with the task of assigning several job applicants to one of these categories on the basis of the APQ scores.

The sample sizes and the group means on the three scales were as follows:

		Means on:		
Group k	n_k	X_1	X_2	X_3
1. Passenger agents	85	12.5882	24.2235	9.0235
2. Mechanics	93	18.5376	21.1398	10.1398
3. Operations control persons	66	15.5758	15.4545	13.2424

The within-groups covariance matrix, computed by dividing each element of the within-groups SSCP matrix $\mathbf{W}$ shown on page 95 by $\Sigma\, n_k - 3 = 241$, is as follows:

$$\mathbf{C}_w = \begin{bmatrix} 16.4640 & 1.4590 & .3180 \\ 1.4590 & 18.2832 & .9769 \\ .3180 & .9769 & 11.1341 \end{bmatrix}$$

Suppose that four job applicants (already accepted for employment, but as yet unassigned to particular job categories) had thc following scores on the three scales of the APQ:

i	X_{1i}	X_{2i}	X_{3i}
1	15	24	6
2	15	16	11
3	16	23	14
4	19	20	15

The first step for computing the χ^2-value of each individual with respect to each group is to express his or her three scores as deviations from the means of the group. The results are:

i	$\mathbf{x}_{i1}$			$\mathbf{x}_{i2}$		
1	2.4118	−.2235	−3.0235	−3.5376	2.8602	−4.1398
2	2.4118	−8.2235	1.9765	−3.5376	−5.1398	.8602
3	3.4118	−1.2235	4.9765	−2.5376	1.8602	3.8602
4	6.4118	−4.2235	5.9765	.4624	−1.1398	4.8602

i	$\mathbf{x}_{i3}$		
1	−.5758	8.5455	−7.2424
2	−.5758	.5455	−2.2424
3	.4242	7.5455	.7576
4	3.4242	4.5455	1.7576

Next, we need the inverse $\mathbf{C}_w^{-1}$ of the within-groups covariance matrix, which is computed to be

$$\mathbf{C}_w^{-1} = \begin{bmatrix} 6.1191 & -.4812 & -.1325 \\ -.4812 & 5.5331 & -.4717 \\ -.1325 & -.4717 & 9.0266 \end{bmatrix} \times 10^{-2}$$

We may now carry out the computations for the χ^2's in accordance with Eq. 10.1. Thus the χ^2-value for individual 1 with respect to group 1 (passenger agents) is obtained as

$$\chi_{11}^2 = [2.4118 \quad -.2235 \quad -3.0235]$$
$$\times \begin{bmatrix} .061191 & -.004812 & -.001325 \\ -.004182 & .055331 & -.004717 \\ -.001325 & -.004717 & .090266 \end{bmatrix}$$
$$\times \begin{bmatrix} 2.4118 \\ -.2235 \\ -3.0235 \end{bmatrix}$$
$$= [.1527 \quad -.0097 \quad -.2751] \begin{bmatrix} 2.4118 \\ -.2235 \\ -3.0235 \end{bmatrix}$$
$$= 1.2022$$

Similarly, this person's χ^2-values with reference to group 2 (mechanics) and group 3 (operations control persons) are

$$\chi_{12}^2 = [-3.5376 \quad 2.8602 \quad -4.1398]$$
$$\begin{bmatrix} .061191 & -.004812 & -.001325 \\ -.004182 & .055331 & -.004717 \\ -.001325 & -.004717 & .090266 \end{bmatrix}$$
$$\times \begin{bmatrix} -3.5376 \\ 2.8602 \\ -4.1398 \end{bmatrix}$$
$$= 2.9355$$

and

$$\chi_{13}^2 = [-.5758 \quad 8.5455 \quad -7.2424]$$
$$\times \begin{bmatrix} .061191 & -.004812 & -.001325 \\ -.004818 & .055331 & -.004717 \\ -.001325 & -.004717 & .090266 \end{bmatrix}$$
$$\times \begin{bmatrix} -.5758 \\ 8.5455 \\ -7.2424 \end{bmatrix}$$
$$= 1.8712$$

respectively.

Since this person's smallest χ^2-value is that with reference to group 1, our decision according to the minimum chi-square rule would be to assign him or her to the passenger agents group.

On computing the three χ^2-values for the remaining three people, we may summarize the results and decisions for all four people in a table, as follows (where the smallest χ^2-value for each person is shown in italics):

Individual Number (*i*)	χ^2_{i1}	χ^2_{i2}	χ^2_{i3}	**Decision—Assign to**
1	*1.2022*	2.9355	1.8712	Group 1
2	4.7815	2.1690	*.5017*	Group 3
3	3.0836	*1.9339*	3.1276	Group 2
4	7.1235	2.2687	*1.8982*	Group 3

When the K population covariance matrices are not (or cannot be assumed to be) equal, the separate matrices or their respective sample estimates are used in place of $\mathbf{\Sigma}$ or $\mathbf{C}_w$ in computing χ^2. Thus the formula for the chi-square statistic now becomes

$$\chi^2_{ik} = [X_{1i} - \overline{X}_{1k}, X_{2i} - \overline{X}_{2k}, \ldots, X_{pi} - \overline{X}_{pk}]\, \mathbf{C}_k^{-1} \times [X_{1i} - \overline{X}_{1k}, \ldots, X_{pi} - \overline{X}_{pk}]' \qquad (10.2)$$

where $\mathbf{C}_k = \mathbf{S}_k/(n_k - 1)$ is the covariance matrix of the kth sample. At the same time, the classification rule is modified to be based on minimizing, not χ^2 itself, but an adjusted quantity χ'^2 defined as follows:

$$\chi'^2_{ik} = \chi^2_{ik} + \ln|\mathbf{C}_k| \qquad (10.3)$$

which is proportional to the natural logarithm of the multivariate normal density function $\mathrm{N}(\overline{\mathbf{X}}_k, \mathbf{C}_k)$ evaluated at the point $\mathbf{X}'_i = [X_{1i}, X_{2i}, \ldots, X_{pi}]$. That is, for each individual to be classified, we compute the quantity χ'^2_{ik}, defined by Eqs. 10.2 and 10.3 for each of the K groups, and assign him or her to that group for which his or her χ'^2 value is the smallest.[1]

[1]The maximum *centour score* rule proposed by Tiedeman in Tiedeman, Bryan, and Rulon (1953) (also described in Rulon, Tiedeman, Tatsuoka, and Langmuir, 1967, p. 167) is intermediate between the minimum χ^2 and minimum χ'^2 rules stated above. An individual's centour score with respect to group k is found by computing χ^2_{ik} from Eq. 10.2, determining its percentile rank as a chi-square variate with p d.f.'s, and subtracting this value from 100. Thus the larger a person's centour score with respect to a given group, the closer is his or her score point to the centroid of that group.

EXAMPLE 10.2 Continuing with the example given in connection with the minimum chi-square rule, we now discard the assumption that the covariance matrices in the three populations are equal, and use a separate covariance matrix $\mathbf{C}_k$ for each group. These are computed as $\mathbf{C}_k = \mathbf{S}_k/(n_k - 1)$, where $\mathbf{S}_k$ is the SSCP matrix for the kth sample. The dispersion matrices and their respective inverses are as follows:

$$\mathbf{C}_1 = \begin{bmatrix} 20.2451 & 4.6170 & -2.4069 \\ 4.6170 & 18.7947 & 2.5066 \\ -2.4069 & 2.5066 & 9.8804 \end{bmatrix}$$

$$\mathbf{C}_1^{-1} = \begin{bmatrix} .055084 & -.015858 & .017442 \\ -.015858 & .059635 & -.018992 \\ .017442 & -.018992 & .110277 \end{bmatrix}$$

$$\mathbf{C}_2 = \begin{bmatrix} 12.7078 & -1.5760 & 2.5436 \\ -1.5760 & 20.7085 & .1759 \\ 2.5436 & .1759 & 10.5129 \end{bmatrix}$$

$$\mathbf{C}_2^{-1} = \begin{bmatrix} .083571 & .006533 & -.020329 \\ .006533 & .048806 & -.002397 \\ -.020329 & -.002397 & .100080 \end{bmatrix}$$

$$\mathbf{C}_3 = \begin{bmatrix} 16.8942 & 1.6727 & .6890 \\ 1.6727 & 14.1902 & .1343 \\ .6890 & .1343 & 13.6326 \end{bmatrix}$$

$$\mathbf{C}_3^{-1} = \begin{bmatrix} .060011 & -.007046 & -.002963 \\ -.007046 & .071305 & -.000346 \\ -.002963 & -.000346 & .073506 \end{bmatrix}$$

The new χ^2-values for individual 1, computed in accordance with Eq. 10.2, are

$$\chi_{11}^2 = [2.4118 \quad -.2235 \quad -3.0235]$$

$$\times \begin{bmatrix} .055084 & -.015858 & .017442 \\ -.015858 & .059635 & -.018992 \\ .017442 & -.018992 & .110277 \end{bmatrix}$$

$$\times \begin{bmatrix} 2.4118 \\ -.2235 \\ -3.0235 \end{bmatrix}$$

$$= 1.0686$$

$$\chi_{12}^2 = [-3.5376 \quad 2.8602 \quad -4.1398]$$

$$\times \begin{bmatrix} .083571 & .006533 & -.020329 \\ .006533 & .048806 & -.002397 \\ -.020329 & -.002397 & .100080 \end{bmatrix}$$

$$\times \begin{bmatrix} -3.5376 \\ 2.8602 \\ -4.1398 \end{bmatrix}$$

$$= 2.4896$$

$$\chi_{13}^2 = [-.5758 \quad 8.5455 \quad -7.2424]$$

$$\times \begin{bmatrix} .060011 & -.007046 & -.002963 \\ -.007046 & .071305 & -.000346 \\ -.002963 & -.000346 & .073506 \end{bmatrix}$$

$$\times \begin{bmatrix} -.5758 \\ 8.5455 \\ -7.2424 \end{bmatrix}$$

$$= 9.1699$$

Next, following Eq. 10.3, χ'^2 values are obtained by adding to each χ^2 value the natural logarithm of the determinant of the corresponding covariance matrix. We find

$$|\mathbf{C}_1| = 3257.0 \qquad |\mathbf{C}_2| = 2605.0 \qquad |\mathbf{C}_3| = 3223.4$$

and

$$\ln|\mathbf{C}_1| = 8.0886 \qquad \ln|\mathbf{C}_2| = 7.8652 \qquad \ln|\mathbf{C}_3| = 8.0782$$

Thus, for individual 1, we obtain

$$\chi_{11}'^2 = 1.0686 + 8.0886 = 9.1572$$

$$\chi_{12}'^2 = 2.4896 + 7.8652 = 10.3548$$

$$\chi_{13}'^2 = 9.1699 + 8.0782 = 17.2481$$

Similar computations for the remaining three people yield results and decisions as shown in the following table.

Individual Number (i)	χ'^2_{i1}	χ'^2_{i2}	χ'^2_{i3}	Decision—Assign to
1	*9.1572*	10.3548	17.2481	Group 1
2	14.2854	10.6568	*8.4865*	Group 3
3	12.5058	*10.3660*	12.1401	Group 2
4	18.5103	10.2387	*10.2219*	Group 3

We see that the decisions based on minimum χ'^2 are identical to those based on minimum χ^2, made earlier, as far as these four people are concerned.

10.2 TAKING PRIOR PROBABILITIES INTO CONSIDERATION

The preceding classification rules, based on minimum chi-square and the adjusted quantity of Eq. 10.3, do not take into consideration the prior probabilities of group membership—that is, the probability of drawing at random a member of each group from a mixed population of all K groups. Otherwise stated, these procedures assume that the relative frequencies of all groups are equal. Hence the optimal property of minimum probability of misclassifications holds only when the prior probabilities are in fact all equal. When this is not the case, a further modification on the χ'^2 of Eq. 10.3 becomes necessary in order to take into account the different prior probabilities (The term "prior" signifies that these are probabilities of group membership before we know a person's scores on the p predictor variables.)

Let p_k denote the probability that a person selected at random from a mixed population comprising all K groups is a member of the kth group. Then the appropriate modification of the χ'^2-statistic is given by a constant times the natural logarithm of the multivariate normal density function for group k, multiplied by p_k. That is,

$$\chi''^2_{ik} = \chi'^2_{ik} - 2 \ln p_k \qquad (10.4)$$

where χ'^2_{ik} is as defined in Eq. 10.3. Again, the decision rule is to assign the person to that group for which his or her χ''^2 value is the smallest.

Although the definition of χ''^2 is straightforward, the estimation of the p_k's often poses a serious problem. It is seldom the case that the K samples at hand are proportional in size to their respective populations. (In fact, the relative sizes of the populations are sometimes undefined, all K of them being theoretically infinite.)

In applications to personnel classification, however, we often have at least a rough idea of what the relative sizes of the various occupational or curricular groups have been in the past, and these may be taken as estimates of p_k. Such a procedure has sometimes been criticized for its tendency to "perpetuate the status quo" by keeping traditionally large groups large, and small ones small. This criticism would be justified to the extent that the "traditional" sizes of various vocational and curricular groups have been arbitrarily determined by policy or expediency considerations. No doubt such considerations are a factor in determining the relative sizes of occupational and curricular groups. But there are also certain "natural" limits set by the needs of a society. More physicians are needed, for example, than biology professors; the number of physicians needed is, in turn, exceeded by the number of electrical engineers employed by our technological society.

Thus, except at times when the structure of society is undergoing a drastic change, there is some justification in using the traditional relative sizes of the various groups as estimates of the prior probabilities of group membership. But what if society *is* undergoing a rapid transition, as from an agrarian to an industrial society? Then surely the relative sizes of, for example, the agricultural expert population and the mechanical engineer population over the past decade will be meaningless in estimating their relative sizes five years hence.

In the above type of situation, the vocational or educational guidance counselor will have to resort to reasonable forecasts of social demands for various occupational categories in the near future. The estimates of prior probabilities of group membership would then be based on such forecasts instead of on the relative sizes of the groups in the recent past.

Let us now return to Eq. 10.4, defining the χ''^2-statistic, to see what the effect of different p_k values is on this statistic, and hence on the classification procedure. Since the term involving p_k is $-2 \ln p_k$, and p_k is a positive number less than 1, the following conclusions may be drawn: (1) the additive component due to p_k is always positive; and (2) the larger the value of p_k, the smaller this additive component. Consequently, if for some two groups j and k, p_k is larger than p_j, then a person for whom $\chi'^2_{ik} = \chi'^2_{ij}$ will have a *smaller* χ''^2_{ik} than χ''^2_{ij}. Thus, if this person's χ'^2-value had been smaller for these two groups than for any other group (and hence his or her assignment to group j or k on the basis of χ'^2-values would have been a matter of toss-up), the effect of the prior probabilities is to break the tie in favor of the group with the larger prior probability, because $\chi''^2_{ik} < \chi''^2_{ij}$.

Although we considered the case when $\chi'^2_{ik} = \chi'^2_{ij}$ above, it is clear that even if $\chi'^2_{ik} > \chi'^2_{ij}$, we may have $\chi''^2_{ik} < \chi''^2_{ij}$ provided that p_k is sufficiently larger than p_j. That is, a decision based on χ'^2-values (measuring dissimilarity) may be reversed in favor of a group with a large prior probability of membership (i.e., a large group) when χ''^2-values are used as the basis of classification.

We thus see that the role played by prior probabilities is, as it were, to temper our decisions based on resemblance alone with considerations of relative group sizes. Where we might tend to oversupply small groups and undersupply large ones by using resemblance as the sole basis for classification we introduce a corrective effect by taking prior probabilities of group membership into account.

EXAMPLE 10.3. Let us now modify the classification based on χ'^2-values, given in Example 10.2, by introducing prior-probability considerations. As seen from Eq. 10.4, we have merely to add -2 times the natural logarithm of p_k to each χ'^2_{ik}-value in order to get χ''^2_{ik}. For simplicity we use the relative sample sizes $n_k/(n_1 + n_2 + n_3)$ as our estimates of p_k—since this is a fictitious example and we have no other means for estimating the prior probabilities. The values are,

$$p_1 = 85/244 = .34836 \quad p_2 = 93/244 = .38115 \quad p_3 = 66/244 = .27049$$

and -2 times their respective natural logarithms are

$$-2 \ln p_k = 2.1090 \qquad 1.9291 \qquad 2.6150$$

for $k = 1, 2, 3$, respectively. We add the appropriate one of these three numbers to each of the three χ'^2_{ik}-values already computed for each individual i. Thus, for individual 1,

$$\chi''^2_{11} = 9.1572 + 2.1090 = 11.2662$$

$$\chi''^2_{12} = 10.3548 + 1.9291 = 12.2839$$

$$\chi''^2_{13} = 17.2481 + 2.6150 = 19.8631$$

We similarly obtain the χ''^2-values for individuals 2, 3, and 4. The results and decisions for all four people are as shown in the following table.

Individual Number (i)	χ''^2_{i1}	χ''^2_{i2}	χ''^2_{i3}	**Decision—Assign to**
1	*11.2662*	12.2839	19.8631	Group 1
2	16.3944	12.5859	*11.1015*	Group 3
3	14.6148	*12.2951*	14.7551	Group 2
4	20.6193	*12.1678*	12.8369	Group 2

We note that the decisions are unchanged from those based on the minimum χ'^2-rule, made earlier, for the first three individuals, but individual 4, who was classified as a member of group 3 previously, is now assigned to group 2. The larger prior probability of membership in group 2 (0.38, as against 0.27 for group 3) has outweighed this person's greater resemblance (by a narrow margin, to be sure) to the average member of group 3 than to that of group 2, as measured by the χ'^2-criterion.

10.3 PROBABILITY OF GROUP MEMBERSHIP

The quantities χ^2 and χ'^2 utilized in Section 10.1 as measures of dissimilarity are closely related to a certain type of probability, specified below, provided that the variables $X_1, X_2, \ldots, X_p$ follow a multivariate normal distribution in each of the K groups. Considering χ'^2 (the more generally applicable of the two statistics), we see from Eq. 10.3 that

$$e^{-\chi_{ik}'^2/2} = |\mathbf{C}_k|^{-1/2} e^{-\chi_{ik}^2/2}$$

and hence

$$(2\pi)^{-p/2} e^{-\chi_{ik}'^2/2} = (2\pi)^{-p/2} |\mathbf{C}_k|^{-1/2} e^{-\chi_{ik}^2/2}$$

which is simply the multivariate normal density function evaluated at the point corresponding to the observed score combination $\mathbf{X}_i$. Therefore, by definition, the quantity

$$(2\pi)^{-p/2} e^{-\chi_{ik}'^2/2}\, dX_1, dX_2, \ldots, dX_p$$

expresses the probability that a randomly drawn member of group k will have a score combination between

$$(X_{1i}, X_{2i}, \ldots, X_{pi})$$

and

$$(X_{1i} + dX_1, X_{2i} + dX_2, \ldots, X_{pi} + dX_p) \tag{10.5}$$

Let us denote this probability by $p(\mathbf{X}_i \mid H_k)$, where H_k stands for the statement: "individual i is a member of group k."

The probability alluded to at the outset of this section, and its relationship to χ'^2, may now be stated as follows: Given that individual i is a randomly selected member of group k, the probability that his or her score combination lies within the limits displayed in 10.5 is equal to

$$p(\mathbf{X}_i \mid H_k) = (2\pi)^{-p/2} e^{-\chi_{ik}'^2/2}\, dX_1, dX_2, \ldots, dX_p \tag{10.6}$$

where $\chi_{ik}'^2$ is as defined in Eq. 10.3.

On the other hand, the quantity χ''^2 introduced in the preceding section, taking prior probabilities of group membership into consideration, cannot be related to a conditional probability of the type shown in Eq. 10.6. For the prior probability does not (and cannot) play any role once we confine our attention to a particular group k, as we do in the conditional probability $p(\mathbf{X}_i \mid H_k)$. To take the prior probabilities of group membership into account and still produce a relevant probability after the score combination $\mathbf{X}_i$ has been observed, we have to consider a type of probability which is, as it were, the *inverse* of that displayed in Eq. 10.6: the probability that individual i is a member of group k, given that his or her score combination is $\mathbf{X}_i$ (or, more precisely, that it lies between the limits stated in Eq. 10.5). It will presently be seen that this type of probability, which we denote by $p(H_k \mid \mathbf{X}_i)$, is functionally related to χ''^2.

For $p(H_k \mid \mathbf{X}_i)$ to be definable, we must make one further assumption that was not needed in considering probabilities of the type $p(\mathbf{X}_i \mid H_k)$. This is the requirement that the as yet unclassified individual i must definitely be a member of one or another of the K groups under consideration. That is, the eventuality that he or she belongs to *none* of these K groups is prohibited.[2]

Granted the assumption that the set of statements $H_1, H_2, \ldots, H_k$ exhausts all the possibilities with regard to the group membership of individual i, we may compute his or her $p(H_k \mid \mathbf{X}_i)$ by means of Bayes' theorem on "inverse probability," or *posterior* probability, as it is more commonly known today (see, e.g., Hays, 1981). The formula is

$$p(H_k \mid \mathbf{X}_i) = \frac{p_k p(\mathbf{X}_i \mid H_k)}{\sum_{j=1}^{K} p_j p(\mathbf{X}_i \mid H_j)} \qquad k = 1, 2, \ldots, K \tag{10.7}$$

where p_k is the prior probability of membership in group k, and $p(\mathbf{X}_i \mid H_k)$ is as defined in Eq. 10.6. Substituting from this equation and canceling the common factors $(2\pi)^{-p/2}$ and $dX_1, dX_2, \ldots, dX_p$ from numerator and denominator, we may write $p(H_k \mid \mathbf{X}_i)$ explicitly in terms of χ'^2_{ik}, as follows:

$$p(H_k \mid \mathbf{X}_i) = \frac{p_k \exp(-\chi'^2_{ik}/2)}{\sum_{j=1}^{K} p_j \exp(-\chi'^2_{ij}/2)}$$

or, further expressing χ'^2 in terms of χ^2 from Eq. 10.3,

$$p(H_k \mid \mathbf{X}_i) = \frac{p_k |\mathbf{C}_k|^{-1/2} \exp(-\chi^2_{ik}/2)}{\sum_{j=1}^{K} p_j |\mathbf{C}_j|^{-1/2} \exp(-\chi^2_{ij}/2)} \qquad k = 1, 2, \ldots, K \tag{10.8}$$

Alternatively, we may eliminate χ'^2 in the previous equation in favor of χ''^2 by use of Eq. 10.4, in which case we get

[2]In practice, we would be stipulating this requirement even when using probabilies of the type of Eq. 10.6 if we make a forced classification of every person into one or another of the K groups on the basis of the minimum χ^2 (or miniumum χ'^2) rule. But there is no logical necessity to do so. We may decide that a person belongs to none of the K groups if his or her χ^2 (or χ'^2) values for all of them are fairly large—or, equivalently, if his or her $p(\mathbf{X}_i \mid H_k)$ is quite small for all k.

$$p(H_k \mid \mathbf{X}_i) = \frac{\exp(-\chi''^2_{ik}/2)}{\sum_{j=1}^{K} \exp(-\chi''^2_{ij}/2)} \qquad k = 1, 2, \ldots, K \qquad (10.8a)$$

We thus see that whereas $p(\mathbf{X}_i \mid H_k)$ is expressible only in terms of χ^2 or χ'^2, $p(H_k \mid \mathbf{X}_i)$ is related to χ''^2 as well as the two earlier statistics. This is the posterior probability, after observing individual i's score combination $(X_{1i}, X_{2i}, \ldots, X_{pi})$, that he or she is a member of group k. For short, we shall refer to this simply as the probability of group membership, omitting the qualifier "posterior" as understood.

We reiterate the distinction between the probability defined in Eq. 10.6 and that in Eq. 10.8, for a clear understanding of this distinction becomes important in the sequel. The former, that is, $p(\mathbf{X}_i \mid H_k)$, represents the proportion of individuals, among members of group k, who have score combinations in the vicinity of $(X_{1i}, X_{2i}, \ldots, X_{pi})$. The latter, $p(H_k \mid \mathbf{X}_i)$, may be interpreted as the proportion of individuals, among those with the score combination or thereabouts, who are members of group k; here it is assumed that our universal set consists of a mixture of groups $1, 2, \ldots, K$, and nothing else.

In using the probability of group membership for classification purposes, the decision rule is of course to assign each person to that group for which his or her $p(H_k \mid \mathbf{X}_i)$-value is the *largest*. Since this probability is calculated by using Bayes' theorem, the decision rule is often referred to as the *Bayes' decision rule*. This is a reversal from the rules using χ^2, χ'^2, and χ''^2, in which we sought to minimize the values of these statistics. It should be noted, however, that the numerator of expression (10.8a) for $p(H_i \mid \mathbf{X}_i)$ is $\exp(-\chi''^2_{ik}/2)$, which is a monotonically decreasing function of χ''^2_{ik}, and that the denominator does not change with k for any one individual i. Hence the classification based on maximum $p(H_k \mid \mathbf{X}_i)$ is actually identical with that based on minimum χ''^2. The reader may well wonder why we should bother to compute $p(H_k \mid \mathbf{X}_i)$ at all, if identical decisions are reachable by using χ''^2_{ik}, which need to be computed first before getting $p(H_k \mid \mathbf{X}_i)$. The reason will become clear in Section 10.6.

EXAMPLE 10.4. Since the χ''^2-values have already been computed for our four individuals, it is but a short step to getting their $p(H_k \mid \mathbf{X}_i)$ values, which we here abbreviate as $p_{k;i}$ to simplify the notation. Tables of the exponential function usually give e^{-x} as well as e^x as functions of x. If such a table is available, we have only to divide each χ''^2_{ik} by 2 and enter the table with $\chi''^2_{ik}/2$ as argument, making sure to interpolate between tabled values because e^{-x} is quite sensitive to small variations in the argument for small and moderate values of x $(= \chi''^2_{ik}/2)$ such as we shall often encounter. If an exponential-function table is not available, we may compute $\exp(-\chi''^2_{ik}/2)$, the basic ingredients of

$p_{k;i}$, with the aid of a table of common logarithms, utilizing the following sequence of relations:

$$\log [\exp(-\chi''^2_{ik}/2)] = (-\chi''^2_{ik}/2) \log e$$
$$= (-\chi''^2_{ik}/2)(.43429)$$
$$= (-.21715)\chi''^2_{ik} = L_{ik} \quad \text{(say)}$$

Therefore,

$$\exp(-\chi''^2_{ik}/2) = \text{antilog } L_{ik}$$

That is, we need only to multiply each χ''^2_{ik}-value by the constant $-.21715$, and find the number whose common logarithm is equal to this product. (Many hand-held electronic calculators have a built-in exponential function, so the computation of $\exp(-\chi''^2_{ik}/2)$ has been greatly simplified, without having to use a table of logarithms.)

The values of $\exp(-\chi''^2_{ik}/2)$ and their sum (over $k = 1, 2, 3$) for each of the four people are as follows:

i	$\exp(-\chi''^2_{i1}/2)$	$\exp(-\chi''^2_{i2}/2)$	$\exp(-\chi''^2_{i3}/2)$	$\Sigma_k \exp(-\chi''^2_{ik}/2)$
1	.0035777	.0021508	.0000486	.0057771
2	.0002755	.0018493	.0038845	.0060093
3	.0006706	.0021388	.0006251	.0034345
4	.0000333	.0022794	.0016312	.0039439

Then, in accordance with Eq. 10.8a, each person's $p_{k;i}$-values are obtained by dividing the first three entries in his or her row by their sum listed as the rightmost entry. Thus $p_{1;1} = .0035777/.0057771 = .6193$, and so on. The $p_{k;i}$-values thus computed for the four individuals, and the resulting classificatory decisions, are shown in the following table, where the largest probability value in each row is in italics.

Individual Number (i)	$p_{1;i}$	$p_{2;i}$	$p_{3;i}$	Decision—Assign to
1	*.6193*	.3723	.0084	Group 1
2	.0458	.3077	*.6464*	Group 3
3	.1953	*.6227*	.1820	Group 2
4	.0084	*.5780*	.4136	Group 2

There is, of course, no difference between these decisions and those made earlier on the basis of the minimum χ''^2-rule, for the two decision rules are, as we already mentioned, mathematically equivalent.

10.4 REDUCTION OF DIMENSIONALITY BY DISCRIMINANT ANALYSIS

In the foregoing classification procedures, the relevant statistic—χ^2, χ'^2, χ''^2, or $p(H_k \mid \mathbf{X}_i)$, as the case may be—was computed in terms of a set of p predictor variables. Often the number of predictors is quite large (10 or more, for example) and the computation of these statistics becomes time consuming, even with the aid of electronic computers, when the number of individuals to be classified is large—as it usually is. It is therefore desirable to reduce the number of variables to be considered—or, in geometric language, to reduce the dimensionality of the space in which we operate.

As discussed extensively by Rulon et al. (1967, chaps. 7–9), and as one might intuitively expect, the technique of multigroup discriminant analysis is the most appropriate means for achieving this reduction of dimensionality (insofar as reduction by linear transformations, that is, rotation of axes, is concerned). In fact, it can be shown (Tatsuoka, 1956) that the results of classification based on the minimum chi-square rule (i.e., when we assume equal covariance matrices in the K populations) are identical regardless of whether we use the original p variables or the $K - 1$ discriminant functions in computing the χ^2-statistic. Thus if we have 10 predictors and 5 groups, it suffices to compute the χ^2-statistic of Eq. 10.1 from the four discriminant-function scores for each person to be classified, instead of from his or her scores on the original 10 variables. This represents a considerable saving of computational time (even allowing for the fact that the discriminant-function coefficients have first to be computed) if the number of individuals to be classified is very large.

When the K population coveriance matrices are not assumed to be equal—that is, when classification is based on χ'^2, χ''^2, or $p(H_k \mid \mathbf{X}_i)$—the identity of classificatory results in the original p-dimensional space and the reduced $(K - 1)$-dimensional space no longer holds. However, experience shows that the classifications in the two spaces yield closely similar results as long as the covariance matrices are not drastically different. (See Lohnes, 1961, and Rulon et al., 1967, p. 317 for empirical evidence of this "robustness property of discriminant analysis.) Thus it seems justifiable to use the discriminant functions even when classification is to be based on those statistics that take the separate group covariance matrices into account.

Apart from computational expediency, however, there is another argument in favor of using discriminant functions rather than the original predictors for computing the decision statistic. This argument, in fact, calls for using only the statistically significant discriminant functions, thus resulting (generally) in an even more drastic reduction of dimensionality than from p to $K - 1$. The idea is that in using the original p variables (or all $K - 1$ of the discriminant functions), we are doubtless capitalizing to some extent on chance differences that happen to differentiate the K samples at hand but do not reflect real differences in the corresponding populations. It seems reasonable, then, that by confining our attention to those discriminant functions that are statistically significant, we would decrease our reliance on apparent differences due to sampling error. If this standpoint is taken, we would insist on using only the significant discriminant functions even when equal population covariance matrices are assumed. We would do so with the realization that we shall no longer obtain classificatory results identical with those using the original variables, as we could, in this case, get by using all $K - 1$ discriminant functions in computing the χ^2-statistic if our only purpose were that of computational expediency.

The use of discriminant functions, whether we choose to retain all $K - 1$ or only the statistically significant ones, does not entail any essential changes in the classification procedures discussed in the preceding sections. We have merely to replace, in Eqs. 10.1 to 10.8a, the original set of variables (X_1, X_2, . . . , X_p) and their covariance matrices by the discriminant functions Y_1, Y_2, . . . , Y_r and the covariance matrices for these variables. No new formulas need to be given.

EXAMPLE 10.5. The discriminant functions appropriate to the numerical example we have been carrying in this chapter were already computed in Chapter 7 (see p. 222). The normalized coefficients for the two functions (both of which were statistically significant at the .001 level) were there found to be given by the columns of the matrix

$$\mathbf{V} = \begin{bmatrix} .3524 & .9145 \\ -.7331 & .1960 \\ .5818 & -.3540 \end{bmatrix}$$

The two discriminant-function scores Y_1 and Y_2 for each individual are computed in accordance with Eq. 7.1, or, in matrix notation,

$$\mathbf{Y}_i' = \mathbf{X}_i'\mathbf{V}$$

This equation is, of course, used also for computing the group means on the two discriminant functions. The results (given only to two decimal places in Chapter 7) are here shown to four decimals to prevent excessive rounding error in the subsequent series of calculations:

Group 1		Group 2		Group 3	
$\bar{Y}_1$	$\bar{Y}_2$	$\bar{Y}_1$	$\bar{Y}_2$	$\bar{Y}_1$	$\bar{Y}_2$
−8.0723	13.0654	−3.0656	17.5065	1.8636	12.5853

The four individuals' discriminant-function scores, expressed as deviations from the respective group means, are as follows:

i	y_{i1}		y_{i2}		y_{i3}	
1	−.7453	3.2321	−5.7520	−1.2090	−10.6812	3.7122
2	8.0285	−.1059	3.0218	−4.5470	−1.9074	.3742
3	4.9946	1.1186	−.0121	−3.3225	−4.9413	1.5987
4	8.8329	2.9201	3.8262	−1.5210	−1.1030	3.4002

Next, the group covariance matrices are each computed from an equation typified by 7.2, that is

$$\mathbf{C}_k(y) = \mathbf{V}'\mathbf{C}_k\mathbf{V}$$

where $\mathbf{C}_k$ is the covariance matrix of the original predictors for group k. The covariance matrices, their determinants, and their inverses are as shown below.

k	$\mathbf{C}_k(y)$	$\lvert\mathbf{C}_k(y)\rvert$	$\mathbf{C}_k(y)^{-1}$
1	$\begin{bmatrix} 10.4488 & -1.0316 \\ -1.0316 & 21.7570 \end{bmatrix}$	226.27	$\begin{bmatrix} .096155 & .004559 \\ .004559 & .046178 \end{bmatrix}$
2	$\begin{bmatrix} 17.9734 & 1.0041 \\ 1.0041 & 10.5044 \end{bmatrix}$	187.79	$\begin{bmatrix} .055936 & -.005347 \\ -.005347 & .095709 \end{bmatrix}$
3	$\begin{bmatrix} 13.6425 & -.0772 \\ -.0772 & 16.5172 \end{bmatrix}$	225.33	$\begin{bmatrix} .073302 & .000343 \\ .000343 & .060545 \end{bmatrix}$

With the foregoing data and the prior probabilities of group membership (p_k) given earlier, we can compute successively, χ^2, χ'^2, χ''^2, and $p(H_k \mid \mathbf{Y}_i)$ for each person with respect to each group. Since the calculations are exactly the same as those shown in the corresponding numerical examples using the original predictor scores, we illustrate the details only for the first person with respect to the first group.

$$\chi_{11}^2 = [-.7453 \quad 3.2321] \begin{bmatrix} .096155 & .004559 \\ .004559 & .046178 \end{bmatrix} \begin{bmatrix} -.7453 \\ 3.2321 \end{bmatrix}$$

$$= .5140 \tag{10.2}$$

$$\chi_{11}'^2 = \chi_{11}^2 + \ln \lvert\mathbf{C}_k(y)\rvert = .5140 + 5.4217 = 5.59357 \tag{10.3}$$

$$\chi_{11}''^2 = \chi_{11}'^2 - 2 \ln p_k = 5.9357 + 2.1090 = 8.0447 \tag{10.4}$$

$$p(H_1 \mid \mathbf{Y}_1) = \frac{\exp(-\chi_{11}''^2/2)}{\sum_{j=1}^{3} \exp(-\chi_{1j}''^2/2)} = \frac{.017910}{.028764} \tag{10.8a}$$

$$= .6227$$

The results for the four people are as follows:

Individual Number (i)	χ^2_{i1}	χ^2_{i2}	χ^2_{i3}	Decision—Assign to
1	.5140	1.9164	9.1703	—
2	6.1906	2.6362	.2747	—
3	2.5076	1.0560	1.9393	—
4	8.1306	1.1027	.7867	—

Individual Number (i)	χ'^2_{i1}	χ'^2_{i2}	χ'^2_{i3}	Decision—Assign to
1	*5.9357*	7.1517	14.5879	Group 1
2	11.6123	7.8715	*5.6923*	Group 3
3	7.9293	*6.2913*	7.3569	Group 2
4	13.5523	6.3380	*6.2043*	Group 3
	χ''^2_{i1}	χ''^2_{i2}	χ''^2_{i3}	
1	*8.0447*	9.0808	17.2029	Group 1
2	13.7213	9.8006	*8.3073*	Group 3
3	10.0383	*8.2204*	9.9719	Group 2
4	15.6613	*8.2671*	8.8193	Group 2
	$p_{1;i}$	$p_{2;i}$	$p_{3;i}$	
1	*.6227*	.3710	.0064	Group 1
2	.0433	.3076	*.6490*	Group 3
3	.2215	*.5496*	.2289	Group 2
4	.0139	*.5607*	.4254	Group 2

No decisions are made on the basis of the χ^2_{ik}-values, since these are not terminal statistics, but are intermediate to obtaining the χ'^2_{ik}-values, having been computed in accordance with Eq. 10.2. As may be ascertained by comparing with previously given results, the decisions based on χ'^2, χ''^2, and $p(H_k \mid \mathbf{Y}_i)$, are respectively, identical with those made of the basis of the corresponding quantities computed from the original variables. In particular, the last-mentioned criterion—probability of group membership—not only yields identical decisions but also shows numerical values fairly similar to the $p(H_k \mid \mathbf{X}_i)$ obtained before. This is another instance of verification of the empirical generalization stated earlier, that classificatory results in the original predictor space and the reduced discriminant space tend to be closely similar even when the separate group covariance matrices are used, provided that they are not drastically different.

10.5 NOTE ON TERMINOLOGY: TWO SENSES OF THE TERM "DISCRIMINANT FUNCTION"

The term "discriminant function" has come to be used in two different senses in the literature. One is the sense in which it was defined in Chapter 7, and the other is more closely related to the various classification-decision measures discussed in this chapter. This fact was noted by Tatsuoka (1971) and by Green (1979), and since a confusion of the two senses can have serious consequences—especially in using computer packages—we devote this section to clarifying the distinction. From Eqs. 10.2 and 10.3 it is evident that the minimum-χ'^2 classification rule is equivalent to the following:

For each person i to be classified, compute the quantity

$$\begin{aligned} Q_{ik} &= (-\tfrac{1}{2})[\ln|\mathbf{C}_k| + (\mathbf{X}_i - \overline{\mathbf{X}}_k)'\mathbf{C}_k^{-1}(\mathbf{X}_i - \overline{\mathbf{X}}_k)] \\ &= (-\tfrac{1}{2})[\ln|\mathbf{C}_k| + \mathbf{X}_i'\mathbf{C}_k^{-1}\mathbf{X}_i - 2\mathbf{X}_i'\mathbf{C}_k^{-1}\overline{\mathbf{X}}_k + \overline{\mathbf{X}}_k'\mathbf{C}_k^{-1}\overline{\mathbf{X}}_k] \qquad (10.9) \\ & \qquad k = 1, 2, \ldots, K \end{aligned}$$

and classify him or her into the group for which the value of Q_{ik} is largest. The expression Q_{ik} may be called the quadratic *classification-function score* for individual i with respect to group k. If, instead of a particular person's scores on the p variables, as in the vector $\mathbf{X}_i$, the generic vector $\mathbf{X}$ is used in the right-hand side of Eq. 10.9, we get the quadratic classification function Q_k for group k.

If the K population covariance matrices $\mathbf{\Sigma}_k$ ($k = 1, 2, \ldots, K$) can be assumed equal and hence can be estimated by the pooled within-groups covariance matrix $\mathbf{C}_w$, the first two terms of the expanded form of the expression for Q_{ik} become identical for all k. These two terms, $\ln|\mathbf{C}_w| + \mathbf{X}_i'\mathbf{C}_w^{-1}\mathbf{X}_i$, therefore have no bearing on which one of an individual's Q_{ik}'s becomes the largest, and hence may be omitted from the classification function. We then get a *linear* classification function,

$$L_k = \mathbf{X}'\mathbf{C}_w^{-1}\overline{\mathbf{X}}_k - \tfrac{1}{2}\overline{\mathbf{X}}_k'\mathbf{C}_w^{-1}\overline{\mathbf{X}}_k \qquad k = 1, 2, \ldots, K \qquad (10.10)$$

This is the expression that the BMD manual (Dixon, 1975) and the SAS User's Guide (SAS Institute, 1979) refer to as the "discriminating function" and the "linearized discriminant function," respectively.

The reader can readily see that expression (10.10) is quite different from what was defined as a discriminant function in Chapter 7. The coefficients of the linear combinations that we call discriminant functions are the elements of the successive eigenvectors of the matrix $\mathbf{W}^{-1}\mathbf{B}$, while the coefficients of the linear function (10.10) (which also has a constant term, it should be noted) are the elements of the vector $\mathbf{C}_w^{-1}\overline{\mathbf{X}}_k$ for each k. It appears that most authors in the behavioral and social sciences use the terms "discriminant function" and "classification function" in the same way as we do in this book, while authors in the biological sciences refer to what we labeled "linear classification function" in Eq. 10.10 as (linear) discriminant functions. Of course, there is no question of which usage is "right" and which, "wrong," but the reader should be aware of the existence of these different usages, especially when using the BMD or SAS packages. (The SPSS package, on the other hand, adopts the same usage as we do.)

In the special case when $K = 2$, there is a simple relationship between the two linear classification functions and the (single) discriminant function. In this case, we have

$$L_1 = \mathbf{X}'\mathbf{C}_w^{-1}\overline{\mathbf{X}}_1 - \tfrac{1}{2}\overline{\mathbf{X}}_1'\mathbf{C}_1^{-1}\overline{\mathbf{X}}_1$$

and

$$L_2 = \mathbf{X}'\mathbf{C}_w^{-1}\overline{\mathbf{X}}_2 - \tfrac{1}{2}\overline{\mathbf{X}}_2'\mathbf{C}_w^{-1}\overline{\mathbf{X}}_2$$

Hence

$$\begin{aligned} L_1 - L_2 &= \mathbf{X}'\mathbf{C}_w^{-1}(\overline{\mathbf{X}}_1 - \overline{\mathbf{X}}_2) - \tfrac{1}{2}\left[\overline{\mathbf{X}}_1'\mathbf{C}_w^{-1}\overline{\mathbf{X}}_1 - \overline{\mathbf{X}}_2'\mathbf{C}_w^{-1}\overline{\mathbf{X}}_2\right] \\ &= \mathbf{X}'\mathbf{C}_w^{-1}(\overline{\mathbf{X}}_1 - \overline{\mathbf{X}}_2) - \tfrac{1}{2}(\overline{\mathbf{X}}_1 + \overline{\mathbf{X}}_2)\mathbf{C}_w^{-1}(\overline{\mathbf{X}}_1 - \overline{\mathbf{X}}_2) \end{aligned}$$

Comparing the coefficient $\mathbf{C}_w^{-1}(\overline{\mathbf{X}}_1 - \overline{\mathbf{X}}_2)$ of $\mathbf{X}'$ in this expression with the $\mathbf{v}$ displayed in Eq. 7.22 (and remembering that the $\mathbf{d}$ there stands for $\overline{\mathbf{X}}_1 - \overline{\mathbf{X}}_2$), we see that they differ only in that the $\mathbf{T}^{-1}$ in Eq. 7.22 has been replaced by $\mathbf{C}_w^{-1}$. However, it can be shown that, in the two-group case, $\mathbf{T}^{-1}\mathbf{d}$ and $\mathbf{C}_w^{-1}\mathbf{d}$ are proportional to each other. Thus we have shown that in the two-group case, the *difference* between the two linear classification functions and the single discriminant function are equivalent to each other. (By "equivalent" we mean that they differ only by a linear transformation and hence will yield identical classification results.)

The relationship between the K linear classification functions and the $K - 1$ discriminant functions in the general case of K (>2) groups is more complicated, and we shall not pursue it here. Interested readers are referred to Green (1979), who indicates in outline the relationship between the $p \times K$ matrix $\mathbf{C}_w^{-1}\overline{\mathbf{X}}$ of coefficients in the classification functions and the $p \times (K - 1)$ matrix $\mathbf{V}$ of coefficients for the discriminant functions. Again, they are related by a linear transformation,

$$\mathbf{C}_w^{-1}\overline{\mathbf{X}} = \mathbf{VH} \tag{10.11}$$

where $\mathbf{H}$ is a $(K - 1) \times K$ transformation matrix. This means that the classification results using the maximum-L_k rule and those using discriminant functions in the manner described in Section 10.4 will be identical, as Green indicates.

The last-mentioned fact should not mislead the reader into thinking that the situation is just the same in the K-group case as it is in the two-group case. Even though linear transformations are involved in both cases in the relation between classification and discriminant function, we have a linear transformation between *scalar* quantities L_1–L_2 and Y in the two-group case, which means that the combining weights of the original p variable are *proportional* across these two linear combinations. On the other hand, the matrix linear transformation (10.11) does not make the two sets of weights proportional. Hence the relative weights assigned to the several variables are different in the two sets of linear combinations in the K-group case. Consequently, even though the classification functions provide information about the group structures just as the discriminant functions do, the two sets of functions given different kinds of information, as Green points out.

10.6 JOINT PROBABILITY OF GROUP MEMBERSHIP AND SUCCESS

We now turn to the situation in which considerations of success or productivity in a group (occupational or educational) are of concern besides sheer ''belongingness'' in the group. This is not a pure classification problem as was defined at the outset of this chapter and was subsequently broadened to take prior probabilities of group membership into account besides resemblance. But it is a natural extension of the classification problem as far as applications to vocational and educational guidance are concerned. Just as it is unreasonable to counsel a person to enter some occupation on the sole basis of his or her ''resemblance'' to current members of that group without regard to societal demands (which lead to different group sizes), it would be remiss to ignore information on the possible degree of success that he or she might enjoy in various occupations, if such information is available.

The customary method for seeking to predict a person's possible degree of success in various occupations has been to use a series of multiple regression equations of some measure of success on a set of predictor variables, one equation having been constructed for each occupational group. The counselee is then advised to consider entering that group for which his or her predicted degree of success is the highest.[3] But this approach obviously suffers from the fallacy of using regression equations that may not be valid for a given person. It must be remembered that each regression equation has been constructed on a sample

[3]This, of course, is an oversimplified description of the process of vocational or educational guidance. In practice, other considerations besides possible degree of success are no doubt taken into account by the counselor. But it seems that such weighting by other factors has been done largely on an intuitive basis, so that the adequacy of the outcome depends heavily on the expertise of the particular counselor.

from a particular occupational (or curricular) group, and hence cannot be validly used for anyone who is not a member of that group. In other words, we cannot take at face value the predicted degrees of success estimated from the several regression equations—one for each group—for a person of unknown group membership. In fact, only one of these equations should, strictly speaking, yield a valid prediction for any one person—namely, the equation for the particular group of which he or she is indeed a member.[4]

A possible approach to overcoming the above difficulty lies in utilizing the joint probability of membership and success in a group (Tatsuoka, 1956). In this method we relinquish the idea of predicting the possible *degree* of success as such, and seek instead to estimate a person's chances of "succeeding" at all, in the sense of exceeding some specified score on the criterion variable. The reason for adopting this more modest goal will presently become clear.

We have already seen, in Section 10.3, how we may compute the probability $p(H_k \mid \mathbf{X}_i)$ that a person with a predictor-score combination $\mathbf{X}_i$ is a member of group k. To obtain the joint probability of membership and success, we now need to compute the probability that a member of group k who has this predictor-score combination $\mathbf{X}_i$ will be "successful"—that is, that he or she will have a criterion score exceeding an acceptable cutoff point.

Let U_k be a criterion variable appropriate for measuring success in group k. This may be a measure of productivity, a supervisor rating of efficiency on the job, annual income of a self-employed person, grade-point average in college, or whatever else may be reasonable and available for assessing the degree of success of a member of the kth vocational or curricular group. (We need not have the same measure for all K groups.) Suppose that we have constructed, in the usual manner, a multiple regression equation of U_k on $X_1, X_2, \ldots, X_p$ (the same predictor variables as before) for each group. These equations may be symbolized by

$$\hat{U}_k = a_k + b_{1k}X_2 + \cdots + b_{pk}X_p \qquad (k = 1, 2, \ldots, K)$$

In the course of constructing these equations, we will have computed, among other things, the sum of squares $\Sigma\, u_k^2$ of the criterion variable (in deviation scores) for the kth sample, and the coefficient of multiple correlation R_{uk} between U_k and $X_1, X_2, \ldots, X_p$. From these, we compute the standard errors of estimate

$$s_{u.x(k)} = \left[\frac{\Sigma\, u_k^2}{n_k - 2}(1 - R_{uk}^2)\right]^{1/2} \qquad k = 1, 2, \ldots, K \qquad (10.12)$$

Next, for each group, we must decide on a suitable cutoff point U_k^* such that

[4]Even if we grant that the several occupations or curricula being considered by a person would often be sufficiently similar to render the predictions from all of the relevant regression equations at least approximately valid, there are other difficulties with using these predictions as the sole bases for counseling. For a discussion of this issue, see Rulon, Tiedeman, Tatusoka, and Langmuir (1967, pp. 323–336).

a member of group k will be called "successful" if and only if his or her criterion score exceeds this value. Then, with the assumption that the predictors and criterion jointly follow a $(p + 1)$-variate normal distribution in each group, we can estimate the probability that a member of group k who has a predictor-score combination $\mathbf{X}_i = (X_{1i}, X_{2i}, \ldots, X_{pi})$ will be "successful." The formula is as follows:

$$p(U_{ki} > U_k^* \mid H_k \;\&\; \mathbf{X}_i) = (2\pi)^{-1/2} \int_{z_{ki}^*}^{\infty} \exp(-z^2/2)\, dz \qquad (10.13)$$

with

$$z_{ki}^* = \frac{U_k^* - \hat{U}_k(\mathbf{X}_i)}{s_{u.x(k)}} \qquad (10.14)$$

where $\hat{U}_k(\mathbf{X}_i)$ is the value of U_k predicted from the regression equation for the predictor-score combination $\mathbf{X}_i$. The presence of an integral sign in Eq. 10.13 need not discourage readers who are unfamiliar with the calculus. It is merely a convenient way of indicating that we are to use a table of normal-curve areas to find the area to the right of the point $z = z_{ki}^*$ on the abscissa. So the only actual calculations needed are those involved in determining z_{ki}^* for the given predictor-score combination.

It should be noted that the probability given by Eq. 10.13 is conditioned just as much on the premise that the person to whom it applies is a member of group k as it is on the fact that his or her predictor-score combination is $\mathbf{X}_i$. We are not really using the kth group's regression equation for a particular person i whose group membership is unknown. Rather, we are estimating the proportion, *among group k members who have a score combination in the vicinity of* $\mathbf{X}_i$, of those who may be expected to have a criterion score exceeding U_k^*.

Now that we have seen how, given a particular predictor-score combination, we may compute for each k $(= 1, 2, \ldots, K)$ the probability of membership in group k, and the probability of "success" in that group *given* membership therein, it remains only to apply the general multiplication theorem of probability theory in order to obtain the desired probability of the joint event: membership *and* success in a group, given a particular predictor-score combination.

The general form of the multiplication theorem is

$$p(A \;\&\; B \mid C) = p(A \mid C)p(B \mid A \;\&\; C)$$

In the context of our problem, we let A stand for the event of membership in group k (i.e., the occurrence of H_k), B for "success" in that group (that is, $U_{ki} > U_k^*$), and C for the observation of a particular predictor-score combination $\mathbf{X}_i$. Thus the desired joint probability may be computed as

$$p(H_k \;\&\; U_{ki} > U_k^* \mid \mathbf{X}_i) = p(H_k \mid \mathbf{X}_i)p(U_{ki} > U_k^* \mid H_k \;\&\; \mathbf{X}_i) \qquad (10.15)$$

where the two factors on the right are given by Eq. 10.8 and 10.13, respectively.

Once these probabilities have been computed for each person to be counseled, the counselor would advise each person to consider entering that group for which his or her probability of membership and success is the largest. An evaluation of the soundness of this procedure by cross-validation methods admittedly poses certain problems, which it would take us too far afield to discuss here. Suggestions for coping with these problems are discussed in Tatsuoka (1956) and Rulon et al. (1967, chap. 10).

EXAMPLE 10.6[5] To illustrate the calculations for obtaining the joint probability of membership and success, we now suppose that scores on a suitable measure U_k of success were available in addition to the APQ scores for all members of each of the three normative samples. We would then be able to construct a multiple regression equation of U_k on X_1, X_2, and X_3 for each group. Suppose that the equations were as follows:

$$\hat{U}_1 = 1.0603 - .0799X_1 + .1691X_2 + .1443X_3$$

$$\hat{U}_2 = -.7587 + .2506X_1 + .1232X_2 - .0781X_3$$

$$\hat{U}_3 = 2.3426 + .1643X_1 - .1440X_2 + .1942X_3$$

The standard errors of estimate associated with these equations, calculated in accordance with Eq. 10.12, were as follows:

$$s_{u.x(1)} = 1.6849 \qquad s_{u.x(2)} = 1.5557$$

and

$$s_{u.x(3)} = 1.5968$$

The predicted success score for a member of each group having a predictor-score combination equal to that of each of our four individuals is then computed from the relevant regression equation. The results are as follows:

i	$\mathbf{X}_i$	$\hat{U}_1$	$\hat{U}_2$	$\hat{U}_3$
1	[15, 24, 6]	4.7860	5.4885	2.5163
2	[15, 16, 11]	4.1547	4.1124	4.6393
3	[16, 23, 14]	5.6914	4.9911	4.3782
4	[19, 20, 15]	5.0887	5.2952	5.4973

[5]The author is indebted to Dr. James Kraatz of the University of Illinois Computer-Based Education Research Laboratory for doing the calculations for this example, including the generation of suitable criterion scores. All results pertaining only to X_1, X_2, and X_3 first appeared in Rulon, Tiedeman, Langmuir, and Tatsuoka (1954, parts 3-4). The scores on U_1, U_2, and U_3 and results involving these variables appeared in Rulon, Tiedeman, Tatsuoka, and Langmuir (1967, chap. 10). The scores were used by permission of John Wiley and Sons, Inc.: the calculations were re-done expressly for this book.

(Note that these are not predictions for our four individuals themselves, but for members of the relevant groups with the stated predictor-score combinations, which are equal to those of our four individuals.) The next step is to decide on the cutoff point U^*_k for "success" in each group. Although the values may, in general, differ from group to group, we here take $U^*_k = 4.5$ for all three groups. We can now determine the z^*_{ki} values in accordance with Eq. 10.14. Thus, for $i = 1$,

$$z^*_{11} = [4.5 - \hat{U}_1(\mathbf{X}_1)]/1.6849 = (4.5 - 4.7860)/1.6849$$

$$= -.1697$$

$$z^*_{21} = [4.5 - \hat{U}_2(\mathbf{X}_1)]/1.5557 = (4.5 - 5.4885)/1.5557$$

$$= -.6354$$

$$z^*_{31} = [4.5 - \hat{U}_3(\mathbf{X}_1)]/1.5968 = (4.5 - 2.5163)/1.5968$$

$$= 1.2423$$

Collecting the results for all four predictor-score combinations $\mathbf{X}_i$ in a single table, we have the following:

i	z^*_{1i}	z^*_{2i}	z^*_{3i}
1	−.1697	−.6354	1.2423
2	.2049	.2491	−.0872
3	−.7071	−.3157	.0763
4	−.3494	−.5112	−.6246

Next, we determine $p(U_{ki} > U^*_k \mid H_k \;\&\; \mathbf{X}_i)$ in accordance with Eq. 10.13 by entering a table of normal-curve areas with each z^*_{ki} value and finding the area to the right of that abscissa point. The results are:

$p(U_{ki} > U^*_k \mid H_k \;\&\; \mathbf{X}_i)$

i \ k	1	2	3
1	.5674	.7374	.1071
2	.4188	.4017	.5348
3	.7602	.6239	.4696
4	.6366	.6954	.7339

The final step, following Eq. 10.15, is to multiply each of these probabilities of success (given membership in group k and the predictor-score combination) by the corresponding probability of group membership, $p(H_k \mid \mathbf{X}_i)$, obtained previously from Eq. 10.8a. The resulting products are the joint probabilities of membership and success in the various groups, given predictor-score combination $\mathbf{X}_i$. For this example, the values of the joint probability, and the "classification" decisions based thereon, are as follows:

	$p(H_k \,\&\, U_{ki} > U_k^* \mid X_i)$			
i \ k	1	2	3	Decision—Assign to
1	*.3514*	.2745	.0009	Group 1
2	.0192	.1236	*.3457*	Group 3
3	.1485	*.3885*	.0855	Group 2
4	.0053	*.4019*	.3035	Group 2

We see that the decisions for these four individuals would be exactly the same as those based on the probability of group membership alone. This, of course, need not be true in general.

10.7 OTHER CLASSIFICATION-DECISION RULES

For the sake of completeness, we describe very briefly in this section three other decision rules that are sometimes encountered in the literature.

Minimizing the Total Loss Due to Misclassification

In some fields of application, it is often possible to associate a *cost* or loss incurred by misclassifying an individual that actually belongs to population k to some other population. The cost may be a monetary one, as in a financial decision, or the loss in terms of waste of human resources when we classify a person capable of functioning in a higher-level occupation into a lower-level one, and so on. We assume that such costs can be quantified as a set of numbers c_k $(k = 1, 2, \ldots, K)$ whose sum is equal to 1. Then the classification index $C(H_k \mid \mathbf{X}_i)$ involves just a slight modification of the probability $p(H_k \mid \mathbf{X}_i)$ of membership in group k, given in Eq. 10.7. Namely,

$$C(H_k \mid \mathbf{X}_i) = \frac{c_k p_k p(\mathbf{X}_i \mid H_k)}{\sum_{j=1}^{K} c_j p_j p(\mathbf{X}_i \mid H_j)} \qquad k = 1, 2, \ldots, K \qquad (10.16)$$

The classification rule, just as in the case of using the probability-of-group-membership rule, is to classify each person to the group for which his or her $C(H_k \mid \mathbf{X}_i)$ is the largest. This will have the effect of *minimizing* the total cost of misclassification, because each person is classified to the group for which his or her probability of membership, *weighted* by the loss that would be incurred if he or she were indeed a member of that group and had not been so classified, is the greatest. Thus the difference between using the maximum-probability rule and the minimum-loss rule is this: If an individual's probability, based on his or her pattern X_i of scores, is slightly larger for belonging to group 3 than to group 5 (say), but the cost of failing to classify him or her to group 5 if he or she were actually a member of that group is considerably greater than failing to classify a group 3 member to that group, then the loss consideration will outweigh the pure probability consideration and lead to his or her being classified in group 5 instead of group 3.

The Coefficient of Profile Similarity, r_p

In 1949, Cattell developed several types of coefficients of profile similarity, mainly for use in conjunction with his then recently published Sixteen Personality-Factors Questionnaire, the 16 PF (Cattell, 1949b, Cattell & Eber, 1970). The types differ in their purpose: for comparing pairs of individual profiles, pairs of group profiles, and an individual to a group profile. It is the last type, for comparing individuals versus groups, that is relevant for classification problems. We discuss here two slightly modified forms of r_p—one to allow the variables to have arbitrary standard deviations instead of the predetermined 2.0 of the scale used for the 16 PF, the other to allow for nonnegligible correlations among the variables. Assuming, as before, that we have p variables and K groups, the coefficient of profile similarity of the ith individual to the kth group's mean profile is

$$r_{p(k;i)} = \frac{C_p - \sum_{j=1}^{p} [(X_{ji} - \overline{X}_{jk})/s_{jk}]^2}{C_p + \sum_{j=1}^{p} [(X_{ji} - \overline{X}_{jk})/s_{jk}]^2} \qquad k = 1, 2, \ldots, K \qquad (10.17)$$

Here C_p is the median of the chi-square distribution with p degrees of freedom, and the other symbols should be obvious from the context. The decision rule is to classify the individual in the group for which his or her $r_{p(k;i)}$ has the largest value. Roughly speaking, the rationale is as follows. If we assume that each variable X_j is approximately normally distributed with mean $\overline{X}_{jk}$ and variance s_{jk}^2 in the kth group, and that all pairs of variables are uncorrelated, then the sum appearing in both numerator and denominator of expression (10.17) will be an approximate chi-square with p degrees of freedom. Hence for a person randomly sampled from population k, the sum will exceed C_p with probability

$\frac{1}{2}$, and hence $r_{p(k;i)}$ will be positive or negative with a 50:50 chance, and will most often be close to zero. The closer a person's score vector $\mathbf{X}_i$ is to the kth group centroid $\mathbf{X}_k$ (i.e., the more he or she resembles the "typical" member of group k), the closer to zero the sum will be, and hence the closer to 1 $r_{p(k;i)}$ will be. The farther $\mathbf{X}_i$ is from $\overline{\mathbf{X}}_k$, on the other hand, the larger the sum will be compared to C_k and the closer $r_{p(k;i)}$ will be to -1. Thus r_p is an index that is somewhat like the correlation coefficient: It can range from -1 to 1, with -1 representing "complete dissimilarity" of the person's profile from the group-mean profile, $+1$ representing identity of the individual and group profiles, and 0 indicating "just a chance resemblance."

Going now to the second modification of taking the correlations between pairs of variables into account, it should not be difficult to see that we need only replace the sum in both numerator and denominator of Eq. 10.17 by either the χ^2_{ik} of Eq. 10.1 or that of Eq. 10.2, depending on whether the K population covariance matrices are regarded as homogeneous or not. The expression for the second modified r_p for individual i with respect to group k thus becomes

$$r^*_{p(k;i)} = \frac{C_p - \chi^2_{ik}}{C_p + \chi^2_{ik}} \qquad k = 1, 2, \ldots, K \tag{10.18}$$

The interpretation is exactly the same as for the $r_{p(k;i)}$ of Eq. 10.17. This second modified r_p, however, has the additional property that classification of an individual to the group with the largest $r^*_{p(k;i)}$ value is *equivalent* to using the previously discussed minimum-χ^2_{ik} classification rule.

The m-Nearest-Neighbor Rule

This is a simple, model-free approach first proposed by Cover and Hart (1967) and subsequently developed in greater detail by Fukunaga (1972), among others. The first step is to compute the (squared) distance between the individual to be classified and each of the m closest neighbors to him or her in the p-dimensional space, where m is some prespecified number. The distance may be defined either in the ordinary Euclidean sense,

$$D^2_{ij} = (\mathbf{X}_i - \mathbf{X}_j)'(\mathbf{X}_i - \mathbf{X}_j) \tag{10.19}$$

or the Mahalanobis generalized distance sense,

$$D^{*2}_{ij} = (\mathbf{X}_i - \mathbf{X}_j)'\mathbf{C}_w^{-1}(\mathbf{X}_i - \mathbf{X}_j) \tag{10.20}$$

which is just like the χ^2_{ik} defined in Eq. 10.1 except that, instead of the deviation of an individual's score vector from a group centroid, the difference between two individual score vectors is involved. If i is the person to be classified, the D^2_{ij} or D^{*2}_{ij} is computed from Eq. 10.19 or 10.20 for each j corresponding to individuals that are "reasonably close" to individual i. (We may actually have to compute them for all persons in the K groups at hand.) We then retain the smallest m D_{ij}^2's (or D_{ij}^{*2}'s) and determine how many of these are associated with members of each of the K groups. Let n_{i1} be the number of D_{ij}^2's, among the m smallest of these quantities, that are associated with members of group 1, n_{i2} be the number of those associated with group 2 members, and so on through n_{iK}.

We then compute the quantities

$$p(k \mid m) = \frac{p_k(n_{ik}/m)}{\sum_{h=1}^{K} p_j(n_{ih}/m)} = \frac{p_k n_{ik}}{\sum_{h=1}^{K} p_h n_{ih}} \qquad k = 1, 2, \ldots, K \qquad (10.21)$$

where p_k is the prior probability of membership in group k, first introduced in Eq. 10.4. The decision rule is to classify individual i in that group for which $p(k \mid m)$ has the largest value. [The middle member of Eq. 10.21 is included merely to show that the numerator (and each term in the denominator) is the product of a prior probability and a *sort of* conditional probability that if individual i is indeed a member of group k, the observed number of members of group k will be found among his or her m closest neighbors. The expression is thus analogous to that in Eq. 10.7. The m, of course, cancels out and the last member of Eq. 10.21 is obtained, which is used for the calculations.]

The SAS package contains a program for carrying out classification using this method, there called the NEIGHBOR procedure.

10.8 RESEARCH EXAMPLES

Gribbons and Lohnes (1968, p. 40–2) approached a pair of research issues by means of a comparison of the classification validities of scales from two interview protocols. They had scaled eight Readiness for Vocational Planning (RVP) variables from an eighth-grade interview protocol collected on a sample of 110 junior high school students in five cities around Boston, Massachusetts in 1958. A similar set of eight RVP scales were scored from tenth-grade interview protocols collected on the same students in 1960. Also collected in 1960 were the high school curriculum group placements of the students. Curriculum formed a categorical variable in three groups: (1) college, (2) business, (3) industrial arts and general. The first research issue was concerned with the hypothesis that both RVP assessments possessed classification validity for the prediction of curriculum group placement in tenth-grade. The second research issue was concerned with the hypothesis that the classification validity of the 10th-grade RVP assessment was greater than the classification validity of the 8th-grade RVP assessment.

Preliminary discriminant analyses indicated that the two MANOVA F-ratios for differences in group centroids were quite similar (for 8th RVP, F = 2.23

while for 10th RVP, $F = 2.34$; n.d.f. = 16 and 214 in each case). However, the two discriminant functions for the three groups in the spaces of the 8th RVP and the 10th RVP were rather different, and the centroids of the three groups mapped into the two separate two-dimensional discriminant spaces rather differently, as shown by the Centroids in Discriminant Spaces Table (Table 10.1). The authors believed that these multivariate analyses were somewhat overpowering for the small sample size (in the sense that the profiles of discriminant function weights could be expected to capitalize on chance to a considerable extent), and that the classification hits and misses would provide the best test of their research hypotheses.

Classification hits and misses for the two assessments can be compared in the Classifications from Membership Probabilities Table (Table 10.2), in which hits are printed in boldface and misses in ordinary face. Each subject was assigned to the group for which she had the highest computed probability of membership in the particular RVP assessment (8th or 10th grade). The table reveals a 7:4 hits/misses ratio for each of the two RVP assessments. The first hypothesis seemed to be confirmed by the data, in that both the 8th-grade and the 10th-grade RVP assessments showed impressive classification validity. The strength of the information from the interview protocols resided in its ability to identify college preparatory students. The weakness resided in its inability to identify many of the business students and most of the industrial arts and general curriculum students. However, the tendency of many of these students to describe themselves in the same terms as those employed by college students may have important educational guidance implications. Was it possible that undue external pressures caused some students who thought of themselves the way most college preparatory students thought of themselves to nevertheless accept assignment into business, industrial art, or general curricula? Alternatively, was it possible that their schools had failed to assist some students who were not potential college material to modify their unrealistic self-assessments?

The second hypothesis failed on the data, and the surprising result was that the classification validity of the 10th-grade RVP assessment was no greater than that of the 8th-grade RVP assessment, despite the fact that the former was a concurrent validity and the latter a true predictive validity.

In any event, the authors derived considerable satisfaction from this classification outcome. They concluded that it "would seem to be evidence that the

TABLE 10.1 Centroids in Discriminant Spaces

	8th RVP		10th RVP	
Group	d.f.1	d.f.2	d.f.1	d.f.2
College	.82	−.02	.77	.28
Business	−.95	.45	−1.42	.43
Industrial arts and general	−1.15	−.69	−.18	−1.71

TABLE 10.2 Classifications from Membership Probabilities

Predicted group	College		Business		Industrial arts and general		Total	
	8 RVP	10 RVP	8 RVP	10 RVP	8 RVP	10 RVP	8 RVP	10 RVP
College	**51**	**55**	14	18	12	14	77	87
Business	10	5	**17**	**13**	4	3	31	21
Industrial arts and general	0	1	0	0	**2**	**1**	2	2
Total	61	61	31	31	18	18	110	110

Readiness for Vocational Planning traits are well-defined for eighth-grade students and may be reliably estimated in the first semester of eighth grade'' (p. 42).

Gribbons and Lohnes (1982, p. 78–9) reported research on the classification validity over an 11-year time span of four measurements collected in 1958 on their sample of 110 8th-grade students for a categorical variable collected in 1969. Unfortunately, by 1969 two of their subjects were deceased (both while in the Armed Services, one in a jeep accident in Alaska, the other in Viet Nam), so only 108 were available for classification. The categorical variable represented the four branch tips of a career development tree structure, and combined information about education beyond high school with occupational field. The cells of the criterion variable were: (1) College Science (CS), (2) Non-college Technology (NT), (3) Non-college Business (NB), (4) College Business and Cultural (CB). Three of the four 1958 predictor variables were standard educational research variables (Gender, Socioeconomic Stratus, Otis IQ), but the fourth was an innovative measure of Readiness for Career Planning (RCP). RCP was essentially the first principal component of the 8-scale RVP assessment discussed in the example above.

Preliminary multivariate analysis yielded a MANOVA F with n.d.f. of 30 and ∞ equal to .7 for the hypothesis of equality of dispersions of the four groups in the space of the four predictors, and a MANOVA F with n.d.f. of 12 and 267 equal to 5.0 for the hypothesis of equality of centroids. The MANOVA η^2 was .41 (but remember that this is an upwardly-biased estimator). Two useful discriminant functions were found to separate the four groups in the 4-space of the measurements. The first discriminant separated the two college-trained groups from the two non-college groups, while the second discriminant separated the predominantly male non-college technology group from the other three groups. The MANOVA for Career Tree Groups Table (Table 10.3) reports the means for the groups on the measurements. Needless to say, the authors were disappoined in the comparatively weak separation of the groups on the RCP variable.

TABLE 10.3 MANOVA Table for Career Tree Groups

1958 scales	Career Tree Group Means				Pooled	
	CS ($N=9$)	NT ($N=22$)	NB ($N=59$)	CB ($N=18$)	S.D.	F^3_{104}
Gender	1.6	1.2	1.7	1.3	.5	6.8
SES (1 = high)	2.2	4.3	4.4	3.0	1.5	8.9
Otis IQ	115.6	103.2	106.6	113.6	8.7	7.4
RCP	33.3	32.6	30.9	37.8	9.1	2.0

The classification of 1969 groups in the space of 1958 scales in Table 10.4 (hits in boldface) reveals that 54% of the sample was correctly classified. Only the college business and cultural group had a majority of its actual members misclassified. More students were predicted into the science and technology groups than actually arrived in them, and fewer students were predicted into the business and technology groups than actually arrived in them. Regarding the students who were predicted into the college science group but who actually arrived in the college business and cultural group, the authors remarked, "Why some of these college sociocultural and business subjects, who 'look more like' college science subjects, are not pursuing science or technology careers is a nice question" (p. 78), which indeed it is.

TABLE 10.4 Classifications of 1969 Groups in Space of 1958 Scales

Actual 1969 Group	Predicted 1969 Group				
	CS	NT	NB	CB	Total
College science	**6**	0	2	1	9
Non-college technical	1	**14**	4	3	22
Non-college business	8	16	**31**	4	59
College business-cultural	5	4	2	**7**	18
Total	20	34	39	15	$N=108$

EXERCISES

1. Using each of the methods discussed in Sections 10.1 to 10.5, arrive at classification decisions for members of the same job-applicant group as that in the numerical examples, whose APQ scores are as follows:

$$\mathbf{X}_i$$

Individual 5: [17, 19, 15]

Individual 6: [16, 20, 11]

2. Suppose, in the context of the numerical examples carried through this chapter and Exercise 1, that we further assume the costs of misclassifying a member of group k ($k = 1, 2, 3$) to either of the other groups are

$$c_1 = .2 \qquad c_2 = .3 \qquad c_3 = .5$$

Use the cost function $C(H_k \mid \mathbf{X}_i)$ given in Eq. 10.16 to classify each of individuals 1, 2, . . . , 6 described in the examples and in Exercise 1 by the minimum-cost rule.

3. Using Eq. 10.18 with χ^2_{ik} defined by Eq. 10.2, calculate the coefficient of profile similarity $r^*_{p(k;i)}$ for each combination of $k = 1, 2, 3$ and $i = 5, 6$.

4. In Chapter 7, Exercise 9, four composite career-choice groups were defined (two of them already being composite groups in the HSB data set itself). They were:

Group I:	clerical, operative, and sales	$n_1 = 86$
Group II:	craftsman, farmer, and technical	$n_2 = 86$
Group III:	professional 1	$n_3 = 161$
Group IV:	professional 2	$n_4 = 94$

Using the three affective and five cognitive variables, determine the four linear classification functions based on the eight variables. (Recall, as pointed out in Section 10.5, that the SAS package calls these the "discriminant functions.") Be sure to exclude from your analysis the 173 students that do not fall into any of these four groups.

APPENDIX A

DETERMINANTS

A.1 DEFINITION AND BASIC COMPUTATIONAL RULES

A determinant of order n is a *number* associated with a square matrix of order n, computed as a function of the latter's elements (which are also called the *elements* of the determinant) in accordance with the following rules.

Given an $n \times n$ matrix

$$\mathbf{A} = \begin{bmatrix} a_{11} & a_{12} & \cdots & a_{1n} \\ a_{21} & a_{22} & \cdots & a_{2n} \\ \vdots & \vdots & & \vdots \\ a_{n1} & a_{n2} & \cdots & a_{nn} \end{bmatrix}$$

the determinant of $\mathbf{A}$, symbolized by $|\mathbf{A}|$ or det ($\mathbf{A}$) and written in full as

$$\begin{vmatrix} a_{11} & a_{12} & \cdots & a_{1n} \\ a_{21} & a_{22} & \cdots & a_{2n} \\ \vdots & \vdots & & \vdots \\ a_{n1} & a_{n2} & \cdots & a_{nn} \end{vmatrix}$$

is computed as follows:

Step 1. Form all possible products of n factors such that each factor is an element of $\mathbf{A}$, no two of which are from the same row or same column of $\mathbf{A}$. There are $n!$ such products possible. For example, when $n = 3$ the required products are the following six ($= 3!$) quantities:

$$a_{11}a_{22}a_{33} \qquad a_{12}a_{23}a_{31} \qquad a_{13}a_{21}a_{32} \qquad a_{13}a_{22}a_{31} \qquad a_{11}a_{23}a_{32} \quad a_{12}a_{21}a_{33}$$

(The requirement that no two factors of a given product be elements of the same row or same column of **A** is reflected in the fact that each of the numerals 1, 2, and 3 occurs once and only once as a row subscript, and also once and only once as a column subscript, among the elements that form each product.)

Step 2. Within each product, arrange the factors so that the row subscripts are in natural order $(1, 2, \ldots, n)$. (Each of the six products in the example above with $n = 3$ has already been written in this manner.) Then examine the order in which the column subscripts appear. Specifically, note how many times a larger number precedes a smaller one in the sequence of column subscripts in each product. This number is called the *number of inversions* (hereafter abbreviated as NI) associated with each product. Thus, for the six products for $n = 3$ in the order listed above, the NIs are

$$0, 2, 2, 3, 1, \quad \text{and} \quad 1$$

respectively. (Note that NI is the number of *instances* of larger numbers preceding smaller ones, *not* the number of larger *numbers* that precede smaller ones. Thus the NI for $a_{13}a_{21}a_{32}$ is 2, because *3* precedes a smaller number in two instances—preceding *1* and preceding *2*—even though *3* is the only number that precedes a smaller one.)

Step 3. Upon determining the NIs for all $n!$ products, multiply each product with an odd-numbered NI by -1 (i.e., change its sign); leave as is all the products with even-numbered NIs (counting 0 as even). Thus, among the six products listed above for $n = 3$, the first three (with NI $= 0, 2, 2$, respectively) will be unaltered, while the last three (with NI $= 3, 1, 1$) will have their signs reversed. (For any n, exactly one-half of the $n!$ products will have their signs reversed).

Step 4. Form the algebraic sum of the products as partially modified in sign in Step 3. This sum is the value of $|\mathbf{A}|$. Thus, for $n = 3$, we have

$$|\mathbf{A}| = \begin{vmatrix} a_{11} & a_{12} & a_{13} \\ a_{21} & a_{22} & a_{23} \\ a_{31} & a_{32} & a_{33} \end{vmatrix} = \begin{array}{l} a_{11}a_{22}a_{33} + a_{12}a_{23}a_{31} + a_{13}a_{21}a_{32} \\ -\; a_{13}a_{22}a_{31} - a_{11}a_{23}a_{32} - a_{12}a_{21}a_{33} \end{array} \tag{A.1}$$

EXAMPLE A.1. Find the value of the following determinant:

$$\begin{vmatrix} 3 & 1 & -2 \\ 2 & 0 & 3 \\ -4 & 2 & 5 \end{vmatrix}$$

Substituting in Eq. A.1, we find

$$\begin{vmatrix} 3 & 1 & -2 \\ 2 & 0 & 3 \\ -4 & 2 & 5 \end{vmatrix} = \begin{array}{l} (3)(0)(5) + (1)(3)(-4) + (-2)(2)(2) \\ -\ (-2)(0)(-4) - (3)(3)(2) - (1)(2)(5) \end{array}$$

$$= 0 - 12 - 8 - 0 - 18 - 10 = -48$$

A.2 SPECIAL ROUTINE FOR THIRD-ORDER DETERMINANTS: SARRUS' RULE

For a third-order determinant, the application of Steps 1 to 4 may be mechanized as follows:

1. To the right of the determinant, copy the first two columns over again:

$$\begin{vmatrix} a_{11} & a_{12} & a_{13} \\ a_{21} & a_{22} & a_{23} \\ a_{31} & a_{32} & a_{33} \end{vmatrix} \begin{matrix} a_{11} & a_{12} \\ a_{21} & a_{22} \\ a_{31} & a_{32} \end{matrix}$$

2. In the extended three-row, five-column configuration constructed above, connect, by a solid line, each of the three first-row elements of $|\mathbf{A}|$ with the two numbers located "southeast" of it. Similarly, connect by a dashed line, each of the three third-row elements of $|\mathbf{A}|$ with the two numbers located "northeast" of it. Thus

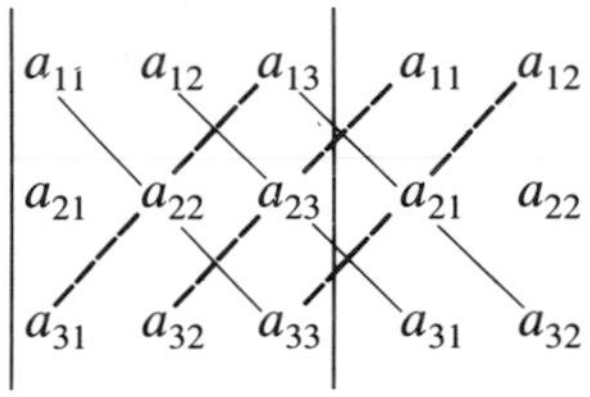

3. Form the product of each triplet of elements connected by a solid line, and prefix the product with a plus sign. Form the product of each triplet of elements connected by a dashed line, and prefix the product with a minus sign:

$$+ (a_{11}a_{22}a_{33}) + (a_{12}a_{23}a_{31}) + (a_{13}a_{21}a_{32})$$
$$- (a_{13}a_{22}a_{31}) - (a_{11}a_{23}a_{32}) - (a_{12}a_{21}a_{33})$$

The resulting algebraic sum gives the value of $|\mathbf{A}|$.

Remember that this method is applicable *only* to determinants of order 3.

EXERCISES

Find the value of each of the following determinants. In the case of third-order determinants, evaluate both by explicit use of the definitional rules (Steps 1 to 4) and by Sarrus' rule.

1. $\begin{vmatrix} 3 & 5 \\ 2 & 4 \end{vmatrix}$ **2.** $\begin{vmatrix} 2 & -3 \\ 1 & 4 \end{vmatrix}$ **3.** $\begin{vmatrix} 1 & 2 & 3 \\ 3 & 1 & 2 \\ 2 & 3 & 1 \end{vmatrix}$ **4.** $\begin{vmatrix} 1 & 0 & 0 \\ 3 & 1 & 2 \\ 2 & 3 & 1 \end{vmatrix}$

5. $\begin{vmatrix} 0 & 2 & 0 \\ 3 & 1 & 2 \\ 2 & 3 & 1 \end{vmatrix}$ **6.** $\begin{vmatrix} 0 & 0 & 3 \\ 3 & 1 & 2 \\ 2 & 3 & 1 \end{vmatrix}$

7. $\begin{vmatrix} 2 & -1 & 3 \\ 3 & 4 & -2 \\ 5 & 3 & 1 \end{vmatrix}$ **8.** $\begin{vmatrix} 1 & 2 & 0 & -1 \\ 0 & 3 & 1 & 2 \\ -2 & 1 & 4 & 0 \\ 3 & 0 & 1 & 1 \end{vmatrix}$

[Answers: **(1)** 2, **(2)** 11, **(3)** 18, **(4)** −5, **(5)** 2 **(6)** 21, **(7)** 0, **(8)** 72.]

A.3 EXPANSION OF A DETERMINANT IN TERMS OF COFACTORS

Evaluating a determinant of order higher than 3 is exceedingly tedious, as the reader will have found in doing Exercise 8. Fortunately, the task can be somewhat simplified by using a derived rule: expanding in terms of cofactors. This rule can be developed, for $n = 3$, by algebraic manipulation of the expression on the right-hand side of Eq. A.1. The extensions to cases when $n > 3$ will be evident.

First rearrange the terms of the expression in question by placing next to each other those products that involve the same first-row element of $|\mathbf{A}|$, thus:

$$|\mathbf{A}| = a_{11}a_{22}a_{33} - a_{11}a_{23}a_{32} + a_{12}a_{23}a_{31} - a_{12}a_{21}a_{33} + a_{13}a_{21}a_{32} - a_{13}a_{22}a_{31}$$

Next, from each adjacent pair of terms we factor out the common first-row element:

$$|\mathbf{A}| = a_{11}(a_{22}a_{33} - a_{23}a_{32}) + a_{12}(a_{23}a_{31} - a_{21}a_{33}) + a_{13}(a_{21}a_{32} - a_{22}a_{31})$$

We then rewrite the expressions in parentheses in the form of determinants, thus:

$$|\mathbf{A}| = a_{11}\begin{vmatrix} a_{22} & a_{23} \\ a_{32} & a_{33} \end{vmatrix} + a_{12}\begin{vmatrix} a_{23} & a_{21} \\ a_{33} & a_{31} \end{vmatrix} + a_{13}\begin{vmatrix} a_{21} & a_{22} \\ a_{31} & a_{32} \end{vmatrix} \quad \text{(A.2)}$$

From the definition of the products constituting the terms in evaluating a determinant (Step 1 on p. 384), it is evident that none of the second-order determinants in (A.2) contains elements from the row or column of which the associated multiplier (a_{11}, a_{12}, and a_{13}, respectively) is an element. Thus the first determinant (whose multiplier is a_{11}) comprises those elements of $|\mathbf{A}|$ that remain after deleting the first row and first column of $|\mathbf{A}|$; the second determinant (with multiplier a_{12}) contains elements that remain on deleting the first row and second column of $|\mathbf{A}|$; and the third (with multiplier a_{13}), those elements that remain after the first row and third column have been deleted.

Next, we pay attention not only to the elements occurring in the three second-order determinants in (A.2), but to the configurations in which they occur. We find that in the first and third determinants, the configuration is just as it would be if we were actually to cross out the appropriate row and column of $|\mathbf{A}|$ and to collect the remaining elements as they stood. In the second determinant, however, the columns are *reversed* from the order in which they occur in $|\mathbf{A}|$. But this anomaly may easily be removed by noting that

$$\begin{vmatrix} a_{23} & a_{21} \\ a_{33} & a_{31} \end{vmatrix} = -\begin{vmatrix} a_{21} & a_{23} \\ a_{31} & a_{33} \end{vmatrix}$$

Making this substitution, the expansion (A.2) of $|\mathbf{A}|$ becomes

$$|\mathbf{A}| = a_{11}\begin{vmatrix} a_{22} & a_{23} \\ a_{32} & a_{33} \end{vmatrix} - a_{12}\begin{vmatrix} a_{21} & a_{23} \\ a_{31} & a_{33} \end{vmatrix} + a_{13}\begin{vmatrix} a_{21} & a_{22} \\ a_{31} & a_{32} \end{vmatrix} \tag{A.3}$$

The determinant of order $n - 1$ obtained by crossing out the ith row and jth column of an nth-order determinant $|\mathbf{A}|$ is known as the *minor determinant* associated with element a_{ij} of $|\mathbf{A}|$. We will denote this as M_{ij}. Then the expansion (A.3) of $|\mathbf{A}|$ may be written as

$$|\mathbf{A}| = a_{11}M_{11} - a_{12}M_{12} + a_{13}M_{13} \tag{A.4}$$

It can be shown, by appropriate regroupings of the right-hand side of Eq. A.1, that expansions of $|\mathbf{A}|$ may be made in terms of the minors associated with elements in any row or any column. For example,

$$|\mathbf{A}| = -a_{21}M_{21} + a_{22}M_{22} - a_{23}M_{23}$$

(where the minors are those associated with the second-row elements) and

$$|\mathbf{A}| = a_{13}M_{13} - a_{23}M_{23} + a_{33}M_{33}$$

(using the third-column minors) are equally qualified with (A.4) as expansions of $|\mathbf{A}|$; they all yield the same numerical value. The rule for attaching a plus or minus sign to each term in any expansion is as follows: If the sum of the row and column subscripts of the element appearing in a term is an even number, attach a plus sign; if this sum is an odd number, attach a minus sign. Symbolically, the sign attached to the term $a_{ij}M_{ij}$ in any expansion is equal to that of

$(-1)^{i+j}$. Consequently, we may incorporate $(-1)^{i+j}$ as part of each term, and precede every such augmented term with a plus sign. Thus

$$\begin{aligned}|\mathbf{A}| &= (-1)^2a_{11}M_{11} + (-1)^3a_{12}M_{12} + (-1)^4a_{13}M_{13}\\ &= (-1)^3a_{21}M_{21} + (-1)^4a_{22}M_{22} + (-1)^5a_{23}M_{23}\\ &= (-1)^4a_{13}M_{13} + (-1)^5a_{23}M_{23} + (-1)^6a_{33}M_{33}\end{aligned}$$

Alternatively, we may let the minors themselves "absorb" the sign of $(-1)^{i+j}$ and denote such "signed minors" by a symbol different from M_{ij}. This, in fact, is how the *cofactors* A_{ij} (introduced on p. 26) are defined. That is,

$$A_{ij} = (-1)^{i+j}M_{ij}$$

is called the cofactor (in $\mathbf{A}$ or $|\mathbf{A}|$) associated with the element a_{ij}. In words, the cofactor of a_{ij} is determined as follows: Cross out the ith row and jth column of $|\mathbf{A}|$, and write the remaining elements, in their original configuration, as a determinant of one smaller order than $|\mathbf{A}|$. If $i + j$ is even, this determinant itself is A_{ij}; if $i + j$ is odd, this determinant multiplied by -1 is A_{ij}. For example, for the fourth-order determinant in Exercise 8,

$$A_{31} = \begin{vmatrix} 2 & 0 & -1 \\ 3 & 1 & 2 \\ 0 & -1 & 1 \end{vmatrix} \qquad \text{while} \qquad A_{32} = (-1)\begin{vmatrix} 1 & 0 & -1 \\ 0 & 1 & 2 \\ 3 & -1 & 1 \end{vmatrix}$$

EXERCISE

1. Determine A_{33} and A_{34} for the example above, and evaluate the fourth-order determinant by expansion along the third row.

To summarize, the method of expanding a determinant in terms of the cofactors of its elements along some row or some column converts the task of evaluating a determinant of order n into one of evaluating n determinants of order $n - 1$, and forming a linear combination of these n determinants. The general rule may be stated as:

$$\begin{aligned}|\mathbf{A}| &= \sum_{j=1}^{n} a_{ij}A_{ij} \qquad \text{(expansion along } i\text{th row)}\\ &= \sum_{i=1}^{n} a_{ij}A_{ij} \qquad \text{(expansion along } j\text{th column)}\end{aligned} \tag{A.5}$$

It will be noted that the method described above does not, in general, offer much actual computational saving from the definitional rules except when $n = 4$ (in which case Sarrus' rule for third-order determinants may be used to facilitate computation of the four cofactors needed for an expansion). What it does do for any n, however, is to systematize the computations so that the task of explicitly writing out all $n!$ products required in Step 1 of the definitional rules can be averted.

It should also be evident that a genuine computational saving is effected if the determinant to be evaluated happens to contain one or more zero elements in any row or column. We would then expand it along that row or column which has the largest number of zeros. (In Exercise 8, each row and column contained one zero, so it is immaterial which row or column we choose to expand along. Using the third row, as suggested above, it was really unnecessary to determine A_{34} in order to evaluate $|\mathbf{A}|$; A_{34} does not contribute to the expansion because it is multiplied by $a_{34} = 0$.)

A.4 SELECTED PROPERTIES OF DETERMINANTS

As we pointed out above, expansion of a determinant in terms of cofactors represents a real computational saving if there are any zero elements in some row or column. Actually, it is always possible to transform a determinant into an equivalent one having at most one nonzero element in some row or column. (*All* elements can be made to vanish, in some row or column, if and only if the value of the determinant is zero.) We shall not describe such transformations in detail, since they are cumbersome to apply to determinants that are likely to be encountered in real problems in multivariate statistical analysis because the elements of such determinants usually are numbers involving many digits. However, some of the properties of determinants on which the transformations are based are important to know for other than computational purposes. We state these properties below without complete proofs, indicating only how they may be proved. In some of the examples given, the determinant of Exercise 3, page 387, is referred to as $|\mathbf{C}|$.

Property 1. If two rows (or two columns) of a determinant are interchanged, the absolute value of the determinant remains unchanged, but the sign is reversed.

BASIS FOR PROOF: Any such interchange will reverse the oddness or evenness of the NI associated with each product in the definition of a determinant.

EXAMPLE A.2. The determinant obtained by interchanging the second and third rows of $|\mathbf{C}|$ is as follows:

$$\begin{vmatrix} 1 & 2 & 3 \\ 2 & 3 & 1 \\ 3 & 1 & 2 \end{vmatrix}$$

Verify that the value of this determinant is -18. (See on page 387 that $|\mathbf{C}| = 18$.)

Property 2. If two rows (or two columns) of a determinant are identical, the value of the determinant is zero.

OUTLINE OF PROOF: If we interchange the two identical rows (or columns), no change is made in the determinant. But, according to Property 1, the sign should be reversed. The only number for which $x = -x$ is $x = 0$.

EXAMPLE A.3. Verify that

$$\begin{vmatrix} 1 & 2 & 3 \\ 3 & 1 & 2 \\ 1 & 2 & 3 \end{vmatrix} = 0$$

Property 3. If every element in any row (or any column) of a determinant is multiplied by a constant, the value of the determinant is multiplied by that same constant.

BASIS FOR PROOF: Expand the determinant along the row (or column) whose elements are multiplied by the constant.

EXAMPLE A.4. If we multiply each element of the second column of $|\mathbf{C}|$ by 2, we obtain the following determinant:

$$\begin{vmatrix} 1 & 4 & 3 \\ 3 & 2 & 2 \\ 2 & 6 & 1 \end{vmatrix}$$

Verify that the value of this determinant is $(2)(18) = 36$.

Property 4. If the elements of one row (or column) of a determinant are proportional to those of another row (or column, respectively), the value of the determinant is zero.

The proof follows directly from combined use of Properties 2 and 3.

EXAMPLE A.5. The third column of the following determinant is proportional to its first column:

$$\begin{vmatrix} 1 & 2 & 2 \\ 3 & 1 & 6 \\ 2 & 3 & 4 \end{vmatrix}$$

Verify that the value of this determinant is zero.

Property 5. If the elements of one row (or column) of a determinant are multiplied by a constant, and the results are added to or subtracted from the corresponding elements of another row (or column, respectively), the value of the determinant is unchanged.

OUTLINE OF PROOF: Let

$$|\mathbf{A}| = \begin{vmatrix} a_{11} & a_{12} & a_{13} \\ a_{21} & a_{22} & a_{23} \\ a_{31} & a_{32} & a_{33} \end{vmatrix}$$

Then

$$\begin{vmatrix} a_{11} & a_{12} & a_{13} \\ a_{21} & a_{22} & a_{23} \\ a_{31} + ka_{11} & a_{32} + ka_{12} & a_{33} + ka_{13} \end{vmatrix}$$

$$= (a_{31} + ka_{11})A_{31} + (a_{32} + ka_{12})A_{32} + (a_{33} + ka_{13})A_{33}$$
(by expansion along third row)

$$= a_{31}A_{31} + a_{32}A_{32} + a_{33}A_{33} + (ka_{11}A_{31} + ka_{12}A_{32} + ka_{13}A_{33})$$

$$= |\mathbf{A}| + \begin{vmatrix} a_{11} & a_{12} & a_{13} \\ a_{21} & a_{22} & a_{23} \\ ka_{11} & ka_{12} & ka_{13} \end{vmatrix}$$
(by using Rule A.5 in reverse)

$$= |\mathbf{A}| + 0$$
(by Property 4)

EXAMPLE A.6. Multiplying each element of the second column of $|\mathbf{C}|$ by 2 and subtracting from the corresponding element of the third column, we obtain

$$\begin{vmatrix} 1 & 2 & -1 \\ 3 & 1 & 0 \\ 2 & 3 & -5 \end{vmatrix}$$

Verify that this determinant has the value 18 (= $|\mathbf{C}|$). Further, multiplying the first row by 5 and subtracting from the third row of this new determinant, we get

$$\begin{vmatrix} 1 & 2 & -1 \\ 3 & 1 & 0 \\ -3 & -7 & 0 \end{vmatrix}$$

Verify that this determinant also has the value 18.

It is by repeated use of Property 5 that we can successively transform a given determinant to get an equivalent determinant with $n - 1$ zeros in some row or some column, as seen in Example A.6.

Property 6. If a linear combination is formed of the cofactors of elements along a given row (or column), using the corresponding elements of a *different* row (or column) as coefficients, then the value of this linear combination is 0.

OUTLINE OF PROOF: The expression in parentheses in the third last line of the proof of Property 5 exemplifies such a linear combination, if we take $k = 1$.

EXAMPLE A.7. For

$$|\mathbf{C}| = \begin{vmatrix} 1 & 2 & 3 \\ 3 & 1 & 2 \\ 2 & 3 & 1 \end{vmatrix}$$

we have

$$C_{12} = -\begin{vmatrix} 3 & 2 \\ 2 & 1 \end{vmatrix} = 1 \qquad C_{22} = \begin{vmatrix} 1 & 3 \\ 2 & 1 \end{vmatrix} = -5$$

and

$$C_{32} = -\begin{vmatrix} 1 & 3 \\ 3 & 2 \end{vmatrix} = 7$$

as the second column cofactors. Linearly combining these cofactors with the first-column elements as coefficients, we obtain

$$(1)(1) + (3)(-5) + (2)(7) = 0$$

Or, using the third-column elements as coefficients, we have

$$(3)(1) + (2)(-5) + (1)(7) = 0$$

Property 6 forms the basis for the method for finding the inverse of a matrix, described on pages 26–28. It can readily be verified that the (i, j)-element of the product $\mathbf{A}\ \mathbf{adj}(\mathbf{A})$ [where $\mathbf{adj}(\mathbf{A})$ is the adjoint of $\mathbf{A}$] is a linear combination of the cofactors of the jth row of $|\mathbf{A}|$, using the elements of the ith row as combining weights (coefficients). Hence the off-diagonal elements (where $i \neq j$) all vanish by virtue of Property 6, while each diagonal element ($j = i$) has the value $|\mathbf{A}|$ because it represents an expansion of $|\mathbf{A}|$ along its ith row.

Property 6 also plays a prominent role in the determination of eigenvectors of a matrix $\mathbf{A}$, described in Chapter 5. If the elements of a vector $\mathbf{v}$ are proportional to any column of the adjoint of $\mathbf{A} - \mu\mathbf{I}$ (where μ is an eigenvalue of $\mathbf{A}$, that is, a root of the characteristic equation, $|\mathbf{A} - \mu\mathbf{I}| = 0$), it follows that each element of the vector $(\mathbf{A} - \mu\mathbf{I})\mathbf{v}$ is equal to zero. All but one of these equalities hinges on Property 6, and the exceptional one is true because the determinant $|\mathbf{A} - \mu\mathbf{I}|$ has the value 0 by definition of μ.

Property 7. The determinant of the product of two square matrices is equal to the product of their respective determinants. That is,

$$|\mathbf{AB}| = |\mathbf{A}||\mathbf{B}| \tag{A.6}$$

provided that $\mathbf{A}$ and $\mathbf{B}$ are square matrices of the same order.

The proof is lengthy but not difficult. We first form the matrix product in accordance with Eq. 2.5. In evaluating the determinant of this product, we utilize the decomposition rule, as in the proof of Property 5, in combination with Property 4.

EXAMPLE A.8. Let

$$\mathbf{A} = \begin{bmatrix} 1 & 2 & 3 \\ 3 & 1 & 2 \\ 2 & 3 & 1 \end{bmatrix} \quad \text{and} \quad \mathbf{B} = \begin{bmatrix} 1 & 0 & 0 \\ 3 & 2 & -1 \\ 5 & 2 & 4 \end{bmatrix}$$

Then

$$\mathbf{AB} = \begin{bmatrix} 22 & 10 & 10 \\ 16 & 6 & 7 \\ 16 & 8 & 1 \end{bmatrix} \quad \text{and} \quad |\mathbf{AB}| = 180$$

while

$$|\mathbf{A}| = 18 \quad \text{and} \quad |\mathbf{B}| = 10$$

Therefore,

$$|\mathbf{A}||\mathbf{B}| = (18)(10) = 180 = |\mathbf{AB}|$$

APPENDIX B

PIVOTAL CONDENSATION METHOD OF MATRIX INVERSION

Most of the practical matrix inversion routines consist in systematic arrangements of sequences of elementary row operations. By an "elementary row operation" we mean that a given row of a matrix is multiplied by some constant, and the result is subtracted from each of the other rows. (The constant is ordinarily such that the subtraction will make one of each minuend row's elements equal to zero after the operation. Thus, a different multiplier is used for each minuend row.) Bearing this general principle in mind will make it easier to follow the description of the pivotal condensation method, given below with reference to a specific numerical example.

The $p \times p$ matrix **A** whose inverse is sought is written in the upper left-hand $p \times p$ sector of a worksheet such as that in Table B.1 ($p = 4$ in this example). The worksheet should be large enough to accommodate $2p + (p - 1)(p + 2)/2$ rows and $2p + 1$ columns of figures with one or two more significant digits than the elements of **A**. (More significant digits than needed or warranted in the final result must be carried in the intermediate calculations to allow for cumulative rounding errors.)

To the right of **A**, we write the identity matrix of order p (i.e., the same order as **A**). We now have $2p$ columns of p rows each. In each row of the last column, we enter the sum of the $2p$ numbers in that row (i.e., the elements of that row of **A** *plus* 1 *plus* $p - 1$ zeros). This column is introduced for the purpose of having a summational check on our calculations, step by step. Henceforth, the numbers in this column are treated in exactly the same way as the numbers to their left in each computational step. Then, as each new row is generated, we check to see if the sum of the numbers in columns (1)–($2p$) of

TABLE B.1 Worksheet for Computing the Inverse of a Matrix

	(1)	(2)	(3)	(4)	(5)	(6)	(7)	(8)	(9)
(0.1)	6.4570	2.8248	2.6712	2.5704	1	0	0	0	15.5234
(0.2)	3.4912	12.1606	4.1503	6.1378	0	1	0	0	26.9399
(0.3)	2.4961	4.8126	4.9151	3.2634	0	0	1	0	16.4872
(0.4)	3.7817	5.6247	3.8276	8.1125	0	0	0	1	22.3645
(1.0)	1	.437479	.413691	.398080	.154871	0	0	0	2.404120
(1.1)	0	10.633273	2.706022	4.748023	−.540686	1	0	0	18.546633
(1.2)	0	3.720609	3.882486	2.269753	−.386574	0	1	0	10.486274
(1.3)	0	3.988286	2.263145	6.607081	−.585676	0	0	1	13.272836
(2.0)	0	1	.254486	.446525	−.050849	.094044	0	0	1.744207
(2.1)	0	0	2.935643	.608408	−.197385	−.349901	1	0	3.996762
(2.2)	0	0	1.248182	4.826212	−.382876	−.375074	0	1	6.316440
(3.0)	0	0	1	.207249	−.067237	−.119191	.340641	0	1.361462
(3.1)	0	0	0	4.567528	−.298952	−.226302	−.425182	1	4.617092
(4. 0)	0	0	0	1	−.065452	−.049546	−.093088	.218937	1.010851
(4. −1)	0	0	1	0	−.053672	−.108923	.359933	−.045374	1.151964
(4. −2)	0	1	0	0	−.007964	.143887	−.050032	−.086214	.999678
(4. −3)	1	0	0	0	.206614	.001836	−.089957	−.030667	1.087827

$|\mathbf{A}| = (6.4570)(10.6333)(2.9356)(4.5675) = 920.624$

that row is equal (within rounding error) to the column-($2p + 1$) entry; if not, some arithmetical error has occurred, and we must correct it before proceeding.

1. Each of the $2p + 1$ numbers in row (0.1) is divided by the leftmost number (called the first "pivotal element"), 6.4570, which is underscored in the worksheet. The resulting quotients are written in the corresponding columns of row (1.0), which is called the first "pivotal row."
2. The numbers in row (1.0) are multiplied by the column-(1) number in row (0.2) (3.4912 here), and the products are subtracted from the corresponding numbers in row (0.2) to yield the elements of row (1.1); similarly, row (1.0) is multiplied by the first element of row (0.3) (2.4961 here), and the result subtracted from row (0.3) to yield row (1.2); and so on through row ($0.p$).

We now have a set of $p - 1$ rows, (1.1), (1.2), . . . , ($1.p - 1$), the first element of each of which is 0 by design. [We say that we have "swept out" the first column of **A**, using row (1.0) as pivot.] Steps 1 and 2 are now repeated on these rows, exclusive of column (1). That is, each element of row (1.1) is divided by the second pivotal element 10.633273 to yield the second pivotal row (2.0). This is then used to sweep out column (2) of rows (1.2), (1.3), . . . , ($1.p - 1$), and the results are recorded as rows (2.1), (2.2), . . . , ($2.p - 2$), respectively.

The cycle is repeated until we come to row ($p - 1.1$), in which the first $p - 1$ elements are all 0, as in row (3.1) here. The next step still follows rule (1); that is, we divide each element of row (3.1) by the fourth pivotal element (4.567528 here) to get row (4.0), or (p.0) in general. The numbers in columns (5)–(8) [or ($p + 1$)–($2p$) in general] in this row form the *last* row of $\mathbf{A}^{-1}$. The computation of rows (4. − 1), (4. − 2), and so on, is somewhat different from the preceding steps. This phase is known as the "backward solution."

To generate row (4. − 1), we eliminate the column-(4) [or column-p) in general] element of row (3.0) by subtracting .207249 times row (4.0) from row (3.0). Symbolically,

$$(4. - 1) = (3.0) - (3.0)_4 \times (4.0)$$

where $(3.0)_4$ denotes the fourth element of row (3.0). [In the p-dimensional case, this generalizes to: $(p. - 1) = (p - 1.0) - (p - 1.0)_p \times (p.0)$.]

Row (4. − 2) is obtained by subtracting .446525 times row (4.0) *and* .254486 times row (4. − 1) from row (2.0); that is,

$$(4. - 2) = (2.0) - (2.0)_4 \times (4.0) - (2.0)_3 \times (4. - 1)$$

which generalizes to

$$(p. - 2) = (p - 2.0) - (p - 2.0)_p \times (p.0) - (p - 2.0)_{p-1} \times (p. - 1)$$

Thus, the backward solution involves subtracting multiples of successively more and more rows, $(p.0)$, $(p.-1)$, $(p.-2)$, and so on, from earlier and earlier pivotal rows. The elements in columns $(p+1)$ through $(2p)$ of the successively generated rows $(p.0)$, $(p.-1)$, $(p.-2)$, . . . , $(p.-(p-1))$ form rows p, $p-1, p-2, \ldots, 1$, respectively, of $\mathbf{A}^{-1}$. That is, the successive rows of $\mathbf{A}^{-1}$ appear in *reverse order* in the lower right-hand $p \times p$ sector of the worksheet. Thus, for this example we have, on rounding the numbers to five decimal places (i.e., a maximum of five significant digits),

$$\mathbf{A}^{-1} = \begin{bmatrix} .20661 & .00184 & -.08996 & -.03067 \\ -.00796 & .14389 & -.05003 & -.08621 \\ -.05367 & -.10892 & .35993 & -.04537 \\ -.06545 & -.04955 & -.09309 & .21894 \end{bmatrix}$$

Even though we have carried a step-by-step summational check of our calculations, it is always wise to perform a final, positive check by multiplying the result by the original matrix $\mathbf{A}$ to see that we obtain the identity matrix. This is left for the reader to carry out.

B.1 OTHER USES OF PIVOTAL CONDENSATION

The method described above may be used for two other purposes besides computing the inverse of a matrix.

1. Solving a system of linear equations $\mathbf{Ax} = \mathbf{c}$, where $\mathbf{A}$ is a square, nonsingular matrix of coefficients, $\mathbf{x}$ is the column vector of unknowns, and $\mathbf{c}$ is a column vector of known constants. For this purpose, we write the vector $\mathbf{c}$ in column $(p+1)$, rows (0.1) through $(0.p)$, of our worksheet instead of writing the identity matrix in columns $(p+1)$ through $(2p)$. Thus the worksheet will now have only $p+2$ columns, the last being the summational check column as before. All computations are done exactly as described above, and the column-$(p+1)$ entries in the final rows $(p.0)$, $(p.-1)$, $(p.-2)$, . . . , $(p.-(p-1))$ give the values of $x_p, x_{p-1}, x_{p-2}, \ldots, x_1$, respectively, of the solution.

2. Evaluating a determinant $|\mathbf{A}|$. The value of $|\mathbf{A}|$ is readily obtained as a by-product of the process of inverting the matrix $\mathbf{A}$, or of solving the system of linear equations $\mathbf{Ax} = \mathbf{c}$. All we have to do is to find the product of all the pivotal elements as used in the process, namely, the first nonzero element of rows (0.1), (1.1), (2.1), . . . , $(p-1.1)$ in the worksheet. (The computation of $|\mathbf{A}|$ for our numerical example is shown at the foot of Table B.1, where the pivotal elements have been rounded to four decimal places.) Consequently, if our sole purpose is to evaluate a given determinant, we need only enter its elements in the upper left-hand $p \times p$ sector of our worksheet, and the row

sums in column $(p + 1)$ for carrying out the summational checks. The computations will now terminate with row $(p - 1.1)$, whose only nonzero entries will be the last (i.e., pth) pivotal element and the entry in the check column, which should equal the former within rounding error.

B.2 EXCEPTIONAL CASES

For the sake of completeness, we indicate a slight modification of the procedure described in the foregoing, which becomes necessary in some exceptional situations—even though these are hardly likely to arise when we seek the inverses of matrices that are of interest in multivariate analysis. The anomaly in question is the occurrence of a 0 in the diagonal position in some step; that is, when an element which should, according to the foregoing instructions, be used as a pivotal element happens to be 0—so that we cannot divide by it to generate the next pivotal row. This may happen (a) when the matrix is nonsingular, in which case we need the modification described below in order to proceed; or (b) when the matrix is singular, in which case we have to abandon our attempt to find its inverse, for none exists. A simple example of each case should suffice to illustrate how to proceed in the first case and how to recognize an instance of the second case.

1. Suppose that we set out to find the inverse of

$$\mathbf{B} = \begin{bmatrix} 2.0 & 2.0 & 2.0 \\ 1.5 & 1.5 & 1.0 \\ 2.0 & 3.0 & 1.0 \end{bmatrix}$$

in accordance with the procedure described above. The worksheet is shown in Table B.2.

TABLE B.2 Worksheet for Computing Inverse of B

	(1)	(2)	(3)	(4)	(5)	(6)	(7)
(0.1)	2.0	2.0	2.0	1	0	0	7.0
(0.2)	1.5	1.5	1.0	0	1	0	5.0
(0.3)	2.0	3.0	1.0	0	0	1	7.0
(1.0)	1	1.0	1.0	.5	0	0	3.5
(1.1)	0	0.0	−.50	−.75	1	0	−.25
(1.2)	0	1.0	1.0	−1.0	0	1	0.0
(2.0)	0	1	−1.0	−1.0	0	1.0	0.0
(2.1)	0	0	−.50	−.75	1	0	−.25
(3. 0)	0	0	1	1.5	−2.0	0.0	.5
(3.−1)	0	1	0	.5	−2.0	1.0	.5
(3.−2)	1	0	0	−1.5	4.0	−1.0	2.5

In row (1.1), obtained by subtracting 1.5 times row (1.0) from row (0.2), we find that not only the column (1) entry is zero (by design), but the column (2) entry is also zero (by peculiarity). [Throughout Table B.2, elements that are zero by design (i.e., will be zero in inverting any matrix) are written as "0" without a decimal point, while those that are zero because of a peculiarity of the particular matrix is written as "0.0"; a similar distinction is made between 1 and 1.0.] Ordinarily, we would divide each element of row (1.1) by its column (2) element in order to get the second pivotal row, (2.0). We cannot do so in this case because the column (2) element is zero.

In such a situation, provided that there is at least one nonzero element among the first p (= 3 here) in that row, we adopt the following modification: Use the next row after the anomalous one(s) for generating the next pivotal row. We shall call this step a "row interchange." Thus, in the present example, we divide row (1.2) by its column (2) element (which happens to be 1.0 here) in order to obtain row (2.0), the second pivotal row. (The divisor 1.0 is underscored as the second pivotal element.) Then, row (2.1) is obtained by multiplying row (2.0) by 0.0 (the column (2) element of row (1.1)) and subtracting the result from row (1.1); in other words, row (1.1) itself is recorded as row (2.1). We then proceed as usual. Of course, if we encounter another anomalous row in the process of inverting a larger matrix, we perform a row interchange again before proceeding.

The desired inverse emerges, as usual, with the rows in reverse order, in columns (p + 1) through ($2p$) of rows (p.0), (p.−1), . . . , (p.−(p − 1)). Thus for this example,

$$\mathbf{B}^{-1} = \begin{bmatrix} -1.5 & 4.0 & -1.0 \\ .5 & -2.0 & 1.0 \\ 1.5 & -2.0 & 0.0 \end{bmatrix}$$

The value of the determinant $|\mathbf{B}|$, however, may need to have its sign adjusted depending on how many row interchanges were performed. This value is given by the product of all p pivotal elements if an *even* number of row interchanges were made (just as in the case of no interchange); but if an *odd* number of row interchanges were performed, the sign of this product is reversed to get the value of the determinant. In other words, the determinant is equal to $(-1)^m$ times the product of pivotal elements, where m is the number of row interchanges that were made. In the present example, with $m = 1$, we have

$$|\mathbf{B}| = (-1)[(2.0)(1.0)(-.50)] = 1.0$$

2. The case of a singular matrix. As implied above in the proviso that an anomalous row must nevertheless have at least one nonzero element among its first p before we decide to make a row interchange, a singular matrix of order

TABLE B.3 Worksheet Showing That a Matrix Is Singular

	(1)	(2)	(3)	(4)	(5)	(6)	(7)
(0.1)	1.0	−4.0	−3.0	1	0	0	−5.0
(0.2)	−2.0	5.0	4.0	0	1	0	8.0
(0.3)	3.0	3.0	1.0	0	0	1	8.0
(1.0)	1	−4.0	−3.0	1.0	0	0	−5.0
(1.1)	0	−3.0	−2.0	2.0	1	0	−2.0
(1.2)	0	15.0	10.0	−3.0	0	1	23.0
(2.0)	0	1	2/3	−2/3	−1/3	0	2/3
(2.1)	0	0	0.0	7.0	5.0	1	13.0

$p \times p$ is identified by the fact that at some stage or other, there will emerge a row whose first p elements are *all* equal to zero. In the example shown in Table B.3, where we set out to find the inverse of the 3×3 matrix in the upper left-hand sector, we see that the first three elements of row (2.1) are all zeros. We therefore conclude that the matrix is singular and abandon our attempt to find its inverse. Furthermore, we can assert that the rank of the matrix is equal to the total number of pivotal rows obtained, or 2 in this case.

In this example we were able to determine the rank of the matrix as soon as a row with first p elements zero was found, because that row (2.1) was in fact the last possible one before we would normally start the backward solution. This is not true in general, however. We may encounter a row with p initial zeros at an earlier stage, in which case we must proceed further before we can tell what the rank of the matrix is. (Of course, we already know at that point that the rank is smaller than p, and hence that the matrix is singular.) The further steps to be taken are as follows: Delete the row with p initial zeros from further consideration, and proceed "as far as we can go," using the cycle of steps (1) and (2) described earlier. By "as far as we can go" is meant that we continue until either we get a single pivotal row with no other rows to be operated on by it, or when the sweeping out by some pivotal row results in *all* the other rows at that stage becoming "null rows" (through the pth column). In either case, no further pivotal rows can be constructed, so our computation is necessarily terminated. The rank of the matrix is then equal to the total number of pivotal rows, (1.0), (2.0), and so on, that were constructed in all. Table B.4 illustrates the procedure with reference to a 4×4 matrix. (We have condensed the worksheet by retaining only the first four columns. The reader should supply the check column in retracing the calculations for practice.)

Observe that row (2.1), obtained by operating on row (1.2) with the second pivotal row, (2.0), is a null row. However, we do not stop here and conclude that the rank of the matrix is 2 (which is the number of pivotal rows constructed

TABLE B.4 Determining the Rank of a Matrix

	(1)	(2)	(3)	(4)
(0.1)	2.0	−1.0	4.0	2.0
(0.2)	1.0	0.0	2.0	3.0
(0.3)	2.0	−1/2	4.0	4.0
(0.4)	3.0	2.0	−1.0	2.0
(1.0)	1	−1/2	2.0	1.0
(1.1)	0	1/2	0.0	2.0
(1.2)	0	1/2	0.0	2.0
(1.3)	0	7/2	−7.0	−1.0
(2.0)	0	1	0.0	4.0
(2.1)	0	0	0.0	0.0
(2.2)	0	0	−7.0	−15.0
(3.0)	0	0	1	15/7

thus far). We would do so only if row (2.2) were also a null row. But since it is not, we use row (2.2) for generating the next pivotal row, (3.0). Now we have no other row on which to operate with this new pivotal row. Our computation necessarily ends at this point, and we have determined that the rank of our matrix is 3.

APPENDIX C

SYMBOLIC DIFFERENTIATION BY VECTORS OR MATRICES

At several points in this book we made use of what is known as a *symbolic derivative* with respect to a vector (Dwyer and MacPhail, 1948). The process of symbolic differentiation by a vector involves no more (and no less) than finding the partial derivative of a given function with respect to each element of the vector, and then arranging the results in the form of a vector of the same type. Thus, if $f(\mathbf{x})$ is a function of the elements of a vector $\mathbf{x}$, the symbolic derivative $\frac{\partial}{\partial \mathbf{x}} f(\mathbf{x})$ is, by definition, the column vector whose elements are

$$\frac{\partial f}{\partial x_1}, \frac{\partial f}{\partial x_2}, \frac{\partial f}{\partial x_3}, \ldots, \frac{\partial f}{\partial x_p}$$

Similarly,

$$\frac{\partial}{\partial \mathbf{x}'} f(\mathbf{x})$$

is defined to be the row vector with these same elements.

The type of function of vector elements that occurs most frequently in multivariate analysis is the quadratic form $\mathbf{x}'\mathbf{A}\mathbf{x}$, or composite functions of this function,such as $(\mathbf{x}'\mathbf{A}\mathbf{x})/(\mathbf{x}'\mathbf{B}\mathbf{x})$, $\log(\mathbf{x}'\mathbf{A}\mathbf{x})$, $\exp[-(\mathbf{x}'\mathbf{A}\mathbf{x})/2]$, and so on. It is therefore useful to have a general formula for writing out the symbolic derivative of $\mathbf{x}'\mathbf{A}\mathbf{x}$ with respect to $\mathbf{x}$. We derive the formula for the simplest case, when $\mathbf{x}$ is a two-dimensional vector and $\mathbf{A}$ is a 2×2 matrix. It will readily be seen that the derivation generalizes to any dimensionality.

Since

$$\mathbf{x}'\mathbf{A}\mathbf{x} = [x_1, x_2]\begin{bmatrix} a_{11} & a_{12} \\ a_{21} & a_{22} \end{bmatrix}\begin{bmatrix} x_1 \\ x_2 \end{bmatrix} = a_{11}x_1^2 + (a_{12} + a_{21})x_1x_2 + a_{22}x_2^2$$

it follows that

$$\frac{\partial}{\partial x_1}(\mathbf{x}'\mathbf{A}\mathbf{x}) = 2a_{11}x_1 + (a_{12} + a_{21})x_2$$

and

$$\frac{\partial}{\partial x_2}(\mathbf{x}'\mathbf{A}\mathbf{x}) = (a_{12} + a_{21})x_1 + 2a_{22}x_2$$

On collecting these partial derivatives into a column vector, we obtain

$$\begin{aligned}\frac{\partial}{\partial \mathbf{x}}(\mathbf{x}'\mathbf{A}\mathbf{x}) &= \begin{bmatrix} 2a_{11}x_1 + (a_{12} + a_{21})x_2 \\ (a_{12} + a_{21})x_1 + 2a_{22}x_2 \end{bmatrix} \\ &= \begin{bmatrix} 2a_{11} & a_{12} + a_{21} \\ a_{12} + a_{21} & 2a_{22} \end{bmatrix}\begin{bmatrix} x_1 \\ x_2 \end{bmatrix} \\ &= \left(\begin{bmatrix} a_{11} & a_{12} \\ a_{21} & a_{22} \end{bmatrix} + \begin{bmatrix} a_{11} & a_{21} \\ a_{12} & a_{22} \end{bmatrix}\right)\begin{bmatrix} x_1 \\ x_2 \end{bmatrix} = (\mathbf{A} + \mathbf{A}')\mathbf{x}\end{aligned}$$

An exactly parallel derivation can be made when $\mathbf{x}$ is a vector of any dimensionality, say p, and $\mathbf{A}$ is a $p \times p$ matrix, by noting that

$$\mathbf{x}'\mathbf{A}\mathbf{x} = \sum_i \sum_j a_{ij}x_ix_j = \sum_{i=1}^{p} a_{ii}x_i^2 + \sum\sum_{i<j} (a_{ij} + a_{ji})x_ix_j$$

We have thus shown that

$$\frac{\partial}{\partial \mathbf{x}}(\mathbf{x}'\mathbf{A}\mathbf{x}) = (\mathbf{A} + \mathbf{A}')\mathbf{x} \tag{C.1}$$

Similarly, by arranging the partial derivatives with respect to x_i in the form of a row vector, we get

$$\frac{\partial}{\partial \mathbf{x}'}(\mathbf{x}'\mathbf{A}\mathbf{x}) = \mathbf{x}'(\mathbf{A} + \mathbf{A}') \tag{C.1'}$$

When the matrix $\mathbf{A}$ of the quadratic form is symmetrical, so that $\mathbf{A}' = \mathbf{A}$, Eqs. C.1 and C.1′ reduce, respectively, to the simpler forms

$$\frac{\partial}{\partial \mathbf{x}}(\mathbf{x}'\mathbf{A}\mathbf{x}) = 2\mathbf{A}\mathbf{x} \tag{C.1s}$$

and

$$\frac{\partial}{\partial \mathbf{x}'}(\mathbf{x}'\mathbf{A}\mathbf{x}) = 2\mathbf{x}'\mathbf{A} \tag{C.1's}$$

A special case of a quadratic form with a symmetric matrix is the squared norm $\mathbf{x}'\mathbf{x}$ of a vector $\mathbf{x}$. We have merely to let $\mathbf{A} = \mathbf{I}$ in Eqs. C.1s and C.1′s to obtain

$$\frac{\partial}{\partial \mathbf{x}}(\mathbf{x}'\mathbf{x}) = 2\mathbf{x} \tag{C.2}$$

and

$$\frac{\partial}{\partial \mathbf{x}'}(\mathbf{x}'\mathbf{x}) = 2\mathbf{x}' \tag{C.2′}$$

Another related type of expression whose symbolic (partial) derivative is often needed is the *bilinear form* $\mathbf{x}'\mathbf{A}\mathbf{y}$ or $\mathbf{y}'\mathbf{A}\mathbf{x}$. It may readily be verified that

$$\frac{\partial}{\partial \mathbf{x}}(\mathbf{x}'\mathbf{A}\mathbf{y}) = \frac{\partial}{\partial \mathbf{x}}(\mathbf{y}'\mathbf{A}\mathbf{x}) = \mathbf{A}\mathbf{y} \tag{C.3}$$

and

$$\frac{\partial}{\partial \mathbf{x}'}(\mathbf{x}'\mathbf{A}\mathbf{y}) = \frac{\partial}{\partial \mathbf{x}'}(\mathbf{y}'\mathbf{A}\mathbf{x}) = \mathbf{y}'\mathbf{A} \tag{C.3′}$$

Although the symbolic differentiation of a quadratic form by a vector (together with the usual "chain rule" for differentiation of a composite function) is the most that is used, in the way of symbolic derivatives, in this book, we present some additional results for the benefit of those readers who wish to pursue the mathematics of multivariate analysis further. The formulas given below are especially useful in connection with obtaining maximum likelihood estimates of the parameters of multivariate normal distributions.

C.1 SYMBOLIC DERIVATIVE OF A QUADRATIC FORM WITH RESPECT TO ITS MATRIX

Consider now the situation in which the vector $\mathbf{x}$ of a quadratic form is fixed, and we wish to find the partial derivatives of $\mathbf{x}'\mathbf{A}\mathbf{x}$ with respect to the elements a_{ij} of $\mathbf{A}$. The matrix whose (i, j) element is $\frac{\partial}{\partial a_{ij}}(\mathbf{x}'\mathbf{A}\mathbf{x})$, is called the *symbolic derivative* of $\mathbf{x}'\mathbf{A}\mathbf{x}$ with respect to $\mathbf{A}$. We illustrate the derivation with the two-dimensional case. For short, let us denote $\mathbf{x}'\mathbf{A}\mathbf{x}$ by Q. (We temporarily assume $\mathbf{A}$ to be nonsymmetric.)

$$Q = a_{11}x_1^2 + (a_{12} + a_{21})x_1x_2 + a_{22}x_2^2$$

Therefore,

$$\frac{\partial Q}{\partial a_{11}} = x_1^2 \qquad \frac{\partial Q}{\partial a_{12}} = x_1x_2 \qquad \frac{\partial Q}{\partial a_{21}} = x_1x_2 \qquad \frac{\partial Q}{\partial a_{22}} = x_2^2$$

Collecting these partial derivatives into a 2 × 2 matrix, we have

$$\frac{\partial Q}{\partial \mathbf{A}} = \begin{bmatrix} x_1^2 & x_1x_2 \\ x_2x_1 & x_2^2 \end{bmatrix} = \begin{bmatrix} x_1 \\ x_2 \end{bmatrix} [x_1, x_2] = \mathbf{xx}'$$

It may readily be verified that an exactly parallel derivation holds for the p-dimensional case. We thus have, as a general rule,

$$\frac{\partial}{\partial \mathbf{A}} (\mathbf{x}'\mathbf{A}\mathbf{x}) = \mathbf{xx}' \qquad \text{provided that } \mathbf{A} \text{ is nonsymmetric} \qquad \text{(C.4)}$$

To see that Eq. C.4 does not hold when $\mathbf{A}$ is symmetrical, consider again the two dimensional case, with $a_{12} = a_{21} = b$ (say). The quadratic form now becomes

$$Q = a_{11}x_1^2 + 2bx_1x_2 + a_{22}x_2^2$$

so that

$$\frac{\partial Q}{\partial a_{11}} = x_1^2 \qquad \frac{\partial Q}{\partial a_{12}} = \frac{\partial Q}{\partial a_{21}} = \frac{\partial Q}{\partial b} = 2x_1x_2 \qquad \frac{\partial Q}{\partial a_{22}} = x_2^2$$

Hence, collecting these resuts into a matrix, the symbolic derivative becomes

$$\frac{\partial Q}{\partial \mathbf{A}} = \begin{bmatrix} x_1^2 & 2x_1x_2 \\ 2x_2x_1 & x_2^2 \end{bmatrix} = \begin{bmatrix} x_1 \\ x_2 \end{bmatrix} [x_1, x_2] + \begin{bmatrix} 0 & x_1x_2 \\ x_2x_1 & 0 \end{bmatrix}$$

A somewhat neater form of this expression may be obtained by doubling $\mathbf{xx}'$ and subtracting x_1^2 and x_2^2 from the respective diagonal elements. Generalizing to p dimensions, we may write

$$\frac{\partial}{\partial \mathbf{A}} (\mathbf{x}'\mathbf{A}\mathbf{x}) = 2\mathbf{xx}' - \mathbf{D}(x_i^2) \qquad \text{when } \mathbf{A} \text{ is symmetric} \qquad \text{(C.4s)}$$

where $\mathbf{D}(x_i^2)$ denotes the diagonal matrix with $x_1^2, x_2^2, \ldots, x_p^2$ as its diagonal elements.

C.2 SYMBOLIC DERIVATIVE OF A DETERMINANT WITH RESPECT TO ITS MATRIX

Next, let us consider the partial derivative of the determinant $|\mathbf{A}|$ with respect to each of its elements. Expanding the determinant along its ith row, we have, in accordance with Eq. A.5 of Appendix A,

$$|\mathbf{A}| = a_{i1}A_{i1} + a_{i2} + \cdots + a_{ij}A_{ij} + \cdots + a_{ip}A_{ip} \tag{C.5}$$

where A_{ij} is the cofactor of a_{ij}. Now, by definition of a cofactor, *none* of the cofactors in the expansion above contains the element a_{ij} (since the ith row and some column of $|\mathbf{A}|$ have been deleted in getting each of these cofactors), provided that $\mathbf{A}$ is nonsymmetric. Hence, for any particular j, the only term in the expansion above which contains a_{ij} is $a_{ij}A_{ij}$. Therefore, the partial derivative of $|\mathbf{A}|$ with respect to a_{ij} is

$$\frac{\partial}{\partial a_{ij}} |\mathbf{A}| = A_{ij} \tag{C.6}$$

Hence, on collecting these partial derivatives for all (i, j) pairs into a $p \times p$ matrix, we have

$$\frac{\partial |\mathbf{A}|}{\partial \mathbf{A}} = (A_{ij}) = [\mathbf{adj}(\mathbf{A})]' \qquad \text{when } \mathbf{A} \text{ is nonsymmetric} \tag{C.7}$$

where $\mathbf{adj}(\mathbf{A})$ denotes the adjoint of $\mathbf{A}$. (Note that the result is the transpose of $\mathbf{adj}(\mathbf{A})$ rather than $\mathbf{adj}(\mathbf{A})$ itself, because by definition, $[\mathbf{adj}(\mathbf{A})]_{ij} = A_{ji}$.)

When $\mathbf{A}$ is a symmetric matrix, $a_{ij}A_{ij}$ is not the only term in expansion (C.5) that contains the element a_{ij} (or, equivalently, a_{ji}). In fact, *every* cofactor except A_{ii} contains a_{ij} (in the "guise" of a_{ji}). Hence the partial derivative of $|\mathbf{A}|$ with respect to a_{ij} becomes more complicated than when $\mathbf{A}$ is nonsymmetric. Let us examine the outcome for the determinant of a symmetric 3×3 matrix $\mathbf{A}$. To indicate the equality of elements in symmetrical positions, we use b, c, and d for the off-diagonal elements. Expanding $|\mathbf{A}|$ along its first row, we have

$$\begin{vmatrix} a_{11} & b & c \\ b & a_{22} & d \\ c & d & a_{33} \end{vmatrix} = a_{11} \underbrace{\begin{vmatrix} a_{22} & d \\ d & a_{33} \end{vmatrix}}_{A_{11}} - b \underbrace{\begin{vmatrix} b & d \\ c & a_{33} \end{vmatrix}}_{-A_{12}} + c \underbrace{\begin{vmatrix} b & a_{22} \\ c & d \end{vmatrix}}_{A_{13}}$$

The partial derivative of $|\mathbf{A}|$ with respect to $a_{13} = a_{31} = c$ is, therefore,

$$\begin{aligned} \frac{\partial}{\partial c} |\mathbf{A}| &= (-b) \frac{\partial}{\partial c} (ba_{33} - dc) + \frac{\partial}{\partial c} [c(bd - a_{22}c)] \\ &= (-b)(-d) + (bd - a_{22}c) + c(-a_{22}) = 2(bd - a_{22}c) = 2A_{13} \end{aligned}$$

This result generalizes to symmetrical determinants of any order, although not quite as obviously as in the case of (C.4s); that is,

$$\frac{\partial}{\partial a_{ij}} |\mathbf{A}| = 2A_{ij} \qquad \text{for } i \neq j \qquad \text{when } \mathbf{A} \text{ is symmetric} \tag{C.6s}$$

It should be clear from the expansion displayed above that, for a diagonal element a_{ii}, its occurrence is only the explicit one in $a_{ii}A_{ii}$ (in the expansion of

$|\mathbf{A}|$ along row i), even when $\mathbf{A}$ is symmetric. Hence the partial derivative of $|\mathbf{A}|$ with respect to a diagonal element a_{ii} retains the expression in (C.6) with j replaced by i; that is, the factor 2 does not appear as in (C.6s). Collecting the partial derivatives into a matrix, we have the awkward result,

$$\frac{\partial}{\partial \mathbf{A}} |\mathbf{A}| = \begin{bmatrix} A_{11} & 2A_{12} & \cdots & 2A_{1p} \\ 2A_{21} & A_{22} & \cdots & 2A_{2p} \\ \vdots & & & \vdots \\ 2A_{p1} & 2A_{p2} & \cdots & A_{pp} \end{bmatrix} = 2\,\mathbf{adj}(\mathbf{A}) - \mathbf{D}(A_{ii}) \qquad \text{(C.7s)}$$

where $\mathbf{D}(A_{ii})$ denotes the diagonal matrix with $A_{11}, A_{22}, \ldots, A_{pp}$ as its diagonal elements, when $\mathbf{A}$ is a symmetric matrix. Note, however, that we need not write $[\mathbf{adj}(\mathbf{A})]'$ here, since $\mathbf{A}' = \mathbf{A}$ implies that $[\mathbf{adj}(\mathbf{A})]' = \mathbf{adj}(\mathbf{A})$.

C.3 SYMBOLIC "PSEUDODERIVATIVES"

It was seen in Eqs. C.4s and C.7s that symbolic differentiation with respect to a matrix yields a more complicated result when the matrix is symmetric than when it is not (C.4 and C.7). This is unfortunate in view of the fact that most of the matrices that play the role of (symbolic) variable of differentiation in multivariate analysis are symmetrical. However, two observations suggest a way to avoid this complication.

First, the complexity of the symbolic derivative with respect to a symmetric matrix consists only in the fact that the diagonal elements retain the same expression as in the case of a nonsymmetric matrix, while the off-diagonal elements are *twice* the corresponding expression for the nonsymmetric case. (This was substantiated above only when the function to be differentiated was a quadratic form or the determinant of the matrix, but the same holds true for any scalar function of a matrix.)

Second, the situations in which we seek symbolic derivatives with respect to matrices are mostly (if not exclusively) when we will subsequently set the result equal to a null matrix—as when we seek the necessary conditions for maximizing or minimizing the function. To set the symbolic derivative equal to a null matrix is, of course, equivalent to setting the partial derivative with respect to each element of the matrix, individually, equal to zero. Consequently, the presence of a factor of 2 in the off-diagonal elements of the symbolic derivative with respect to a symmetric matrix (as compared to the corresponding expression in the nonsymmetric case) is really immaterial—as long as we are setting the whole thing equal to a null matrix anyway.

We are thus led to the conclusion that—provided that our purpose is to set the result equal to a null matrix—we may, without any distortion, construct a matrix whose diagonal elements are the actual partial derivatives of the function with respect to the diagonal elements of the matrix of differentiation **A**, but whose off-diagonal elements are *one-half* of the partial derivatives with respect to the off-diagonal elements a_{ij}. Let us call the matrix thus constructed, the symbolic "pseudoderivative" of $f(\mathbf{A})$ with respect to **A**, and denote it by $\frac{\partial^*}{\partial \mathbf{A}} f(\mathbf{A})$ to distinguish it from the true symbolic derivative $\frac{\partial}{\partial \mathbf{A}} f(\mathbf{A})$. It then follows that the formulas for the symbolic pseudoderivatives of $\mathbf{x}'\mathbf{A}\mathbf{x}$ and $|\mathbf{A}|$ with respect to **A** are identical with Eqs. C.4 and C.7, respectively (except that, in the latter, we need not take the transpose of **adj**(**A**)). Thus

$$\frac{\partial^*}{\partial \mathbf{A}} (\mathbf{x}'\mathbf{A}\mathbf{x}) = xx' \qquad \text{when } \mathbf{A} \text{ is symmetric} \tag{C.4*s}$$

and

$$\frac{\partial^*}{\partial \mathbf{A}} |\mathbf{A}| = (A_{ij}) = (A_{ji}) = \mathbf{adj}(\mathbf{A}) \qquad \text{when } \mathbf{A} \text{ is symmetric} \tag{C.7*s}$$

EXAMPLE C.1. The problem of finding the maximum-likelihood estimates of the centroid $\boldsymbol{\mu}$ and variance–covariance matrix $\boldsymbol{\Sigma}$ of a p-variate normal population on the basis of a random sample of n observation vectors $\mathbf{x}_i$ ($i = 1, 2, \ldots, n$) amounts to the following: We are to determine those "values" of $\boldsymbol{\mu}$ and $\boldsymbol{\Sigma}$ (treated as indeterminates) that simultaneously maximize the likelihood of the observed sample—that is, the n-fold joint density function evaluated for the sample at hand. In practice, we maximize the natural logarithm of the likelihood function (which is a monotonically increasing function of the likelihood itself), which is given by

$$\begin{aligned} L(\boldsymbol{\mu}, \boldsymbol{\Sigma}) = (-np/2) \ln 2\pi + (n/2) \ln |\mathbf{H}| \\ - (1/2) \sum_{i=1}^{n} (\mathbf{x}_i - \boldsymbol{\mu})'\mathbf{H}(\mathbf{x}_i - \boldsymbol{\mu}) \end{aligned} \tag{C.8}$$

where $\mathbf{H} = \boldsymbol{\Sigma}^{-1}$, the inverse of the population variance-covariance matrix. The necessary conditions for maximizing $L(\boldsymbol{\mu}, \boldsymbol{\Sigma})$ are obtained by setting its symbolic derivatives with respect to $\boldsymbol{\mu}$ and $\boldsymbol{\Sigma}$ (or, equivalently with respect to $\boldsymbol{\mu}$ and **H**), respectively, equal to a null vector and a null matrix.

Since $\boldsymbol{\mu}$ is involved only in the last term of L, which is a sum of n quadratic forms, we obtain, by applying Eq. C.1′s to each summand,

$$\frac{\partial}{\partial \boldsymbol{\mu}'} L(\boldsymbol{\mu}, \boldsymbol{\Sigma}) = \tfrac{1}{2} \sum_{i=1}^{n} 2(\mathbf{x}_i - \boldsymbol{\mu})'\mathbf{H} = \left[\sum_{i=1}^{n} (\mathbf{x}_i - \boldsymbol{\mu})' \right] \mathbf{H}$$

On setting this expression equal to the p-dimensional null vector $\mathbf{0}'$, we get (since $\mathbf{H}$ is nonsingular)

$$\sum_{i=1}^{n} (\mathbf{x}_i - \boldsymbol{\mu})' = \mathbf{0}'$$

from which we obtain

$$\boldsymbol{\mu} = \frac{1}{n} \sum_{i=1}^{n} \mathbf{x}_i = \bar{\mathbf{x}} \qquad \text{(sample centroid)} \tag{C.9}$$

as an estimate of $\boldsymbol{\mu}$.

Next, we find the symbolic pseudoderivative of expression (C.8) with respect to $\mathbf{H}$:

$$\begin{aligned} \frac{\partial^*}{\partial \mathbf{H}} L(\boldsymbol{\mu}, \boldsymbol{\Sigma}) &= (n/2) \frac{1}{|\mathbf{H}|} \frac{\partial^*}{\partial \mathbf{H}} |\mathbf{H}| \\ &\quad - (1/2) \sum_{i=1}^{n} \left[\frac{\partial^*}{\partial \mathbf{H}} (\mathbf{x}_i - \boldsymbol{\mu})' \mathbf{H} (\mathbf{x}_i - \boldsymbol{\mu}) \right] \\ &= (n/2) \frac{\mathbf{adj(H)}}{|\mathbf{H}|} - (1/2) \sum_{i=1}^{n} (\mathbf{x}_i - \boldsymbol{\mu})(\mathbf{x}_i - \boldsymbol{\mu})' \end{aligned}$$

where we have used Eqs. C.7*s and C.4*s, respectively, for evaluating the symbolic pseudoderivatives of the determinant and quadratic forms with respect to the matrix $\mathbf{H}$. Setting the last expression equal to $\mathbf{0}(p \times p)$ after replacing $\mathbf{adj(H)}/|\mathbf{H}|$ by $\mathbf{H}^{-1}$ to which it is equal by definition, we get

$$n\mathbf{H}^{-1} - \sum_{i=1}^{n} (\mathbf{x}_i - \boldsymbol{\mu})(\mathbf{x}_i - \boldsymbol{\mu})' = \mathbf{0} \tag{C.10}$$

But since we have already found an estimate (C.9) for $\boldsymbol{\mu}$ that is necessary in order to maximize $L(\boldsymbol{\mu}, \boldsymbol{\Sigma})$, we may substitute this estimate, namely $\bar{\mathbf{x}}$, in place of $\boldsymbol{\mu}$ in Eq. C.10. Noting also that the $\mathbf{H}^{-1}$, which satisfies (C.10), is really the desired estimate $\hat{\boldsymbol{\Sigma}}$ of $\boldsymbol{\Sigma}$ (because $\mathbf{H}^{-1} = (\boldsymbol{\Sigma}^{-1})^{-1} = \boldsymbol{\Sigma}$), we finally get

$$\hat{\boldsymbol{\Sigma}} = (1/n) \sum_{i=1}^{n} (\mathbf{x}_i - \bar{\mathbf{x}})(\mathbf{x}_i - \bar{\mathbf{x}})' = (1/n)\mathbf{S} \tag{C.11}$$

where $\mathbf{S}$ is the sample SSCP matrix.

Thus the estimates C.9 and C.11 for $\boldsymbol{\mu}$ and $\boldsymbol{\Sigma}$ are necessary for maximizing $L(\boldsymbol{\mu}, \boldsymbol{\Sigma})$. It can be shown that they are also sufficient. Therefore, the sample centroid and $\mathbf{S}/n$ are the maximum-likelihood estimates for $\boldsymbol{\mu}$ and $\boldsymbol{\Sigma}$, respectively.

APPENDIX D

PRINCIPAL COMPONENTS (OR FACTORS) BY HOTELLING'S ITERATIVE PROCEDURE FOR SOLVING EIGENVALUE PROBLEMS

D.1 BASIC METHOD

In his 1933 paper on principal component analysis, Hotelling developed an iterative procedure for calculating the eigenvalues and vectors of any symmetric matrix. By an iterative method is meant one in which we start with some initial trial value (here a trial vector), perform a certain operation on it (here, premultiplying it by the given matrix) to obtain a second trial value (vector), repeat the same operation on it to get a third trial value, and so on until two successive trial values agree with each other to a specified degree of accuracy. When this occurs, the iteration is said to have converged on the value (vector) sought.

Let **R** be the given $p \times p$ symmetric matrix whose eigenvalues and eigenvectors are sought. The computational steps are as follows:

1. Take any p-dimensional column vector $\mathbf{v}_{(0)}$ as the initial trial vector.
2. Form the product $\mathbf{R}\mathbf{v}_{(0)} = \mathbf{u}_{(1)}$, say.
3. Divide $\mathbf{u}_{(1)}$ by its largest (in absolute value) element, and denote the result by $\mathbf{v}_{(1)}$.
4. Use $\mathbf{v}_{(1)}$ as the second trial vector and repeat steps 2 and 3 to obtain $\mathbf{v}_{(2)}$, which then becomes the third trial vector.

5. The foregoing cycle is repeated (iterated) until two successive trial vectors $\mathbf{v}_{(k)}$ and $\mathbf{v}_{(k+1)}$, for instance, are identical within a specified number of decimal places. (That this convergence will eventually occur was proven by Hotelling.)
6. The number by which $\mathbf{Rv}_{(k)} = \mathbf{u}_{(k+1)}$ was divided in order to get $\mathbf{v}_{(k+1)}$ (that is, the largest element of $\mathbf{u}_{(k+1)}$) is the largest eigenvalue λ_1 of $\mathbf{R}$.
7. Normalize the terminal trial vector $\mathbf{v}_{(k)}$ to unity by dividing it by the square root of the sum of the squares of its elements. The result is $\mathbf{v}_1$, the unit eigenvector of $\mathbf{R}$ corresponding to λ_1.
8. Form the matrix product $\mathbf{v}_1\mathbf{v}_1'$, multiply it by λ_1, and subtract the result from $\mathbf{R}$. The resulting difference matrix is called the first residual matrix, denoted $\mathbf{R}_1$. That is,

$$\mathbf{R}_1 = \mathbf{R} - \lambda_1\mathbf{v}_1\mathbf{v}_1' \tag{D.1}$$

9. Carry out steps 1–7 using $\mathbf{R}_1$ in place of $\mathbf{R}$. The output in steps 6 and 7 are the largest eigenvalue and associated eigenvector of $\mathbf{R}_1$. These are, respectively, the *second* largest eigenvalue λ_2 and its associated eigenvector $\mathbf{v}_2$ of the original matrix $\mathbf{R}$. (This is a consequence of the relation

$$\mathbf{A} = \lambda_1\mathbf{v}_1\mathbf{v}_1' + \lambda_2\mathbf{v}_2\mathbf{v}_2' + \cdots + \lambda_r\mathbf{v}_r\mathbf{v}_r'$$

proved in Exercise 3, p. 178.)
10. Compute the second residual matrix,

$$\mathbf{R}_2 = \mathbf{R}_1 - \lambda_2\mathbf{v}_2\mathbf{v}_2'$$

and again carry out steps 1 through 7 to obtain λ_3 and $\mathbf{v}_3$.
The process is continued until the desired number of eigenvalue-eigenvector pairs of $\mathbf{R}$ have been found. (See the discussion at the end of Section 5.6*a*.)

D.2 ACCELERATING THE CONVERGENCE

Although the basic method outlined above will always converge, the number of iterations required may sometimes be forbiddingly large, especially when two successive eigenvalues of $\mathbf{R}$ happen to be close to each other in numerical value. Hotelling showed in a subsequent paper (1936) that the rate of convergence may be greatly accelerated by using a suitable power of $\mathbf{R}$ instead of $\mathbf{R}$ itself. That this will lead to the desired results, and will do so more rapidly, may be seen as follows:

Theorem 11 on page 159 shows that if λ_i and $\mathbf{v}_i$ are the ith eigenvalue and vector of any matrix $\mathbf{A}$, then $\mathbf{v}_i$ is also the ith eigenvector of any power $\mathbf{A}^m$ of $\mathbf{A}$, with associated eigenvalue λ_i^m. Hence if we use $\mathbf{R}^4$ (for example) instead of

R in the basic method, the eigenvectors we get are immediately the eigenvectors of **R** too, and the fourth roots of the eigenvalues will be the corresponding eigenvalues of **R**. Furthermore, since forming the product $\mathbf{R}^4\mathbf{v}_{(j)}$ is the same as operating on any trial vector $\mathbf{v}_{(j)}$ four times by **R** itself, it follows that one iteration of steps 2 and 3 using $\mathbf{R}^4$ is equivalent to *four* iterations using **R**; hence the acceleration in convergence.

It might seem that the economy thus gained would be offset by the labor of having to raise the successive residual matrices $\mathbf{R}_1$, $\mathbf{R}_2$, and so on to a high power for each successive eigenvalue–eigenvector pair. This is not the case. For once we have raised the original **R** to a given power, the same power of the subsequent residual matrices can be obtained without any "powering," as shown below.

Squaring both sides of Eq. D.1, we get

$$\begin{aligned}\mathbf{R}_1^2 &= (\mathbf{R} - \lambda_1\mathbf{v}_1\mathbf{v}_1')^2 \\ &= \mathbf{R}^2 - 2\lambda_1\mathbf{R}(\mathbf{v}_1\mathbf{v}_1') + \lambda_1^2(\mathbf{v}_1\mathbf{v}_1')(\mathbf{v}_1\mathbf{v}_1') \\ &= \mathbf{R}^2 - 2\lambda_1(\mathbf{R}\mathbf{v}_1)\mathbf{v}_1' + \lambda_1^2\mathbf{v}_1(\mathbf{v}_1'\mathbf{v}_1)\mathbf{v}_1'\end{aligned}$$

where we have used the distributive and associative laws, respectively, in going from the first to the second, and from the second to the third expressions on the right. But, in the last expression, we may replace $\mathbf{R}\mathbf{v}_1$ by $\lambda_1\mathbf{v}_1$ (since $\mathbf{v}_1$ is the eigenvector of **R** associated with λ_1), and $\mathbf{v}_1'\mathbf{v}_1$ by 1 (since $\mathbf{v}_1$ is a unit vector). We then have

$$\begin{aligned}\mathbf{R}_1^2 &= \mathbf{R}^2 - 2\lambda_1^2(\mathbf{v}_1\mathbf{v}_1') + \lambda_1^2(\mathbf{v}_1\mathbf{v}_1') \\ &= \mathbf{R}^2 - \lambda_1^2(\mathbf{v}_1\mathbf{v}_1')\end{aligned}$$

Similarly, it may be shown that for any positive integer m,

$$\mathbf{R}_1^m = \mathbf{R}^m - \lambda_1^m(\mathbf{v}_1\mathbf{v}_1') \tag{D.2}$$

and, more generally, for the jth residual matrix,

$$\mathbf{R}_j^m = \mathbf{R}_{j-1}^m - \lambda_j^m(\mathbf{v}_j\mathbf{v}_j') \tag{D.3}$$

EXAMPLE D.1. Given the following reduced correlation matrix for three variables (estimated communalities in the diagonal), let us determine as many eigenvalue–eigenvector pairs as are necessary "essentially" to account for the observed matrix, and thereby construct the principal-factor matrix.

$$\mathbf{R} = \begin{bmatrix} .6 & -.4 & .1 \\ -.4 & .9 & .3 \\ .1 & .3 & .3 \end{bmatrix}$$

In order to speed up the convergence of the iterations, we first raise $\mathbf{R}$ to the fourth power by squaring twice:

$$\mathbf{R}^4 = (\mathbf{R}^2)^2 = \begin{bmatrix} .6067 & -.9159 & -.2040 \\ -.9159 & 1.5509 & .4171 \\ -.2040 & .4171 & .1394 \end{bmatrix}$$

As our initial trial vector $\mathbf{v}_{(0)}$ (step 1), we take the three-dimensional column vector whose elements are equal to the row totals of $\mathbf{R}^4$:

$$\mathbf{v}_{(0)} = \begin{bmatrix} -.5132 \\ 1.0521 \\ .3525 \end{bmatrix}$$

(For those who wonder what basis there could be for this choice of $\mathbf{v}_{(0)}$, it may be pointed out that this is equivalent to having taken $[1, 1, 1]'$ as a trial vector one iteration before $\mathbf{v}_{(0)}$. Lacking all grounds for a logical choice, this vector with uniform elements should—on the average—be the best bet.)

Next, we form the product $\mathbf{R}^4\mathbf{v}_{(0)}$ in accordance with step 2:

$$\mathbf{R}^4\mathbf{v}_{(0)} = [-1.3469, 2.2488, .5927]' = \mathbf{u}_{(1)}$$

where, in order to save space, we have indicated the column vector $\mathbf{u}_{(1)}$ by exhibiting a row vector flagged with the transposition sign. This notation will be used hereunder.

Since the largest absolute-valued element of $\mathbf{u}_{(1)}$ is 2.2488, we divide each element by this number to obtain $\mathbf{v}_{(1)}$, as prescribed in step 3:

$$[-1.3469, 2.2488, .5927]' \div 2.2488 = [-.5989, 1, .2635]' = \mathbf{v}_{(1)}$$

This becomes the second trial vector with which to carry out steps 2 and 3 as stated in 4. For further economy of space, we display the subsequent iterations without explicitly labelling each $\mathbf{u}_{(i)}$, and indicate step 3 (division by the largest element of $\mathbf{u}_{(i)}$) by an arrow ($\rightarrow$) leading to $\mathbf{v}_{(i)}$:

$$\mathbf{R}^4\mathbf{v}_{(1)} = [-1.3330, 2.2094, .5760]' \rightarrow [-.6033, 1, .2607]' = \mathbf{v}_{(2)}$$

$$\mathbf{R}^4\mathbf{v}_{(2)} = [-1.3351, 2.2122, .5765]' \rightarrow [-.6035, 1, .2606]' = \mathbf{v}_{(3)}$$

$$\mathbf{R}^4\mathbf{v}_{(3)} = [-1.3352, 2.2123, .5765]' \rightarrow [-.6035, 1, .2606]' = \mathbf{v}_{(4)}$$

We terminate our iterations at this point, following instruction 5, since $\mathbf{v}_{(3)} = \mathbf{v}_{(4)}$ to four decimal places, which is the limit of accuracy to be expected, because we have only four significant digits in the elements of $\mathbf{R}^4$. Hence, according to statement 6, the largest element 2.2123 of $\mathbf{u}_{(4)}$ is the largest eigenvalue of $\mathbf{R}^4$, and its fourth root $(2.2123)^{1/4} = 1.2196$ is the largest eigenvalue λ_1 of $\mathbf{R}$.

Step 7 consists in normalizing $\mathbf{v}_{(3)}$ to unity by dividing each of its elements by $[(-.6035)^2 + (1)^2 + (.2606)^2]^{1/2} = 1.1967$. The result is

$$\mathbf{v}_1 = [-.5043, .8356, .2178]'$$

the unit-length eigenvector of $\mathbf{R}$ associated with its largest eigenvalue $\lambda_1 = 1.2196$.

We are now ready to form the fourth power of the first residual matrix $\mathbf{R}_1$ in accordance with step 8 as modified by replacing Eq. D.1 by D.2, and thence to determine the second eigenvalue–eigenvector pair as instructed in 9. However, let us first consider the question of whether it is necessary to do so. That is, we want to test whether the first factor alone will adequately account for the observed correlation matrix. Several criteria for this purpose were mentioned in Section 5.6. We will here use the simplest and most intuitive of these: the proportion of common variance accounted for by the first factor. This is assessed by calculating the ratio

$$\lambda_i/\mathrm{tr}\,(\mathbf{R}) = 1.2196/(.6 + .9 + .3) = .68$$

Thus, only about 68% of the common variance is explained by the first principal factor, so we should certainly extract at least one more factor.

Following Eq. D.2 with $m = 4$, we compute

$$\mathbf{R}_1^4 = \mathbf{R}^4 - \lambda_1^4(\mathbf{v}_1\mathbf{v}_1')$$

$$= \begin{bmatrix} .6067 & -.9159 & -.2040 \\ -.9159 & 1.5509 & .4171 \\ -.2040 & .4171 & .1394 \end{bmatrix}$$

$$-(2.2123)\begin{bmatrix} -.5043 \\ .8356 \\ .2178 \end{bmatrix}[-.5043, .8356, .2178]$$

$$= \begin{bmatrix} .4411 & .1636 & .3891 \\ .1636 & .0627 & .1446 \\ .3891 & .1446 & .3454 \end{bmatrix} \times 10^{-1}$$

Using the vector with elements equal to the row totals of $10\mathbf{R}_1^4$ as the initial trial vector, that is,

$$\mathbf{v}_{(0)} = [.9938, .3709, .8791]'$$

we carry out steps 1 to 7 as instructed in 9, thus:

$$10\mathbf{R}_1^4\mathbf{v}_{(0)} = [.8411, .3130, .7440]' \rightarrow [1, .3721, .8845]' = \mathbf{v}_{(1)}$$

$$10\mathbf{R}_1^4\mathbf{v}_{(1)} = [.8461, .3148, .7484]' \rightarrow [1, .3718, .8845]' = \mathbf{v}_{(2)}$$

$$10\mathbf{R}_1^4\mathbf{v}_{(2)} = [.8461, .3148, .7484]' \rightarrow [1, .3718, .8845]' = \mathbf{v}^{(3)}$$

The iteration is terminated at this point since $\mathbf{v}_{(2)} = \mathbf{v}_{(3)}$ to four places. Note, however, that the fourth power of the largest eigenvalue of $\mathbf{R}_1$ is not .8461 itself, but one-tenth of this number, or .08461, since we have used $10\mathbf{R}_1^4$ in our iterations. Hence the second largest eigenvalue λ_2 of the original correlation matrix $\mathbf{R}$ is $(.08461)^{1/4} = .5393$, and its associated eigenvector is

$$\begin{aligned}\mathbf{v}_2 = \mathbf{v}_{(2)}/\sqrt{\mathbf{v}_{(2)}'\mathbf{v}_{(2)}} &= [1, .3718, .8845]'/1.3858 \\ &= [.7216, .2683, .6382]'\end{aligned}$$

We now examine what proportion of the common variance is accounted for by the first two principal factors. The relevant index is

$$(1.2196 + .5393)/1.8 = .978$$

which shows that about 98% of the common variance has already been accounted for. Hence we may safely conclude that computation of the third factor is unnecessary.

To get a better idea of what it means to say that the first two factors account for practically all of the common variance, let us construct the factor matrix. This is obtained by rescaling each eigenvector to have a length equal to the square root of the corresponding eigenvalue instead of unity. Thus the factor matrix $\mathbf{F}$ is given by

$$\mathbf{F} = \mathbf{V}\mathbf{\Lambda}^{1/2}$$

where $\mathbf{V}$ is the matrix whose columns are $\mathbf{v}_1$ and $\mathbf{v}_2$ (i.e., the principal-components transformation matrix), and $\mathbf{\Lambda}$ is the diagonal matrix with diagonal elements λ_1 and λ_2. For the present example,

$$\mathbf{F} = \begin{bmatrix} -.5043 & .7216 \\ .8356 & .2683 \\ .2178 & .6382 \end{bmatrix} \begin{bmatrix} \sqrt{1.2196} & 0 \\ 0 & \sqrt{.5393} \end{bmatrix} = \begin{bmatrix} -.5569 & .5299 \\ .9228 & .1970 \\ .2405 & .4687 \end{bmatrix}$$

This factor matrix postmultiplied by its transpose should essentially reproduce the correlation matrix $\mathbf{R}$. Calculating the product, we find

$$\mathbf{F}\mathbf{F}' = \begin{bmatrix} .5909 & -.4095 & .1144 \\ -.4095 & .8904 & .3143 \\ .1144 & .3143 & .2775 \end{bmatrix}$$

whose difference from **R** is

$$\mathbf{R} - \mathbf{F}\mathbf{F}' = \begin{bmatrix} .0091 & .0095 & -.0144 \\ .0095 & .0096 & -.0143 \\ -.0144 & -.0143 & .0225 \end{bmatrix}$$

(This is, of course, the same as the second residual matrix $\mathbf{R}_2$ computed as $\mathbf{R}_2 = \mathbf{R}_1 - \lambda_2\mathbf{v}_2\mathbf{v}_2'$.) The discrepancies, or residuals, are all seen to be of the order of .01, except in the (3, 3)-element, where the difference is about .02. Surely no one would deem it meaningful to extract a third factor from this residual matrix!

APPENDIX E

STATISTICAL AND NUMERICAL TABLES

TABLE E.1 Normal Curve Areas[a]

z	.00	.01	.02	.03	.04	.05	.06	.07	.08	.09
0.0	.0000	.0040	.0080	.0120	.0160	.0199	.0239	.0279	.0319	.0359
0.1	.0398	.0438	.0478	.0517	.0557	.0596	.0636	.0675	.0714	.0753
0.2	.0793	.0832	.0871	.0910	.0948	.0987	.1026	.1064	.1103	.1141
0.3	.1179	.1217	.1255	.1293	.1331	.1368	.1406	.1443	.1480	.1517
0.4	.1554	.1591	.1628	.1664	.1700	.1736	.1772	.1808	.1844	.1879
0.5	.1915	.1950	.1985	.2019	.2054	.2088	.2123	.2157	.2190	.2224
0.6	.2257	.2291	.2324	.2357	.2389	.2422	.2454	.2486	.2517	.2549
0.7	.2580	.2611	.2642	.2673	.2704	.2734	.2764	.2794	.2823	.2852
0.8	.2881	.2910	.2939	.2967	.2995	.3023	.3051	.3078	.3106	.3133
0.9	.3159	.3186	.3212	.3238	.3264	.3289	.3315	.3340	.3365	.3389
1.0	.3413	.3438	.3461	.3485	.3508	.3531	.3554	.3577	.3599	.3621
1.1	.3643	.3665	.3686	.3708	.3729	.3749	.3770	.3790	.3810	.3830
1.2	.3849	.3869	.3888	.3907	.3925	.3944	.3962	.3980	.3997	.4015
1.3	.4032	.4049	.4066	.4082	.4099	.4115	.4131	.4147	.4162	.4177
1.4	.4192	.4207	.4222	.4236	.4251	.4265	.4279	.4292	.4306	.4319
1.5	.4332	.4345	.4357	.4370	.4382	.4394	.4406	.4418	.4429	.4441
1.6	.4452	.4463	.4474	.4484	.4495	.4505	.4515	.4525	.4535	.4545
1.7	.4554	.4564	.4573	.4582	.4591	.4599	.4608	.4616	.4625	.4633
1.8	.4641	.4649	.4656	.4664	.4671	.4678	.4686	.4693	.4699	.4706
1.9	.4713	.4719	.4726	.4732	.4738	.4744	.4750	.4756	.4761	.4767
2.0	.4772	.4778	.4783	.4788	.4793	.4798	.4803	.4808	.4812	.4817
2.1	.4821	.4826	.4830	.4834	.4838	.4842	.4846	.4850	.4854	.4857
2.2	.4861	.4864	.4868	.4871	.4875	.4878	.4881	.4884	.4887	.4890
2.3	.4893	.4896	.4998	.4901	.4904	.4906	.4909	.4911	.4913	.4916
2.4	.4918	.4920	.4922	.4925	.4927	.4929	.4931	.4932	.4934	.4936
2.5	.4938	.4940	.4941	.4943	.4945	.4946	.4948	.4949	.4951	.4952
2.6	.4953	.4955	.4956	.4957	.4959	.4960	.4961	.4962	.4963	.4964
2.7	.4965	.4966	.4967	.4968	.4969	.4970	.4971	.4972	.4973	.4974
2.8	.4974	.4975	.4976	.4977	.4977	.4978	.4979	.4979	.4980	.4981
2.9	.4981	.4982	.4982	.4983	.4984	.4984	.4985	.4985	.4986	.4986
3.0	.4987	.4987	.4987	.4988	.4988	.4989	.4989	.4989	.4990	.4990

[a]Abridged from Table 1 of *Statistical Tables and Formulas,* by A. Hald (New York: John Wiley & Sons, Inc., 1952). Reproduced by permission of A. Hald and the publishers.

TABLE E.2 Upper Percentage Points of the *t*-Distribution[a]

n \ $1-\alpha$	.75	.90	.95	.975	.99	.995	.9995
1	1.000	3.078	6.314	12.706	31.821	63.657	636.619
2	.816	1.886	2.920	4.303	6.965	9.925	31.598
3	.765	1.638	2.353	3.182	4.541	5.841	12.941
4	.741	1.533	2.132	2.776	3.747	4.604	8.610
5	.727	1.476	2.015	2.571	3.365	4.032	6.859
6	.718	1.440	1.943	2.447	3.143	3.707	5.959
7	.711	1.415	1.895	2.365	2.998	3.499	5.405
8	.706	1.397	1.860	2.306	2.896	3.355	5.041
9	.703	1.383	1.833	2.262	2.821	3.250	4.781
10	.700	1.372	1.812	2.228	2.764	3.169	4.587
11	.697	1.363	1.796	2.201	2.718	3.106	4.437
12	.695	1.356	1.782	2.179	2.681	3.055	4.318
13	.694	1.350	1.771	2.160	2.650	3.012	4.221
14	.692	1.345	1.761	2.145	2.624	2.977	4.140
15	.691	1.341	1.753	2.131	2.602	2.947	4.073
16	.690	1.337	1.746	2.120	2.583	2.921	4.015
17	.689	1.333	1.740	2.110	2.567	2.898	3.965
18	.688	1.330	1.734	2.101	2.552	2.878	3.922
19	.688	1.328	1.729	2.093	2.539	2.861	3.883
20	.687	1.325	1.725	2.086	2.528	2.845	3.850
21	.686	1.323	1.721	2.080	2.518	2.831	3.819
22	.686	1.321	1.717	2.074	2.508	2.819	3.792
23	.685	1.319	1.714	2.069	2.500	2.807	3.767
24	.685	1.318	1.711	2.064	2.492	2.797	3.745
25	.684	1.316	1.708	2.060	2.485	2.787	3.725
26	.684	1.315	1.706	2.056	2.479	2.779	3.707
27	.684	1.314	1.703	2.052	2.473	2.771	3.690
28	.683	1.313	1.701	2.048	2.467	2.763	3.674
29	.683	1.311	1.699	2.045	2.462	2.756	3.659
30	.683	1.310	1.697	2.042	2.457	2.750	3.646
40	.681	1.303	1.684	2.021	2.423	2.704	3.551
60	.679	1.296	1.671	2.000	2.390	2.660	3.460
120	.677	1.289	1.658	1.980	2.358	2.617	3.373
∞	.674	1.282	1.645	1.960	2.326	2.576	3.291

[a]Taken from Table III of R. A. Fisher and F. Yates, *Statistical Tables for Biological, Agricultural, and Medical Research,* published by Oliver & Boyd Ltd., Edinburgh, by permission of the authors and publishers.

TABLE E.3 Percentage Points of the Chi-square Distribution[a,b]

ν \ $1-\alpha$	.005	.010	.025	.050	.100	.250	.500
1	$.0^4393$	$.0^3157$	$.0^3982$	$.0^2393$	.0158	.102	.455
2	.0100	.0201	.0506	.103	.211	.575	1.386
3	.0717	.115	.216	.352	.584	1.213	2.366
4	.207	.297	.484	.711	1.064	1.923	3.357
5	.412	.554	.831	1.145	1.610	2.675	4.351
6	.676	.872	1.237	1.635	2.204	3.455	5.348
7	.988	1.239	1.690	2.167	2.833	4.255	6.346
8	1.344	1.647	2.180	2.733	3.490	5.071	7.344
9	1.735	2.088	2.700	3.325	4.168	5.899	8.343
10	2.156	2.558	3.247	3.940	4.865	6.737	9.342
11	2.603	3.053	3.816	4.575	5.578	7.584	10.341
12	3.074	3.571	4.404	5.226	6.304	8.438	11.340
13	3.565	4.107	5.009	5.892	7.042	9.299	12.340
14	4.075	4.660	5.629	6.571	7.790	10.165	13.339
15	4.601	5.229	6.262	7.261	8.547	11.037	14.339
16	5.142	5.812	6.908	7.962	9.312	11.912	15.339
17	5.697	6.408	7.564	8.672	10.085	12.792	16.338
18	6.265	7.015	8.231	9.390	10.865	13.675	17.338
19	6.844	7.633	8.907	10.117	11.651	14.562	18.338
20	7.434	8.260	9.591	10.851	12.443	15.452	19.337
21	8.034	8.897	10.283	11.591	13.240	16.344	20.337
22	8.643	9.542	10.982	12.338	14.042	17.240	21.337
23	9.260	10.196	11.689	13.091	14.848	18.137	22.337
24	9.886	10.856	12.401	13.848	15.659	19.037	23.337
25	10.520	11.524	13.120	14.611	16.473	19.939	24.337
26	11.160	12.198	13.844	15.379	17.292	20.843	25.337
27	11.808	12.879	14.573	16.151	18.114	21.749	26.336
28	12.461	13.565	15.308	16.928	18.939	22.657	27.336
29	13.121	14.257	16.047	17.708	19.768	23.567	28.336
30	13.787	14.954	16.791	18.493	20.599	24.478	29.336
40	20.707	22.164	24.433	26.509	29.051	33.660	39.335
50	27.991	29.707	32.357	34.764	37.689	42.942	49.335
60	35.535	37.485	40.482	43.188	46.459	52.294	59.335
z	−2.576	−2.326	−1.960	−1.645	−1.282	−.675	.000

[a]For $\nu > 60$, use the approximation $\chi^2 = \frac{1}{2}[z + \sqrt{2\nu - 1}]^2$, where z is the corresponding percent point of the unit normal distribution, shown in the last line.

[b]Abridged from Catherine M. Thompson, Tables of percentage points of the incomplete beta function and of the chi-square distribution, *Biometrika, 32* (1941), pp. 187–191, and is published here with the permission of the editor of *Biometrika*.

TABLE E.3 (continued) Percentage Points of the Chi-square Distribution

ν \ $1-\alpha$	.750	.900	.950	.975	.990	.995	.999
1	1.323	2.706	3.841	5.024	6.635	7.879	10.828
2	2.773	4.605	5.991	7.378	9.210	10.597	13.816
3	4.108	6.251	7.815	9.348	11.345	12.838	16.266
4	5.385	7.779	9.488	11.143	13.277	14.860	18.467
5	6.626	9.236	11.071	12.833	15.086	16.750	20.515
6	7.841	10.645	12.592	14.449	16.812	18.548	22.458
7	9.037	12.017	14.067	16.013	18.475	20.278	24.322
8	10.219	13.362	15.507	17.535	20.090	21.955	26.125
9	11.389	14.684	16.919	19.023	21.666	23.589	27.877
10	12.549	15.987	18.307	20.483	23.209	25.188	29.588
11	13.701	17.275	19.675	21.920	24.725	26.757	31.264
12	14.845	18.549	21.026	23.337	26.217	28.300	32.909
13	15.984	19.812	22.362	24.736	27.688	29.820	34.528
14	17.117	21.064	23.685	26.119	29.141	31.319	36.123
15	18.245	22.307	24.996	27.488	30.578	32.801	37.697
16	19.369	23.542	26.296	28.845	32.000	34.267	39.252
17	20.489	24.769	27.587	30.191	33.407	35.719	40.790
18	21.605	25.989	28.869	31.526	34.805	37.157	42.312
19	22.718	27.204	30.144	32.852	36.191	38.582	43.820
20	23.828	28.412	31.410	34.170	37.566	39.997	45.315
21	24.935	29.615	32.671	35.479	38.932	41.401	46.797
22	26.039	30.813	33.924	36.981	40.289	42.796	48.268
23	27.141	32.007	35.173	38.076	41.638	44.181	49.728
24	28.241	33.196	36.415	39.364	42.980	45.559	51.179
25	29.339	34.382	37.653	40.647	44.314	46.928	52.618
26	30.435	35.563	38.885	41.923	45.642	48.290	54.052
27	31.528	36.741	40.113	43.195	46.963	49.645	55.476
28	32.621	37.916	41.337	44.461	48.278	50.993	56.892
29	33.711	39.088	42.557	45.722	49.588	52.336	58.301
30	34.800	40.256	43.773	46.979	50.892	53.672	59.703
40	45.616	51.805	55.759	59.342	63.691	66.766	73.402
50	56.334	63.167	67.505	71.420	76.154	79.490	86.661
60	66.982	74.397	79.082	83.298	88.379	91.952	99.607
z	.675	1.282	1.645	1.960	2.326	2.576	3.090

TABLE E.4 Table of *F* for .05 (roman), .01 (*italic*), and .001 (boldface) Levels of Significance[a]

n_2 \ n_1	1	2	3	4	5	6	8	12	24	∞
	161	200	216	225	230	234	239	244	249	254
1	*4052*	*4999*	*5403*	*5625*	*5724*	*5859*	*5981*	*6106*	*6234*	*6366*
	405284	**500000**	**540379**	**562500**	**576405**	**585937**	**598144**	**610667**	**623497**	**636619**
	18.51	19.00	19.16	19.25	19.30	19.33	19.37	19.41	19.45	19.50
2	*98.49*	*99.01*	*99.17*	*99.25*	*99.30*	*99.33*	*99.36*	*99.42*	*99.46*	*99.50*
	998.5	**999.0**	**999.2**	**999.2**	**999.3**	**999.3**	**999.4**	**999.4**	**999.5**	**999.5**
	10.13	9.55	9.28	9.12	9.01	8.94	8.84	8.74	8.64	8.53
3	*34.12*	*30.81*	*29.46*	*28.71*	*28.24*	*27.91*	*27.49*	*27.05*	*26.60*	*26.12*
	167.5	**148.5**	**141.1**	**137.1**	**134.6**	**132.8**	**130.6**	**128.3**	**125.9**	**123.5**
	7.71	6.94	6.59	6.39	6.26	6.16	6.04	5.91	5.77	5.63
4	*21.20*	*18.00*	*16.69*	*15.98*	*15.52*	*15.21*	*14.80*	*14.37*	*13.93*	*13.46*
	74.14	**61.25**	**56.18**	**53.44**	**51.71**	**50.53**	**49.00**	**47.41**	**45.77**	**44.05**
	6.61	5.79	5.41	5.19	5.05	4.95	4.82	4.68	4.53	4.36
5	*16.26*	*13.27*	*12.06*	*11.39*	*10.97*	*10.67*	*10.27*	*9.89*	*9.47*	*9.02*
	47.04	**36.61**	**33.20**	**31.09**	**29.75**	**28.84**	**27.64**	**26.42**	**25.14**	**23.78**
	5.99	5.14	4.76	4.53	4.39	4.28	4.15	4.00	3.84	3.67
6	*13.74*	*10.92*	*9.78*	*9.15*	*8.75*	*8.47*	*8.10*	*7.72*	*7.31*	*6.88*
	35.51	**27.00**	**23.70**	**21.90**	**20.81**	**20.03**	**19.03**	**17.99**	**16.89**	**15.75**
	5.59	4.74	4.35	4.12	3.97	3.87	3.73	3.57	3.41	3.23
7	*12.25*	*9.55*	*8.45*	*7.85*	*7.46*	*7.19*	*6.84*	*6.47*	*6.07*	*5.65*
	29.22	**21.69**	**18.77**	**17.19**	**16.21**	**15.52**	**14.63**	**13.71**	**12.73**	**11.69**
	5.32	4.46	4.07	3.84	3.69	3.58	3.44	3.28	3.12	2.93
8	*11.26*	*8.65*	*7.59*	*7.01*	*6.63*	*6.37*	*6.03*	*5.67*	*5.28*	*4.86*
	25.42	**18.49**	**15.83**	**14.39**	**13.49**	**12.86**	**12.04**	**11.19**	**10.30**	**9.34**

	5.12	4.26	3.86	3.63	3.48	3.37	3.23	3.07	2.90	2.71
9	*10.56*	*8.02*	*6.99*	*6.42*	*6.06*	*5.80*	*5.47*	*5.11*	*4.73*	*4.31*
	22.86	**16.39**	**13.90**	**12.56**	**11.71**	**11.13**	**10.37**	**9.57**	**8.72**	**7.81**
	4.96	4.10	3.71	3.48	3.33	3.22	3.07	2.91	2.74	2.54
10	*10.04*	*7.56*	*6.55*	*5.99*	*5.64*	*5.39*	*5.06*	*4.71*	*4.33*	*3.91*
	21.04	**14.91**	**12.55**	**11.28**	**10.48**	**9.92**	**9.20**	**8.45**	**7.64**	**6.76**
	4.84	3.98	3.59	3.36	3.20	3.09	2.95	2.79	2.61	2.40
11	*9.65*	*7.20*	*6.22*	*5.67*	*5.32*	*5.07*	*4.74*	*4.40*	*4.02*	*3.60*
	19.69	**13.81**	**11.56**	**10.35**	**9.58**	**9.05**	**8.35**	**7.63**	**6.85**	**6.00**
	4.75	3.88	3.49	3.26	3.11	3.00	2.85	2.69	2.50	2.30
12	*9.33*	*6.93*	*5.95*	*5.41*	*5.06*	*4.82*	*4.50*	*4.16*	*3.78*	*3.36*
	18.64	**12.97**	**10.80**	**9.63**	**8.89**	**8.38**	**7.71**	**7.00**	**6.25**	**5.42**
	4.67	3.80	3.41	3.18	3.02	2.92	2.77	2.60	2.42	2.21
13	*9.07*	*6.70*	*5.74*	*5.20*	*4.86*	*4.62*	*4.30*	*3.96*	*3.59*	*3.16*
	17.81	**12.31**	**10.21**	**9.07**	**8.35**	**7.86**	**7.21**	**6.52**	**5.78**	**4.97**
	4.60	3.74	3.34	3.11	2.96	2.85	2.70	2.53	2.35	2.13
14	*8.86*	*6.51*	*5.56*	*5.03*	*4.69*	*4.46*	*4.14*	*3.80*	*3.43*	*3.00*
	17.14	**11.78**	**9.73**	**8.62**	**7.92**	**7.43**	**6.80**	**6.13**	**5.41**	**4.60**
	4.54	3.68	3.29	3.06	2.90	2.79	2.64	2.48	2.29	2.07
15	*8.68*	*6.36*	*5.42*	*4.89*	*4.56*	*4.32*	*4.00*	*3.67*	*3.29*	*2.87*
	16.59	**11.34**	**9.34**	**8.25**	**7.57**	**7.09**	**6.47**	**5.81**	**5.10**	**4.31**
	4.49	3.63	3.24	3.01	2.85	2.74	2.59	2.42	2.24	2.01
16	*8.53*	*6.23*	*5.29*	*4.77*	*4.44*	*4.20*	*3.89*	*3.55*	*3.18*	*2.75*
	16.12	**10.97**	**9.00**	**7.94**	**7.27**	**6.81**	**6.19**	**5.55**	**4.85**	**4.06**
	4.45	3.59	3.20	2.96	2.81	2.70	2.55	2.38	2.19	1.96
17	*8.40*	*6.11*	*5.18*	*4.67*	*4.34*	*4.10*	*3.79*	*3.45*	*3.08*	*2.65*
	15.72	**10.66**	**8.73**	**7.68**	**7.02**	**6.56**	**5.96**	**5.32**	**4.63**	**3.85**

[a]Reprinted, in rearranged form, from Table V of R. A. Fisher and F. Yates, *Statistical Tables for Biological, Agricultural, and Medical Research,* published by Oliver & Boyd Ltd., Edinburgh, by permission of the authors and publishers.

TABLE E.4 (continued) Table of *F* for .05 (roman), .01 (*italic*), and .001 (boldface) Levels of Significance[a]

n_2 \ n_1	1	2	3	4	5	6	8	12	24	∞
18	4.41	3.55	3.16	2.93	2.77	2.66	2.51	2.34	2.15	1.92
	8.28	*6.01*	*5.09*	*4.58*	*4.25*	*4.01*	*3.71*	*3.37*	*3.00*	*2.57*
	15.38	**10.39**	**8.49**	**7.46**	**6.81**	**6.35**	**5.76**	**5.13**	**4.45**	**3.67**
19	4.38	3.52	3.13	2.90	2.74	2.63	2.48	2.31	2.11	1.88
	8.18	*5.93*	*5.01*	*4.50*	*4.17*	*3.94*	*3.63*	*3.30*	*2.92*	*2.49*
	15.08	**10.16**	**8.28**	**7.26**	**6.61**	**6.18**	**5.59**	**4.97**	**4.29**	**3.52**
20	4.35	3.49	3.10	2.87	2.71	2.60	2.45	2.28	2.08	1.84
	8.10	*5.85*	*4.94*	*4.43*	*4.10*	*3.87*	*3.56*	*3.23*	*2.86*	*2.42*
	14.82	**9.95**	**8.10**	**7.10**	**6.46**	**6.02**	**5.44**	**4.82**	**4.15**	**3.38**
21	4.32	3.47	3.07	2.84	2.68	2.57	2.42	2.25	2.05	1.81
	8.02	*5.78*	*4.87*	*4.37*	*4.04*	*3.81*	*3.51*	*3.17*	*2.80*	*2.36*
	14.59	**9.77**	**7.94**	**6.95**	**6.32**	**5.88**	**5.31**	**4.70**	**4.03**	**3.26**
22	4.30	3.44	3.05	2.82	2.66	2.55	2.40	2.23	2.03	1.78
	7.94	*5.72*	*4.82*	*4.31*	*3.99*	*3.76*	*3.45*	*3.12*	*2.75*	*2.31*
	14.38	**9.61**	**7.80**	**6.81**	**6.19**	**5.76**	**5.19**	**4.58**	**3.92**	**3.15**
23	4.28	3.42	3.03	2.80	2.64	2.53	2.38	2.20	2.00	1.76
	7.88	*5.66*	*4.76*	*4.26*	*3.94*	*3.71*	*3.41*	*3.07*	*2.70*	*2.26*
	14.19	**9.47**	**7.67**	**6.69**	**6.08**	**5.65**	**5.09**	**4.48**	**3.82**	**3.05**
24	4.26	3.40	3.01	2.78	2.62	2.51	2.36	2.18	1.98	1.73
	7.82	*5.61*	*4.72*	*4.22*	*3.90*	*3.67*	*3.36*	*3.03*	*2.66*	*2.21*
	14.03	**9.34**	**7.55**	**6.59**	**5.98**	**5.55**	**4.99**	**4.39**	**3.74**	**2.97**
25	4.24	3.38	2.99	2.76	2.60	2.49	2.34	2.16	1.96	1.71
	7.77	*5.57*	*4.68*	*4.18*	*3.86*	*3.63*	*3.32*	*2.99*	*2.62*	*2.17*
	13.88	**9.22**	**7.45**	**6.49**	**5.88**	**5.46**	**4.91**	**4.31**	**3.66**	**2.89**

	4.22	3.37	2.98	2.74	2.59	2.47	2.32	2.15	1.95	1.69
26	*7.22*	*5.53*	*4.64*	*4.14*	*3.82*	*3.59*	*3.29*	*2.96*	*2.58*	*2.13*
	13.74	**9.12**	**7.36**	**6.41**	**5.80**	**5.38**	**4.83**	**4.24**	**3.59**	**2.82**
	4.21	3.35	2.96	2.73	2.57	2.46	2.30	2.13	1.93	1.67
27	*7.68*	*5.49*	*4.60*	*4.11*	*3.78*	*3.56*	*3.26*	*2.93*	*2.55*	*2.10*
	13.61	**9.02**	**7.27**	**6.33**	**5.73**	**5.31**	**4.76**	**4.17**	**3.52**	**2.75**
	4.20	3.34	2.95	2.71	2.56	2.44	2.29	2.12	1.91	1.65
28	*7.64*	*5.45*	*4.57*	*4.07*	*3.75*	*3.53*	*3.23*	*2.90*	*2.52*	*2.06*
	13.50	**8.93**	**7.19**	**6.25**	**5.66**	**5.24**	**4.69**	**4.11**	**3.46**	**2.70**
	4.18	3.33	2.93	2.70	2.54	2.43	2.28	2.10	1.90	1.64
29	*7.60*	*5.42*	*4.54*	*4.04*	*3.73*	*3.50*	*3.20*	*2.87*	*2.49*	*2.03*
	13.39	**8.85**	**7.12**	**6.19**	**5.59**	**5.18**	**4.64**	**4.05**	**3.41**	**2.64**
	4.17	3.32	2.92	2.69	2.53	2.42	2.27	2.09	1.89	1.62
30	*7.56*	*5.39*	*4.51*	*4.02*	*3.70*	*3.47*	*3.17*	*2.84*	*2.47*	*2.01*
	13.29	**8.77**	**7.05**	**6.12**	**5.53**	**5.12**	**4.58**	**4.00**	**3.36**	**2.59**
	4.08	3.23	2.84	2.61	2.45	2.34	2.18	2.00	1.79	1.51
40	*7.31*	*5.18*	*4.31*	*3.83*	*3.51*	*3.29*	*2.99*	*2.66*	*2.29*	*1.80*
	12.61	**8.25**	**6.60**	**5.70**	**5.13**	**4.73**	**4.21**	**3.64**	**3.01**	**2.23**
	4.00	3.15	2.76	2.52	2.37	2.25	2.10	1.92	1.70	1.39
60	*7.08*	*4.98*	*4.13*	*3.65*	*3.34*	*3.12*	*2.82*	*2.50*	*2.12*	*1.60*
	11.97	**7.76**	**6.17**	**5.31**	**4.76**	**4.37**	**3.87**	**3.31**	**2.69**	**1.90**
	3.92	3.07	2.68	2.45	2.29	2.17	2.02	1.83	1.61	1.25
120	*6.85*	*4.79*	*3.95*	*3.48*	*3.17*	*2.96*	*2.66*	*2.34*	*1.95*	*1.38*
	11.38	**7.31**	**5.79**	**4.95**	**4.42**	**4.04**	**3.55**	**3.02**	**2.40**	**1.56**
	3.84	2.99	2.60	2.37	2.21	2.09	1.94	1.75	1.52	1.00
∞	*6.64*	*4.60*	*3.78*	*3.32*	*3.02*	*2.80*	*2.51*	*2.18*	*1.79*	*1.00*
	10.83	**6.91**	**5.42**	**4.62**	**4.10**	**3.74**	**3.27**	**2.74**	**2.13**	**1.00**

[a]Reprinted, in rearranged form, from Table V of R. A. Fisher and F. Yates, *Statistical Tables for Biological, Agricultural, and Medical Research,* published by Oliver & Boyd Ltd., Edinburgh, by permission of the authors and publishers.

TABLE E.5 Four-Place Logarithms

n	0	1	2	3	4	5	6	7	8	9
10	0000	0043	0086	0128	0170	0212	0253	0294	0334	0374
11	0414	0453	0492	0531	0569	0607	0645	0682	0719	0755
12	0792	0828	0864	0899	0934	0969	1004	1038	1072	1106
13	1139	1173	1206	1239	1271	1303	1335	1367	1399	1430
14	1461	1492	1523	1553	1584	1614	1644	1673	1703	1732
15	1761	1790	1818	1847	1875	1903	1931	1959	1987	2014
16	2041	2068	2095	2122	2148	2175	2201	2227	2253	2279
17	2304	2330	2355	2380	2405	2430	2455	2480	2504	2529
18	2553	2577	2601	2625	2648	2672	2695	2718	2742	2765
19	2788	2810	2833	2856	2878	2900	2923	2945	2967	2989
20	3010	3032	3054	3075	3096	3118	3139	3160	3181	3201
21	3222	3243	3263	3284	3304	3324	3345	3365	3385	3404
22	3424	3444	3464	3483	3502	3522	3541	3560	3579	3598
23	3617	3636	3655	3674	3692	3711	3729	3747	3766	3784
24	3802	3820	3838	3856	3874	3892	3909	3927	3945	3962
25	3979	3997	4014	4031	4048	4065	4082	4099	4116	4133
26	4150	4166	4183	4200	4216	4232	4249	4265	4281	4298
27	4314	4330	4346	4362	4378	4393	4409	4425	4440	4456
28	4472	4487	4502	4518	4533	4548	4564	4579	4594	4609
29	4624	4639	4654	4669	4683	4698	4713	4728	4742	4757
30	4771	4786	4800	4814	4829	4843	4857	4871	4886	4900
31	4914	4928	4942	4955	4969	4983	4997	5011	5024	5038
32	5051	5065	5079	5092	5105	5119	5132	5145	5159	5172
33	5185	5198	5211	5224	5237	5250	5263	5276	5289	5302
34	5315	5328	5340	5353	5366	5378	5391	5403	5416	5428
35	5441	5453	5465	5478	5490	5502	5514	5527	5539	5551
36	5563	5575	5587	5599	5611	5623	5635	5647	5658	5670
37	5682	5694	5705	5717	5729	5740	5752	5763	5775	5786
38	5798	5809	5821	5832	5843	5855	5866	5877	5888	5899
39	5911	5922	5933	5944	5955	5966	5977	5988	5999	6010
40	6021	6031	6042	6053	6064	6075	6085	6096	6107	6117
41	6128	6138	6149	6160	6170	6180	6191	6201	6212	6222
42	6232	6243	6253	6263	6274	6284	6294	6304	6314	6325
43	6335	6345	6355	6365	6375	6385	6395	6405	6415	6425
44	6435	6444	6454	6464	6474	6484	6493	6503	6513	6522
45	6532	6542	6551	6561	6571	6580	6590	6599	6609	6618
46	6628	6637	6646	6656	6665	6675	6684	6693	6702	6712
47	6721	6730	6739	6749	6758	6767	6776	6785	6794	6803
48	6812	6821	6830	6839	6848	6857	6866	6875	6884	6893
49	6902	6911	6920	6928	6937	6946	6955	6964	6972	6981
50	6990	6998	7007	7016	7024	7033	7042	7050	7059	7067

TABLE E.5 (continued) Four-Place Logarithms

n	0	1	2	3	4	5	6	7	8	9
51	7076	7084	7093	7101	7110	7118	7126	7135	7143	7152
52	7160	7168	7177	7185	7193	7202	7210	7218	7226	7235
53	7243	7251	7259	7267	7275	7284	7292	7300	7308	7316
54	7324	7332	7340	7348	7356	7364	7372	7380	7388	7396
55	7404	7412	7419	7427	7435	7443	7451	7459	7466	7474
56	7482	7490	7497	7505	7513	7520	7528	7536	7543	7551
57	7559	7566	7574	7582	7589	7597	7604	7612	7619	7627
58	7634	7642	7649	7657	7664	7672	7679	7686	7694	7701
59	7709	7716	7723	7731	7738	7745	7752	7760	7767	7774
60	7782	7789	7796	7803	7810	7818	7825	7832	7839	7846
61	7853	7860	7868	7875	7882	7889	7896	7903	7910	7917
62	7924	7931	7938	7945	7952	7959	7966	7973	7980	7987
63	7993	8000	8007	8014	8021	8028	8035	8041	8048	8055
64	8062	8069	8075	8082	8089	8096	8102	8109	8116	8122
65	8129	8136	8142	8149	8156	8162	8169	8176	8182	8189
66	8195	8202	8209	8215	8222	8228	8235	8241	8248	8254
67	8261	8267	8274	8280	8287	8293	8299	8306	8312	8319
68	8325	8331	8338	8344	8351	8357	8363	8370	8376	8382
69	8388	8395	8401	8407	8414	8420	8426	8432	8439	8445
70	8451	8457	8463	8470	8476	8482	8488	8494	8500	8506
71	8513	8519	8525	8531	8537	8543	8549	8555	8561	8567
72	8573	8579	8585	8591	8597	8603	8609	8615	8621	8627
73	8633	8639	8645	8651	8657	8663	8669	8675	8681	8686
74	8692	8698	8704	8710	8716	8722	8727	8733	8739	8745
75	8751	8756	8762	8768	8774	8779	8785	8791	8797	8802
76	8808	8814	8820	8825	8831	8837	8842	8848	8854	8859
77	8865	8871	8876	8882	8887	8893	8899	8904	8910	8915
78	8921	8927	8932	8938	8943	8949	8954	8960	8965	8971
79	8976	8982	8987	8993	8998	9004	9009	9015	9020	9025
80	9031	9036	9042	9047	9053	9058	9063	9069	9074	9079
81	9085	9090	9096	9101	9106	9112	9117	9122	9128	9133
82	9138	9143	9149	9154	9159	9165	9170	9175	9180	9186
83	9191	9196	9201	9206	9212	9217	9222	9227	9232	9238
84	9243	9248	9253	9258	9263	9269	9274	9279	9284	9289
85	9294	9299	9304	9309	9315	9320	9325	9330	9335	9340
86	9345	9350	9355	9360	9365	9370	9375	9380	9385	9390
87	9395	9400	9405	9410	9415	9420	9425	9430	9435	9440
88	9445	9450	9455	9460	9465	9469	9474	9479	9484	9489
89	9494	9499	9504	9509	9513	9518	9523	9528	9533	9538
90	9542	9547	9552	9557	9562	9566	9571	9576	9581	9586

TABLE E.5 (continued) Four-Place Logarithms

n	0	1	2	3	4	5	6	7	8	9
91	9590	9595	9600	9605	9609	9614	9619	9624	9628	9633
92	9638	9643	9647	9652	9657	9661	9666	9671	9675	9680
93	9685	9689	9694	9699	9703	9708	9713	9717	9722	9727
94	9731	9736	9741	9745	9750	9754	9759	9763	9768	9773
95	9777	9782	9786	9791	9795	9800	9805	9809	9814	9818
96	9823	9827	9832	9836	9841	9845	9850	9854	9859	9863
97	9868	9872	9877	9881	9886	9890	9894	9899	9903	9908
98	9912	9917	9921	9926	9930	9934	9939	9943	9948	9952
99	9956	9961	9965	9969	9974	9978	9983	9987	9991	9996

Table E.6 Squares, Square Roots, and Reciprocals

n	n^2	$\sqrt{n}$	$\sqrt{10n}$	$1/n$	n	n^2	$\sqrt{n}$	$\sqrt{10n}$	$1/n$
1	1	1.000	3.162	1.00000	31	961	5.568	17.607	.03226
2	4	1.414	4.472	.50000	32	1024	5.657	17.889	.03125
3	9	1.732	5.477	.33333	33	1089	5.745	18.166	.03030
4	16	2.000	6.325	.25000	34	1156	5.831	18.439	.02941
5	25	2.236	7.071	.20000	35	1225	5.916	18.708	.02857
6	36	2.449	7.746	.16667	36	1296	6.000	18.974	.02778
7	49	2.646	8.367	.14286	37	1369	6.083	19.235	.02703
8	64	2.828	8.944	.12500	38	1444	6.164	19.494	.02632
9	81	3.000	9.487	.11111	39	1521	6.245	19.748	.02564
10	100	3.162	10.000	.10000	40	1600	6.325	20.000	.02500
11	121	3.317	10.488	.09091	41	1681	6.403	20.248	.02439
12	144	3.464	10.954	.08333	42	1764	6.481	20.494	.02381
13	169	3.606	11.402	.07692	43	1849	6.557	20.736	.02326
14	196	3.742	11.832	.07143	44	1936	6.633	20.976	.02273
15	225	3.873	12.247	.06667	45	2025	6.708	21.213	.02222
16	256	4.000	12.649	.06250	46	2116	6.782	21.448	.02174
17	289	4.123	13.038	.05882	47	2209	6.856	21.679	.02128
18	324	4.243	13.416	.05556	48	2304	6.928	21.909	.02083
19	361	4.359	13.784	.05263	49	2401	7.000	22.136	.02041
20	400	4.472	14.142	.05000	50	2500	7.071	22.361	.02000
21	441	4.583	14.491	.04762	51	2601	7.141	22.583	.01961
22	484	4.690	14.832	.04545	52	2704	7.211	22.804	.01923
23	529	4.796	15.166	.04348	53	2809	7.280	23.022	.01887
24	576	4.899	15.492	.04167	54	2916	7.348	23.238	.01852
25	625	5.000	15.811	.04000	55	3025	7.416	23.452	.01818
26	676	5.099	16.125	.03846	56	3136	7.483	23.664	.01786
27	729	5.196	16.432	.03704	57	3249	7.550	23.875	.01754
28	784	5.292	16.733	.03571	58	3364	7.616	24.083	.01724
29	841	5.385	17.029	.03448	59	3481	7.681	24.290	.01695
30	900	5.477	17.321	.03333	60	3600	7.746	24.495	.01667

Table E.6 (continued) Squares, Square Roots, and Reciprocals

n	n^2	$\sqrt{n}$	$\sqrt{10n}$	$1/n$	n	n^2	$\sqrt{n}$	$\sqrt{10n}$	$1/n$
61	3721	7.810	24.698	.01639	81	6561	9.000	28.460	.01235
62	3844	7.874	24.900	.01613	82	6724	9.055	28.636	.01220
63	3969	7.937	25.100	.01587	83	6889	9.110	28.810	.01205
64	4096	8.000	25.298	.01562	84	7056	9.165	28.983	.01190
65	4225	8.062	25.495	.01538	85	7225	9.220	29.155	.01176
66	4356	8.124	25.690	.01515	86	7396	9.274	29.326	.01163
67	4489	8.185	25.884	.01493	87	7569	9.327	29.496	.01149
68	4624	8.246	26.077	.01471	88	7744	9.381	29.665	.01136
69	4761	8.307	26.268	.01449	89	7921	9.434	29.833	.01124
70	4900	8.367	26.458	.01429	90	8100	9.487	30.000	.01111
71	5041	8.426	26.646	.01408	91	8281	9.539	30.166	.01099
72	5184	8.485	26.833	.01389	92	8464	9.592	30.332	.01087
73	5329	8.544	27.019	.01370	93	8649	9.644	30.496	.01075
74	5476	8.602	27.203	.01351	94	8836	9.695	30.659	.01064
75	5625	8.660	27.386	.01333	95	9025	9.747	30.822	.01053
76	5776	8.718	27.568	.01316	96	9216	9.798	30.984	.01042
77	5929	8.775	27.749	.01299	97	9409	9.849	31.145	.01031
78	6084	8.832	27.928	.01282	98	9604	9.899	31.305	.01020
79	6241	8.888	28.107	.01266	99	9801	9.950	31.464	.01010
80	6400	8.944	28.284	.01250	100	10000	10.000	31.623	.01000

APPENDIX F

DATA FROM PROJECT "HIGH SCHOOL AND BEYOND"[a]

Variable Number	Variable Name	Variable Columns	Variable Label	Coded Responses[b]
1	ID	1-3	Student identification number	(Three-digit consecutive numbers)
2	SEX	6	Sex	1 = Male 2 = Female
3	RACE	10	Race or ethnicity	1 = Hispanic 2 = Asian 3 = Black 4 = White
4	SES	15	Socioeconomic status	1 = Low 2 = Medium 3 = High
5	SCTYP	20	School type	1 = Public 2 = Private
6	HSP	25	High school program	1 = General 2 = Academic preparatory 3 = Vocational/technical
7	LOCUS	28-32	Locus of control	(Standardized to mean of 0 and SD of 1)
8	CONCPT	35-39	Self-concept	(Standardized to mean of 0 and SD of 1)
9	MOT	41-44	Motivation	(Average of three motivation items)

Variable Number	Variable Name	Variable Columns	Variable Label	Coded Responses[b]
10	CAR	46-47	Career choice	1 = Clerical 2 = Craftsman 3 = Farmer 4 = Homemaker 5 = Laborer 6 = Manager 7 = Military 8 = Operative 9 = Professional 1 10 = Professional 2 11 = Proprietor 12 = Protective 13 = Sales 14 = School 15 = Service 16 = Technical 17 = Not working
11	RDG	50-53	Standardized reading score	(Standardized to mean of 50 and SD of 10)
12	WRTG	55-58	Standardized writing score	(Standardized to mean of 50 and SD of 10)
13	MATH	60-63	Standardized math score	(Standardized to mean of 50 and SD of 10)
14	SCI	65-68	Standardized science score	(Standardized to mean of 50 and SD of 10)
15	CIV	70-73	Standardized civics score	(Standardized to mean of 50 and SD of 10)

[a]"High School and Beyond" is a large-scale national longitudinal study conducted by the National Opinion Research Center (1980) under contract with the National Center for Education Statistics.
[b]"Professional 1" refers to accountant, artist, nurse, engineer, librarian, writer, social worker, actor, athlete, politician, etc.; "professional 2" refers to clergyman, dentist, physician, lawyer, scientist, college teacher, etc.

ID	SEX	RACE	SES	SCTYP	HSP	LOCUS	CONCPT	MOT	CAR	RDG	WRTG	MATH	SCI	CIV
1	2	1	1	1	3	0.29	0.88	0.67	10	33.6	43.7	40.2	39.0	40.6
2	1	1	1	1	1	-0.42	0.03	0.33	2	46.9	35.9	41.9	36.3	45.6
3	2	1	1	1	1	0.71	0.03	0.67	9	41.6	59.3	41.9	44.4	45.6
4	2	1	2	1	3	0.06	0.03	0.00	15	38.9	41.1	32.7	41.7	40.6
5	2	1	2	1	3	0.22	-0.28	0.00	1	36.3	48.9	39.5	41.7	45.6
6	1	1	2	1	1	0.46	0.03	0.00	11	49.5	46.3	46.2	41.7	35.6
7	1	1	1	1	2	0.44	-0.47	0.33	10	62.7	64.5	48.0	63.4	55.6
8	2	1	1	1	2	0.68	0.25	1.00	9	44.2	51.5	36.9	49.8	55.6
9	1	1	2	1	2	0.06	0.56	0.33	9	46.9	41.1	45.3	47.1	55.6
10	2	1	1	1	2	0.05	0.15	1.00	11	44.2	49.5	40.5	39.0	50.6
11	2	1	1	1	2	0.25	0.34	1.00	9	49.5	61.9	42.9	41.7	50.6
12	1	1	1	1	1	0.06	-0.59	0.00	8	44.2	41.1	59.8	44.4	35.6
13	1	1	1	1	2	-1.10	0.03	0.00	7	46.9	40.4	42.8	45.0	30.6
14	2	1	2	1	1	0.52	-0.59	0.67	4	41.6	59.3	38.6	36.3	50.6
15	1	1	2	1	3	-1.28	1.19	1.00	9	41.6	41.1	45.9	44.4	40.6
16	2	1	1	1	2	0.06	0.03	1.00	10	46.9	41.1	46.2	39.6	40.6
17	2	1	3	1	2	0.26	0.03	1.00	10	60.1	64.5	55.7	63.4	60.5
18	1	1	2	1	3	0.21	0.94	0.00	12	38.9	48.9	41.8	58.5	45.6
19	1	1	2	1	2	0.06	0.03	1.00	6	57.4	54.1	58.9	47.1	50.6
20	1	1	3	1	2	0.28	0.03	1.00	10	73.3	64.5	68.0	63.4	70.5
21	1	1	2	1	1	-0.44	-0.84	0.33	2	46.9	46.3	45.9	37.9	45.6
22	2	1	1	1	2	-0.43	0.13	0.33	7	38.9	44.3	51.6	43.9	48.1
23	2	1	1	1	3	-0.22	0.32	0.33	1	35.2	38.5	39.9	34.7	40.6
24	2	1	1	1	3	0.73	-1.42	0.33	15	38.9	41.1	41.0	36.3	45.6
25	1	1	2	1	3	0.26	0.03	0.33	11	48.4	48.9	52.2	43.9	50.6
26	2	1	2	1	2	1.12	-0.74	0.67	15	31.0	41.1	36.0	36.9	45.6
27	2	1	3	1	1	0.46	0.59	1.00	16	52.1	62.5	53.6	56.3	65.5
28	2	1	2	1	1	-0.60	-0.47	0.67	9	46.9	54.1	49.0	52.6	60.5
29	2	1	2	1	2	-0.90	0.03	0.67	4	36.3	44.3	36.1	33.6	40.6
30	2	1	2	1	2	0.31	1.19	1.00	9	46.9	44.3	41.9	39.0	50.6
31	1	1	2	1	2	-0.21	-1.38	0.00	9	34.2	46.3	44.5	39.0	35.6
32	2	1	2	1	2	0.69	0.34	1.00	9	52.1	56.7	53.4	60.7	45.6
33	2	1	2	1	1	-0.19	-1.67	0.00	13	42.1	39.8	42.2	38.5	40.6
34	1	1	2	1	3	-1.78	0.56	0.33	8	37.3	43.7	45.4	39.0	45.6
35	1	1	2	1	3	0.30	0.56	1.00	16	44.2	33.3	37.7	30.9	41.9
36	1	1	2	1	2	1.36	0.44	0.67	9	70.7	58.0	65.4	63.4	60.5
37	2	1	2	1	3	-1.05	-1.65	0.00	9	46.9	46.3	38.6	47.1	60.5
38	1	1	2	1	3	0.29	0.39	0.33	7	41.6	33.3	37.6	28.2	45.6
39	1	1	2	1	2	-0.83	0.13	1.00	9	41.6	35.9	36.7	36.3	55.6
40	1	1	3	1	2	0.21	-0.28	0.67	12	46.9	41.1	53.6	41.7	55.6
41	1	1	3	1	2	-0.03	0.63	0.67	9	52.1	54.1	48.2	55.3	60.5
42	1	1	2	1	2	0.03	0.28	0.33	8	52.1	54.1	52.0	55.3	50.6
43	1	1	3	1	3	-0.83	1.19	0.67	9	38.9	38.5	44.4	26.0	41.9
44	2	1	1	1	3	0.22	-0.76	1.00	15	31.0	41.1	49.2	33.6	35.6
45	1	1	3	1	2	1.16	0.32	1.00	9	60.1	54.1	48.6	58.0	30.6
46	1	1	1	1	3	-0.22	-1.34	0.67	16	46.9	31.3	43.6	36.3	35.6
47	1	1	1	1	1	-0.18	0.15	0.00	15	38.9	28.1	36.0	36.3	35.6
48	2	1	3	1	3	0.02	0.03	0.33	4	41.6	35.9	37.7	41.7	45.6
49	2	1	2	1	2	-1.06	0.03	0.67	6	46.9	56.7	48.0	44.4	40.6
50	1	1	2	1	1	-1.33	0.03	1.00	9	41.6	48.9	33.7	44.4	45.6
51	1	1	1	1	3	-1.55	0.03	0.00	9	36.3	41.1	43.5	33.6	40.6
52	1	1	2	1	3	-1.58	0.56	0.67	2	49.5	33.3	48.7	44.4	35.6
53	2	1	3	1	3	0.47	-0.09	0.33	9	33.6	33.9	38.8	39.6	30.6
54	1	1	3	1	3	0.46	0.03	0.67	9	54.8	54.7	56.9	58.0	40.6
55	2	1	1	1	1	-1.78	0.56	1.00	9	28.3	46.3	42.8	44.4	50.6
56	1	1	1	1	2	-0.84	-1.09	0.33	2	41.6	41.7	45.1	36.3	40.6
57	2	1	2	1	1	-0.89	-0.18	0.67	10	44.2	43.7	44.5	41.7	50.6
58	1	1	3	1	2	-0.68	-0.26	0.00	2	60.1	51.5	56.8	60.7	60.5
59	2	1	1	1	3	-0.60	0.32	1.00	1	38.9	56.7	41.2	33.6	55.6
60	1	1	2	1	3	-0.59	1.19	0.67	6	36.3	36.5	35.1	33.6	35.6

ID	SEX	RACE	SES	SCTYP	HSP	LOCUS	CONCPT	MOT	CAR	RDG	WRTG	MATH	SCI	CIV
61	1	1	2	1	1	-0.04	0.65	1.00	16	44.2	44.3	61.1	49.8	45.6
62	2	1	1	1	1	0.52	-0.28	0.67	9	38.9	41.7	33.7	30.9	50.6
63	2	1	1	1	1	-0.34	0.59	1.00	7	38.9	33.9	35.1	44.4	40.6
64	1	1	2	1	3	-0.44	0.03	0.33	9	41.6	38.5	38.7	56.3	45.6
65	1	1	2	1	3	-0.21	0.03	1.00	6	41.6	43.7	51.1	49.8	40.6
66	2	2	3	1	2	0.32	-1.17	0.33	9	52.1	54.1	58.1	47.1	60.5
67	2	2	1	1	2	-0.71	-2.29	0.33	9	65.4	64.5	63.7	58.0	70.5
68	1	2	1	1	1	0.45	0.03	0.67	2	41.6	25.5	44.3	36.3	35.6
69	1	2	1	1	1	0.71	0.03	1.00	10	46.9	61.9	46.2	60.7	45.6
70	1	2	2	1	2	0.06	0.44	0.67	9	52.1	61.9	66.3	47.1	45.6
71	2	2	2	1	3	0.06	0.28	1.00	1	44.2	54.1	47.1	58.0	50.6
72	1	2	3	1	2	0.91	0.34	1.00	10	52.1	54.1	58.1	55.8	50.6
73	2	2	2	1	1	-1.09	0.03	0.33	2	46.9	43.7	41.9	41.7	35.6
74	2	2	3	1	2	0.02	0.03	0.67	10	65.4	56.7	62.3	58.5	55.6
75	1	2	3	2	2	-1.05	0.28	0.67	9	62.7	64.5	66.3	58.0	60.5
76	2	2	3	1	2	0.11	-1.42	1.00	16	60.1	59.3	62.1	60.7	50.6
77	1	2	2	1	3	0.89	0.32	0.67	10	46.9	44.3	48.7	53.1	55.6
78	1	2	1	1	3	-0.45	-0.64	0.33	9	65.4	56.7	55.2	52.6	55.6
79	1	2	2	1	2	0.46	-0.30	1.00	10	53.2	60.6	61.2	56.9	55.6
80	2	2	2	1	2	0.46	-1.17	0.00	9	68.0	62.5	65.4	60.7	60.5
81	2	2	3	1	2	0.28	0.94	1.00	10	54.8	56.0	43.5	49.8	50.6
82	2	2	2	1	1	0.06	0.90	0.67	9	38.9	52.8	54.4	49.8	40.6
83	1	2	2	1	2	0.06	0.94	1.00	7	52.1	51.5	57.9	60.7	35.6
84	1	2	3	1	2	-0.17	0.31	1.00	10	65.4	67.1	75.5	71.5	70.5
85	1	2	1	1	1	-0.38	0.03	0.00	2	52.1	43.7	49.4	55.3	40.6
86	2	2	3	1	2	0.23	-0.60	1.00	10	57.4	67.1	68.0	59.1	65.5
87	1	2	2	1	2	1.36	0.59	1.00	10	62.7	67.1	69.6	68.8	55.6
88	2	2	3	1	2	-0.39	1.19	1.00	9	40.5	59.3	41.9	33.6	50.6
89	1	2	2	1	1	-1.28	-1.05	0.33	9	57.4	43.7	57.8	60.7	45.6
90	2	2	1	1	3	-0.65	-1.17	0.00	8	36.3	30.7	45.9	36.3	40.6
91	2	2	2	2	1	-0.40	-0.47	1.00	10	54.8	59.3	51.8	41.7	55.6
92	2	2	3	1	3	0.06	-0.29	0.67	1	49.5	56.7	68.1	49.8	55.6
93	2	2	2	1	2	0.55	0.90	1.00	9	62.7	61.9	59.6	60.7	65.5
94	2	2	3	1	3	-0.35	-0.89	0.00	1	49.5	67.1	66.4	66.1	55.6
95	2	2	2	1	2	0.30	-1.09	0.00	9	51.1	56.7	55.3	47.7	56.8
96	2	2	1	1	3	-0.93	0.65	1.00	5	47.4	45.6	49.4	42.3	45.6
97	2	2	1	1	2	1.16	-0.72	0.67	10	56.9	64.5	72.2	54.2	55.6
98	1	2	3	1	2	0.00	0.15	0.33	10	73.3	64.5	75.5	60.7	50.6
99	1	2	2	1	2	0.52	0.34	1.00	9	52.1	54.1	51.4	49.8	50.6
100	2	1	3	2	2	0.42	-0.47	0.33	10	73.3	61.2	57.4	54.7	65.5
101	1	1	2	2	1	-0.67	-0.47	0.00	8	44.2	48.9	43.1	55.3	45.6
102	2	1	3	2	2	0.45	0.65	1.00	9	57.4	56.7	46.9	52.6	55.6
103	2	1	1	2	1	0.06	0.32	0.00	15	60.1	54.1	49.9	49.8	50.6
104	1	1	2	2	2	0.26	0.88	1.00	9	65.4	59.3	62.3	60.7	65.5
105	2	1	3	2	2	0.46	0.34	1.00	9	57.4	54.1	59.6	60.7	70.5
106	2	3	1	1	1	-0.89	-1.23	1.00	10	44.2	48.9	43.8	35.2	50.6
107	2	3	2	1	3	0.57	-2.62	0.00	11	38.9	28.1	38.4	44.4	35.6
108	2	3	3	1	3	0.26	0.94	0.00	17	54.8	38.5	46.8	36.3	45.6
109	2	3	1	1	3	0.31	1.19	1.00	1	40.5	46.9	40.4	39.0	50.6
110	1	3	1	1	3	0.93	-1.63	0.67	8	38.9	41.1	40.3	34.1	45.6
111	1	3	3	1	2	0.68	0.94	1.00	9	62.7	52.1	66.1	68.8	50.6
112	1	3	1	1	2	0.03	0.32	1.00	6	44.7	56.7	50.3	30.9	55.6
113	1	3	3	1	2	-0.03	0.87	1.00	7	53.2	46.3	43.0	41.7	50.6
114	1	3	2	1	2	0.24	-0.35	1.00	8	30.5	35.9	36.9	33.6	40.6
115	2	3	3	1	2	1.14	1.19	1.00	9	65.9	67.1	67.1	60.7	65.5
116	2	3	2	1	2	0.27	0.34	1.00	7	49.5	48.9	48.0	41.7	60.5
117	2	3	1	1	3	-1.34	1.19	1.00	9	44.2	41.1	44.2	44.4	40.6
118	1	3	1	1	1	0.45	-0.91	0.33	2	41.6	41.1	42.8	49.8	40.6
119	2	3	1	1	2	0.28	1.19	1.00	1	56.4	59.3	38.5	55.3	56.8
120	2	3	1	1	3	-1.33	-0.26	0.67	15	38.4	37.2	33.4	30.9	36.9

ID	SEX	RACE	SES	SCTYP	HSP	LOCUS	CONCPT	MOT	CAR	RDG	WRTG	MATH	SCI	CIV
121	1	3	2	1	2	0.49	0.94	1.00	9	49.5	39.8	44.5	55.3	55.6
122	2	3	1	1	2	0.10	0.32	0.67	9	40.0	40.4	42.4	48.8	60.5
123	2	3	2	1	1	1.11	-1.05	0.33	9	70.7	67.1	63.0	63.4	65.5
124	2	3	2	1	3	0.22	-0.76	0.67	16	45.8	52.1	54.9	44.4	55.6
125	2	3	3	1	1	0.75	-0.72	0.67	14	42.6	46.3	41.2	28.2	45.6
126	1	3	2	1	2	0.89	0.15	0.67	10	68.0	59.3	56.6	50.4	60.5
127	2	3	1	1	2	-0.49	0.03	0.00	8	46.9	37.2	42.8	41.7	45.6
128	1	3	2	1	2	0.49	0.44	0.67	10	53.7	43.7	51.1	52.0	43.1
129	1	3	2	1	3	-0.38	0.34	0.67	7	38.9	28.1	35.3	39.0	40.6
130	2	3	1	1	3	-0.60	0.90	0.67	1	46.9	61.9	45.4	33.6	45.6
131	2	3	2	1	2	1.36	0.65	0.67	10	59.0	56.7	52.9	48.8	56.8
132	2	3	1	1	2	-0.24	-0.16	0.33	16	62.7	61.9	52.4	47.1	60.5
133	2	3	1	1	3	-0.23	0.44	0.33	14	33.6	35.2	40.9	29.3	25.7
134	2	3	2	1	2	0.51	-1.29	0.33	1	39.4	41.7	46.9	35.2	50.6
135	1	3	2	1	3	-0.60	-1.43	1.00	10	38.9	51.5	38.6	39.0	35.6
136	2	3	1	1	2	0.06	0.94	0.67	9	45.3	54.7	44.3	33.6	40.6
137	1	3	2	1	2	-0.37	-1.90	0.67	6	54.8	36.5	37.7	49.8	60.5
138	1	3	2	1	2	0.46	0.94	0.33	5	36.3	30.7	44.8	39.6	40.6
139	2	3	1	1	2	0.44	0.65	1.00	8	46.9	46.3	48.5	32.5	40.6
140	2	3	1	1	3	0.10	1.19	1.00	15	38.9	35.9	46.8	39.0	40.6
141	1	3	2	1	3	-2.23	1.19	1.00	8	36.3	38.5	39.3	39.0	45.6
142	1	3	2	1	2	1.13	0.87	1.00	10	57.4	54.7	51.7	49.8	50.6
143	1	3	2	1	2	0.26	1.19	1.00	9	52.1	52.8	37.6	52.6	50.6
144	2	3	1	1	1	-0.04	-0.29	0.67	10	41.6	38.5	40.2	33.6	45.6
145	1	3	3	1	3	0.69	0.03	0.67	9	49.5	39.8	38.6	49.3	46.8
146	2	3	1	1	1	-1.76	-0.93	0.67	6	36.3	38.5	36.4	36.3	50.6
147	1	3	3	1	3	-0.66	-0.08	1.00	12	44.7	33.3	33.7	39.0	35.6
148	1	3	2	1	1	-0.44	-0.31	0.67	9	49.5	59.3	42.1	52.6	60.5
149	1	3	1	1	2	1.16	1.19	0.33	9	55.3	46.9	49.1	54.2	51.8
150	2	3	2	1	2	0.91	-0.28	1.00	7	60.1	67.1	56.2	37.4	50.6
151	2	3	3	1	1	0.75	-0.26	0.67	15	41.6	35.9	42.0	31.4	39.4
152	1	3	2	1	2	-1.14	-0.06	1.00	6	36.3	46.3	36.9	41.7	35.6
153	2	3	2	1	1	0.46	0.03	1.00	14	52.1	56.7	62.8	47.1	60.5
154	2	3	1	1	2	0.45	0.03	1.00	1	49.5	46.3	52.9	52.6	65.5
155	2	3	1	1	3	-0.54	-0.59	1.00	4	38.9	35.9	38.4	33.6	35.6
156	1	3	3	1	3	-0.42	-0.60	1.00	14	54.8	56.7	64.7	58.0	60.5
157	1	3	2	1	3	-0.38	0.56	1.00	2	33.6	36.5	45.7	39.0	30.6
158	2	3	1	1	1	1.16	0.59	1.00	10	36.3	45.0	44.3	39.0	40.6
159	1	3	1	1	1	-0.68	0.32	0.67	9	36.3	44.3	34.4	47.1	45.6
160	2	3	1	2	1	0.70	0.87	1.00	1	46.9	54.1	46.4	49.8	55.6
161	2	3	2	2	2	0.48	0.63	1.00	10	54.8	48.2	58.7	51.5	45.6
162	2	3	3	2	2	-0.40	-1.80	0.33	1	50.0	45.0	42.9	41.7	45.6
163	2	3	2	2	2	0.66	0.00	1.00	10	52.1	48.9	48.5	44.4	60.5
164	1	4	3	1	2	0.06	0.03	1.00	9	62.7	64.5	71.3	55.3	70.5
165	1	4	1	1	3	0.46	-1.17	0.33	5	46.9	30.7	40.3	58.0	35.6
166	1	4	2	1	3	-0.40	0.88	0.33	3	54.8	45.0	45.9	58.0	50.6
167	2	4	3	1	2	0.06	0.03	0.67	10	54.8	54.1	47.1	49.8	50.6
168	2	4	2	1	1	-0.69	-1.13	0.67	1	46.9	54.1	45.4	44.4	45.6
169	2	4	2	1	2	1.16	1.19	1.00	10	70.7	64.5	72.2	66.1	55.6
170	1	4	1	1	1	0.30	-0.59	1.00	16	54.8	46.3	45.5	58.0	60.5
171	2	4	2	1	2	0.08	0.59	1.00	9	57.4	56.7	53.9	60.7	60.5
172	1	4	2	1	3	-1.30	0.03	1.00	7	54.8	41.1	40.2	44.4	40.6
173	2	4	1	1	1	-0.61	0.03	0.33	11	44.2	54.1	40.3	52.6	40.6
174	1	4	2	1	2	-0.16	0.65	1.00	7	68.0	61.9	69.7	71.5	60.5
175	2	4	2	1	2	1.16	0.65	1.00	1	65.4	67.1	63.1	55.3	70.5
176	2	4	3	1	1	0.91	0.65	0.33	11	54.8	48.9	41.9	52.6	55.6
177	1	4	2	1	2	0.68	0.03	0.67	10	71.2	63.2	60.2	65.5	65.5
178	1	4	2	1	2	0.06	-0.29	0.67	3	57.4	64.5	51.1	63.4	60.5
179	2	4	2	1	2	1.16	0.03	1.00	6	70.7	67.1	61.2	60.7	70.5
180	2	4	2	1	3	-0.59	0.03	0.33	14	33.6	54.1	41.0	41.7	50.6

ID	SEX	RACE	SES	SCTYP	HSP	LOCUS	CONCPT	MOT	CAR	RDG	WRTG	MATH	SCI	CIV
181	2	4	3	1	2	1.13	0.87	1.00	10	76.0	62.5	60.0	67.2	65.5
182	1	4	3	1	2	0.42	0.56	0.67	17	70.7	56.0	62.0	67.7	63.0
183	2	4	1	1	2	1.16	0.63	0.67	13	58.5	54.1	46.3	48.2	53.1
184	1	4	3	1	1	1.16	-0.81	0.67	9	65.4	64.5	48.0	63.4	65.5
185	2	4	2	1	1	0.45	0.56	1.00	5	52.1	43.7	41.9	47.1	45.6
186	1	4	2	1	3	-0.16	-0.53	0.67	5	62.7	61.9	67.0	66.1	45.6
187	2	4	1	1	1	0.68	-0.60	0.33	14	52.1	64.5	59.7	55.8	50.6
188	2	4	3	1	2	0.06	0.03	1.00	10	41.6	54.1	41.2	41.7	40.6
189	2	4	3	1	2	1.11	0.90	0.33	6	62.7	64.5	61.4	58.0	60.5
190	2	4	3	1	3	-0.41	0.03	0.67	15	49.5	51.5	45.1	58.0	35.6
191	1	4	3	1	2	0.49	0.94	1.00	16	54.8	56.7	63.8	60.7	65.5
192	2	4	2	1	2	0.67	0.03	0.67	9	52.1	56.7	51.1	55.3	55.6
193	2	4	2	1	2	0.51	0.03	1.00	14	54.8	54.1	66.4	41.7	55.6
194	1	4	3	1	1	-0.84	0.34	0.00	10	41.6	46.3	59.6	63.4	65.5
195	1	4	3	1	2	0.27	0.03	1.00	10	60.1	51.5	48.1	52.6	55.6
196	2	4	2	1	3	0.06	0.03	1.00	1	68.0	61.9	55.5	49.8	50.6
197	2	4	1	1	3	-1.15	-1.38	1.00	17	49.5	56.7	46.1	53.6	40.6
198	2	4	2	1	2	0.66	-0.47	0.67	9	54.8	59.3	50.6	60.7	50.6
199	1	4	1	1	3	-0.82	0.63	1.00	12	36.8	36.5	41.5	33.1	31.9
200	1	4	3	1	2	-0.27	0.88	1.00	13	52.1	64.5	60.6	60.7	45.6
201	2	4	2	1	3	0.31	0.63	1.00	1	38.9	56.7	46.3	58.0	55.6
202	1	4	2	1	2	1.13	-0.60	0.67	16	73.3	67.1	71.3	63.4	65.5
203	2	4	2	1	2	-0.44	-0.78	1.00	10	58.0	48.9	43.8	44.4	65.5
204	2	4	2	1	2	0.44	0.03	1.00	9	55.3	51.5	48.0	58.0	65.5
205	2	4	1	1	3	-0.90	-2.54	0.00	4	44.2	43.7	40.2	39.6	30.6
206	2	4	1	1	3	-0.18	0.28	0.33	4	44.2	51.5	46.1	47.1	50.6
207	2	4	2	1	1	1.11	0.90	0.33	4	55.3	50.2	41.7	58.5	55.6
208	1	4	1	1	1	0.05	-0.60	0.00	8	57.4	51.5	40.6	46.6	56.8
209	2	4	2	1	1	-0.63	0.44	0.33	4	41.6	43.7	46.8	36.3	50.6
210	2	4	2	1	1	0.06	0.03	1.00	9	54.8	59.3	46.2	47.1	65.5
211	2	4	2	1	1	0.71	-0.29	0.67	9	57.4	61.9	55.5	58.0	65.5
212	1	4	2	1	1	0.26	0.03	1.00	2	57.4	51.5	55.3	60.7	60.5
213	2	4	1	1	1	0.04	-0.47	0.67	1	33.6	59.3	44.7	47.1	45.6
214	2	4	2	1	3	-1.81	-0.64	1.00	15	41.6	54.1	46.9	47.1	45.6
215	2	4	2	1	2	-0.40	-0.22	0.67	9	52.1	38.5	36.1	44.4	45.6
216	2	4	3	1	2	0.68	0.03	1.00	13	65.4	67.1	64.5	60.7	70.5
217	2	4	2	1	2	0.00	-0.42	0.67	11	49.5	51.5	52.9	39.0	55.6
218	2	4	2	1	3	0.46	0.03	1.00	1	57.4	67.1	47.7	55.8	50.6
219	2	4	2	1	3	0.10	0.03	0.33	6	49.5	56.7	48.0	47.1	60.5
220	2	4	2	1	2	0.66	0.34	1.00	4	56.9	49.5	49.5	50.9	58.1
221	1	4	1	1	1	-0.36	-1.67	1.00	3	44.2	43.7	56.4	58.0	60.5
222	1	4	3	1	1	0.93	0.34	1.00	10	46.9	54.1	54.6	55.3	50.6
223	1	4	2	1	3	0.48	0.03	1.00	10	60.1	46.3	51.4	52.6	60.5
224	2	4	2	1	1	-1.15	-0.47	0.67	4	44.2	56.7	52.2	44.4	55.6
225	1	4	2	1	2	0.89	0.59	0.67	6	60.6	56.7	70.5	58.0	65.5
226	1	4	3	1	2	0.49	1.19	1.00	10	46.9	51.5	50.6	49.8	55.6
227	1	4	2	1	2	0.44	0.65	1.00	3	52.1	54.1	54.6	41.7	55.6
228	2	4	2	1	2	0.00	-0.82	0.67	10	54.8	67.1	63.8	49.8	50.6
229	2	4	1	1	2	0.91	0.03	1.00	1	60.6	59.3	49.4	44.4	65.5
230	2	4	3	1	1	-0.20	1.19	1.00	15	41.6	46.9	41.7	47.1	60.5
231	2	4	1	1	1	0.00	-1.09	0.67	9	62.7	51.5	45.9	47.1	45.6
232	2	4	2	1	2	0.20	-0.47	0.33	11	38.9	54.1	53.7	52.6	40.6
233	2	4	2	1	3	0.06	0.03	0.67	15	62.7	61.9	58.8	60.7	60.5
234	2	4	3	1	2	0.96	0.87	1.00	9	57.4	64.5	48.0	47.1	50.6
235	2	4	2	1	2	0.20	-0.55	0.67	9	60.1	61.9	48.7	49.8	50.6
236	1	4	2	1	2	0.08	0.94	0.67	10	73.3	61.9	60.5	71.5	55.6
237	1	4	2	1	2	1.36	-0.80	0.67	13	52.1	59.3	59.6	59.1	40.6
238	1	4	3	1	2	0.46	0.94	1.00	10	65.4	61.9	67.9	66.1	65.5
239	2	4	1	1	3	-0.84	-0.18	0.00	1	33.6	46.3	38.4	36.3	45.6
240	1	4	2	1	2	0.46	0.03	0.67	9	62.7	54.1	54.8	55.3	65.5

ID	SEX	RACE	SES	SCTYP	HSP	LOCUS	CONCPT	MOT	CAR	RDG	WRTG	MATH	SCI	CIV
241	1	4	1	1	2	0.48	0.69	0.67	16	62.7	43.0	58.9	65.0	44.3
242	1	4	3	1	2	0.75	1.19	1.00	10	68.0	61.9	52.2	60.7	65.5
243	2	4	2	1	2	-0.40	0.03	0.00	1	54.8	49.5	55.3	47.1	51.8
244	2	4	3	1	2	0.71	0.34	0.67	9	68.0	61.9	64.5	68.8	60.5
245	1	4	2	1	1	0.46	0.34	0.67	9	62.2	56.7	51.9	54.7	58.1
246	2	4	2	1	3	0.43	-0.47	1.00	1	42.6	50.8	44.6	39.0	55.6
247	2	4	2	1	3	0.46	-0.47	0.00	9	49.5	61.9	41.4	55.3	30.6
248	1	4	2	1	1	-0.49	0.03	0.33	9	65.4	59.3	56.8	66.1	60.5
249	2	4	2	1	2	-1.56	0.03	0.33	15	62.7	54.1	53.0	55.3	45.6
250	1	4	2	1	1	-0.17	-0.84	0.67	15	62.7	56.7	54.3	58.0	50.6
251	1	4	3	1	1	0.66	0.34	1.00	7	60.1	51.5	53.0	63.4	55.6
252	2	4	1	1	2	0.48	0.32	1.00	14	49.5	54.1	38.7	49.8	50.6
253	1	4	2	1	1	-0.14	-0.24	0.67	9	54.8	38.5	57.0	52.6	45.6
254	2	4	3	1	2	0.48	-0.47	0.33	9	52.1	61.9	55.5	60.7	60.5
255	1	4	2	1	2	-1.13	-0.55	0.33	12	44.2	46.9	45.5	39.0	50.6
256	1	4	3	1	1	0.06	0.03	0.00	10	44.2	33.3	53.9	58.0	30.6
257	1	4	2	1	1	0.20	-0.47	0.33	3	38.9	38.5	53.0	52.6	50.6
258	2	4	2	1	3	-0.64	-0.51	0.67	4	49.5	56.7	47.7	44.4	60.5
259	2	4	2	1	1	0.23	0.44	1.00	2	49.5	51.5	45.5	49.8	55.6
260	2	4	2	1	2	0.10	1.19	1.00	9	52.1	56.7	62.3	66.1	70.5
261	1	4	2	1	3	0.28	0.03	1.00	9	54.8	59.3	49.5	63.4	56.8
262	2	4	3	1	2	0.52	0.59	1.00	9	68.0	59.9	64.0	68.8	65.5
263	1	4	2	1	2	0.46	0.03	0.67	8	54.8	64.5	58.6	55.3	60.5
264	2	4	1	1	1	-0.80	0.15	0.33	15	41.6	41.1	39.5	47.1	60.5
265	2	4	1	1	3	-1.33	-0.60	0.33	1	34.7	35.2	40.2	50.9	33.1
266	2	4	1	1	1	-1.33	0.34	0.00	4	44.2	41.1	36.9	41.7	45.6
267	1	4	3	1	2	0.68	0.34	1.00	16	57.4	56.7	62.8	58.0	50.6
268	2	4	3	1	2	0.46	-0.55	0.33	10	41.6	54.1	50.3	49.8	51.8
269	1	4	2	1	3	-0.44	-0.22	0.00	2	44.2	33.3	40.9	47.1	45.6
270	2	4	2	1	2	0.23	-0.76	1.00	9	65.4	64.5	59.6	66.1	60.5
271	2	4	3	1	3	0.29	0.03	1.00	4	49.5	48.9	56.2	47.1	45.6
272	2	4	1	1	1	0.46	0.03	0.67	16	52.1	56.7	53.0	47.1	50.6
273	2	4	2	1	1	0.46	0.03	0.33	13	44.7	38.5	45.9	44.4	35.6
274	2	4	3	1	1	0.33	-0.26	1.00	4	52.1	67.1	57.0	63.4	60.5
275	2	4	3	1	1	-0.44	-0.47	1.00	9	54.8	65.1	66.1	49.8	60.5
276	1	4	2	1	3	-1.05	-0.34	0.67	1	44.2	38.5	41.4	42.3	35.6
277	2	4	3	1	2	1.11	0.34	1.00	10	73.3	67.1	62.3	58.0	65.5
278	1	4	1	1	2	0.46	-0.18	1.00	10	52.1	46.3	48.3	58.0	60.5
279	1	4	1	1	3	0.08	0.94	0.33	2	49.5	52.8	50.6	48.8	66.8
280	1	4	3	1	2	0.31	0.03	1.00	10	54.8	48.9	61.2	60.7	55.6
281	1	4	3	1	3	-0.18	0.03	0.67	2	44.2	48.9	56.3	41.7	60.5
282	2	4	2	1	1	0.71	-0.05	0.67	1	52.7	47.6	52.2	45.0	50.6
283	1	4	3	1	2	0.02	0.13	0.33	9	73.3	59.9	70.5	60.7	70.5
284	1	4	3	1	2	1.36	1.19	1.00	9	65.4	48.9	66.3	58.0	35.6
285	2	4	2	1	2	0.00	0.03	0.67	1	61.1	61.2	62.2	57.4	53.1
286	2	4	3	1	2	0.52	0.34	1.00	10	65.4	54.1	61.4	58.0	55.6
287	2	4	3	1	2	0.66	0.34	0.67	9	54.8	56.7	61.9	63.4	55.6
288	1	4	1	1	1	0.25	0.65	0.00	3	49.5	48.9	50.4	63.4	55.6
289	1	4	3	1	2	-0.38	0.03	1.00	9	60.1	54.1	58.0	58.0	60.5
290	2	4	3	1	1	0.46	-1.11	0.33	9	65.4	67.1	67.0	66.1	70.5
291	2	4	2	1	1	0.23	0.03	1.00	13	65.4	56.7	65.4	58.0	70.5
292	2	4	1	1	3	-1.09	-0.90	0.67	9	57.4	59.9	50.5	52.6	36.9
293	2	4	2	1	2	0.71	-0.55	1.00	9	60.1	59.3	64.3	60.7	65.5
294	1	4	1	1	2	-0.82	-0.76	0.00	1	57.4	43.7	59.6	52.6	50.6
295	2	4	3	1	1	1.36	0.03	0.67	14	46.9	59.3	56.3	66.1	60.5
296	2	4	3	1	3	-0.14	-0.86	0.00	9	38.9	46.3	50.1	36.3	45.6
297	2	4	3	1	2	0.46	1.19	0.67	10	65.9	64.5	61.8	60.7	60.5
298	2	4	3	1	2	0.06	0.32	1.00	1	62.7	64.5	70.5	68.8	70.5
299	1	4	3	1	3	-0.51	-1.98	0.00	17	60.1	57.3	58.8	58.0	50.6
300	1	4	3	1	2	0.68	0.03	0.33	9	73.3	64.5	64.0	60.7	65.5

ID	SEX	RACE	SES	SCTYP	HSP	LOCUS	CONCPT	MOT	CAR	RDG	WRTG	MATH	SCI	CIV
301	2	4	3	1	2	0.75	1.19	1.00	9	60.1	61.9	67.1	49.8	55.6
302	1	4	2	1	2	0.69	0.05	0.67	10	60.1	59.3	57.9	40.6	65.5
303	2	4	1	1	1	-0.84	-0.24	1.00	4	54.8	64.5	44.5	52.6	50.6
304	1	4	3	1	2	0.01	0.32	0.00	16	52.1	41.1	50.6	52.6	55.6
305	2	4	2	1	1	-0.16	-0.45	0.67	4	49.5	54.1	61.9	52.6	50.6
306	2	4	3	1	2	-0.60	0.28	1.00	9	33.6	41.1	42.9	41.7	50.6
307	1	4	3	1	2	0.46	0.34	1.00	9	76.0	52.1	64.1	63.9	60.5
308	1	4	1	1	2	0.00	0.65	0.67	7	46.9	33.3	50.4	47.1	55.6
309	2	4	3	1	1	0.32	0.90	0.67	10	33.6	51.5	41.9	49.8	50.6
310	1	4	3	1	2	0.52	0.65	0.33	2	54.3	62.5	56.6	54.7	45.6
311	1	4	3	1	3	0.46	0.63	0.67	2	46.9	52.8	49.3	53.1	35.6
312	2	4	2	1	2	0.66	0.03	1.00	1	52.1	56.7	41.9	52.6	60.5
313	2	4	2	1	2	0.46	-0.47	0.67	10	50.0	41.1	45.1	44.4	55.6
314	1	4	2	1	1	0.26	0.03	1.00	2	49.5	35.9	47.8	52.6	35.6
315	2	4	3	1	2	0.91	-0.47	0.67	10	57.4	59.3	57.2	59.1	65.5
316	2	4	2	1	3	1.36	0.87	1.00	4	36.3	43.7	37.2	41.7	40.6
317	2	4	2	1	2	0.02	-1.05	0.33	9	60.1	64.5	61.2	60.7	50.6
318	2	4	1	1	1	-0.44	-0.47	0.33	9	38.9	46.3	41.9	41.7	45.6
319	1	4	1	1	3	-0.44	-0.55	0.33	4	46.9	38.5	47.1	41.7	25.7
320	1	4	1	1	3	0.22	0.03	0.67	16	60.1	41.7	52.0	52.6	50.6
321	1	4	1	1	3	-0.91	-0.59	0.67	8	44.2	39.1	46.3	36.3	40.6
322	1	4	2	1	1	-0.84	-0.57	0.33	2	33.6	33.3	41.0	36.3	35.6
323	1	4	2	1	3	0.66	0.34	0.33	2	33.6	38.5	42.9	47.1	45.6
324	2	4	3	1	2	0.94	0.03	1.00	10	68.0	56.7	59.6	58.0	25.7
325	2	4	2	1	1	-0.65	0.03	0.33	9	41.6	39.1	42.3	41.7	40.6
326	1	4	1	1	1	1.13	0.03	0.33	5	45.8	41.7	43.1	53.6	36.9
327	2	4	1	1	3	1.36	0.63	1.00	1	68.0	64.5	58.9	63.4	45.6
328	2	4	2	1	3	0.22	0.03	0.33	1	52.1	54.7	49.5	53.6	60.5
329	1	4	3	1	2	-0.43	0.03	1.00	9	52.1	44.3	53.1	58.0	45.6
330	1	4	1	1	1	0.28	-0.06	0.33	2	44.2	30.7	35.3	47.1	45.6
331	2	4	1	1	1	-1.99	0.03	0.00	9	39.4	54.1	38.7	47.1	35.6
332	1	4	2	1	2	-0.40	0.03	0.00	10	44.2	33.3	47.8	44.4	45.6
333	1	4	2	1	1	-0.03	-0.16	0.00	6	33.6	43.7	42.6	47.1	35.6
334	2	4	2	1	2	0.08	-0.24	0.33	10	52.1	59.3	48.0	55.3	60.5
335	2	4	2	1	1	0.46	0.34	0.67	9	62.7	61.9	52.9	44.4	40.6
336	1	4	3	1	1	0.06	0.03	0.67	11	41.6	39.1	56.3	45.0	50.6
337	1	4	2	1	2	-0.39	-0.28	0.67	9	44.2	51.5	51.1	63.4	60.5
338	2	4	1	1	3	0.26	-0.26	0.00	6	49.5	56.7	43.8	47.1	40.6
339	1	4	2	1	2	0.46	0.03	1.00	7	70.7	56.7	51.3	71.5	55.6
340	1	4	3	1	2	0.47	0.01	1.00	10	68.0	64.5	61.5	55.3	60.5
341	2	4	3	1	2	0.71	0.03	1.00	14	68.0	61.9	72.2	66.1	56.8
342	2	4	2	1	2	-0.16	-0.47	0.67	9	36.3	61.9	52.2	39.0	45.6
343	1	4	1	1	1	-0.37	-0.47	0.67	9	41.6	48.9	42.6	49.8	55.6
344	2	4	1	1	1	0.27	-2.52	0.33	8	38.9	32.0	35.3	46.6	51.8
345	1	4	2	1	2	0.71	0.34	0.67	9	52.1	41.1	51.3	60.7	65.5
346	2	4	2	1	2	0.46	0.03	0.67	9	57.4	59.3	53.9	49.8	55.6
347	2	4	3	1	2	0.28	0.03	0.67	9	62.7	67.1	65.6	61.8	65.5
348	2	4	2	1	1	0.46	0.03	1.00	10	65.4	51.5	61.4	60.7	50.6
349	1	4	3	1	3	-1.50	0.03	0.67	9	33.6	48.9	38.6	42.3	55.6
350	1	4	3	1	2	-0.41	0.28	0.33	16	46.9	35.9	42.6	58.0	55.6
351	2	4	2	1	2	0.32	0.90	0.67	1	52.1	59.3	58.1	47.1	45.6
352	2	4	2	1	1	0.32	0.65	1.00	9	54.8	61.9	58.1	58.0	60.5
353	1	4	2	1	2	-0.21	-1.13	0.00	2	38.9	41.1	43.6	55.3	45.6
354	2	4	2	1	2	-0.60	-1.18	0.67	1	54.8	59.3	68.0	49.3	65.5
355	2	4	1	1	1	-0.40	-0.89	0.33	8	41.6	56.7	45.0	50.4	43.1
356	2	4	1	1	3	-0.89	-1.29	0.33	9	60.1	64.5	41.0	52.6	50.6
357	2	4	2	1	3	0.06	-0.28	1.00	1	65.9	64.5	54.6	60.7	60.5
358	2	4	3	1	2	0.45	0.03	1.00	4	62.7	51.5	54.4	49.8	50.6
359	1	4	2	1	2	0.22	0.32	1.00	14	49.5	56.7	62.8	49.8	70.5
360	2	4	3	1	2	0.71	0.28	1.00	11	60.1	64.5	63.0	68.8	60.5

ID	SEX	RACE	SES	SCTYP	HSP	LOCUS	CONCPT	MOT	CAR	RDG	WRTG	MATH	SCI	CIV
361	2	4	2	1	3	0.91	0.34	1.00	1	68.0	59.3	53.0	63.4	60.5
362	1	4	2	1	1	0.00	0.65	1.00	6	54.8	64.5	51.1	49.8	55.6
363	1	4	1	1	3	-0.38	0.03	1.00	8	38.9	41.7	37.7	44.4	45.6
364	2	4	2	1	2	-0.16	-2.58	1.00	10	52.1	59.3	58.1	52.6	65.5
365	2	4	1	1	1	-1.10	-0.28	0.67	4	38.9	41.1	45.9	47.1	45.6
366	2	4	2	1	1	0.93	0.03	1.00	10	65.9	67.1	70.5	52.6	65.5
367	1	4	3	1	1	0.70	-0.16	0.33	9	68.0	59.3	55.7	63.4	65.5
368	1	4	2	1	3	-0.45	-0.26	0.00	2	38.9	35.9	44.2	55.3	55.6
369	2	4	2	1	3	0.02	-2.03	0.67	1	38.9	59.3	39.5	39.0	50.6
370	2	4	1	1	3	-0.19	-2.58	0.00	1	41.6	54.1	41.0	41.7	40.6
371	1	4	2	1	1	0.91	-0.47	0.67	8	44.2	41.1	50.5	58.0	40.6
372	1	4	2	1	1	0.09	-1.42	0.33	11	52.1	61.9	59.6	58.0	55.6
373	2	4	1	1	2	0.00	0.65	1.00	16	68.0	64.5	58.3	58.5	55.6
374	1	4	1	1	2	0.70	-0.16	1.00	10	49.5	61.9	50.9	46.1	66.8
375	1	4	1	1	1	-0.60	0.56	0.33	2	36.3	43.7	44.3	47.1	30.6
376	1	4	2	1	1	-1.33	0.03	0.67	2	41.6	30.7	56.9	47.1	50.6
377	2	4	2	1	2	-0.16	0.03	1.00	8	57.4	57.3	64.0	58.0	65.5
378	1	4	3	1	2	0.46	0.65	1.00	9	68.0	54.1	63.0	63.4	60.5
379	1	4	3	1	2	-0.33	0.38	0.67	6	62.7	59.3	56.5	55.3	55.6
380	2	4	3	1	1	0.25	0.03	0.67	4	49.5	51.5	55.5	44.4	55.6
381	2	4	2	1	2	0.22	-0.57	1.00	1	61.1	67.1	64.5	50.4	55.6
382	1	4	3	1	2	0.22	1.19	0.33	13	38.9	33.3	37.7	47.1	40.6
383	2	4	2	1	1	0.27	0.03	0.67	1	46.9	56.7	63.8	55.3	55.6
384	1	4	2	1	3	-2.23	1.19	0.33	5	44.2	38.5	41.2	49.8	45.6
385	2	4	1	1	1	-0.38	0.37	0.67	4	44.2	43.7	46.1	47.1	50.6
386	2	4	2	1	2	0.06	0.65	0.33	11	47.4	64.5	53.9	55.8	50.6
387	2	4	3	1	2	0.89	0.88	1.00	9	52.1	59.3	64.0	61.2	55.6
388	2	4	3	1	1	1.13	0.03	0.33	9	43.2	54.1	54.8	55.3	45.6
389	2	4	1	1	1	-1.02	-1.67	0.67	5	45.3	43.0	52.3	60.7	56.8
390	2	4	3	1	2	0.45	0.03	0.67	9	60.1	61.9	51.9	53.1	58.1
391	2	4	3	1	2	0.06	0.65	0.00	9	65.4	59.3	56.8	46.1	65.5
392	1	4	1	1	3	-1.74	0.61	0.33	17	45.8	34.6	37.9	39.0	45.6
393	1	4	2	1	3	-1.30	0.13	0.33	11	44.2	41.1	51.8	47.1	40.6
394	1	4	2	1	2	0.46	0.03	0.00	8	73.3	61.9	73.1	68.8	65.5
395	2	4	2	1	2	0.55	-0.24	0.67	13	46.9	56.7	54.6	47.1	55.6
396	2	4	2	1	3	-0.68	0.03	0.33	17	38.9	38.5	41.2	39.0	35.6
397	1	4	2	1	3	0.25	-0.26	0.33	2	49.5	30.7	40.3	33.6	30.6
398	2	4	1	1	3	0.96	0.94	0.67	9	46.9	50.2	46.5	47.1	51.8
399	2	4	1	1	1	-0.24	0.03	0.67	9	45.8	54.1	49.5	47.1	55.6
400	1	4	1	1	1	-0.40	0.65	0.67	5	44.2	43.7	38.6	33.6	45.6
401	2	4	1	1	2	-0.40	0.03	1.00	15	44.2	54.1	59.3	58.0	50.6
402	2	4	2	1	2	0.93	0.65	1.00	6	62.7	59.3	71.3	68.8	60.5
403	2	4	1	1	2	0.00	0.34	1.00	15	62.7	59.9	65.4	53.6	65.5
404	2	4	2	1	1	-0.38	-0.47	0.67	4	62.7	43.7	44.7	52.6	41.9
405	1	4	2	1	2	0.20	0.56	0.67	12	60.1	52.8	51.5	58.0	50.6
406	1	4	2	1	2	0.73	0.03	1.00	16	65.4	59.3	69.6	63.4	50.6
407	2	4	2	1	3	0.06	0.03	1.00	4	43.2	47.6	45.6	47.1	45.6
408	1	4	3	1	2	-0.45	-0.60	0.67	10	52.1	59.9	49.5	48.8	60.5
409	2	4	3	1	2	1.13	0.03	0.67	9	62.7	64.5	64.5	52.6	60.5
410	2	4	1	1	2	-0.18	-1.09	0.67	15	62.7	56.7	58.8	71.5	60.5
411	1	4	3	1	2	0.32	0.28	0.67	16	62.7	63.2	68.9	74.2	70.5
412	2	4	2	1	3	0.66	-0.60	0.00	15	42.6	56.7	40.3	49.8	50.6
413	2	4	2	1	1	0.51	0.03	0.33	6	60.1	51.5	53.9	63.4	50.6
414	2	4	2	1	1	0.08	0.03	1.00	13	54.8	67.1	47.4	49.8	65.5
415	2	4	2	1	2	0.46	-0.86	1.00	14	68.0	59.3	60.5	55.3	70.5
416	1	4	1	1	2	1.11	-0.09	0.00	9	44.2	48.9	48.0	49.8	55.6
417	2	4	2	1	1	0.07	0.34	1.00	9	65.4	59.3	51.2	60.7	55.6
418	1	4	2	1	3	-0.89	0.56	0.33	2	44.2	41.1	40.3	49.8	25.7
419	1	4	2	1	3	-0.19	0.03	0.33	2	54.8	51.5	42.8	60.7	50.6
420	1	4	1	1	2	-0.27	0.94	0.00	17	44.7	54.1	50.4	45.0	50.6

ID	SEX	RACE	SES	SCTYP	HSP	LOCUS	CONCPT	MOT	CAR	RDG	WRTG	MATH	SCI	CIV
421	1	4	3	1	3	-0.40	-1.09	0.67	7	62.7	43.7	47.1	52.6	55.6
422	2	4	2	1	2	0.00	0.03	0.67	9	62.7	64.5	38.3	55.8	55.6
423	1	4	2	1	2	0.46	-0.47	0.67	10	68.0	64.5	70.6	68.8	59.3
424	2	4	2	1	3	0.75	0.34	1.00	9	47.4	41.7	51.6	38.5	50.6
425	2	4	1	1	1	1.36	0.65	1.00	15	46.9	51.5	57.0	49.8	40.6
426	1	4	1	1	2	0.68	0.87	1.00	10	54.8	56.7	56.9	58.0	70.5
427	1	4	2	1	3	0.05	-0.88	0.67	2	62.7	63.2	74.6	71.5	65.5
428	2	4	2	1	3	-0.84	0.03	0.00	9	36.3	59.3	47.7	39.0	55.6
429	1	4	3	1	3	1.13	0.56	1.00	6	60.1	48.9	50.4	55.3	65.5
430	1	4	3	1	1	1.13	0.63	0.67	10	60.1	64.5	57.9	60.7	65.5
431	2	4	2	1	1	0.55	-0.60	1.00	1	54.8	54.1	51.3	41.7	45.6
432	1	4	3	1	1	0.06	0.03	1.00	9	44.2	54.1	51.4	39.0	45.6
433	2	4	2	1	3	-0.64	0.03	1.00	4	41.6	46.3	38.4	36.3	45.6
434	2	4	3	1	2	0.48	0.03	1.00	4	65.4	61.9	60.4	63.4	60.5
435	2	4	1	1	1	-0.19	-0.85	0.00	15	54.8	60.6	55.5	58.0	45.6
436	1	4	3	1	2	0.20	-2.54	0.00	16	54.8	61.9	63.6	63.4	65.5
437	2	4	2	1	3	-0.60	-0.26	0.67	4	49.5	41.1	38.4	52.6	35.6
438	1	4	2	1	2	-1.33	0.03	1.00	16	36.3	28.1	53.7	33.6	25.7
439	2	4	1	1	2	0.68	0.88	1.00	9	46.9	61.9	53.0	52.6	60.5
440	2	4	3	1	2	1.36	0.94	1.00	4	52.1	48.9	51.3	41.7	45.6
441	2	4	2	1	1	0.89	0.65	0.67	9	68.0	48.9	56.3	63.4	55.6
442	2	4	2	1	3	-0.14	0.13	0.67	9	41.6	56.7	51.3	47.1	60.5
443	1	4	2	1	3	-0.43	-1.22	0.00	5	41.6	46.9	55.5	44.4	40.6
444	1	4	2	1	2	0.04	0.03	0.67	9	65.4	51.5	61.2	68.8	60.5
445	1	4	1	1	1	0.44	-0.16	0.67	9	62.7	48.9	48.8	66.1	45.6
446	1	4	1	1	2	0.65	-0.30	1.00	10	60.1	61.9	74.6	58.0	60.5
447	1	4	2	1	3	-1.33	0.65	0.00	2	40.5	38.5	47.1	44.4	30.6
448	1	4	2	1	3	0.68	1.19	0.33	14	41.6	41.1	57.2	71.5	30.6
449	1	4	2	1	3	0.10	0.03	0.00	3	62.7	48.9	42.1	66.1	70.5
450	2	4	3	1	2	0.50	0.03	0.67	17	68.0	59.3	58.8	66.1	65.5
451	2	4	2	1	3	0.02	-0.32	1.00	4	46.9	46.3	52.0	47.7	45.6
452	2	4	2	1	1	0.06	0.03	1.00	17	41.6	46.3	46.2	39.0	45.6
453	2	4	2	1	1	-0.83	0.65	0.33	4	44.2	59.3	55.5	36.3	55.6
454	2	4	3	1	2	0.91	0.15	0.00	9	54.8	56.7	55.5	58.0	60.5
455	2	4	1	1	1	0.06	0.59	1.00	1	52.1	51.5	45.4	39.0	50.6
456	1	4	2	1	3	-1.57	0.65	0.33	6	52.1	56.7	45.2	55.3	50.6
457	1	4	2	1	3	-1.33	0.03	0.33	5	38.9	30.7	40.4	39.0	50.6
458	1	4	2	1	2	0.46	0.65	1.00	16	49.5	48.9	60.5	55.3	55.6
459	2	4	2	1	1	0.03	0.03	1.00	11	41.6	43.0	44.5	39.0	50.6
460	1	4	3	1	2	-0.93	-0.22	0.67	10	65.4	64.5	65.5	60.7	65.5
461	1	4	1	1	2	0.24	-0.43	0.33	16	70.7	43.7	58.8	66.1	60.5
462	1	4	2	1	1	0.33	0.28	0.67	9	52.1	46.3	48.1	55.3	40.6
463	1	4	2	1	1	-0.11	0.25	1.00	9	44.2	44.3	45.6	39.0	50.6
464	2	4	2	1	2	0.94	-0.30	1.00	10	60.1	67.1	52.4	55.3	50.6
465	1	4	3	1	1	0.75	-0.49	0.33	16	41.6	41.1	41.2	44.4	35.6
466	2	4	2	1	2	0.30	-0.60	1.00	9	49.5	59.3	53.1	60.7	60.5
467	1	4	2	1	1	1.13	0.03	1.00	6	38.9	54.1	47.5	49.8	40.6
468	1	4	3	1	1	-0.16	0.03	0.33	11	44.2	59.3	49.5	55.3	45.6
469	1	4	2	1	1	0.46	0.65	1.00	16	68.0	59.3	57.9	74.2	65.5
470	2	4	3	1	2	1.36	0.94	1.00	10	57.4	64.5	49.9	55.3	55.6
471	2	4	3	1	2	0.46	0.32	0.67	16	49.5	54.1	47.8	60.7	50.6
472	2	4	2	1	1	-0.40	-0.26	0.33	15	52.1	54.1	55.3	52.6	50.6
473	2	4	2	1	3	0.06	0.28	1.00	13	60.1	59.3	56.2	60.7	60.5
474	2	4	1	1	3	-0.61	0.36	0.00	4	44.2	48.9	46.1	47.1	55.6
475	1	4	3	1	2	0.71	0.03	1.00	10	54.8	61.2	53.7	48.8	60.5
476	2	4	3	1	1	0.23	-0.26	0.00	9	52.1	61.9	56.3	58.0	55.6
477	2	4	2	1	2	0.46	-0.28	1.00	9	57.4	56.7	56.4	63.4	60.5
478	2	4	2	1	2	0.67	-0.59	0.67	9	54.8	64.5	55.3	49.8	60.5
479	2	4	2	1	2	0.46	0.03	1.00	10	60.1	64.5	56.3	63.4	60.5
480	1	4	1	1	3	-0.88	-0.29	0.33	8	44.2	54.1	41.0	49.8	45.6

ID	SEX	RACE	SES	SCTYP	HSP	LOCUS	CONCPT	MOT	CAR	RDG	WRTG	MATH	SCI	CIV
481	2	4	1	1	2	0.89	0.34	0.33	10	57.4	61.9	72.2	60.7	60.5
482	2	4	2	1	3	0.06	-1.07	0.00	9	38.9	48.9	45.2	44.4	50.6
483	2	4	1	1	3	0.06	-0.60	0.67	9	62.7	56.0	57.3	60.7	56.8
484	2	4	2	1	3	-0.22	-2.27	0.00	1	57.4	51.5	40.3	60.7	55.6
485	2	4	2	1	2	0.71	0.34	0.00	2	62.7	64.5	57.4	60.7	70.5
486	2	4	2	1	2	0.53	0.81	0.67	9	54.8	59.3	61.4	47.1	55.6
487	2	4	1	1	2	-0.47	-0.47	0.00	1	52.1	57.3	63.7	58.0	55.6
488	1	4	3	1	3	0.43	-0.05	0.67	12	54.8	45.6	51.8	60.7	45.6
489	2	4	2	1	2	-0.44	-1.13	1.00	9	54.8	61.9	69.6	60.7	55.6
490	1	4	2	1	3	-1.28	0.34	0.33	9	31.0	35.9	46.1	39.0	45.6
491	2	4	2	1	1	-0.14	0.56	1.00	4	54.8	61.9	54.6	47.1	45.6
492	2	4	1	1	2	-0.14	-1.05	1.00	9	70.7	65.1	66.4	63.4	60.5
493	1	4	1	1	3	-0.86	0.28	1.00	3	36.3	48.9	54.4	60.7	35.6
494	2	4	1	1	3	0.30	0.03	0.67	1	36.8	59.3	40.7	49.8	35.6
495	1	4	2	1	1	0.68	0.03	1.00	10	65.4	64.5	51.3	66.1	65.5
496	2	4	2	1	2	0.23	0.03	1.00	9	52.1	59.3	52.9	60.7	50.6
497	2	4	2	1	3	0.00	-0.18	0.00	9	62.7	59.3	55.5	45.5	46.8
498	1	4	2	1	3	0.26	0.03	1.00	8	62.7	59.3	53.2	63.4	55.6
499	1	4	2	1	1	-1.58	-0.26	0.33	7	62.7	48.9	35.3	66.1	40.6
500	2	4	1	1	3	0.06	0.28	1.00	3	65.9	63.2	50.3	60.1	55.6
501	1	4	2	1	2	0.91	0.59	1.00	10	65.4	67.1	67.1	66.1	61.8
502	1	4	2	1	2	0.28	0.32	1.00	10	52.1	54.1	56.5	55.3	50.6
503	1	4	3	1	2	-0.19	-0.73	0.67	10	73.3	60.6	64.7	66.1	65.5
504	1	4	2	1	2	1.36	0.63	1.00	10	62.7	61.9	68.6	68.8	55.6
505	1	4	1	1	1	0.71	1.19	0.33	3	54.8	59.3	62.5	68.8	45.6
506	2	4	2	1	2	0.25	-0.47	0.33	10	76.0	67.1	72.2	66.1	65.5
507	1	4	2	1	3	-0.40	-0.76	0.67	7	52.1	35.9	50.3	47.1	50.6
508	1	4	3	1	2	0.68	-0.47	0.33	7	46.9	61.9	60.5	68.8	65.5
509	1	4	2	1	3	-0.41	0.34	0.67	16	52.1	38.5	57.9	52.6	50.6
510	1	4	2	1	3	-0.86	1.19	0.33	8	33.6	28.1	31.8	39.6	35.6
511	1	4	2	1	2	-0.44	1.19	1.00	2	60.1	54.1	59.6	55.3	65.5
512	1	4	1	1	3	-1.10	0.03	1.00	8	38.9	38.5	42.8	41.7	40.6
513	1	4	3	1	2	0.06	0.32	0.67	12	49.5	43.7	55.5	68.8	65.5
514	1	4	2	1	2	0.06	0.03	0.00	11	46.9	51.5	57.2	52.6	60.5
515	1	4	2	1	2	0.68	-0.26	1.00	10	62.7	61.9	56.2	47.1	45.6
516	1	4	2	1	2	0.66	0.03	0.33	6	65.4	59.3	67.1	66.1	65.5
517	2	4	1	1	1	0.26	0.87	1.00	3	49.5	61.9	61.4	63.4	50.6
518	1	4	2	1	1	0.06	0.03	1.00	2	54.8	41.1	42.8	47.1	40.6
519	1	4	2	2	2	0.32	1.19	1.00	9	57.4	61.9	64.0	55.8	55.6
520	1	4	2	2	2	0.06	0.03	0.33	9	68.0	59.3	71.3	66.1	55.6
521	1	4	3	2	2	0.46	0.34	1.00	16	52.1	46.3	50.4	52.6	50.6
522	1	4	2	2	2	0.00	0.65	1.00	7	52.1	61.9	62.1	58.0	60.5
523	2	4	3	2	1	0.68	0.32	1.00	15	36.3	56.7	41.9	49.8	40.6
524	2	4	2	2	2	-0.66	-1.07	0.67	1	49.5	61.9	60.4	47.1	50.6
525	2	4	3	2	2	0.48	-0.82	0.33	9	62.7	64.5	58.2	61.8	61.8
526	1	4	2	2	2	0.71	0.03	0.33	11	46.9	46.9	41.0	41.7	50.6
527	2	4	2	2	2	1.36	-1.18	1.00	9	62.7	61.9	63.1	60.7	60.5
528	2	4	3	2	2	-0.19	1.19	1.00	9	60.1	61.9	54.6	60.7	65.5
529	1	4	2	2	2	0.55	0.63	1.00	16	54.8	59.3	62.3	58.0	50.6
530	2	4	3	2	2	0.26	0.94	1.00	10	60.1	59.3	51.8	53.1	55.6
531	2	4	3	2	2	0.68	0.06	0.67	1	49.5	51.5	41.2	41.7	45.6
532	1	4	2	2	3	0.00	0.34	0.33	2	46.9	56.7	57.2	58.0	45.6
533	2	4	1	2	2	-0.17	0.03	0.67	15	44.2	59.3	45.7	55.3	50.6
534	1	4	2	2	2	0.06	0.56	1.00	10	49.5	56.7	46.2	55.3	50.6
535	2	4	2	2	2	0.52	0.34	1.00	10	46.9	64.5	59.6	49.8	55.6
536	2	4	3	2	2	0.91	-1.67	0.33	9	57.4	54.1	54.8	60.7	60.5
537	1	4	2	2	1	-0.40	0.65	0.00	10	52.1	46.3	51.3	58.0	55.6
538	2	4	3	2	2	0.46	-0.22	1.00	10	70.7	64.5	68.7	58.0	70.5
539	2	4	2	2	2	-0.40	-1.34	1.00	8	57.4	61.9	55.5	49.8	55.6
540	1	4	2	2	2	0.71	1.19	0.67	2	54.8	48.9	52.4	58.0	60.5

ID	SEX	RACE	SES	SCTYP	HSP	LOCUS	CONCPT	MOT	CAR	RDG	WRTG	MATH	SCI	CIV
541	1	4	2	2	2	-0.14	0.34	0.33	9	49.5	46.3	44.8	58.0	60.5
542	2	4	2	2	2	0.52	0.03	1.00	14	46.9	51.5	53.7	41.7	40.6
543	2	4	3	2	2	-0.47	0.28	1.00	10	65.4	61.9	47.1	60.7	60.5
544	2	4	2	2	2	0.96	0.03	0.67	11	44.2	51.5	43.2	44.4	50.6
545	2	4	2	2	2	0.91	-0.47	1.00	9	46.9	64.5	55.8	61.8	60.5
546	2	4	3	2	2	0.02	-0.14	0.00	9	60.1	64.5	67.0	58.5	66.8
547	1	4	2	2	2	0.47	0.03	0.67	13	62.7	59.3	48.8	55.3	70.5
548	2	4	2	2	2	0.47	0.34	0.67	10	33.6	43.0	41.0	49.8	35.6
549	2	4	3	2	2	0.46	0.03	1.00	9	41.6	64.5	47.1	53.1	60.5
550	2	4	2	2	3	1.13	0.88	0.33	1	49.5	51.5	52.7	55.3	55.6
551	1	4	2	2	2	0.96	0.63	1.00	12	65.4	64.5	70.3	66.1	65.5
552	2	4	2	2	2	0.50	-0.22	1.00	9	54.8	61.9	61.3	60.7	50.6
553	1	4	2	2	2	0.71	0.28	0.67	16	62.7	56.7	54.7	58.0	40.6
554	2	4	1	2	2	-0.14	-0.60	0.33	15	44.2	48.9	54.3	49.8	45.6
555	1	4	3	2	2	0.06	1.19	1.00	11	65.4	64.5	67.9	63.4	55.6
556	2	4	2	2	2	0.87	0.87	0.67	10	57.4	61.9	62.8	55.3	40.6
557	2	4	2	2	2	1.16	1.19	1.00	16	60.1	59.3	55.3	58.0	45.6
558	2	4	2	2	3	0.46	0.56	0.67	9	44.2	54.1	52.2	44.4	45.6
559	2	4	2	2	1	0.27	0.03	0.67	4	57.4	41.1	57.0	55.3	51.8
560	2	4	1	2	2	-0.19	-1.73	1.00	1	44.7	47.6	50.3	39.0	50.6
561	2	4	3	2	2	0.68	0.03	1.00	6	44.2	48.9	46.2	47.7	45.6
562	2	4	3	2	2	0.23	0.03	1.00	1	62.7	61.9	55.5	55.3	60.5
563	2	4	3	2	2	0.91	0.03	1.00	9	57.4	59.3	55.5	55.3	50.6
564	1	4	3	2	2	0.32	-0.88	0.33	9	57.4	54.1	42.2	44.4	55.6
565	1	4	2	2	2	0.71	0.03	0.33	5	46.9	59.3	63.0	52.6	45.6
566	1	4	3	2	2	1.11	0.34	1.00	10	65.4	64.5	64.1	66.1	70.5
567	2	4	2	2	2	0.26	-0.57	1.00	10	41.6	58.6	54.6	55.3	60.5
568	2	4	2	2	2	0.00	0.34	0.33	9	46.9	59.3	53.7	58.0	45.6
569	1	4	1	2	2	0.20	-0.47	0.67	2	56.4	49.5	51.9	47.1	48.1
570	2	4	2	2	2	-0.84	-2.60	0.33	15	44.2	64.5	53.0	55.3	60.5
571	2	4	3	2	2	1.16	0.09	1.00	9	46.9	52.1	42.5	47.7	60.5
572	1	4	2	2	3	0.23	0.15	0.33	2	49.5	51.5	48.0	52.6	50.6
573	2	4	1	2	1	-0.23	0.69	0.67	15	46.9	61.9	48.0	39.0	50.6
574	1	4	3	2	2	0.03	0.56	0.67	2	65.4	67.1	63.0	66.1	70.5
575	2	4	2	2	1	0.68	0.03	0.00	10	44.2	35.9	43.6	47.1	40.6
576	2	4	2	2	3	0.43	-1.09	0.67	14	63.3	64.5	52.1	58.0	65.5
577	2	4	2	2	2	0.47	0.34	0.67	16	44.2	48.9	48.0	39.0	50.6
578	2	4	2	2	3	-0.59	0.28	1.00	14	46.9	51.5	48.5	49.8	50.6
579	2	4	1	2	3	-0.38	0.03	1.00	9	45.3	38.5	43.5	40.1	45.6
580	2	4	3	2	2	0.73	0.03	1.00	9	62.7	62.5	68.9	60.7	60.5
581	1	4	2	2	3	-0.66	-0.47	1.00	9	52.7	41.1	40.2	58.0	60.5
582	1	4	3	2	2	0.10	-0.16	0.33	9	49.5	59.3	51.0	47.1	45.6
583	1	4	2	2	1	-1.33	0.03	0.67	16	57.4	56.7	59.8	58.0	55.6
584	1	4	3	2	2	0.22	1.19	1.00	11	57.4	54.1	48.8	55.3	40.6
585	1	4	3	2	2	0.10	0.56	0.67	9	54.8	61.9	60.4	55.3	65.5
586	1	4	3	2	2	0.71	-0.06	0.67	9	44.2	37.8	49.2	39.0	45.6
587	1	4	3	2	2	0.68	-0.53	1.00	11	48.4	48.2	60.5	60.1	49.3
588	1	4	2	2	1	-0.93	-0.80	0.33	9	52.1	51.5	49.5	55.3	45.6
589	1	4	3	2	2	-0.60	0.34	0.00	2	49.5	41.7	50.3	36.3	60.5
590	1	4	2	2	2	0.06	0.03	0.67	16	46.9	56.7	56.1	60.7	55.6
591	2	4	3	2	2	1.14	0.63	1.00	15	55.3	53.4	44.5	53.6	48.1
592	2	4	3	2	2	0.06	0.03	0.67	14	46.9	60.6	51.3	63.4	30.6
593	1	4	2	2	2	-1.09	-0.26	0.33	11	44.2	41.1	45.1	47.1	45.6
594	1	4	3	2	2	0.02	0.28	0.67	9	44.2	51.5	52.9	52.6	45.6
595	1	4	3	2	2	0.00	0.03	0.33	10	52.1	59.3	50.4	60.7	60.5
596	1	4	3	2	2	0.27	-1.05	0.33	9	60.1	54.1	56.3	55.3	55.6
597	1	4	2	2	2	0.55	0.34	1.00	16	62.7	61.9	72.9	63.4	55.6
598	1	4	2	2	2	-0.61	0.34	0.67	16	68.0	54.1	74.6	66.1	65.5
599	2	4	2	2	2	0.23	0.94	1.00	10	57.4	67.1	57.9	60.7	55.6
600	1	4	2	2	2	0.02	0.03	0.33	9	62.7	54.1	64.7	58.0	50.6

REFERENCES

Anderson, T. W. (1958). *An introduction to multivariate statistical analysis*. New York: Wiley.

Astin, H. S., and Myint, T. (1971). Career development of young women during the post-high school years. *Journal of Counseling Psychology, 18,* 369–393. (Monograph).

Bartlett, M. S. (1947). Multivariate analysis. *Journal of the Royal Statistical Society, Series B, 9,* 176–197.

Bartlett, M. S. (1951). A further note on tests of significance in factor analysis. *British Journal of Psychology, Statistical Section, 4,* 1–2.

Bean, J. P. (1985). Interaction effects based on class level in an explanatory model of college student dropout syndrome. *American Educational Research Journal, 22,* 35–64.

Beckenbach, E. F., Drooyan, I., and Wooton, W. (1978). *College algebra* (4th ed.). Belmont, CA: Wadsworth.

Bentler, P. M. (1980). Multivariate analysis with latent variables: Causal modeling. *Annual Review of Psychology, 31,* 419–456.

Bentler, P. M. (1985). *Theory and implementation of EQS: A structural equations program*. Los Angeles: BMDP Statistical Software, Inc.

Biniaminov, I., and Glasman, N. S. (1983). School determinants of student achievement in secondary education. *American Educational Research Journal, 20,* 251–268.

Blalock, H. M., Jr. (1964). *Causal inferences in nonexperimental research*. Chapel Hill: University of North Carolina Press.

Blalock, H. M., Jr. (ed.). (1971). *Causal models in the social sciences*. Chicago: Aldine-Atherton.

Blalock, H. M., Jr. (1979). *Social statistics*. New York: McGraw-Hill.

Bock, R. D. (1963). Programming univariate and multivariate analysis of variance. *Technometrika, 5,* 95–117.

Bock, R. D. (1965). A computer program for univariate and multivariate analysis of variance. In *Proceedings of the IBM scientific computing symposium on statistics*. White Plains, NY: IBM Data Processing Division.

Bock, R. D. (1966). Contributions of multivariate statistical methods to educational research. In R. B. Cattell (Ed.), *Handbook of multivariate experimental psychology*. Skokie, IL: Rand McNally.

Bock, R. D. (1975). *Multivariate statistical methods in behavioral research*. New York: McGraw-Hill.

Bock, R. D., and Haggard, E. A. (1968). The use of multivariate analysis of variance in behavioral research. In D. K. Whitla (Ed.), *Handbook of measurement and assessment in behavioral sciences*. Reading, MA: Addison-Wesley.

Boli, J., Allen, M. L., and Payne, A. (1985). High-ability women and men in undergraduate mathematics and chemistry courses. *American Educational Research Journal, 22,* 605–626.

Box, G. E. P. (1949). A general distribution theory for a class of likelihood criteria. *Biometrika, 36,* 317–346.

Brown, G. W. (1947). Discriminant functions. *Annals of Mathematical Statistics, 18,* 514–528.

Bryan, J. G. (1950). *A method for the exact determination of the characteristic equation of a matrix with applications to the discriminant functions for more than two groups*. Unpublished doctoral dissertation, Harvard University.

Carroll, J. B. (1953). An analytical solution for approximating simple structure in factor analysis. *Psychometrika, 18,* 23–28.

Carter, L. F. (1984). The sustaining effects study of compensatory and elementary education. *Educational Researcher, 13,* (7), 4–13.

Cattell, R. B. (1949a). r_p and other coefficients of pattern similarity. *Psychometrika, 14,* 279–298.

Cattell, R. B. (1949b). The sixteen personality factor questionnaire. Champaign, IL: Institute for Personality and Ability Testing.

Cattell, R. B. (1966). The scree test for the number of factors. *Multivariate Behavioral Research, 1,* 245–276.

Cattell, R. B., and Coulter, M. A. (1966). Principles of behavioral taxonomy and the mathematical basis of the taxonome computer program. *The British Journal of Mathematical and Statistical Psychology, 19,* 237–269.

Cattell, R. B., and Eber, H. W. (1970). *The sixteen personality factor questionnaire*. Champaign, IL: Institute for Personality and Ability Testing.

Clyde, D. J., Cramer, E. M., and Sherin, R. J. (1966). *Multivariate statistical programs*. Coral Gables, FL: Biometric Laboratory of the University of Miami.

Cooley, W. W., and Lohnes, P. R. (1962). *Multivariate procedures for the behavioral sciences*. New York: John Wiley.

Cooley, W. W., and Lohnes, P. R. (1968). *Predicting development of young adults*. Pittsburgh: University of Pittsburgh, Project TALENT.

Cooley, W. W., and Lohnes, P. R. (1971). *Multivariate data analysis*. New York: Wiley.

Cooley, W. W., and Lohnes, P. R. (1976). *Evaluation Research in Education*. New York: Irvington-Halsted-Wiley.

Cooper, R. A., and Weeks, A. J. (1983). *Data, models and statistical analysis*. Totowa, NJ: Barnes and Noble.

Cover, T. M., and Hart, P. E. (1967). Nearest neighbor pattern classification. *Institute of Electrical and Electronics Engineers Transactions: Information Theory, IT13,* 21–27.

Cronbach, L. J., and Gleser, G. C. (1953). Assessing profile similarity. *Psychological Bulletin, 50,* 456–473.

Dixon, W. J. (ed.). (1985) *BMDP statistical software*. Berkeley: University of California Press.

Du Mas, F. M. (1949). The coefficient of profile similarity. *Journal of Clinical Psychology, 5,* 121–131.

Duncan, O. D. (1966). Path analysis: Sociological examples. *American Journal of Sociology, 72,* 1–16.

Duncan, O. D. (1975). *Introduction to structural equations models*. New York: Academic Press.

Dwyer, P. S., and MacPhail, M. S. (1948). Symbolic matrix derivatives. *Annals of Mathematical Statistics, 19,* 517–534.

Dykstra, R. (1978). Summary of the second-grade phase of the Cooperative Research Program in primary reading instruction. *Reading Research Quarterly, 4,* 49–700.

Edwards, A. L. (1968). *Experimental design in psychological research* (3rd ed.). New York: Holt, Rinehart and Winston.

Entwisle, D. R., Alexander, K. L., Cadigan, D., and Pallas, A. (1986). The schooling process in first grade: two samples a decade apart. *American Educational Research Journal, 23,* 587–613.

Finn, J. D. (1972a). *MULTIVARIANCE: Univariate and multivariate analysis of variance, covariance and regression*. Ann Arbor, MI: National Educational Resources, Inc.

Finn, J. D. (1972b). Expectations and the educational environment. *Review of Educational Research, 42,* 387–410.

Finn, J. D. (1974). *A general model for multivariate analysis*. New York: Holt, Rinehart and Winston.

Finn, J. D. (1980). Sex differences in educational outcomes: a cross-national study. *Sex Roles, 6,* 9–26.

Finn, J. D. and Bock, R. D. (1985). *Multivariance VII User's Guide*. Mooresville, IN: Scientific Software.

Fisher, R. A. (1936). The use of multiple measurements in taxonomic problems. *Annals of Eugenics, 7,* 179–188.

Fukunaga, K. (1972). *Introduction to statistical pattern recognition*. New York: Academic Press.

Garrett, H. E. (1943). The discriminant function and its use in psychology. *Psychometrika, 8,* 65–79.

Glass, G. V, and Hopkins, K. D. (1984). *Statistical methods in education and pschology* (2nd ed.). Englewood Cliffs, NJ: Prentice-Hall.

Glass, G. V, and Stanley, J. C. (1970). *Statistical methods in education and psychology*. Englewood Cliffs, NJ: Prentice-Hall.

Goldberger, A. S. and Duncan, O.D. (eds.). (1973). *Structural equation models in the social sciences*. New York: Academic Press.

Greeley, A. M. (1982). *Catholic high schools and minority students.* New Brunswick, NJ: Transaction.

Green, B. F. (1979). The two kinds of linear discriminant functions and their relationship. *Journal of Educational Statistics, 4,* 247–263.

Green, P. E., and Carroll, J. D. (1976). *Mathematical tools for applied multivariate analysis.* New York: Academic Press.

Gribbons, W. D., and Lohnes, P. R. (1968). *Emerging careers.* New York: Teachers College Press.

Gribbons, W. D., and Lohnes, P. R. (1969). Eighth-grade vocational maturity in relation to nine-year career patterns. *Journal of Counseling Psychology, 16,* 557–562.

Gribbons, W. D., and Lohnes, P. R. (1982). *Careers in theory and experience: A twenty-year longitudinal study.* Albany, NY: State University of New York Press.

Harman, H. H. (1967). *Modern factor analysis* (rev. ed.). Chicago: University of Chicago Press.

Harris, R. J. (1985). *A primer of multivariate statistics* (2nd ed.). New York: Academic Press.

Harris, W. C., and Kaiser, H. F. (1964). Oblique factor analytic solutions by orthogonal transformations. *Psychometrika, 29,* 347–362.

Hays, W. L. (1963). *Statistics for psychologists* (2nd ed.). New York: Holt, Rinehart and Winston.

Hays, W. L. (1981). *Statistics for psychologists* (3rd ed.). New York: Holt, Rinehart, and Winston.

Heck, D. L. (1960). Charts of some upper percentage points of the distribution of the largest characteristic root. *Annals of Mathematical Statistics, 31,* 625–642.

Heise, D. R. (1975). *Causal analysis,* New York: John Wiley.

Hoerl, A. E., and Kennard, R. W. (1970). Ridge regression: Biased estimation for non-orthogonal problems. *Technometrics, 12,* 55–67.

Horst, P. (1963). *Matrix algebra for social scientists.* New York: Holt, Rinehart and Winston.

Hotelling, H. (1931). The generalization of Student's ratio. *Annals of Mathematical Statistics, 2,* 360–378.

Hotelling, H. (1933). Analysis of a complex of statistical variables into principal components. *Journal of Educational Psychology, 24,* 417–441, 498–520.

Hotelling, H. (1935). The most predictable criterion. *Journal of Educational Psychology, 26,* 139–142.

Hotelling, H. (1951). A generalized T-test and measure of multivariate dispersion. *Proceedings of the Second Berkeley Symposium of Mathematical Statistics and Probability, 2,* 23–41.

Humphreys, L. G. (1969). Note on a criterion for the number of common factors. *Educational and Psychological Measurement, 29,* 571–578.

Jones, L. V. (1966). Analysis of variance in its multivariate developments. In R. B. Cattell (Ed.), *Handbook of multivariate experimental psychology.* Skokie, IL: Rand McNally.

Jöreskog, K. G. (1969). A general approach to confirmatory factor analysis. *Psychometrika, 34,* 183–202.

Jöreskog, K. G. (1971). Simultaneous factor analysis in several populations. *Psychometrika, 36,* 409–426.

Jöreskog, K. G., and Sörbom, D. (1979). *Advances in factor analysis and structural equation models*. Cambridge, MA: Abt Books.

Jöreskog, K. G., and Sörbom, D. (1983). *LISREL VI: Analysis of linear structural relationships by maximum likelihood and least squares methods*. Chicago: National Educational Resources.

Kaiser, H. F. (1956). *The varimax method of factor analysis*. Unpublished doctoral dissertation, University of California, Berkeley.

Kaiser, H. F. (1958). The varimax criterion for analytic rotation in factor analysis. *Psychometrika, 23,* 187–200.

Kaiser, H. F. (1960). The application of electronic computers to factor analysis. *Educational and Psychological Measurement, 20,* 141–151.

Kaiser, H. F. (1976). *SEARCH: A new program for oblique rotation to simple structure*. Unpublished manuscript, University of California, Berkeley.

Keeves, J. P. (1974). The performance cycle: Motivation and attention as mediating variables in school performance. *Home Environment and School Study Report 5*. Melbourne: Australian Council for Educational Research.

Keeves, J. P. (ed.). (1986). Aspiration, motivation and achievement: Different methods of analysis and different results. *International Journal of Educational Research, 10,* entire.

Keith, T. Z. and Page, E. B. (1985). Do Catholic high schools improve minority student achievement? *American Educational Research Journal, 22,* 337–349.

Kelley, T. L. (1935). *Essential traits of mental life*. Cambridge, MA: Harvard University Press.

Kendall, M. G. (1957). *A course in multivariate analysis*. London: Charles Griffin. (Reprinted (1961) New York: Hafner.)

Kendall, M. G. and Stuart, A. (1966). *The advanced theory of statistics, volume 3: Design and analysis, and time-series*. New York: Hafner.

Kenny, D. A. (1979). *Correlation and causation*. New York: Wiley-Interscience.

Kerlinger, F. N. (1979). *Behavioral research: A conceptual approach*. New York: Holt, Rinehart and Winston.

Laughlin, J. E. (1978). Comment on "Estimating coefficients in linear models: It don't make no nevermind." *Psychological Bulletin, 85,* 247–253.

Lawley, D. N., and Maxwell, A. E. (1963). *Factor analysis as a statistical method*. London: Butterworth.

Lohnes, P. R. (1961). Test space and discriminant space classification and related significance tests. *Educational and Psychological Measurement, 21,* 559–574.

Lohnes, P. R. (1966). *Measuring adolescent personality*. Pittsburgh: University of Pittsburgh, Project TALENT.

Lohnes, P. R. (1986). Factorial modeling. *International Journal of Educational Research, 10,* 181–189.

Lohnes, P. R. and Gray, M. M. (1972). Intelligence and the Cooperative Reading Studies. *Reading Research Quarterly, 7,* 466–476.

Lohnes, P. R. and Marshall, T. O. (1965). Redundancy in student records. *American Educational Research Journal, 2,* 19–23.

Lomax, R. G. (1985). A structural model of public and private schools. *Journal of Experimental Education, 53,* 216–226.

Lunneborg, C. E., and Abbott, R. D. (1983). *Elementary multivariate analysis for the behavioral sciences: Applications of basic structure*. New York: North-Holland.

Lyson, T. A. and Falk, W. W. (1984). Recruitment to school teaching: The relationship between high school plans and early adult attainments. *American Educational Research Journal, 21,* 181–193.

Mahalanobis, P. C. (1936). On the generalized distance in statistics. *Proceedings of the National Institute of Science, India, 12,* 49–55.

Mallows, C. L. (1973). Some comments on C^p. *Technometrics, 15,* 661–675.

Marquardt, D. W., and Snee, R. D. (1975). Ridge regression in practice. *The American Statistician, 29,* 3–20.

Mayeske, G. W., Okada, T., Cohen, W. M., Beaton, A. E., Jr., and Wisler, C. E. (1973). *A study of the achievement of our nation's students*. Washington, D.C.: U.S. Government Printing Office.

McKeon, J. J. (1964). Canonical analysis: Some relations between canonical correlation, factor analysis, discriminant function analysis, and scaling theory. *Psychometric Monograph* (Whole No. 13).

McQuitty, L. L. (1955). *A method of pattern analysis for isolating typological and dimensional constructs*. Rep. No. TN-55-62. Lackland AFB, TX: Air Force Personnel and Training Center.

Mendenhall, W. (1968). *Introduction to linear models and the design and analysis of experiments*. Belmont, CA: Wadsworth.

Mijares, T. A. (1964). *Percentage points of the sum $V_{1(s)}$ of s roots (s = 1 − 50)*. Manila: University of the Philippines, Statistical Center.

Miller, J. K. (1969). *The development and application of bi-multivariate correlation: A measure of statistical association between multivariate measurement sets*. Unpublished doctoral dissertation, State University of New York at Buffalo.

Mood, A. M. (1971). Partitioning variance in multiple regression analyses as a tool for developing learning models. *American Educational Research Journal, 8,* 191–202.

Morrison, D. F. (1976). *Multivariate statistical methods* (2nd ed.). San Francisco: McGraw-Hill.

Mosteller, F., and Tukey, J. W. (1977). *Data analysis and regression*. Reading, MA: Addison-Wesley.

Mulaik, S. A. (1972). *The foundations of factor analysis*. New York: McGraw-Hill.

National Opinion Research Center (1980). *High school and beyond: Information for users, Base year 1980. (Contract No. 300-78-0208)*. Chicago: National Opinion Research Center.

Neuhaus, J. O., and Wrigley, C. (1954). The quartimax method: An analytical approach to orthogonal simple structure. *British Journal of Statistical Psychology, 7,* 81–91.

Newman, A. P. (1972). Later achievement study of pupils underachieving in reading in first grade. *Reading Research Quarterly, 7,* 477–508.

Nichols, R. C. (1985). *MAPS: Multiple Analysis Program System*. Amherst, NY: Maps & Co.

Olson, C. L. (1976). On choosing a test statistic in multivariate analysis of variance. *Psychological Bulletin, 83,* 579–586.

Overall, J. E., and Klett, C. J. (1972). *Applied multivariate analysis*. New York: McGraw-Hill.

3. $$(\mathbf{AB})\mathbf{C}' = \begin{bmatrix} 2 & -4 & 4 \\ 2 & -4 & 2 \\ -3 & 6 & 1 \end{bmatrix} \begin{bmatrix} 2 & 1 \\ 3 & -2 \\ -1 & 1 \end{bmatrix} = \begin{bmatrix} -12 & 14 \\ -10 & 12 \\ 11 & -14 \end{bmatrix}$$

$$\mathbf{A}(\mathbf{BC}') = \begin{bmatrix} 3 & 1 \\ 2 & 0 \\ -1 & 2 \end{bmatrix} \begin{bmatrix} -5 & 6 \\ 3 & -4 \end{bmatrix} = \begin{bmatrix} -12 & 14 \\ -10 & 12 \\ 11 & -14 \end{bmatrix}$$

Verifies associative law of matrix multiplication (Eq. 2.9).

4. $$[(\mathbf{AB})\mathbf{C}']' = \begin{bmatrix} -12 & -10 & 11 \\ 14 & 12 & -14 \end{bmatrix}$$ (from answer to Exercise 3)

$$(\mathbf{CB}')\mathbf{A}' = \begin{bmatrix} -5 & 3 \\ 6 & -4 \end{bmatrix} \begin{bmatrix} 3 & 2 & -1 \\ 1 & 0 & 2 \end{bmatrix} = \begin{bmatrix} -12 & -10 & 11 \\ 14 & 12 & -14 \end{bmatrix}$$

Verifies Eqs 2.2 and 2.12.

5. $$[1 \quad 2 \quad -1] \begin{bmatrix} 2 & -4 & 4 \\ 2 & -4 & 2 \\ -3 & 6 & 1 \end{bmatrix} \begin{bmatrix} 1 \\ 2 \\ -1 \end{bmatrix} = [9 \quad -18 \quad 7] \begin{bmatrix} 1 \\ 2 \\ -1 \end{bmatrix}$$

Using Eq. 2.10, $(2)1^2 + (-4)2^2 + (1)(-1)^2 + (-4 + 2)(1)(2) + (4 - 3)(1)(-1) + (2 + 6)(2)(-1) = -34$

6. $$\mathbf{DC} = \begin{bmatrix} 4 & 6 & -2 \\ 3 & -6 & 3 \end{bmatrix}$$ The first diagonal element of **D** multiplies the first row of **C**; the second diagonal element of **D** multiplies the second row of **C**.

7. $$\mathbf{AD} = \begin{bmatrix} 6 & 3 \\ 4 & 0 \\ -2 & 6 \end{bmatrix}$$ The first diagonal element of **D** multiplies the first column of **A**; the second diagonal element of **D** multiplies the second column of **A**.

8. $$\mathbf{EA} = \begin{bmatrix} 15 & 5 \\ 10 & 0 \\ -5 & 10 \end{bmatrix}$$ The common diagonal element of **E** multiplies each row of **A**. hence every element of **A** gets multiplied by the common diagonal element 5.

9 **1.** (a) $$\begin{bmatrix} 2 & -\frac{3}{2} \\ -3 & \frac{5}{2} \end{bmatrix}$$ (b) $$\begin{bmatrix} -3 & -2 \\ -2 & -1 \end{bmatrix}$$

(c) singular (d) $$\begin{bmatrix} \frac{1}{3} & 0 \\ 0 & -\frac{1}{5} \end{bmatrix}$$

Pearson, K. (1894). Contributions to the mathematical theory of evolution: I. On the dissection of asymmetrical frequency-curves. *Philosophical Transactions of the Royal Society, Series A, 185,* 71–90.

Pearson, K. (1901). The lines of closest fit to a system of points. *Philosophic Magazine, 2,* 559–572.

Pillai, K. C. S. (1960). *Statistical tables for tests of multivariate hypotheses.* Manila: University of the Philippines, Statistical Service Center.

Pillai, K. C. S. (1965). On the distribution of the largest characteristic root in multivariate analysis. *Biometrika, 52,* 405–415.

Pruzek, R. M., and Frederick, B. C. (1978). Weighting predictors in linear models: Alternatives to least squares and limitations of equal weights. *Psychological Bulletin, 85,* 254–266.

Ralston, A. (1978). *A first course in numerical analysis.* New York: Wiley.

Rao, C. R. (1952). *Advanced statistical methods in biometric research.* New York: Wiley.

Rao, C. R. (1965). *Linear statistical inference and its applications.* New York: Wiley.

Richardson, M. (1966). *College algebra* (3rd ed.). Englewood Cliffs, NJ: Prentice-Hall.

Roy, S. N. (1957). *Some aspects of multivariate analysis.* New York: Wiley.

Rulon, P. J., Tiedeman, D. V., Langmuir, C. R., and Tatsuoka, M. M. (1954). *The profile problem.* Cambridge, MA: Educational Research Corporation.

Rulon, P. J., Tiedeman, D. V., Tatsuoka, M. M., and Langmuir, C. R. (1967). *Multivariate statistics for personnel classification.* New York: Wiley.

Saunders, D. R. (1953). *An analytic method for rotation to orthogonal simple structure* (Research Bulletin 53-10). Princeton, NJ: Educational Testing Service.

Schatzoff, M. (1964). *Exact distributions of Wilks' likelihood ratio criterion and comparisons with competitive tests.* Unpublished doctoral dissertation, Harvard University.

Schatzoff, M. (1966a). Exact distributions of Wilks' likelihood ratio criterion. *Biometrika, 53,* 347–358.

Schatzoff, M. (1966b). Sensitivity comparisons among tests of the general linear hypothesis. *Journal of the American Statistical Association, 61,* 415–435.

Searle, S. R. (1966). *Matrix algebra for the biological sciences (including applications in statistics).* New York: Wiley.

Searle, S. R. (1971). *Linear models.* New York: Wiley.

Seber, G. A. F. (1966). *The linear hypothesis: A general theory.* New York: Hafner Press.

Smith, G., and Campbell, F. (1980). A critique of some ridge regression methods. *Journal of the American Statistical Association, 75,* 74–81.

Stein, C. (1960). Multiple regression. In I. Olkin et al. (Eds.), *Contributions to probability and statistics.* Stanford, CA: Stanford University Press.

Stephenson, W. (1963). *The study of behavior: Q-technique and its methodology.* Chicago: University of Chicago Press.

Stevens, J. P. (1979). Comment on Olson: *Choosing a test statistic in multivariate analysis of variance. Psychological Bulletin, 86,* 355–360.

Stewart, D. K., and Love, W. A. (1968). A general canonical correlation index. *Psychological Bulletin, 70,* 160–163.

Takeuchi, K., Yanai, H., and Mukherjee, B. N. (1982). *The foundations of multi-*

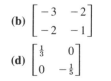

variate analysis: A unified approach by means of projection onto linear subspaces. New Delhi: Wiley Eastern Ltd.

Tatsuoka, M. M. (1953). *The relationship between canonical correlation and discriminant analysis*. Cambridge, MA: Educational Research Corporation.

Tatsuoka, M. M. (1956). *Joint probability of membership in a group and success therein: An index which combines the information from discriminant and regression analyses*. Unpublished doctoral dissertation, Harvard University.

Tatsuoka, M. M. (1970). *Discriminant analysis: The study of group differences*. Champaign, IL: Institute for Personality and Ability Testing.

Tatsuoka, M. M. (1971a). *"Discriminant Functions" à la BMD vs. à la SOUPAC*. Unpublished manuscript, University of Illinois at Urbana-Champaign.

Tatsuoka, M. M. (1971b) *Multivariate analysis: Techniques for educational and psychological research,* New York: John Wiley.

Tatsuoka, M. M. and Tiedeman, D. V. (1954). Discriminant analysis. *Review of Educational Research, 25,* 402–420.

Tatsuoka, M. M., and Tiedeman, D. V. (1963). Statistics as an aspect of scientific method in research on teaching. In Gage, N. L. (Ed.), *Handbook of research on teaching*. Chicago: Rand McNally.

Thurstone, L. L. (1935). *The vectors of the mind*. Chicago: University of Chicago Press.

Thurstone, L. L. (1947). *Multiple factor analysis*. Chicago: University of Chicago Press.

Tiedeman, D. V., Bryan, J. G., and Rulon, P. J. (1953). *The utility of the Airman Classification Battery for assignment of airmen to eight air force specialties*. Cambridge, MA: Educational Research Corporation.

Tildesley, M. L. (1921). A first study of the Burmese skull. *Biometrika, 13,* 176–262.

Tucker, L. R (1938). A graphical method for determining the number of factors. In L. L. Thurstone, *Primary mental abilities*. *Psychometric Monograph* (Whole No. 1).

Tucker, L. R, and Finkbeiner, C. T. (1981). *Transformation of factors by artificial personal probability functions* (Research Rep. No. 81-58). Princeton, NJ: Educational Testing Service.

Wainer, H. (1976). Estimating coefficients in linear models: It don't make no nevermind. *Psychological Bulletin, 83,* 213–217.

Welch, W. W., Anderson, R. E., and Harris, L. J. (1982). The effects of schooling on mathematics achievement. *American Educational Research Journal, 19,* 145–153.

Wherry, R. J. (1947). Multiple bi-serial and multiple point bi-serial correlation. *Psychometrika, 12,* 189–195.

Wilks, S. S. (1932). Certain generalizations in the analysis of variance. *Biometrika. 24,* 471–494.

Winer, B. J. (1962). *Statistical principles in experimental design*. New York: McGraw-Hill.

Wolfle, L. M. (1980). The enduring effects of education on verbal skills. *Sociology of Education, 53,* 104–114.

Zimmerman, B. J., and Pons, M. M. (1986). Development of a structured interview for assessing student use of self-regulated learning strategies. *American Educational Research Journal, 23,* 614–628.

ANSWERS TO EXERCISES

Chapter 2

p. 13 1. (a) $\begin{bmatrix} 5 & 0 & -\frac{15}{2} \\ 5 & -4 & \frac{1}{2} \end{bmatrix}$ **(c)** $\begin{bmatrix} 6 & 0 & -9 \\ \frac{11}{2} & -\frac{29}{6} & \frac{2}{3} \end{bmatrix}$

(d) $\begin{bmatrix} -1 & -\frac{1}{2} \\ 0 & \frac{5}{6} \\ \frac{3}{2} & -\frac{1}{6} \end{bmatrix}$

Parts (b) and (e) are meaningless because the two matrices are of different orders.

2. $\mathbf{D} = \begin{bmatrix} 2 & 0 & -3 \\ 1 & -\frac{5}{3} & \frac{1}{3} \end{bmatrix}$ $\quad \mathbf{E} = \begin{bmatrix} -2 & 0 & 3 \\ 4 & 2 & -1 \end{bmatrix}$

3. $\mathbf{D} + \mathbf{E} = \begin{bmatrix} 0 & 0 & 0 \\ 5 & \frac{1}{3} & -\frac{2}{3} \end{bmatrix} = 2\begin{bmatrix} 0 & 0 & 0 \\ \frac{5}{2} & \frac{1}{6} & -\frac{1}{3} \end{bmatrix} = 2\mathbf{C}'$

p. 22 1. $\mathbf{A}(\mathbf{B} + \mathbf{C}) = \begin{bmatrix} 3 & 1 \\ 2 & 0 \\ -1 & 2 \end{bmatrix} \begin{bmatrix} 3 & 1 & 0 \\ 0 & 0 & 2 \end{bmatrix} = \begin{bmatrix} 9 & 3 & 2 \\ 6 & 2 & 0 \\ -3 & -1 & 4 \end{bmatrix}$

$\mathbf{AB} = \begin{bmatrix} 2 & -4 & 4 \\ 2 & -4 & 2 \\ -3 & 6 & 1 \end{bmatrix} \quad \mathbf{AC} = \begin{bmatrix} 7 & 7 & -2 \\ 4 & 6 & -2 \\ 0 & -7 & 3 \end{bmatrix}$

2. $(\mathbf{B} + \mathbf{C})\mathbf{A} = \begin{bmatrix} 3 & 1 & 0 \\ 0 & 0 & 2 \end{bmatrix} \begin{bmatrix} 3 & 1 \\ 2 & 0 \\ -1 & 2 \end{bmatrix} = \begin{bmatrix} 11 & 3 \\ -2 & 4 \end{bmatrix}$

$\mathbf{BA} = \begin{bmatrix} -2 & 3 \\ 0 & 1 \end{bmatrix} \quad \mathbf{CA} = \begin{bmatrix} 13 & 0 \\ -2 & 3 \end{bmatrix}$

2. **(a)** $\frac{1}{2}\begin{bmatrix} 2 & 4 & 3 \\ -3 & 2 & 1 \\ -1 & 3 & 2 \end{bmatrix}$ **(b)** $\frac{1}{6}\begin{bmatrix} 3 & 1 & 1 \\ 0 & 2 & -4 \\ 3 & -3 & 3 \end{bmatrix}$

(c) singular

3. The inverse of a diagonal matrix is the diagonal matrix whose diagonal elements are each equal to the reciprocal of the corresponding diagonal element of the original. Proof follows from Eqs. 2.13a and 2.13b (pre- and postmultiplying by a diagonal matrix).

4.

$$(\mathbf{A}^{-1})(\mathbf{A}^{-1})^{-1} = \mathbf{I}$$ (definition of inverse matrix)

$$\mathbf{A}(\mathbf{A}^{-1})(\mathbf{A}^{-1})^{-1} = \mathbf{AI} = \mathbf{A}$$ (premultiplying by **A** and definition of identity matrix)

$$(\mathbf{AA}^{-1})(\mathbf{A}^{-1})^{-1} = \mathbf{A}$$ (associative law)

$$(\mathbf{A}^{-1})^{-1} = \mathbf{A}$$ (definitions of inverse and identity)

Equation 2.18 follows immediately on multiplying $c\mathbf{A}$ by $(1/c)\mathbf{A}^{-1}$.

5. $\mathbf{A}^{-1} = \frac{1}{4}\begin{bmatrix} 4 & -2 \\ -8 & 5 \end{bmatrix}$; therefore, $\mathbf{v}'\mathbf{A}^{-1}\mathbf{v} = (\tfrac{1}{4})[-2 \quad 6]\begin{bmatrix} 4 & -2 \\ -8 & 5 \end{bmatrix}\begin{bmatrix} -2 \\ 6 \end{bmatrix}$

$= 79$

6.

$$(\mathbf{AB})(\mathbf{B}^{-1}\mathbf{A}^{-1}) = \mathbf{A}(\mathbf{BB}^{-1})\mathbf{A}^{-1}$$ (associative law)

$$= \mathbf{A}(\mathbf{I})\mathbf{A}^{-1}$$ (definition of inverse)

$$= \mathbf{A}^{-1} = \mathbf{I}$$ (associative law, definitions of identity and inverse)

9. **(a)** $\mathbf{S} = \begin{bmatrix} 269.11579 & 48.49093 & 33.73143 & 1401.46745 \\ 48.49093 & 298.15100 & 41.79604 & 79.94174 \\ 33.73143 & 41.79604 & 70.36058 & 507.68054 \\ 1401.46745 & 79.94174 & 507.68054 & 56{,}667.74869 \end{bmatrix}$

(b) $\mathbf{C} = \frac{\mathbf{S}}{599} = \begin{bmatrix} .44928 & .08095 & .05631 & 2.33968 \\ .08095 & .49775 & .06978 & .13346 \\ .05631 & .06978 & .11746 & .84755 \\ 2.33968 & .13346 & .84755 & 94.60392 \end{bmatrix}$

10. $\mathbf{R} = \begin{bmatrix} 1.00000 & .17119 & .24513 & .35888 \\ .17119 & 1.00000 & .28857 & .01945 \\ .24513 & .28857 & 1.00000 & .25425 \\ .35888 & .01945 & .25425 & 1.00000 \end{bmatrix}$

Chapter 3

p. 63 **1.** $\hat{Y} = -7.79 + .373X_1 + .302X_2 + .297X_3$ and $R_{y.123} = .558.$

2. The validity coefficients are found to be $r_{1y} = .3121$, $r_{2y} = .4231$, $r_{3y} = .3787$. Since r_{1y} is the smallest of the three, we delete X_1. That is, we delete the entire first row and first column of **S** and use the resulting 3 × 3 SSCP matrix as our starting point and obtain $\hat{Y} = 23.4 + .325X_2 + .308X_3$ and $R_{y.23} = .493$.

3. Substituting in Eq. 3.23, we get $F = 4.57$ with 1 and 46 degrees of freedom. From Table E.4, $F^1_{46;.95} = 4.05$, $F^1_{46;.99} = 7.25$. The difference between $R_{y.123}$ and $R_{y.23}$ is significant at the 5% level but not at the 1% level.

4. **(a)** Computing the SSCP matrix for the 10-case cross-validation sample by use of Eq. 2.8, we get

$$\mathbf{S} = \begin{bmatrix} 894.9 & 142.1 & 315.0 & 246.6 \\ 142.1 & 1372.9 & -338.0 & 39.4 \\ 315.0 & -338.0 & 812.0 & -8.0 \\ 246.6 & 39.4 & -8.0 & 624.4 \end{bmatrix}$$

Partitioning this appropriately and substituting the parts in Eq. 3.21 together with the regression weights obtained in Exercise 1, we get

$$R^2_{cv} = .04554, \qquad \text{hence} \quad R_{cv} = .2134$$

(b) The multiple-R obtained in Exercise 1, corrected for shrinkage by Eq. 3.24, is

$$R_{\text{corr},s} = \sqrt{1 - \tfrac{49}{46}\tfrac{48}{45}\tfrac{51}{50}(1 - .3111)} = .449$$

Fisher's Z-transform of this and of the cross-validation R found in part (b) are, respectively, $Z_{\text{corr},s} = .483$ and $Z_{cv} = .216$. Hence $z = (.213 - .483)\sqrt{7} = -.706$, which is clearly not significant.

5. The equal-weights predictor composite is, of course, $\tilde{z}_y = z_1 + z_2 + z_3$. The correlation $r_{\tilde{z}_y z_y(1)}$ between $\tilde{z}_y$ and the actual z_y in the first sample is calculated from Eq. 3.27 (where $\mathbf{R}_{pp}$ and $\mathbf{R}_{pc}$ are the appropriate sectors of the **R** computed from the given **S** in accordance with Eq. 2.20a, resulting in the value .557. To calculate the correlation $r_{\tilde{z}_y z_y(2)}$ in the cross-validation sample, we must first convert the equal-weights composite into raw-score form by using Eq. 3.26; the result is

$$\tilde{Y}_{(1)} = -240.1 + 1.418X_1 + .973X_2 + 1.152X_3$$

Then applying Eq. 3.18, we get $r_{\tilde{z}_y z_y} = .218$. The shrinkage is .557 − .218 = .339. The shrinkage of the multiple-R was .558 − .213 = .345. Thus the shrinkage is very slightly smaller for the equal-weights composite than it was for the optimized multiple-R.

6. **(a)** $\hat{Y} = 37.178 + 4.031X_1 + 2.686X_2 + 3.777X_3 - .9825X_4 + 4.757X_5$
$R_{y.12345} = .4838$

(b) All five variables are included, the order being LOCUS, SEX, SES, MOT, and CONCPT. Since all the variables are in, the final multiple regression equation and the multiple-R are the same as in part (a).

7. The multiple-R of WRTG with the three affective variables is $R_{y.345}$ = .4078. With the other two variables, sex and SES added, the multiple-R is as found in Exercise 6(a): $R_{y.12345}$ = .4838. The F for testing the significance of the incremental R^2 is, from Eq. 3.23,

$$F = \frac{[(.4838)^2 - (.4078)^2]/2}{[1 - (.4838)^2]/594} = 26.28 > 4.79 = F^2_{120;.99} > F^2_{594;.99}$$

The addition of sex and SES to the three affective variables significantly increases the predictability of the writing test score.

8. $\hat{Y} = 39.078 + 4.530X_1 + 1.189X_2 + 4.465X_3 - .1038X_4 + 4.722X_5$; $R_{y.12345} = .5123$; $R_s = .4843$; $R_{cv} = .4244$; $Z_{R_s} = .5286$; $Z_{R_{cv}} = .4530$; $z = -1.301$, $P = .097$. The cross-validation R is not significantly smaller than what would be expected on the basis of shrinkage.

Chapter 4

p. 75 1. Find $|\boldsymbol{\Sigma}|$ and $\boldsymbol{\Sigma}^{-1}$, substitute these in Eq. 4.13, and show that the expression factors as indicated in the problem.

2. $(X_1 - 15)^2 + (X_2 - 20)^2 - 1.2(X_1 - 15)(X_2 - 20) = 44.37$

3. $[(5 - 15)^2 + (8 - 20)^2 - 1.2(5 - 15)(8 - 20)]/16 = 6.25$

which is slightly greater than the 95th centile of the chi-square distribution with 2 d.f. Hence, slightly more than 95% of the population lies within the ellipse. (Interpolation in Table E.3 gives about 95.5%.)

p. 79 1. $\chi^2_2(.95) = 5.991$. Therefore, by Eq. 4.19, the equation of the desired ellipse is

$$[Y_1 - 18, Y_2 - 15]\begin{bmatrix} 25 & 12 \\ 12 & 16 \end{bmatrix}^{-1}\begin{bmatrix} Y_1 - 18 \\ Y_2 - 15 \end{bmatrix} = 5.991/20$$

Or, in scalar notation,

$$(1.25)(Y_1 - 18)^2 + (1.9531)(Y_2 - 15)^2 - (1.875)(Y_1 - 18)(Y_2 - 15) = 5.991$$

The ellipse may be plotted by computing pairs of Y_2 values, for each of several Y_1 values, from this equation. A better method will be described in Chapter Five.

p. 107 1. $\mathbf{d} = \begin{bmatrix} .7 \\ 2.6 \end{bmatrix}$ $\mathbf{S}_d = \begin{bmatrix} 22.1 & 2.8 \\ 2.8 & 90.4 \end{bmatrix}$ $\mathbf{S}_d^{-1} = \begin{bmatrix} 4.5427 & -.1407 \\ -.1407 & 1.1106 \end{bmatrix} \times 10^{-2}$

Hence

$$\mathbf{T}^2 = (10)(9)[.7 \quad 2.6]\begin{bmatrix} 4.5427 & -.1407 \\ -.1407 & 1.1106 \end{bmatrix}\begin{bmatrix} .7 \\ 2.6 \end{bmatrix} \times 10^{-2} = 8.299$$

The final test statistic is $F = 3.688 < {}_{.95}F^2_8 = 4.46$. The null hypothesis cannot be rejected at the 5% level.

2. $$\mathbf{T}^2 = [(20)(30)(48)/50][-.9 \quad .6]\begin{bmatrix} 95.4 & 11.6 \\ 11.6 & 29.2 \end{bmatrix}^{-1}\begin{bmatrix} -.9 \\ .6 \end{bmatrix} = 15.32$$

$$F = [47/(48)(2)](15.32) = 7.50 > {}_{.99}F^2_{47} = 5.11$$

We conclude, at the 1% level, that the population centroids for drug addicts and alcoholics on these two tests are different.

3. The determinants of **W** and **T** are approximately 5903×10^4 and $15{,}152 \times 10^4$. Hence, $\Lambda = 5903/15{,}152 = .3896$. Since $K = 3$ and $p = 2$ in this problem, we may use either the second or the fourth expression in Table 4.2 to obtain $F = 16.86$, which far exceeds the 99th percentile of the F^4_{112} distribution. The null hypothesis of equality of centroids is rejected.

4. **(a)** For H_{01}, $z = 2.2$; for H_{02}, $z = 2.0$. Since both z's are greater than 1.96, both null hypotheses are rejected at the 5% level.
(b) Using Eq. 4.20, $Q = 5.5625 < \chi^2_{2;.95} = 5.991$. Therefore, H_0: $\boldsymbol{\mu} = [15, 20]$ is not rejected at $\alpha = .05$.

6. $T^2_2 = 142.01$; $[594/(598)(5)](142.01) = 28.21 > 2.66 = F^8_{120;.99} > F^{10}_{594;.95}$. The univariate t_j for the significance of $\bar{X}_{j1} - \bar{X}_{j2}$ $(j = 1, 2, \ldots, 5)$ are $t_1 = 1.02$; $t_2 = -6.16^{**}$; $t_3 = 1.17$; $t_4 = 3.38^{**}$; $t_5 = -1.56$; only the two values with double asterisks are significant at $\alpha = .01$. The student should note the discrepancy between the conclusion from the single multivariate test and that from the set of five univariate tests, and should understand the reason for this discrepancy.

7. $$\Lambda = .746 \qquad \frac{1 - \sqrt{.746}}{\sqrt{.746}}\,\frac{593}{5} = 18.71 > 2.66 = F^8_{120;.99} > F^{10}_{1186;.99}$$

8. The determinants of the pooled within-groups covariance matrix **C** and of the individual group covariance matrices $\mathbf{C}_1$, $\mathbf{C}_2$, and $\mathbf{C}_3$ are $|\mathbf{C}| = 4.3353 \times 10^8$, $|\mathbf{C}_1| = 4.8190 \times 10^8$, $|\mathbf{C}_2| = 3.3392 \times 10^8$, and $|\mathbf{C}_3| = 5.0858 \times 10^8$. (The group sizes are $n_1 = 145$, $n_2 = 308$, and $n_3 = 147$.) Box's statistic for testing the equality of the population covariance matrices is $Mh = 41.04 < 43.77 = \chi^2_{30;.95}$; the H_0, $\boldsymbol{\Sigma}_1 = \boldsymbol{\Sigma}_2 = \boldsymbol{\Sigma}_3$, is not rejected at the 5% level. Looking at the sample generalized variances in their own right, the fact that the academic preparatory group has the smallest value agrees with our intuitive expectation: This group would be expected to be more homogeneous in cognitive achievement test scores than are the other two.

Chapter 5

p. 122 1. Substitute $\mathbf{Y}' = \mathbf{X}'\mathbf{V}$ in $\mathbf{Z}' = \mathbf{Y}'\mathbf{U}$ to obtain $\mathbf{Z}' = (\mathbf{X}'\mathbf{V})\mathbf{U} = \mathbf{X}'(\mathbf{VU})$. It remains only to prove that **VU** is an orthogonal matrix. That is, show that $(\mathbf{VU})(\mathbf{VU})' = \mathbf{I}$ and that $|\mathbf{VU}| = 1$. Use Eqs. 2.12 and A.6 (Appendix A).

p. 132 2. The major axis of the ellipse makes an angle of $-36°52'$ (i.e., about 37° in the clockwise direction) with the positive X_1 (horizontal) axis of the reference system.

3. $$\text{var}(\mathbf{Y}) = [.8 \quad -.6]\begin{bmatrix} 23 & -12 \\ -12 & 16 \end{bmatrix}\begin{bmatrix} .8 \\ -.6 \end{bmatrix} = 32.00$$

Axes 10° away from OY in the clockwise and counterclockwise directions make angles of $-46°52'$ and $-26°52'$, respectively, with OX_1. For each of these angles, we look up the cosine and sine in a table of trigonometric functions, and use these as the first and second elements, respectively, of the transformation vector (i.e., in the place .8 and $-.6$ above). We thus find

$$\text{var}(Y') = [.6837 \quad -.7298]\begin{bmatrix} 23 & -12 \\ -12 & 16 \end{bmatrix}\begin{bmatrix} .6837 \\ -.7298 \end{bmatrix} = 31.25$$

$$\text{var}(Y'') = [.8921 \quad -.4519]\begin{bmatrix} 23 & -12 \\ -12 & 16 \end{bmatrix}\begin{bmatrix} .8921 \\ -.4519 \end{bmatrix} = 31.25$$

The variance of Y is greater than that of either Y' or Y''.

4. The minor axis makes an angle of $(-36°52' + 90°) = 53°8'$ with the positive X_1-axis.

5. Calling the variable defined by the minor axis Y_2, and those by axes 10° away from it in the clockwise and counterclockwise directions, Y_2' and Y_2'', respectively, we find (in the same way as in Exercise 3) var $(Y_2) = 7.00$, var $(Y_2') = 7.75$, var $(Y_2'') = 7.76$.

p. 137 1. The characteristic equation is $\lambda^3 - 17\lambda^2 + 86\lambda - 112 = 0$, and the eigenvalues are $\lambda_1 = 8$, $\lambda_2 = 7$, and $\lambda_3 = 2$.

p. 139 1. $\mathbf{v}_2' = [.8944 \quad -.4472 \quad 0] \qquad \mathbf{v}_3' = [-.1825 \quad -.3651 \quad .9128]$

p. 142 1. The eigenvalues are $\lambda_1 = 173.52$, $\lambda_2 = 58.90$, $\lambda_3 = 16.67$. The orthogonal transformation matrix, whose columns are the corresponding eigenvectors $\mathbf{v}_1$, $\mathbf{v}_2$, and $\mathbf{v}_3$, is

$$\mathbf{V} = \begin{bmatrix} .8181 & .4091 & -.4042 \\ .4997 & -.1576 & .8517 \\ -.2846 & .8988 & .3334 \end{bmatrix}$$

2. For the first transformed variable, we have

$$\sigma_{y_1}^2 = [.8181 \ .4997 \ -.2846]\begin{bmatrix} 128.7 & 61.4 & -21.0 \\ 61.4 & 56.9 & -28.3 \\ -21.0 & -28.3 & 63.5 \end{bmatrix}\begin{bmatrix} .8181 \\ .4997 \\ -.2846 \end{bmatrix}$$

$$= 173.52$$

The other two cases are verified similarly.

p. 153 1. Collecting the three vectors of Example 5.7 (as columns) in a matrix, we have

$$\mathbf{A} = \begin{bmatrix} 1 & 3 & -3 \\ 2 & -1 & 15 \\ -1 & 0 & -6 \end{bmatrix}$$

The characteristic equation $|\mathbf{A} - \lambda\mathbf{I}| = 0$ is found to be

$$-\lambda^3 - 6\lambda^2 + 10\lambda = 0$$

which clearly has one zero root. The rank of **A** is 2, so the given set of vectors is linearly dependent.

2. The characteristic equation of the matrix comprising the vectors in Example 5.8 is

$$-\lambda^3 + (7/2)\lambda^2 + (11/2)\lambda = 0$$

The set is therefore linearly dependent.

3. The set of vectors in Example 5.9 form a diagonal matrix, and hence the characteristic equation is readily found to be

$$(d_1 - \lambda)(d_2 - \lambda)(d_3 - \lambda) = 0$$

whose roots are d_1, d_2, and d_3, none of which is zero. Therefore, the given set is linearly independent. This result obviously generalizes to higher dimensions: Any set of vectors which, taken in suitable order, forms a diagonal matrix (with no zero in the diagonal position) is linearly independent.

4. The characteristic equation of the matrix comprising the vectors in Example 5.10 is

$$-\lambda^3 - 2\lambda^2 + 16\lambda + 28 = 0$$

which clearly has no zero root. Therefore, the set is linearly independent.

p. 160 1. $\lambda_1 = 10, \quad \lambda_2 = 8, \quad \lambda_3 = 3 \qquad \mathbf{v}_1' = [.2672 \quad .8018 \quad .5345]$

$\mathbf{v}_2' = [.9487 \quad -.3162 \quad 0] \qquad \mathbf{v}_3' = [-.1690 \quad -.5070 \quad .8452]$

The verifications in parts (a)–(d) are simple. To verify (e) numerically, assume (for simplicity) that the centroid is (0, 0, 0). Then take any $\mathbf{x}'$, for example [1, 2, 3], and compute $\mathbf{x}'\boldsymbol{\Sigma}^{-1}\mathbf{x}$. Next, using the transformation matrix with $\mathbf{v}_1$, $\mathbf{v}_2$, $\mathbf{v}_3$ as columns, transform **x** into **y** and $\boldsymbol{\Sigma}$ into $\boldsymbol{\Sigma}_y$; compute $\mathbf{y}'\boldsymbol{\Sigma}^{-1}\mathbf{y}$ and compare with the value of $\mathbf{x}'\boldsymbol{\Sigma}^{-1}\mathbf{x}$ obtained previously.

2. The principal minor determinants of $\boldsymbol{\Sigma}$ are

$$8, 8, 5, \quad \begin{vmatrix} 8 & 3 \\ 3 & 5 \end{vmatrix} = 31, \quad \begin{vmatrix} 8 & 1 \\ 1 & 5 \end{vmatrix} = 39, \quad \begin{vmatrix} 8 & 0 \\ 0 & 8 \end{vmatrix} = 64, \text{ and } |\boldsymbol{\Sigma}| = 240$$

$\boldsymbol{\Sigma}$ is symmetric. The required factor matrix **B**, following Eq. 5.36, is

$$\mathbf{B} = \begin{bmatrix} .8481 & 2.5355 & 1.6902 \\ 2.6833 & -.8943 & 0 \\ -.2927 & -.8782 & 1.4640 \end{bmatrix}$$

p. 177 1. The eigenvalues are .8, .7, and .2. The eigenvectors are

$$[.4082 \quad .8165 \quad .4082] \qquad [.8944 \quad -.4472 \quad 0]$$
$$[-.1825 \quad -.3651 \quad .9128]$$

5. The two matrices whose common eigenvalues and respective eigenvectors we need to obtain are

$$\mathbf{A}'\mathbf{A} = \begin{bmatrix} 100 & 0 \\ 0 & 50 \end{bmatrix} \quad \text{and} \quad \mathbf{A}\mathbf{A}' = \begin{bmatrix} 73 & 15 & 36 \\ 15 & 25 & -20 \\ 36 & -20 & 52 \end{bmatrix} \qquad \text{with } \lambda_1 = 100, \ \lambda_2 = 50$$

Their eigenvectors are the columns of

$$\mathbf{U} = \begin{bmatrix} 1 & 0 \\ 0 & 1 \end{bmatrix} \quad \text{and} \quad \mathbf{V} = \begin{bmatrix} .8 & 3/\sqrt{50} \\ 0 & 5/\sqrt{50} \\ .6 & -4/\sqrt{50} \end{bmatrix}$$

respectively. Hence, in accordance with Eq. 5.44, the left general inverse of **A** is

$$\mathbf{X} = \mathbf{U}\begin{bmatrix} \sqrt{\lambda_1} & 0 \\ 0 & \sqrt{\lambda_2} \end{bmatrix}^{-1} \mathbf{V}' = \begin{bmatrix} .08 & .0 & .06 \\ .06 & .10 & -.08 \end{bmatrix}$$

$$\textbf{8. } \mathbf{C} = \begin{bmatrix} 102.07 & 61.77 & 64.61 & 67.73 & 58.88 \\ 61.77 & 94.60 & 57.93 & 53.73 & 56.24 \\ 64.61 & 57.93 & 88.64 & 59.35 & 49.68 \\ 67.73 & 53.73 & 59.35 & 94.21 & 49.55 \\ 58.88 & 56.24 & 49.68 & 49.55 & 97.60 \end{bmatrix}$$

λ_j	$\lambda_{j/\Sigma\lambda_j}$			$\mathbf{v}_j'$		
327.95	.687	[.4888	.4418	.4366	.4436	.4224]
50.82	.794	[−.2182	.2161	−.2651	−.4638	.7876]
39.19	.876	[−.2178	.7727	.2307	−.3660	−.4102]
30.02	.939	[.1304	−.3963	.7293	−.5372	.0739]
29.15	1.000	[−.8056	−.0635	.3924	.4068	.1658]

9. $\mathbf{C}_1 = \begin{bmatrix} 105.35 & 71.40 & 68.29 & 72.95 & 66.94 \\ 71.40 & 103.55 & 67.88 & 64.78 & 66.10 \\ 68.29 & 67.88 & 96.18 & 64.10 & 54.10 \\ 72.95 & 64.78 & 64.10 & 101.70 & 56.87 \\ 66.94 & 66.10 & 54.10 & 56.87 & 112.67 \end{bmatrix}$ (male group)

λ_j	$\lambda_{j/\Sigma\lambda_j}$	$\mathbf{V}'_j$					
365.71	.704	[.4731	.4581	.4274	.4407	.4352]	(Only three given
56.04	.812	[−.1188	−.0103	−.3550	−.3609	.8541]	because almost
38.59	.886	[−.2653	.5888	.4246	−.6237	−.1168]	90% of variance accounted for)

$\mathbf{C}_2 = \begin{bmatrix} 99.32 & 55.77 & 61.39 & 62.55 & 52.82 \\ 55.77 & 77.05 & 51.80 & 50.53 & 45.44 \\ 61.39 & 51.80 & 82.24 & 54.46 & 46.68 \\ 62.55 & 50.53 & 54.46 & 84.95 & 45.14 \\ 52.82 & 45.44 & 46.68 & 45.14 & 84.60 \end{bmatrix}$ (female group)

λ_j	$\lambda_{j/\Sigma\lambda_j}$	$\mathbf{v}'_j$					
297.78	.695	[.5062	.4195	.4468	.4492	.4078]	(Only three given
42.65	.795	[−.2398	−.0289	−.1711	−.3421	.8918]	because almost
31.55	.869	[−.5789	.7172	.3057	−.1888	−.1462]	90% of variance accounted for)

Chapter 6

p. 209 1. The eigenvalue of the observed correlation matrix are 3.75, 1.28, .74, .72, .50, .40, .33, and .29. By the Kaiser–Guttman criterion and also by the scree test (when the eigenvalues are plotted against their ordinal numbers), a two-factor solution seems to be adequate. The unrotated, two-principal-factor pattern matrix is:

	Factor	
Variable	**I**	**II**
Locus of control	.46	.22
Self-concept	.11	.47
Motivation	.31	.55
Reading achievement	.85	−.08
Writing achievement	.78	−.03
Math achievement	.80	−.09
Science achievement	.77	−.14
Civics achievement	.69	.07

2. After rotation to an orthogonal simple structure by varimax, the factor matrix (showing also the final communality estimates) is:

Variable	Factor I	Factor II	h^2
Locus of control	.39	.33	.26
Self-concept	−.01	.48	.23
Motivation	.16	.60	.39
Reading achievement	.84	.13	.72
Writing achievement	.76	.16	.60
Math achievement	.80	.11	.65
Science achievement	.78	.05	.61
Civics achievement	.68	.10	.48

Except for the fact that the locus of control variable loads moderately on both factors, factor I is clearly a cognitive factor and factor II, an affective factor. As is to be expected, the cognitive variables have higher communality values than do the affective, reflecting the fact that the cognitive tests have higher reliability coefficients than the affective tests.

3. After rotation to oblique simple structure by the direct oblimin criterion (SSCP), the factor matrix (with the final communality estimates) is:

Variable	Factor I	Factor II	h^2
Locus of control	.36	.27	.26
Self-concept	−.07	.49	.23
Motivation	.09	.60	.39
Reading achievement	.85	−.01	.72
Writing achievement	.77	.04	.60
Math achievement	.81	−.02	.65
Science achievement	.80	−.08	.61
Civics achievement	.70	−.01	.48

($r_{I,II} = .29$)

The loadings in this case are only slightly different from those for the orthogonal rotation to simple structure. (This is reflected in the fact that the factors have a relatively small correlation, $r_{I,II} = .29$.) Nevertheless, there is some improvement in the extent to which the criteria for simple structure are met: the loadings of the cognitive variables on factor II are much closer to zero than they were in the orthogonal simple structure. Similarly, the two affective variables that had small loadings on factor I now have, on the average, even smaller loadings. The locus of control variable continues to have moderate loadings on both factors.

Chapter 7

p. 264 $\mathbf{W}^{-1}\mathbf{B} = \begin{bmatrix} .13762 & .06940 & -.10411 \\ -.00880 & .02493 & .00739 \\ -.11964 & -.05638 & .09056 \end{bmatrix}$

1. The characteristic equation is $\lambda^3 - .25311\lambda^2 + .00672\lambda = 0$, with roots $\lambda_1 = .22297$, $\lambda_2 = .03015$, and $\lambda_3 = 0$. The eigenvectors associated with the two nonzero roots are

$$\mathbf{v}_1' = [.7529 \quad -.0579 \quad -.6556] \qquad \mathbf{v}_2' = [-.2607 \quad .9071 \quad .3304]$$

$$\text{Significance test: } V = (296)(2.3026)[\log 1.223 + \log 1.030]$$

$$= 59.52 + 8.74 = 68.26$$

$$68.26 > \chi_6^2(.95) = 12.59 \qquad \text{and} \qquad 8.74 > \chi_2^2(.95) = 5.99$$

Both discriminant functions are significant at the 5% level. The discriminant-function means for the three groups are:

	Group A	Group B	Group C
Y_1-mean	3.22	3.87	3.08
Y_2-mean	3.76	3.97	4.09

On the first discriminant dimension, group B (physical science) is much higher than the other two groups. Noting that Y_1 has a large positive weight for the SAT mathematics test, and a substantial negative weight for the persuasiveness scale, we may interpret this as an abstract and non-"person-oriented" factor.

The second discriminant function has a high positive weight for the SAT verbal test, and a moderate weight for the persuasiveness scale. It is understandable that the industrial management group should average highest on this dimension. We might call this a verbal and "people-oriented" factor.

2. The characteristic equation of $\mathbf{W}^{-1}\mathbf{B}$ is $\lambda^3 - .3575\lambda^2 + .0216\lambda = 0$, with roots $\lambda_1 = .2807$, $\lambda_2 = .0768$, and $\lambda_3 = 0$. The eigenvectors associated with the two nonzero roots are

$$\mathbf{v}_1' = [.5439 \quad .8361 \quad -.0717] \qquad \mathbf{v}_2' = [-.3633 \quad .8999 \quad -.2415]$$

$$\text{Significance test: } V = (96)(2.3026)[\log 1.2807 + \log 1.0768]$$

$$= 23.75 + 7.10 = 30.85$$

The overall V (as a chi-square with 6 d.f.'s) is significant at the .001 level, and the residual 7.10 after removing the effect of the first discriminant is (as a chi-square with 2 d.f.'s) significant at the .05 level.

3. The SSCP matrix for the three predictor variables and the two group-membership "criterion" variables Y_1 and Y_2 in the total sample is

$$\mathbf{S} = \left[\begin{array}{rrr|rr} 271.34 & 40.38 & 55.43 & -31.00 & 7.80 \\ 40.38 & 69.61 & 5.49 & -15.50 & 12.90 \\ 55.43 & 5.49 & 59.80 & -5.00 & -.60 \\ \hline -31.00 & -15.50 & -5.00 & 25.00 & -15.00 \\ 7.80 & 12.90 & -.60 & -15.00 & 21.00 \end{array}\right]$$

The upper left-hand 3 × 3 submatrix $\mathbf{S}_{pp}$ is the sum of **W** and **B** given in the question. The lower right-hand 2 × 2 submatrix $\mathbf{S}_{cc}$ is computed by noting that $\Sigma\, Y_1 = \Sigma\, Y_1^2 = n_1 = 50$, $\Sigma\, Y_1 Y_2 = 0$, and $\Sigma\, Y_2 = \Sigma\, Y_2^2 = n_2 = 30$, with $N = n_1 + n_2 + n_3 = 100$. For computing the upper right-hand 3 × 2 submatrix $\mathbf{S}_{pc}$, we first have to retrieve the group totals and grand total on X_1, X_2, and X_3 from the group means given. Then note that $\Sigma\, X_i Y_1 = T_{i1}$ and $\Sigma\, X_i Y_2 = T_{i2}$ (the totals of X_i in groups 1 and 2, respectively). The matrix whose eigenvectors yield the discriminant weights is

$$(\mathbf{S}_{pp}^{-1}\mathbf{S}_{pc}\mathbf{S}_{cc}^{-1}\mathbf{S}_{cp} = \begin{bmatrix} .156826 & .043267 & .032550 \\ .138458 & .129894 & .009831 \\ -.000329 & -.018279 & .003687 \end{bmatrix}$$

The two nonzero eigenvalues are $\mu_1^2 = .2193$ and $\mu_2^2 = .0712$. The corresponding eigenvectors are, of course, equal within rounding error to those found in Exercise 2.

6. The eigenvalues of $\mathbf{W}^{-1}\mathbf{B}$ are $\lambda_1 = .33417$ and $\lambda_2 = .004573$, which lead to Bartlett's approximate chi-square statistics

$$V = [599 - (5 + 3)/2](\ln 1.33417 + \ln 1.004573)$$
$$= 174.26 >> \chi^2_{10;.95}$$
$$V - V_1 = (595)(\ln 1.004573) = 2.71 < \chi^2_{4;.95}$$

So only one discriminant function is significant, and its combining weights, normalized to unity, are the elements of $\mathbf{v}_1 = [.3777, .2824, .7174, -.2127, .4666]'$ The standardized weights are proportional to the elements of

$$\mathbf{v}_1^* = [.2920, .2119, .5089, -.1642, .3551]'$$

Thus the variables contributing the most to differentiation between boys and girls are X_3 (math) and X_5 (civics).

7. The sole nonzero eigenvalue of $\mathbf{W}^{-1}\mathbf{B}$ is $\lambda_1 = .28132$, yielding

$$V = [599 - (8 + 2)/2](\ln 1.28132) = 147.25 >> \chi^2_{8;.95}$$

The raw and standardized discriminant weights are proportional to the elements of

$$v = [-.4476, .4753, -.4363, .0279, -.1287, .0419, .0737, -.0058]'$$

and

$$v^* = [-.298, .333, -.149, .282, -1.214, .394, .709, -.0571]'$$

The two most discriminating variables are X_5 (writing) and X_7 (science).

8. The eigenvalues (squared canonical correlation coefficients) are $\mu_1^2 = .20072$, $\mu_2^2 = .02672$, $\mu_3^2 = .00061$. From these, Bartlett's chi-square statistics for testing the ovarall significance and the successive partial canonicals are computed as

$$\begin{aligned} V &= -[600 - 3/2 - (3 + 5)/2] \\ &\quad \times (\ln .79928 + \ln .97328 + \ln .99939) \\ &= -(594.5)[-(.2240 + .02708 + .00061)] \\ &= 149.66 > \chi^2_{15;.95} \\ V - V_1 &= -(594.5)[-(.02708 + .00061)] \\ &= 16.49 >> \chi^2_{8;.95} = 15.51 \\ V - V_1 - V_2 &= -(594.5)(-.00061) = .363 < \chi^2_{3;.95} \end{aligned}$$

Therefore, there are two significant canonical correlations between the set of three affective variables and the set of five cognitive variables. The combining weights that define the two pairs of canonical variates, normalized to unity, are the elements of the eigenvectors

$$\begin{aligned} \mathbf{u}_1 &= [.6899, \; -.1386, \; .7105]' \\ \mathbf{v}_1 &= [.5876, \; .7443, \; .2603, \; -.0778, \; .1641]' \\ \mathbf{u}_2 &= [.2707, \; .3054, \; -.9129]' \\ \mathbf{v}_2 &= [.0842, \; -.4699, \; .0211, \; .8196, \; -.3161]' \end{aligned}$$

9. The three nonzero eigenvalues of $\mathbf{W}^{-1}\mathbf{B}$ and their corresponding eigenvectors are as follows:

$$\lambda_1 = .158187 \qquad \lambda_2 = .101878 \qquad \lambda_3 = .018793$$

$$\mathbf{v}_1 = \begin{bmatrix} .38300 \\ -.28857 \\ .75522 \\ .00018 \\ .11015 \\ -.03763 \\ -.06216 \\ .02125 \end{bmatrix} \qquad \mathbf{v}_2 = \begin{bmatrix} .43947 \\ .43961 \\ .74892 \\ .03471 \\ -.04559 \\ .03829 \\ .03760 \\ .00224 \end{bmatrix} \qquad \mathbf{v}_3 = \begin{bmatrix} -.69532 \\ .66060 \\ -1.58125 \\ .00179 \\ .05068 \\ -.07566 \\ .05716 \\ .04511 \end{bmatrix}$$

Carrying out the sequential significance tests, we first find

$$V = (420)[\ln 1.158187 + \ln 1.101878 + \ln 1.018793] = 110.25$$

which, compared to the chi-square distribution with 24 d.f., far exceeds the 95th percentile. Therefore, at least the first discriminant function is significant. To test the modified H_0 after removing the first eigenvector, we find

$$V - V_1 = (420)[\ln 1.101878 + \ln 1.018793] = 48.47 > \chi^2_{14;.95}$$

Therefore, the second discriminant function also is significant. The test statistic for the modified H_0 after removing the effects of the first two discriminant functions is

$$V - V_1 - V_2 = (420)[\ln 1.018793] = 7.82$$

which is not significant as a chi-square with 6 d.f. Thus we conclude that there are two significant discriminant functions.

Chapter 8

p. 291 1.

Source	SSCP Matrix	n.d.f.	$\lvert \mathbf{S}_h + \mathbf{S}_e \rvert$ or $\lvert \mathbf{S}_e \rvert$	Λ	F-ratio
Education	$\begin{bmatrix} .042 & .255 & -.012 \\ .255 & 1.721 & -.116 \\ -.012 & -.116 & .014 \end{bmatrix}$	2	2.357	.5325	5.68
Sex	$\begin{bmatrix} .510 & .405 & .260 \\ .405 & .322 & .206 \\ .260 & .206 & .132 \end{bmatrix}$	1	2.355	.5330	13.44
Interaction	$\begin{bmatrix} .029 & -.017 & .004 \\ -.017 & .011 & .003 \\ .004 & .003 & .023 \end{bmatrix}$	2	1.361	.9220	.63
Error	$\begin{bmatrix} .782 & .333 & -.057 \\ .333 & 2.150 & .178 \\ -.057 & .178 & .824 \end{bmatrix}$	48	1.255	—	—

(The elements of the SSCP matrices have been rounded to three decimal places to save space, but five places were carried in the calculations.) The F for education (with 6 and 92 d.f.'s), and the F for sex (3 and 46 d.f.'s) are both significant at the .001 level. Discriminant functions for the education and sex effects:

$$\mathbf{S}_e^{-1}\mathbf{S}_{\text{educ}} = \begin{bmatrix} -.00111 & -.05408 & .01171 \\ .12207 & .83614 & -.05836 \\ -.04105 & -.32559 & .03030 \end{bmatrix} \quad \begin{matrix} \lambda_1 = .85070 \\ \lambda_2 = .01463 \end{matrix}$$

$$\mathbf{v}_1' = [\,-.06398 \quad .92867 \quad -.36533\,]$$

$$\mathbf{S}_e^{-1}\mathbf{S}_{\text{sex}} = \begin{bmatrix} .65291 & .51859 & .33205 \\ .05862 & .04656 & .02981 \\ .34748 & .27600 & .17673 \end{bmatrix} \quad \lambda_1 = .87620$$

$$\mathbf{v}_1' = [\,.88001 \quad .07901 \quad .46834\,]$$

2. (a)

Source	SSCP Matrix	n.d.f.	$\lvert S_h + S_e \rvert$ or $\lvert S_e \rvert$ ($\times 10^{-3}$)	Λ	*F*-ratio	*P*
Sex	$\begin{bmatrix} 185.01 & -223.50 \\ -223.50 & 270.00 \end{bmatrix}$	1	2993	.5460	39.93	$< .01$
Method	$\begin{bmatrix} 42.32 & 74.53 \\ 74.53 & 133.85 \end{bmatrix}$	2	1765	.9257	2.23	$> .01$
Interaction	$\begin{bmatrix} 2.51 & 5.47 \\ 5.47 & 17.55 \end{bmatrix}$	2	1656	.9867	.38	$> .01$
Within-cells	$\begin{bmatrix} 1795.39 & 1230.56 \\ 1230.56 & 1753.50 \end{bmatrix}$	114	1634			

(b) Only the sex effect is significant at the 1% level. The discriminant function for this effect is computed as follows:

$$\mathbf{S}_w^{-1}\mathbf{S}_{\text{sex}} = \begin{bmatrix} .3669 & -.4432 \\ -.3849 & .4650 \end{bmatrix} \quad \lambda_1 = .8319$$

$$\mathbf{v}_1 = [\,.6899 \quad -.7238\,]$$

The discriminant-function means are: $\overline{Y}_{\text{boys}} = 1.09$ $\overline{Y}_{\text{girls}} = -2.73$. It will be instructive to plot the centroids for the original variables and the discriminant-function means in the same graph.

3.

Source	SSCP Matrix	n.d.f.	$\lvert S_h + S_e \rvert$ or $\lvert S_e \rvert$	Λ_h
Sex	$\begin{bmatrix} 1541.81 & 241.66 & -524.65 \\ 241.66 & 37.88 & -82.23 \\ -524.65 & -82.23 & 178.53 \end{bmatrix}$	1	$\begin{vmatrix} 14{,}176.40 & 8068.11 & 7231.08 \\ 8068.11 & 11{,}989.90 & 7260.96 \\ 7231.08 & 7260.96 & 13{,}726.00 \end{vmatrix} = 9.12455 \times 10^{11}$	.7750
HSP	$\begin{bmatrix} 3159.72 & 3400.82 & 2711.42 \\ 3400.82 & 3660.72 & 2915.04 \\ 2711.42 & 2915.04 & 2352.53 \end{bmatrix}$	2	$\begin{vmatrix} 15{,}794.31 & 11{,}227.29 & 10{,}467.23 \\ 11{,}227.29 & 15{,}612.74 & 10{,}258.20 \\ 10{,}467.23 & 10{,}258.20 & 15{,}900.01 \end{vmatrix} = 9.55030 \times 10^{11}$	.7405
Sex × HSP	$\begin{bmatrix} 25.05 & -6.49 & 103.88 \\ -6.49 & 79.46 & 35.01 \\ 103.88 & 35.01 & 480.04 \end{bmatrix}$	2	$\begin{vmatrix} 12{,}659.61 & 7819.96 & 7859.62 \\ 7819.96 & 12{,}031.42 & 7378.24 \\ 7859.62 & 7378.22 & 14{,}027.50 \end{vmatrix} = 7.53342 \times 10^{11}$	.9387
Within	$\begin{bmatrix} 12{,}634.59 & 7826.45 & 7755.74 \\ 7826.45 & 11{,}951.98 & 7343.19 \\ 7755.74 & 7343.19 & 13{,}547.46 \end{bmatrix}$	170	$\begin{vmatrix} 12{,}634.57 & 7826.45 & 7755.74 \\ 7826.45 & 11{,}951.98 & 7343.19 \\ 7755.74 & 7343.19 & 13{,}547.46 \end{vmatrix} = 7.07197 \times 10^{11}$	—
Total	$\begin{bmatrix} 17{,}361.17 & 11{,}462.41 & 10{,}046.42 \\ 11{,}462.41 & 15{,}730.04 & 10{,}021.01 \\ 10{,}046.42 & 10{,}211.01 & 16{,}558.57 \end{bmatrix}$	175	—	—

$F_{\text{sex}} = [(1 - .7750)/.7750](168/3) = 16.25 > 5.79 = F^3_{120;.99} > F^3_{168;.99}$

H_0: $\boldsymbol{\mu}_M = \boldsymbol{\mu}_F$ is rejected at the 1% level.

$$F_{\text{HSP}} = [(1 - \sqrt{.7405})/\sqrt{.7405}](168/3) = 9.02 > 4.42 = F^6_{120;.99} > F^6_{336;.99}$$

H_0: $\boldsymbol{\mu}_1 = \boldsymbol{\mu}_2 = \boldsymbol{\mu}_3$ is rejected at the 1% level.

$$F_{\text{int}} = [(1 - \sqrt{.9387})/\sqrt{.9387}](168/3) = 1.80 < F^6_{\infty;.99} < F^6_{336;.99}$$

The null hypothesis of no interaction effect cannot be rejected.

4. The sole discriminant function for the sex effect of Exercise 3 is

$Y = -.8360X_1 + .1546X_2 + .5269X_3$; $\bar{Y}_M = -4.99$ $\bar{Y}_F = -10.90$

The standardized discriminant weights, normalized to unity, are the elements of $\mathbf{v}^* = [-.8280, .1489, .5406]'$. Thus the variable that contributes the most to the differentiation between males and females is writing ($= X_1$), the larger $\bar{X}_1$ for the girls leading to a *smaller* $\bar{Y}$ for them. The second most discrivariable is science ($= X_3$), for which the smaller $\bar{X}_3$ for girls contributes to the smaller $\bar{Y}$ for them.

For the HSP effect, there are two nonzero eigenvalues of $\mathbf{S}_w^{-1}\mathbf{S}_{\text{HSP}}$: $\lambda_1 = .3455$ and $\lambda_2 = .003635$. [Note that $(1.3455)(1.003635) = 1.3504 = 1/.7405$, the reciprocal of Λ_{HSP} found in Exercise 3.] However, Bartlett's chi-square test of the null hypothesis, $\lambda_{1(\text{pop})} \neq 0$, $\lambda_{2(\text{pop})}$ 0 yields $V - V_1 = (169) \ln 1.003635 = .6132$, which, as a chi-square with 2 d.f., is clearly nonsignificant. Therefore, only one discriminant function (whose weights are the elements of the eigenvector $\mathbf{v}_1$ corresponding to the larger eigenvalue λ_1) is significant. This is $Y = .4975X_1 + .8552X_2 + .1454X_3$. The standardized discriminant weights, normalized to unity, are the elements of $\mathbf{v}_1^* = [.5072, .8480, .1535]$, indicating that the math achievement test is the best differentiator among the three high school programs, with reading achievement coming next. The discriminant-function means for the three programs (where 1 = general, 2 = academic, 3 = vocational) are $\bar{Y}_1 = 73.29$, $\bar{Y}_2 = 82.94$, $\bar{Y}_3 = 67.58$.

5.

Source	SSCP Matrix	n.d.f.	$\lvert S_h + S_e \rvert$ or $\lvert S_e \rvert$	Λ_h
SES	$\begin{bmatrix} 2171.1 & 1650.4 & 2122.1 & 2372.1 & 1874.7 \\ 1650.4 & 1254.6 & 1613.6 & 1803.4 & 1425.0 \\ 2122.1 & 1613.6 & 2088.1 & 2325.8 & 1830.0 \\ 2372.1 & 1803.4 & 2325.8 & 2595.4 & 2047.0 \\ 1874.7 & 1425.0 & 1330.0 & 2047.0 & 1619.1 \end{bmatrix}$	2	$\begin{vmatrix} 49{,}791.8 & 27{,}231.9 & 27{,}967.2 & 32{,}669.7 & 25{,}340.6 \\ 27{,}231.9 & 47{,}455.2 & 24{,}920.5 & 24{,}837.2 & 24{,}447.9 \\ 27{,}967.2 & 24{,}920.5 & 42{,}346.1 & 27{,}841.0 & 19{,}910.1 \\ 32{,}669.7 & 24{,}837.2 & 27{,}841.0 & 50{,}248.2 & 22{,}018.3 \\ 25{,}340.6 & 24{,}447.9 & 19{,}910.1 & 22{,}018.3 & 48{,}647.0 \end{vmatrix} = 3.0684 \times 10^{22}$	.9312
HSP	$\begin{bmatrix} 7454.9 & 7067.0 & 7473.5 & 5255.3 & 7126.5 \\ 7067.0 & 6739.6 & 7092.0 & 5049.7 & 6803.8 \\ 7473.5 & 7092.0 & 7493.6 & 5280.7 & 7153.1 \\ 5255.3 & 5049.7 & 5280.7 & 3819.0 & 5104.7 \\ 7126.5 & 6803.8 & 7153.1 & 5104.7 & 6869.9 \end{bmatrix}$	2	$\begin{vmatrix} 55{,}075.5 & 32{,}648.4 & 33{,}318.6 & 35{,}552.9 & 30{,}592.5 \\ 32{,}648.4 & 52{,}940.2 & 30{,}398.9 & 28{,}083.5 & 29{,}826.6 \\ 33{,}318.6 & 30{,}398.9 & 47{,}751.6 & 30{,}796.0 & 25{,}233.1 \\ 35{,}552.9 & 28{,}083.5 & 30{,}796.0 & 51{,}471.8 & 25{,}076.0 \\ 30{,}592.5 & 29{,}826.6 & 25{,}233.1 & 25{,}076.0 & 53{,}897.8 \end{vmatrix} = 3.6170 \times 10^{22}$	.7900
SES × HSP	$\begin{bmatrix} 735.3 & 1.6 & 171.2 & -19.9 & -50.6 \\ 1.6 & 176.0 & 53.9 & -5.9 & -0.9 \\ 171.2 & 53.9 & 232.5 & -181.7 & -90.7 \\ -19.9 & -5.9 & -181.7 & 186.0 & 140.3 \\ -50.6 & -0.9 & -90.7 & 140.3 & 368.9 \end{bmatrix}$	4	$\begin{vmatrix} 48{,}355.9 & 25{,}583.0 & 26{,}016.3 & 30{,}277.7 & 23{,}415.3 \\ 25{,}583.0 & 46{,}376.6 & 23{,}360.8 & 23{,}027.8 & 23{,}022.0 \\ 26{,}016.3 & 23{,}360.8 & 40{,}490.5 & 25{,}333.6 & 17{,}989.3 \\ 30{,}277.7 & 23{,}027.8 & 25{,}333.6 & 47{,}838.8 & 20{,}111.6 \\ 23{,}415.3 & 23{,}022.0 & 17{,}989.3 & 20{,}111.6 & 47{,}396.8 \end{vmatrix} = 3.0620 \times 10^{22}$	.9332

Source	SSCP Matrix	n.d.f.	$\lvert S_h + S_e \rvert$ or $\lvert S_e \rvert$	Λ_h
Within	$\begin{bmatrix} 47{,}620.7 & 25{,}581.4 & 25{,}845.1 & 30{,}297.6 & 23{,}465.9 \\ 25{,}581.4 & 46{,}200.6 & 23{,}306.9 & 23{,}033.8 & 23{,}022.9 \\ 25{,}845.1 & 23{,}306.9 & 40{,}258.0 & 25{,}515.3 & 18{,}080.0 \\ 30{,}297.6 & 23{,}033.8 & 25{,}515.3 & 47{,}652.8 & 19{,}971.3 \\ 23{,}465.9 & 23{,}022.9 & 18{,}080.0 & 19{,}971.3 & 47{,}028.0 \end{bmatrix}$	591	$\begin{vmatrix} 47{,}620.7 & 25{,}581.4 & 25{,}845.1 & 30{,}297.6 & 23{,}465.9 \\ 25{,}581.4 & 46{,}200.6 & 23{,}306.9 & 23{,}033.8 & 23{,}022.9 \\ 25{,}845.1 & 23{,}306.9 & 40{,}258.0 & 25{,}515.3 & 18{,}080.0 \\ 30{,}297.6 & 23{,}033.8 & 25{,}515.1 & 47{,}652.8 & 19{,}971.3 \\ 23{,}465.9 & 23{,}022.9 & 18{,}080.0 & 19{,}971.3 & 47{,}028.0 \end{vmatrix}$ $= 2.8574 \times 10^{22}$	—
Total	$\begin{bmatrix} 57{,}981.9 & 34{,}300.4 & 35{,}611.9 & 37{,}905.1 & 32{,}416.5 \\ 34{,}300.4 & 54{,}370.8 & 32{,}066.3 & 29{,}881.0 & 31{,}250.7 \\ 35{,}611.9 & 32{,}066.3 & 50{,}072.2 & 32{,}940.1 & 26{,}972.4 \\ 37{,}905.1 & 29{,}881.0 & 32{,}940.1 & 54{,}253.2 & 27{,}263.3 \\ 32{,}416.5 & 31{,}250.7 & 26{,}972.4 & 27{,}263.3 & 55{,}885.7 \end{bmatrix}$	599		

$$F_{\text{SES}} = [(1 - \sqrt{.9312})/\sqrt{.9312}](587/5) = 4.26$$

$$> 3.55 = F^{8}_{120;.999} > F^{10}_{1174;.999}$$

$$F_{\text{HSP}} = [(1 - \sqrt{.7900})/\sqrt{.7900}](587/5) = 14.69 > 3.55$$

$$= F^{8}_{120;.999} > F^{10}_{1174;.999}$$

$$F_{\text{SES}\times\text{HSP}} = [(1 - .9332^{1/s})/.9332^{1/s}][(ms - pv_h/2 + 1)/pv_h]$$

with $m = 590$, $s = 3.3166$, yielding $F = 2.05$. By interpolation in Table E.4, 2.05 is found to be greater than $F^{18}_{\infty;.99} = 1.99$. Therefore, it is safe to conclude that $F_{\text{obs}} > F^{20}_{1947.8;.99}$

Chapter 9

p. 349 1. and **2.** Since these are the same problems as Chapter 8, Exercises 3 and 5, respectively, except that they are to be done by the GLM rather than the "classical" approach, the answers are the same as those for the corresponding exercises in Chapter 8. However, in those GLM programs that allow more than one option for "type of SS," only by selecting the appropriate type(s) will we get consonant results. (For SAS PROC GLM, type III SS is the appropriate one.)

Chapter 10

p. 382 1.

Individual Number (i)	χ^2_{i1}	χ^2_{i2}	χ^2_{i3}	Decision—Assign to
5	6.3714	2.6162	*.9844*	Group 3
6	2.2514	*.5198*	1.6884	Group 2
	χ'^2_{i1}	χ'^2_{i2}	χ'^2_{i3}	
5	17.5633	11.0470	*9.2331*	Group 3
6	11.2337	*8.6721*	9.9175	Group 2
	χ''^2_{i1}	χ''^2_{i2}	χ''^2_{i3}	
5	19.6723	12.9761	*11.8481*	Group 3
6	13.3427	*10.6012*	12.5325	Group 2
		$p(H_k \,\&\, U_{ki} > U^*_k \mid \mathrm{x}_i)$		
5	.0080	.1947	*.3743*	Group 3
6	.0869	*.3613*	.1007	Group 2

2. The cost-function values for the six individuals are as follows:

Individual Number (i)	$C(H_1 \mid X_i)$	$C(H_2 \mid X_i)$	$C(H_3 \mid X_i)$	Decision—Assign to
1	.5166	.4659	.0175	Group 1
2	.0216	.2174	.7610	Group 3
3	.1232	.5896	.2872	Group 2
4	.0044	.4540	.5415	Group 3
5	.0059	.2531	.7410	Group 3
6	.0939	.5544	.3517	Group 2

3. The coefficients of profile similarity of individuals 5 and 6 with respect to each of the three groups are as follows:

Individual Number (i)	$r^*_{p(1;i)}$	$r^*_{p(2;i)}$	$r^*_{p(3;i)}$	Decision—Assign to
5	−.4584	−.0502	.4124	Group 3
6	.0248	.6398	.1671	Group 2

4. The coefficients $\mathbf{C}_w^{-1}\overline{X}_k$ and the constant term $(-1/2)\overline{X}_k'\mathbf{C}_w^{-1}\overline{X}_k$ for the four linear classification functions of the form shown in Eq. 10.10 are as follows:

	Group			
	I	II	III	IV
CONSTANT	−27.6593	−28.4295	−29.7231	−32.0149
MOT	3.6717	3.2643	3.5016	4.4956
CONCPT	−0.5954	0.0117	−0.1583	−0.2061
LOCUS	−4.7563	−4.9158	−4.7802	−4.2786
RDG	0.0263	0.0477	0.0438	0.0600
WRTG	0.2933	0.1665	0.2807	0.2692
MATH	0.1997	0.2512	0.1969	0.2288
SCI	0.2427	0.3300	0.2824	0.2695
CIV	0.2211	0.2069	0.2351	0.2278

Actually, the four group covariance matrices were highly significantly different in this problem, and hence a quadratic classification function of the form of Eq. 10.9 should have been computed for each group. We ignored this fact and "forced" the classification function to be linear by using the pooled within-groups covariance matrix instead of the separate group covariance matrices as in Eq. 10.9. The purpose was to offer a basis for comparing classification functions and discriminant functions (computed for Exercise 9 of Chapter 7). The interested reader may wish to verify that a relation of the form of Eq. 10.11 holds between the 8×4 matrix of linear classification-function weights given above and the 8×3 matrix of discriminant function weights found in Chapter 7, Exercise 9. When the 427 students belonging to one or another of the four career-plan groups are classified by using the maximum-L_k rule (see p. 369), the results are shown below.

From Group:	Number of Observations and Percents Classified into Group: 1	2	3	4	Total
1	20	12	48	6	86
	23.26	13.95	55.81	6.98	100.00
2	2	35	43	6	86
	2.33	40.70	50.00	6.98	100.00
3	8	20	118	15	161
	4.97	12.42	73.29	9.32	100.00
4	5	8	60	21	94
	5.32	8.51	63.83	22.34	100.00
Total	35	75	269	48	427
Percent	8.20	17.56	63.00	11.24	100.00
Priors	0.2014	0.2014	0.3770	0.2201	

AUTHOR INDEX

SUBJECT INDEX